Lecture Notes in Civil Engineering

Volume 10

Series editors

Marco di Prisco, Politecnico di Milano, Milano, Italy
Sheng-Hong Chen, School of Water Resources and Hydropower, Wuhan University, Wuhan, China
Giovanni Solari, University of Genoa, Genova, Italy
Ioannis Vayas, National Technical University of Athens, Athens, Greece

Lecture Notes in Civil Engineering (LNCE) publishes the latest developments in Civil Engineering - quickly, informally and in top quality. Though original research reported in proceedings and post-proceedings represents the core of LNCE, edited volumes of exceptionally high quality and interest may also be considered for publication. Volumes published in LNCE embrace all aspects and subfields of, as well as new challenges in, Civil Engineering. Topics in the series include:

- Construction and Structural Mechanics
- Building Materials
- Concrete, Steel and Timber Structures
- Geotechnical Engineering
- Earthquake Engineering
- Coastal Engineering
- Hydraulics, Hydrology and Water Resources Engineering
- Environmental Engineering and Sustainability
- Structural Health and Monitoring
- Surveying and Geographical Information Systems
- Heating, Ventilation and Air Conditioning (HVAC)
- Transportation and Traffic
- Risk Analysis
- Safety and Security

More information about this series at http://www.springer.com/series/15087

Marco di Prisco · Marco Menegotto
Editors

Proceedings of Italian Concrete Days 2016

 Springer

Editors
Marco di Prisco
Dipartimento Ingegneria Civile
Politecnico di Milano
Milan
Italy

Marco Menegotto
AICAP
Rome
Italy

ISSN 2366-2557   ISSN 2366-2565   (electronic)
Lecture Notes in Civil Engineering
ISBN 978-3-319-78935-4   ISBN 978-3-319-78936-1   (eBook)
https://doi.org/10.1007/978-3-319-78936-1

Library of Congress Control Number: 2018937355

Printed on acid-free paper

This Springer imprint is published by the registered company Springer International Publishing AG part of Springer Nature
The registered company address is: Gewerbestrasse 11, 6330 Cham, Switzerland

Preface

Concrete is a very old building material: in Rome, we still have beautiful examples of its use, built around two thousand years ago. However, in the last twenty years it has changed much more than any other building material. The compressive strength, its main performance, is increased more than 4 times, with a reduced increase in the cost of the material, thanks to the progress in technology introduced by innovative products. Structural concrete is the most widely used building material in the world, with a global production estimated between 21 and 31 billion tons per year. In 2016, a world population of 7.4 billion people has been estimated: this means an average production of about 1500 liters per person per year of concrete.

All the structures are expected to deteriorate over time when used: diffuse damage due to cracking often causes a shortening of the expected life. Durability and serviceability play a key role for a sustainable progress of the humanity, decreasing the concrete request of human resources and extending the serviceability life of structures and infrastructures with a proper maintenance. The growing use of additives and fibres in the matrix has required the introduction of other performances like those describing the post-cracking SLS and ULS residual strengths in order to take into account the corresponding tremendous toughness increase in the structural design. The performance-based design is nowadays orienting concrete evolution: it is becoming crucial to quantify further performances aimed at guaranteeing durability. At the same time, a strong effort is in progress to reduce the impact of concrete production on the environment, thus reducing CO_2 emissions and embodied energy. The conference "Italian Concrete Days" for the first time organized jointly by AICAP and CTE since the time of their foundation is an event focused on these subjects and open to the international community.

The ICD16 proceedings, published in Italian language on digital media with ISBN 978-88-99916-02-2, contain over 165 technical and scientific papers aligned to the main topics announced: innovative materials, technologies and constructive techniques, fire resistance, robustness and life cycle assessment, maintenance, upgrade and reuse of structures and infrastructures. They trace a state of the art also open to internationality on concrete construction evolution and mark the actual position on the path of sustainability. In this volume, which represents a synthesis of the proceedings aimed at an international promotion of the event and of several significant scientific innovative contributions presented therein, 40 contributions presented in English language are collected. According to the Scientific Committee, for the quality of the texts, the care of the edition and the specificity of the information gathered in, these contributions deserve a wider international diffusion aimed at tracing the evolution of the technical and scientific topics on the structural concrete of Italian civil engineering.

We would like to take this opportunity to thank all the colleagues of the Scientific Committee for their contribution in reviewing and finalizing the papers for the proceedings in a strict time.

Our sincere appreciation is also extended to the secretary offices of both the cultural associations, AICAP and CTE, who were able to work in cooperation, feeling the need to unify their efforts to give a high-quality product, searching a new format able to consider the international visibility requirements and to involve also international experts and a more interdisciplinary knowledge in the discussed topics. To this aim, a special acknowledgement is addressed to the International Association of Structural Concrete *fib,* for having supported the whole venture.

We hope that this volume could favour the needed dialogue between the various stakeholders, who act in the field of the constructions, according to a new deal in the industrialization process, that driven by digitalization, ushers further progress and a meaningful national and international collaboration.

Marco di Prisco
Marco Menegotto

Organization

Scientific Committee

Maria Antonietta Aiello
Franco Angotti
Luigi Ascione
György L. Balázs
Gianni Bartoli
Beatrice Belletti
Andrea Benedetti
Gabriele Bertagnoli
Francesco Biasioli
Fabio Biondini
Franco Bontempi
Gianmichele Calvi
Giuseppe Campione
Roberto Cerioni
Bernardino Chiaia
Mario Chiorino
Piero Colajanni
Dario Coronelli
Hugo Corres Peiretti
Edoardo Cosenza
Andrea Dall'Asta
Piero D'Asdia
Gianmarco De Felice
Barbara De Nicolo
Achille Devitofranceschi
Luigino Dezi
Pierpaolo Diotallevi
Marco di Prisco
Mauro Dolce

Luigi Evangelista
Giovanni Fabbrocino
Ciro Faella
Alessandro Fantilli
Roberto Felicetti
Liberato Ferrara
Giuseppe Ferro
Dora Foti
Paolo Franchin
Dante Galeota
Pietro G. Gambarova
Natalino Gattesco
Alessandra Gubana
Donatella Guzzoni
Ivo Iori
Lidia La Mendola
Pier Giorgio Malerba
Gaetano Manfredi
Giuseppe Mancini
Enzo Martinelli
Luigi A. Materazzi
Claudio Mazzotti
Alberto Meda
Marco Menegotto
Antonio Migliacci
Claudio Modena
Franco Mola
Pietro Monaco
Giorgio Monti

Contents

Materials, Technologies and Construction Techniques

Multi-criteria Analysis for Sustainable Buildings

A. De Angelis[1(✉)], N. Cheche[1], R. F. De Masi[1], M. R. Pecce[1], and G. P. Vanoli[2]

[1] Department of Engineering, University of Sannio, Benevento, Italy
alessandra.deangelis@unisannio.it
[2] Department of Medicine and Health Sciences, University of Molise, Campobasso, Italy

Abstract. Currently climate and housing standards in parallel with the energy savings govern the performance required to buildings. Consequently, innovative multi-functional components try to satisfy both the requirements of structural safety and thermal performance, but the choice of the designer is difficult due to the complexity of this new market. In this paper, the Multi-Criteria Decision Making (MCDM) analysis is proposed as suitable methodology that can provide adequate support for choosing the best building components between different alternatives. As case study, TOPSIS method (Technique for Order Preference by Similarity to Ideal Solution) has been applied for comparing four different types of floor. The criteria assumed in the case study are referred to different fields as well as thermal, acoustic, air quality building science, structural performances, economic and human impacts. The case study allows to point out that the optimal solution depends on the importance (weight) that the decision maker assigns to each considered criterion.

Keywords: Multi-criteria problem · Structural performance · Sustainability
TOPSIS · Weight assignment · R.C. floor

1 Introduction

In the field of buildings construction the target of energy saving has led to a convulsive evolution of the market for the components of the "building envelope" that is focused only on this aspect, even though further performance has to be assured especially for the structural safety. The availability of multi-functional products with interdisciplinary performances makes necessary the adoption of quantitative models to assist designers in decision making, comparing solutions in term of cost, safety, durability, energy efficiency and so on. The main challenge for the involved decision makers includes the need to consider multiple objectives with different tradeoffs values, therefore a reliable approach to decision analysis is necessary. The wide variety of available solutions for the building envelope, suggests the use of non-dominated optimization to identify a set of feasible design solutions that represent a real compromise between the criteria. This approach, referred to as Pareto optimization, has been extensively applied in the literature concerned with multi-criteria design (Grierson 2008). In the context of building design, Multiple Criteria Decision Making (MCDM) methods provide a valuable tool for supporting the choice of the preferred Pareto optimum. Some examples of MCDA application are available in various engineering fields. In their paper Mardani et al.

© Springer International Publishing AG, part of Springer Nature 2018
M. di Prisco and M. Menegotto (Eds.): ICD 2016, LNCE 10, pp. 3–16, 2018.
https://doi.org/10.1007/978-3-319-78936-1_1

(2015) have documented the exponentially grown interest in the MCDM techniques and they have provided a state-of the-art of the literature regarding applications and methodologies. They have established a list of around 400 articles categorized into 15 fields: energy, environment and sustainability, supply chain management, material, quality management, GIS, construction and project management, safety and risk management, manufacturing systems, technology management, operation research and soft computing, strategic and knowledge management, production management, tourism management and other fields. Kabir et al. (2013) in a recent paper have pointed out that approximately 300 works concern the application of multicriteria decision techniques in the field of infrastructure management. Considering building sector, most of the studies that apply the multicriteria approach are focused separately on energy, environmental or structural engineering criteria. More in detail, several papers discuss how high performance buildings require an integrated approach focusing on architecture conception (Fontanelle and Bastos 2014, Songa et al. 2015), on building envelope solution (Gagliano et al. 2015, Yang et al. 2016), on choice of HVAC system designs (Avgelis and Papadopoulos 2009) and integration of renewable energy and storage technologies (Wimmler et al. 2015, Ascione et al. 2015). As summarized above, the application of MCDA is not new, but, it is generally not related to multidisciplinary problems. In this paper, the potentialities of MCDA application are evidenced for the selection of building envelope components; in particular the application of the procedure is developed for a reinforced concrete (RC) floor; several traditional and innovative solutions are analyzed, considering both the energy efficiency as well as the primary structural performance.

2   A Multicriteria Approach for Roof Slab Technology

The choice of the alternatives to be considered to finalize a MCDA procedure is a key issue that can surely influence the result. Indeed, although the evaluation model is accurate and reliable, it can lead to a poor choice if the alternatives under consideration are weak (Brown 2005). There are several tools that may be employed in the definition of decision alternatives, such as brainstorming techniques (Osborn 1957), cognitive mapping (Eden 1988), dialog maps (Conklin 2006), among others.

More in detail, it has been shown that tools where decision alternatives are created considering the decision-makers' objectives (Keeney 1992) or stakeholders' values (Gregory and Keeney 1994) are the most useful. For example, the analyst can ask to decision makers to propose options that could perform adequately a single objective. In this study, the types of floor have been selected in order to compare two traditional types of reinforce concrete (RC) floors (one casted in situ and the other precast) and two new types of RC floors in which innovative materials are used to improve the energy efficiency.

The first type is a traditional brick-concrete floor, that is a RC floor lightened with hollow clay bricks (named type A1 in the following analysis); other three types of RC floor:

- A2: hollow-core floor;
- A3: expanded polystyrene lightened floor;

- A4: wood-cement lightened floor (Fig. 1).

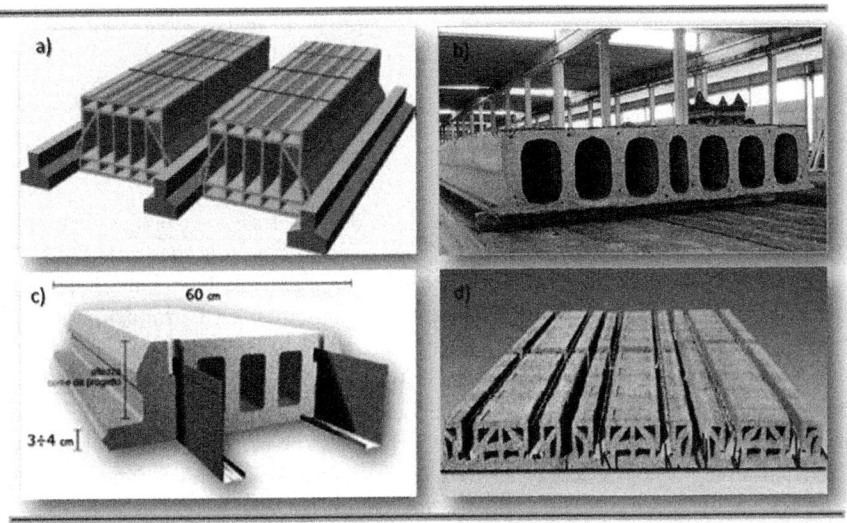

Fig. 1. Types of floor: (a) A1-Brick-concrete floor; (b) A2-Hollow-core floor; (c) A3-Expanded polystyrene lightened floor; (d) A4-Wood-cement lightened floor.

The brick-concrete floor is the most diffuse technology in Europe, but usually, it satisfies only the structural performance. It is completely cast in situ or realized with precast joists; the bricks are spaced to allow the concrete casting of the joists. The description of the properties of the roof layers is shown in Table 1; here, "s" is the thickness, "λ" the thermal conductivity, "c_p" the specific heat and "ρ" the density.

Table 1. Brick-concrete floor

	s [m]	λ [W/mK]	c_p [J/kgK]	ρ [kg/m3]
Interior plaster	0.02	0.700	1000	1800
Ribs and brick	0.20	0.916	878	1008
Slab	0.05	1.600	1000	2500
Screed	0.05	1.060	1000	1800
Waterproofing membrane	0.01	0.170	1000	900
Flooring	0.03	0.700	1000	1000

Instead, the hollow core floor have been created to address the need to reduce the commissioning work time. For this reason, often, it doesn't attempt other functional performance. The element A2 (Table 2) is a precast pre-stressed concrete slab that is completed in-situ with a concrete layer with thickness not less than 4 cm.

Table 2. Hollow-core floor

	s [m]	λ [W/mK]	c_p [J/kgK]	ρ [kg/m3]
Interior plaster	0.02	0.700	1000	1800
Hollow-core floor	0.20	1.120	1000	2400
Slab	0.05	1.600	1000	2500
Screed	0.05	1.060	1000	1800
Waterproofing membrane	0.01	0.170	1000	900
Flooring	0.03	0.700	1000	1000

The EPS-based solution could optimize the energy and environmental performance by operating, at the same time, an increment of the thermal insulation and a reduction of the structural weight.

In particular the third solution, A3, consists of a panel in Expanded Polystyrene (EPS) and a steel 'L' shaped profile embedded inside having a load-carrying function during the casting (generally the self-supporting capacity is guaranteed on a span of 2.50 m for a height floor equal to about 240 mm); two adjacent EPS panel are used as formwork for the casting of the concrete ribs. Therefore, the system is completed by a concrete casting on the top to realize a RC slab. The lower winglet (height ranging from 4 to 8 cm) ensures the continuity of the polystyrene even below the rafters, conferring a low transmittance to the floor and reducing the thermal bridges effect. Table 3 describes the properties of each layer of floor.

Table 3. Expanded polystyrene lightened floor

	s [m]	λ [W/mK]	c_p [J/kgK]	ρ [kg/m3]
Interior plaster	0.02	0.700	1000	1800
Insulating fin	0.04	0.036	1000	35
Ribs and panel	0.20	0.300	1400	413
Slab	0.05	1.600	1000	2300
Screed	0.05	1.060	1000	2000
Waterproofing membrane	0.01	0.170	1000	900
Flooring	0.03	0.700	1000	1000

The element A4, wood-cement lightened floor, consists of pre-assembled panels, of dimensions 100 cm × (20-25-30-39), length up to 6.5–7 m, with horizontal and vertical milling to eliminate thermal bridges and to improve acoustic performance. The floor must be completed on site with supplementary welded steel mesh and finishing casting of the slab. Table 4 shows the characteristic of layers selected for the case study.

Table 4. Wood-cement lightened floor

	s [m]	λ [W/mK]	c_p [J/kgK]	ρ [kg/m3]
Interior plaster	0.02	0.700	1000	1800
Fin insulating wood	0.05	0.103	2200	512
Wood-concrete block	0.20	1.120	1400	2400
Slab	0.05	1.600	1000	2500
Screed	0.05	1.060	1000	1800
Waterproofing membrane	0.01	0.170	1000	900
Flooring	0.03	0.700	1000	1000

3   Parameters for Multi-criteria Analysis

National and international policies devote increasing attention to the reduction of energy consumption and environmental impacts; however in any evaluation process also the economic investment has to be considered. In this section, a general discussion about the selectable criteria for comparing the floor technology is presented. Then, the evaluation system used in the case study is pointed out.

3.1   Structural Properties

The floor is a structural element with significant influence on the realization and behaviour of a building. The floor has the primary structural function of supporting vertical loads but in case of seismic actions it has the further function of distribution of horizontal actions. The design of a floor has to be developed according to the technical codes (in Italy DM 2008 or Eurocode 2). The major issues consist of the fulfilment of the resisting capacity (ultimate limit state) and functional performance (serviceability limit state a deformability limitation). Therefore parameters such as the structural weight, the reduction of the time for the transportation and the mounting operation can to be considered.

3.2   Properties Related to Indoor Comfort and Consumption Issues

Each construction component contributes to thermal and acoustic comfort inside the building, daylight conditions, energy efficiency and sustainability, as evidenced by da Silva and de Almeida (2010). There are a lot of factors that influence the behavior of building components as described below.

Thermal transmittance (UNI EN ISO 6946 2008) is a measure of insulation level of building components; usually, lower value of the thermal transmittance means lower energy consumption for heating. Briefly, according to code approach (Ministerial Decree 2015), this parameter defines the reference building; it is related to the evaluation of global performance index both for refurbished as well as for new design buildings.

In the summer period, thermal performance should be evaluated in dynamic regime; many indicator parameters can be used such as the time lag, the decrement factor, the thermal heat capacity and the periodic thermal transmittance (Y_{IE}) as defined by UNI EN ISO 13786 (2007). Briefly, periodic thermal transmittance represents the heat flow rate through the internal surface of the component when the ambient temperature varies as sinusoidal function. It is evident that a low value of Y_{IE} corresponds to better summer thermal performance of the building component.

Also the spectral properties of finishing layer should be considered. Indeed solar reflectance and thermal emittance significantly affect the temperature of the building components, and these can contribute to reduce the summer energy demand or building overheating as well as the heat island effect. A surface of roof highly reflective and highly emissive allows to minimize the amount of light converted into heat and to maximize the amount of heat that is radiated away. In climatic zones characterized by high energy requirements for cooling, often, this is a suitable technological solution; however also the problem due to reduction of heat gain during the winter period should be evaluated.

Sound insulation is the ability of building elements to reduce sound transmission. It is measured at different frequencies, normally 100–3150 Hz. The airborne sound insulation is expressed by a single value: Weighted Standardized Level Difference ($D_{n,t,w}$), Weighted Sound Reduction Index (R_w) and Weighted Apparent Sound Reduction Index (R'_w). Impact sound insulation can be expressed by a single value between Normalized Impact Sound Pressure Level ($L'_{n,w}$), $L_{nT,w}$ that is the Weighted Standardized Impact Sound Pressure Level and the $L'_{nT,w}$ that is the Weighted Standardized Impact Sound Pressure Level. These are calculated according to the standard EN 12354-1 (2000); EN 12354-2 (2000). In Italy, passive acoustic requirements of buildings are reported by the D.P.C.M. (1997). Moisture in envelope assemblies can cause numerous problems affecting the indoor air quality of a building and the longevity of building components. Moreover moisture can cause corrosion of components and dissolve water soluble constituents damaging structures; the induced degradation could reduce thermal resistance and the strength and/or stiffness of materials.

Vapor permeability is the ability of a material to allow water vapor to pass through it. This is a material property and it is not dependent on size, thickness or shape of the material. Vapor resistance is equivalent to the vapor permeability multiplied by the thickness (ISO 12572 2001). The total value of vapor resistance of building component influences the diffusion of water vapor current density.

3.3 Environmental Performance

Often the energy assessment is related to the need of reducing the environmental impact of building use as well as of construction industry. The sustainability of building sector should take into account the impact on the surrounding environment, human health, consumption of resources and on quality of the ecosystem. The most comprehensive methods to evaluate and reduce environmental impacts is the Life Cycle Assessment (LCA) methodology. According to International Standard ISO 14040 (2010), LCA is a "compilation and evaluation of the inputs, outputs and the potential environmental impacts of a product system throughout its life cycle." By taking into account the construction, assembly, processing and disposal of the component, LCA enables identification of the most significant impacts and stages during its life cycle that need to be targeted for maximum improvements.

3.4   Costs

The cost is surely one of the criteria to be considered in the analysis, since generally a prefixed budget is established in public and private investment, influencing the choice of the builders and buyers. The total cost to be sustained for the realization of each alternative includes: basic resources (manpower and materials), semi-worked (mortar, ready-mixed concrete, etc.) as well as transportation and freight (trucks, cranes, etc.). The works are intended performed in a workman like manner in accordance with applicable laws and regulations.

3.5   Case Study: Evaluation Area and Criteria

For the case study three areas of analysis have been considered:

- Structural Safety;
- Sustainability;
- Economic investment.

These evaluation areas have been structured in categories and evaluation criteria as shown in Fig. 2.

The evaluation area named *Sustainability* includes the *energy* and *indoor quality* categories. More in detail, among the parameters that define the thermal performance of floor thermal transmittance and the periodic thermal transmittance have been chosen. In the same category, the LCA criterion has been added with the aim to evaluate the environmental impact of floor supply chain. The indoor quality referred to building component has been studied in term of acoustic insulation and thus the weighted normalized impact sound pressure level ($L'_{n,w}$) has been considered as criterion. According to EN 12354-2 (2000-09), the limit value for residential kind of use is 63 dB, for tertiary building it varies between 55–58 dB. Finally, the moisture risk is evaluated with the global vapor resistance (R_v). The second evaluation area is named *Structural Safety* and it is divided into two categories: *technology* and *structural performance*. For these, two evaluation criteria have been considered; the commissioning work time (τ_c) and structural weight (S_w). More in detail, the analysed floors have been sized to meet regulatory requirements by adopting the simple supported beam scheme with a span of 6 m. The residential class use has been considered (Cat A), therefore a variable load of 2 kN/m^2 and a permanent load of 2.7 kN/m^2 in addition to the weight have been considered. For all alternatives a steel reinforcement area equivalent to 2Φ16 for each rafter has been imposed considering only the height as design parameter; the design heights obtained for the various types of floors are slightly different for the variation of the self-weight of each one. Finally the third evaluation area is *Investment;* it includes, at this stage of study, only the *cost of construction.*

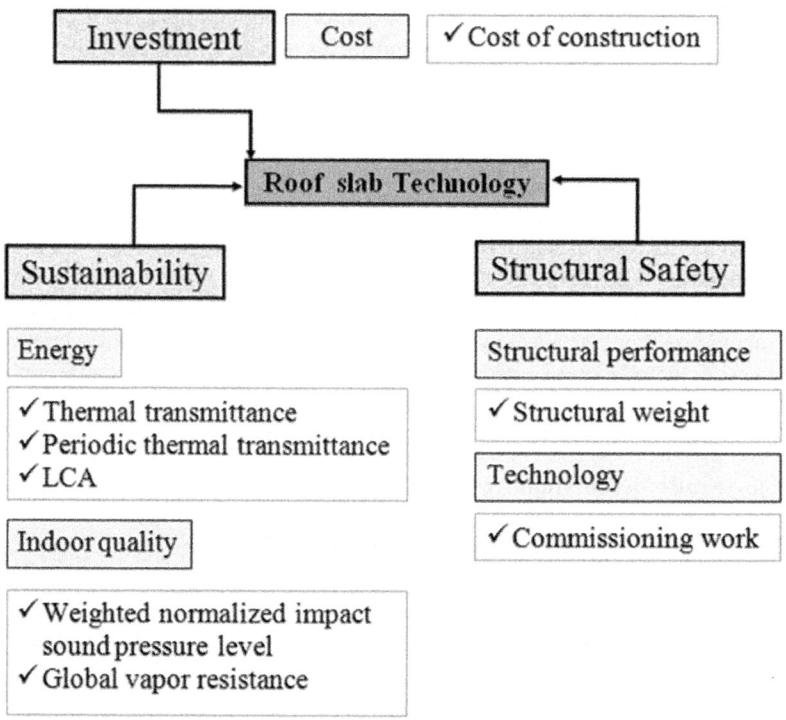

Fig. 2. Hierarchical structure of the evaluation process

The following table shows the values assumed by each criteria for every selected floor (Table 5).

Table 5. Multi-criteria analysis parameters

	U_f W/m^2K	Y_{IE} W/m^2K	LCA P_t	C_I $€/m^2$	$L'_{n,w}$ dB	R_v m^2sPa/kg	S_w kN/m^2	τ_c h/m^2
A1	1.76	0.49	10.12	46.31	51.83	$1200 \cdot 10^9$	3.21	0.42
A2	1.90	0.28	13.19	45.52	51.16	$1200 \cdot 10^9$	3.97	0.21
A3	0.48	0.01	13.25	66.76	56.01	$1440 \cdot 10^9$	2.39	0.36
A4	0.99	0.05	14.37	94.13	49.23	$1150 \cdot 10^9$	3.33	0.70

4   Weighing System Definition

The definition of a weighing system for every item involved in the analysis is another key issue of the MCDM approach. The weight assigned to the criteria, categories and areas of analysis, represents their importance compared to the immediately following level (partial weight) or with respect to the final objective (overall weight). Since there is no unique and standardized method for assigning weights, in this case study different evaluation scenarios are formulated and discussed. In fact the aim of the paper isn't to identify the best design solution of the case study, but to give a road map for developing the evaluation procedure knowing the consequences of the various decisions in a case of interdisciplinary problems. To this aim, 2 scenarios have been considered:

- *Scenario I: Higher weight to Structural Safety category*

In the first scenario, Structural Safety has been characterized by the greater importance. This scenario, therefore, could be representative of an evaluation process for a building construction in a high seismic zone. In seismic areas, indeed, the decision maker or designer attributes inherently greater importance to the structural performance rather than to Sustainability. In this case, a greater weight to the Structural Safety (0.6 in the case study) at the expense of the Sustainability (to which a weight of 0.2 is assigned) has been assigned. The residual weight of 0.2 is assigned to the Economic investment.

- *Scenario II: Higher weight to Sustainability category*

The second scenario is related to new energy efficiency threshold and thus the design of Nearly/Net zero energy buildings (Directive 2010/31/EU). A near-zero energy building is characterized by a very low primary energy needs, covered to a large amount from renewable sources. It is clear that a high performance building component affects the reduction of the final energy needs and therefore it plays a positive role in the global sustainability process. Consequently, greater importance is assigned to sustainability category, favoring the energy, environmental and indoor comfort criteria. Thus, the weight of 0.6 and 0.2 is assigned respectively to sustainability and structural safety area. A weight of 0.2 is assigned to the Investment area as in the previous case.

5   Methodology for Evaluating Criteria Performance

Multi-criteria decision-making (MCDM) methods are formal approaches to structure information and decision evaluation in problems with multiple, conflicting goals. In this paper, the selected solving methodology is TOPSIS based on distances from the ideal solution (Hwang and Yoon 1981). It can be considered a compensatory method because the better solution is the trade-offs between criteria, where a poor result in one criterion can be negated by a good result in another criterion. The best alternative should have the shortest distance from the positive ideal solution and highest distance from the negative ideal solution (also known as anti-ideal solutions). The distances are measured by the Euclidean metric and they are computed for normalized and weighted data. In the following lines, a short description of its application is reported; in particular, seven main steps can be identified.

Firstly, it is necessary to create the evaluation matrix (D) whose generic element a_{ij} expresses the performance of the generic alternative A_i ($i = 1..n$) compared to the generic criterion C_j ($j = 1..m$) and then the normalized matrix (R). Indeed, because the component properties and performance indices have different physical dimensions, the element of the matrix are normalized as:

$$r_{ij} = \frac{a_{ij}}{\sqrt{\sum_{k=1}^{n} a_{kj}^2}} \tag{1}$$

where r_{ij} is in the range of [0-1].

To prescribe the relative priority among the component properties and performance indices, a weight factor (w_{ij}) is given to each of them. This point will be deeply discussed in the next section. Thus, the third step consists in the calculation of the weighted normalized decision matrix (V), where the generic element v_{ij} is obtained as in Eq. (2):

$$v_{ij} = r_{ij} \cdot w_{ij} \tag{2}$$

As further step the ideal (A^-) and negative ideal (A^*) solution have to be calculated. In the matrix R whose elements are normalized and weighted according to Eqs. (1) and (2), the element with the most preferred value (i.e. the highest value in most except for costs) for the j-th component property or performance index is defined as the ideal v_{ij}^*, and the j element with the least preferred value is defined as the non-ideal v_{ij}^-. Hence, the matrices A^* and A^- which consist of v_{ij}^* and v_{ij}^-, respectively, are expressed by Eqs. (3) and (4):

$$A^* = \{(\max_i v_{ij} | j \in J_b), (\min_i v_{ij} | j \in J_c), i = 1, 2, \ldots n\}$$
$$= \{v_{1*}, v_{2*}, \ldots, v_{n*}\} \tag{3}$$

$$A^- = \{(\min_i v_{ij} | j \in J_b), (\max_i v_{ij} | j \in J_c), i = 1, 2, \ldots n\}$$
$$= \{v_{1-}, v_{2-}, \ldots, v_{n-}\} \tag{4}$$

where J_b is associated with the criteria or indices which are regarded as the best, and J_c is associated with the criteria having a negative impact or indices related to price that is, the less, the better. The selection of solution is made upon the distance between the best (S_i^*) and the worst (S_i^-) alternative calculated with the following equations:

$$S_i^* = \sqrt{\sum_{j=1}^{m} \left(v_{ij} - v_j^*\right)^2} \ per \ i = 1, 2, \ldots, n \tag{5}$$

$$S_i^- = \sqrt{\sum_{j=1}^{m} \left(v_{ij} - v_j^-\right)^2} \ per \ i = 1, 2, \ldots, n \tag{6}$$

Then the (C_i^*) of the alternative by S_i^* and S_i^- can be evaluated using the relationship:

$$C_i^* = \frac{S_i^-}{S_i^- + S_i^+} \tag{7}$$

With reference to the value of C_i^*, a preference ranking of alternatives can be made. It is evident that, if A_i coincides with the negative-ideal solution, C_i^* is zero, while for A_i coinciding with A^*, the maximum value of C_i^* is 1. Although, it often happens that the best solution (the one characterized from the highest value of C_i^*) has at the same time the minimum distance and the maximum respectively by A^* and A^-, in certain cases, this condition does not occur.

6   Case Study Results and Discussion

The 4 different RC floors previously introduced have been analysed by the MCDM method: A1-brick-concrete floor; A2- hollow-core floor; A3- expanded polystyrene lightened floor; A4- wood-cement lightened floor.

The 2 scenarios described in Sect. 4 have been identified:

- *Scenario I:* Evaluation process in seismic zone, where the decision maker attributes inherently greater importance to the structural performance;
- *Scenario II:* Projects with energy and environmental performance requirements.

The values of C_i^* are presented in Fig. 3.

The graph of Fig. 3a shows that for scenario I the EPS lightened floor (A3) turns out to be the best alternative (maximum value of C_i^*) together with the precast floor (A2), when higher importance to the structural performance is given. In the first case the lowest self-weight prevails but in the second the success is due to low cost. Indeed the solution A4 (wood-cement floor) has low structural performance both in terms of structural weight and commissioning work; finally the traditional brick floor (A1) results with an intermediate condition.

Moving to analyse the scenario II (Fig. 3b), again the EPS lightened floor appears to be the best solution since it has the lowest value for the added parameters (U_f); the second choice could be the hollow-core floor meanwhile the worst performance competes to alternative A1 about equal to A4.

Finally, the results of proposed case study allow to remark that the adoption of several possible scenarios for definition of the weights can determine a different preference ranking for the alternatives. Thus, it is clear that decision makers can influence differently the final outcome. In the initial phase of decision problem, the final goal and the leading player of the builder process must be identified.

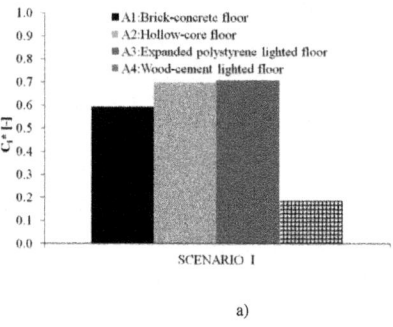

a)

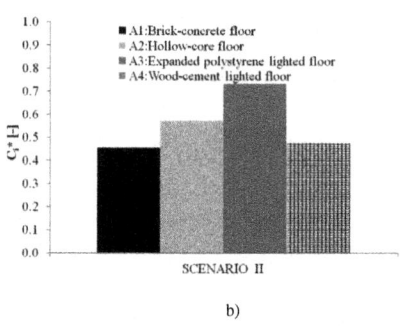

b)

Fig. 3. Results:(a) Scenario I; (b) Scenario II

7   Conclusions

The choice of the best solution for the building envelope components involves the examination of multiple conflicting criteria and objectives, as shown also by the case study of RC floor proposed in this paper. The application of a MCDM method requires at least three complex steps: the definitions of the alternatives and evaluation criteria, and then, the definition of a weighing system.

The application of the procedure to the case study allows to point out that the result changes with decision maker interest. We can conclude that it does not exist the absolute best solution for the building design, but there is a best solution in a particular design context; if the context is clear the MCDM is a useful decision tool.

The findings of the research represented in this paper are expected to be significant in further contributing to MCDM effectiveness in construction and private sectors. In the future, this research could be extended to compare further multifunction technologies as vertical components for the opaque and transparent building envelope, or to evaluate the entire building design configurations.

Acknowledgements. The authors gratefully would like to thank the financial support from the Project Smartcase, MIUR - Italian Ministry of Education, Universities and Research, Managerial Decree n.789 06/03/2014 (ID Number of the Project PON03PE_00093_1).

References

Brown R (2005) Rational choice and judgment: decision analysis for the decider. Wiley, New York

Osborn AF (1957) Applied imagination: principles and procedures of creative problem solving. Charles Scribner's Sons, New York

Eden C (1988) Cognitive mapping: a review. Eur J Oper Res 36(1):1–13

Conklin J (2006) Dialog mapping: building shared understanding of wicked problems. Wiley, Chichester

Keeney RL (1992) Value-focused thinking: a path to creative decision-making. Harvard University Press, Cambridge

Gregory R, Keeney RL (1994) Creating policy alternatives using stakeholder values. Manage Sci 40(8):1035–1048

Min.LL.PP., DM 14 gennaio 2008 Norme tecniche per le costruzioni (NTC). Gazzetta Ufficiale della Repubblica Italiana, no. 29

EN 1992-1-2 2004 Eurocode 2: design of concrete structures - Part 1–2: General rules - Structural fire design

da Silva SM, de Almeida MG (2010) Thermal and acoustic comfort in building. In: Conference paper, Inter.noice, noise and sustainability, 13–16 June 2010, Lisbon (Poertugal)

International Organization for Standardization (ISO) (2008) Building components and building elements—thermal resistance and thermal transmittance—calculation method; International Standard, UNI EN ISO 6946:2008. ISO, Geneva, Switzerland

Italian Government. Ministerial Decree (2015) Decreto requisiti minimi. Ministerial Decree of 26 June 2016

International Organization for Standardization (ISO) (2007) Thermal performance of building components—dynamic thermal characteristics—calculation methods; International Standard, UNI EN ISO 13786:2007. ISO, Geneva, Switzerland

EN 12354-1:2000 Building acoustics. Estimation of acoustic performance in buildings from the performance of elements. Airborne sound insulation between rooms

EN 12354-2:2000-09 Building acoustics - Estimation of acoustic performance of buildings from the performance of elements - Part 2: Impact sound insulation between rooms; German version EN 12354-2:2000

D.P.C.M. (1997) Decreto del Presidente del Consiglio dei Ministri del 5 dicembre 1997. "Determinazione dei requisiti acustici passivi degli edifici" (Gazzetta Ufficiale - Serie generale n. 297)

International Organization for Standardization (ISO). Hygrothermal performance of building materials and products - determination of water vapour transmission properties ISO 12572:2001

International Organization for Standardization (ISO). Environmental management – Life cycle assessment – Principles and framework. International standard ISO 14040:2010

Hwang C, Yoon K (1981) Multiple attribute decision making. In: Lecture notes in economics and mathematical systems, vol. 186. Springer, Berlin

Grierson DE (2008) Pareto multi-criteria decision making. Adv Eng Inform 22(3):371–384

Mardani A, Jusoh A, Nor KMD, Khalifah Z, Zakwan N, Valipour A (2015) Multiple criteria decision-making techniques and their applications – a review of the literature from 2000 to 2014. Econ Res Ekonomska Istraživanja 28(1):516–571. https://doi.org/10.1080/1331677X.2015.1075139

Kabir, G, Sadiq R, Tesfamariam S (2013) A review of multi-criteria decision-making methods for infrastructure management. Struct Infrastruct Eng Maint Manage Life-Cycle Des Perform 10(9):1176–1210. https://doi.org/10.1080/15732479.2013.795978

Fontenelle MR, Bastos L (2014) The multicriteria approach in the architecture conception: defining windows for an office building in Rio de Janeiro. Build Environ 74:96–105

Songa Y, Lia Y, Wanga J, Haoa S, Zhua N, Lina Z (2015) Multi-criteria approach to passive space design in buildings: impact of courtyard spaces on public buildings in cold climates. Build Environ 89:295–307

Gagliano A, Detommaso M, Nocera F, Evola G (2015) A multi-criteria methodology for comparing the energy and environmental behavior of cool, green and traditional roofs. Build Environ 90:71–81

Ascione F, Bianco N, De Masi RF, De Stasio C, Mauro GM, Vanoli GP (2015) Multi-objective optimization of the renewable energy mix for a building. Appl Therm Eng. In Press, Corrected Proof — Note to users. Available online 31 December 2015

Yang M, Lin M, Lin, Y, Tsai K (2016) Multiobjective optimization design of green building envelope material using a non-dominated sorting genetic algorithm. Appl Therm Eng. Available online 13 January 2016 In Press, Accepted Manuscript — Note to users

Avgelis A, Papadopoulos AM (2009) Application of multicriteria analysis in designing HVAC systems. Energy Build 41:774–780

Wimmler C, Hejazi G, Fernandes ED, Moreira C, Connors S (2015) Multi-criteria decision support methods for renewable energy systems on islands. J Clean Energ Technol 3(3):185–195

International Organization for Standardization (ISO). Thermal Performance of Building Components—Dynamic Thermal Characteristics—Calculation Methods; International Standard, UNI EN ISO 13786:2007; ISO: Geneva, Switzerland 2007

Directive 2010/31/EU of the European Parliament and of the Council of 19 May 2010 on the Energy Performance of Buildings (recast). Official Journal of the European Union, 18 June 2010

Turkey – Istanbul - Third Bosporus Bridge

Issues Related to the 305 Meters Tower Erection

G. Fiscina[1], C. J. Garrone[2], R. Sorge[3,4], and M. Mancini[5(✉)]

[1] Astaldi S.p.A., Direttore di Costruzione,
3° Ponte sul Bosforo, Istanbul, Turkey
[2] Astaldi S.p.A., Astaldi, Assistente Direttore di Costruzione,
3° Ponte sul Bosforo, Istanbul, Turkey
[3] Astaldi S.p.A., Responsabile dell'Ingegneria Corporate, Rome, Italy
[4] Astaldi S.p.A., Responsabile della Progettazione,
3° Ponte sul Bosforo, Istanbul, Turkey
[5] Astaldi S.p.A., Servizio Ingegneria,
Responsabile Comunicazione Tecnica di Progetto, Istanbul, Turkey
m.mancini@astaldi.com

Abstract. The aim of the article is to describe the construction method and the difficulties encountered for the execution of an exceptional and unique work within the 3rd Bosphorus Bridge: the erection of the A-shaped towers. The erection of the A-shaped towers was a highly challenging activity due to their geometry and to the two different methods of construction: the slip forming system and an automatic climbing formwork system.

Keywords: Towers' leg · Slip-forming system · Automatic climbing system Mix design · Geometry

1 3rd Bosphorus Bridge

The Northern Marmara Crossing Motorway Project includes the 3rd Bosphorus Bridge (3BB), called Yavus Sultan Selim Bridge.

The 3BB is the widest suspension bridge in the world with a width of 59 m and it has the highest A-shaped towers in the world, with a height of 322 m. The main-span length is 1,408 m, while the total bridge length is 2,250 m.

Moreover, the 3BB is designed as a hybrid bridge with two cable-supporting systems, stay-cables plus two suspension cables and hangers. In this unique project the two cable systems are combined to create a highly rigid solution, called HRSB (Highly Rigid Suspension Bridge). This provides several benefits but the construction phases entail many difficulties. We can affirm that it is the world's best highly-rigid suspension bridge, symbol of modern Turkey (Fig. 1).

© Springer International Publishing AG, part of Springer Nature 2018
M. di Prisco and M. Menegotto (Eds.): ICD 2016, LNCE 10, pp. 17–31, 2018.
https://doi.org/10.1007/978-3-319-78936-1_2

Fig. 1. 3rd Bosphorus Bridge

2   Towers

As stated in the introduction, the 3rd Bosphorus Bridge is a hybrid bridge, therefore, the bridge's A-shaped towers are designed to support the loads transferred by both systems of supporting cables.

The two main cables lay atop the towers, on the tower saddles, while the stay-cables are anchored to specific anchor boxes installed into at the upper portion of the legs (Fig. 2).

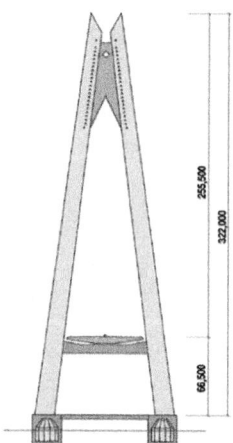

Fig. 2. A-shaped tower

The A-shaped towers' geometry, with two legs having a triangular cross-section, provides enhanced stability. The distance between the two legs of each tower is greater at the bottom level and becomes smaller as height increases. The deck is positioned at one fifth of the total height of the towers, between the two legs.

2.1   Towers' Material

The below table shows the towers' concrete features: strength class, exposure class and the concrete cover thickness (Table 1).

Table 1. Concrete features

Designation	Strenght class (min.)	Exposure class	Concrete cover
Tower's leg (splash zone, up to 30 m)	C50/60	XS3 XC4	50 mm
Tower's leg and concrete crossbeam	C50/60	XS1 XC4	40 mm

Towers' reinforcement steel ratio equals 2%, and a high grade B500B steel has been used.

2.2   Legs

The A-shaped towers' legs have been designed with a triangular hollow cross-section, in order to provide enhanced stability. At the bottom, the distance between the legs are is of 70 m while, at tower's top, the distance decreases up to 8 m.

The below image shows the fixed as well as the variable dimensions of each tower leg throughout its extension (Fig. 3).

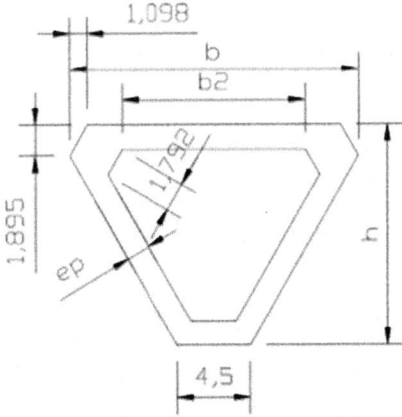

Fig. 3. Tower's leg cross-section

The thickness of concrete decreases as the leg's height increases. At the bottom, the thickness (ep) is of 1.5 m while at the top, concrete thickness is of 0.75 m.

The walls' inclination has two reference angles: 84° on the deck side and 96.75° for the external side (Fig. 4).

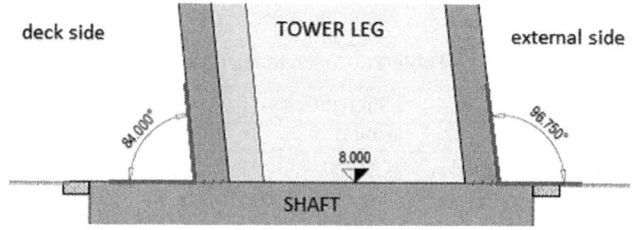

Fig. 4. Walls' inclination

2.3   Crossbeams

In order to further stiffen the whole tower structure, the legs are connected together at three levels: at the deck level (60 m) by a highly pre-stressed concrete crossbeam and at the top by two X-shaped steel bracings, at a height of +266 m and +301 m, respectively. The two X-shaped bracings are lined up with large 'butterfly-shaped' steel panels.

2.4   Diaphragms

Inner intermediate diaphragms are located at different levels in order to increase the global stiffness and also to easily support locally applied forces originating from the temporary struts installed during erection.

2.5   Temporary Struts

During the construction phases, between the to-wer's legs, four temporary steel struts have been installed with the purpose of increasing stability. They were installed approximately every sixty meters, at +60 m, +112 m, +162 m and +208 m.

2.6   Stay-Cables' Anchor Boxes

Stay-cables are anchored into the tower in a series of steel anchorage boxes, embedded in the legs' concrete structure. Each box provides anchorage to one pair of stay-cables, therefore the horizontal action of the main-span cables are transferred to the end-span cables inside the anchorage box (Fig. 5).

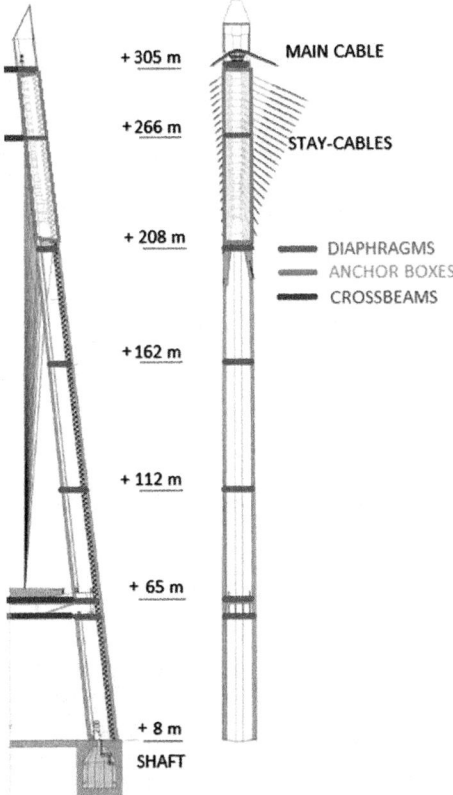

Fig. 5. Leg's elements

3 Construction Method

The erection of the 3rd Bosphorus Bridge's A-shaped towers was a very challenging work.

Several difficulties have been encountered due to the legs' complex geometry; as described, the four tower legs are hollow, tapered and extremely tall. Moreover, also the seasons' temperature range had to be taken into consideration since it can exceed 30 °C.

With the purpose of reducing the duration of construction works, two different construction methods have been adopted: the slip-forming system and an automatic climbing system (ACS).

The sliding system allows having a high casting rate, but it was not possible to use that system for the total height of the legs. At the anchor-boxes level there would be interference between the slip-form and the stay-cable anchorage. Thus, from +208 m on it was necessary to adopt the ACS erection method.

3.1 Slip-Forming System

The slip forming system is a continuous casting process where the concrete is poured into a moving form-shutter.

The slip-form is a machine moved continuously by numerous hydraulic jacks and it has three working platform levels: at the upper level (3) the reinforcement steel frame is installed; at intermediate level (2) the concrete is poured into the form-shutter (1 m high) and it is vibrated; at the lower level (1) the casts' results are checked and the repairs are carried out (Figs. 6, 7, 8 and 9).

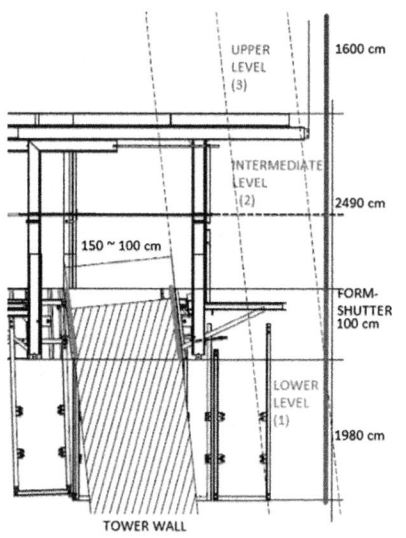

Fig. 6. Slip-form cross section

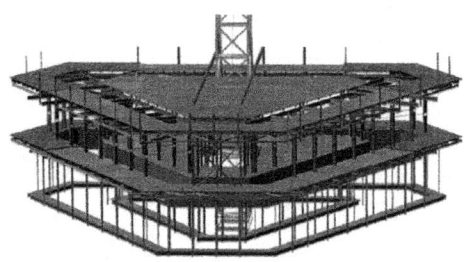

Fig. 7. Slip-form 3D view

Fig. 8. 3BB Slip-form photo

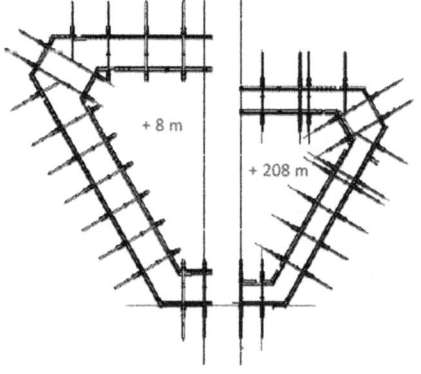

Fig. 9. Slip-form section

In the 3rd Bosphorus Bridge Project the slip-forming system has been used for the towers' legs erection, from the shaft level up to +208 m, the stay-cables' anchor boxes lower level.

The 3BB slip-forms have been designed in order to obtain tapered legs, following a fixed casting angle. The system considers also the concrete thickness changing according to the height, from 150 cm to 100 cm. The form-shutter is equipped with numerous horizontal threaded bars, attached to the inner and outer yoke legs, turning those bars the concrete thickness decreases. The threaded bars are operated in steps of every one meter in height.

During the casting process, concrete is poured in layers of a thickness of 20–30 cm. The form shutter is filled approximately every four hours, when the concrete is sufficiently cured and depending upon the ambient conditions In order to obtain a high casting rate, the slip-form system adopted for the 3rd Bosphorus Bridge project has been arranged for the cold season (the period of expected execution of the works and it has been equipped with heaters and protective tents. By this way, it was possible to cast up to two meters in height per day, which is a very high casting rate.

The concrete is pumped from the shaft area by the use of stationary pumps. For each leg two indepen-dent pipelines are installed, one pipeline is used just in case of the blockage of the active one. The pipelines end in a hopper that uniformly distributes the concrete into the form shutter.

The consistency of the poured concrete is controlled constantly. It is checked by the insertion of a steel bar into the concrete: if the bar can not be pushed down by more than 60 cm then it is considered that the lower concrete layer has reached the required curing and the slip-form may be lifted so as to continue the casting process. Due to the continuous advancement of the slip-forms, concrete was required to be carefully controlled and monitored.

At erection start phase, it was decided to installan 18-meter-tall sliding-guide mast into each hollow leg. The guide mast is supported by a rail system, fixed to the hard concrete, and it is connected to the slip-form by a horizontally controllable joint system, which gives the possibility to adjust the slip-form position (Fig. 10).

Fig. 10. Guide mast

3.2 Automatic Climbing System

The Automatic climbing system (ACS) is a casting process that makes use of a hydraulically operated self-climbing formwork. The formwork is lifted and installed at a higher level once the poured concrete has reached the desired curing grade.

In the 3rd Bosphorus Bridge Project, the ACS is used for the erection of towers' legs from +208 m up to +305 m, along the anchor boxes zone. Just like the slip-forms, the ACS has been designed in order to obtain tapered legs, following a fixed casting angle and, in addition, the possibility to host the anchor boxes' form-tube.

The ACS is organized in six levels: at the upper levels (6 and 5) the reinforcement steel frame is installed; at the intermediate levels (4 and 3) the concrete is poured into the 4.6 m high formwork and it is vibrated; at the lower level (2 and 1) the casts' results are checked and the repairs are carried out. The ACS is anchored to the leg by means of the climbing cones; those elements are removed during erection from the lower levels to the upper levels of the ACS (Figs. 11, 12 and 13).

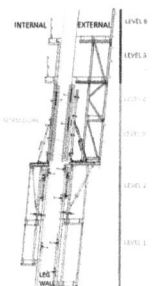 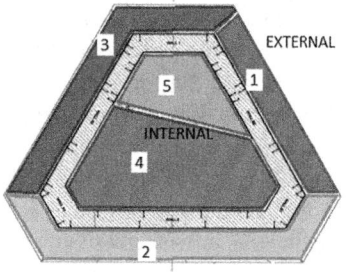

Fig. 11. ACS section   **Fig. 12.** 3BB ACS photo   **Fig. 13.** ACS's modules

Whenever the formwork has to be raised at a new level, the ACS structure is divided in five modules and then lifted up separately by the use of indepen-dent hydraulic cylinders. At the new casting level, the modules are connected and the distance between the formwork is adjusted. The concrete thickness decreases - from 100 cm to 75 cm - as height increases.

With the ACS method, one casting cycle takes five days; hence the casting rate is much lower than the slip-form system. Notwithstanding the low casting rate, the ACS method doesn't present specific problems related to the construction method, thus it can assure good quality results.

4 Difficulties and Solutions

During the erection of towers' legs, some optimizations of the process were required. In this paper, we focused our attention on the issues concerning the mix design and the complex legs' geometry. These optimizations where needed especially during the use of the slip-forming system.

4.1 Mix Design

The towers' specifications require a concrete mix design that provides a strength class C50/60, in accordance to the *TS EN 1992-1-1. Design requirements are set forth here below*:

- Minimum binder content: 300–400 kg/m^3;
- Water/binder ratio: 0.33–0.43;
- Size of coarse aggregates: from 4.75 mm to 25 mm;
- The binder combination shall be OPC/GGBS with high slag content 36–80%.
- Silica fume < 8% of binder content
- Average chloride diffusion coefficient D_{cl^-} (28 days) $\leq 3.0 \times 10$–12 m^2/s.

Consequently, the towers' mix design has been studied in accordance with the above requirements. A particular attention has been drawn to concrete pumpability: due to the extraordinary height to be reached, class S4 turned out to be necessary.

As a result, a mix was designed as set forth in the following tables (Table 2).

Table 2. Concrete mix design

Mix design	Weight (kg/m^3)	Absolute volume (dm^3/m^3)
Air		25
Water	138	138
Cement (CEM I 42,5 R)	135	43
GGBS	265	93
Micro-Silica	0	0
Fly-Ash	0	0
FA 1 (natural sand)	495	187
FA 2 (crushed sand)	374	139
CA 1 (10 mm)	431	159
CA 1 (22 mm)	562	208
Admixture (Optima 100)	6,400	5,565
Admixture (Air + retarder)	1,800	1,667
Tot (kg/m^3)	2407,2	

Before starting the erection of the legs by means of the slip-forms, a small mock-up has been set on site with the intention of controlling the concrete-formwork interaction, verifying the concrete curing period and consequently the casting rate. The first results obtained were not positive: the concrete surface was affected by clear signs of tears, large horizontal cracks and bad surface appearance with holes and big porosities.

When the slip-form panel (form-shutter) slides on a fresh concrete surface, the friction between the moving panel and the concrete increases. Therefore, if the concrete is too sticky, it tends to remain stuck to the panel thus creating evident tears on its surface (Fig. 14).

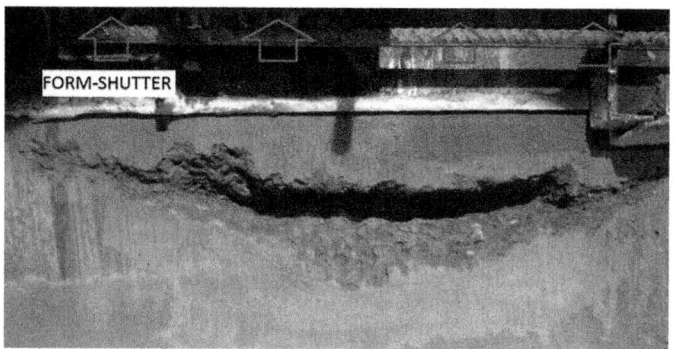

Fig. 14. Moke-up: superficial concrete tear

So, a change in concrete composition turned out to be necessary. The main intention of the analysis was to identify the parameters affecting friction, and consequently determine the slip-form lifting force.

There are several parameters influencing the concrete interaction with the form-shutter. It may be asserted that the amount of the fines is directly related to concrete cohesion and adhesion. Natural round aggregates require less lifting force and the increasing amount of sand reduces friction forces.

As a result of the study, a new concrete mixture was designed as shown in the below table (Table 3).

In addition, in order to minimize friction between the formwork and the concrete, it has been decided to cover the slip-form panel before each cast with a 'lubricant' layer. This layer, made of cement slurry mixed with fine sand, provides low shear strength.

At the end, all the mock-up tests with the adjusted concrete mix design gave positive result.

By comparing the two mix designs it may be observed that water and cement weight increased significantly, by 15% and 61%, respectively. Furthermore, also the content of crushed sand and small coarse aggregates (10 mm) increased, by 7% and 23%, respectively, in spite of other coarse aggregates and natural sand. Moreover, it has been established that admixtures with retarding agent were not needed.

The above changes permitted to increase concrete workability and, therefore, to decrease the concrete stickiness and the friction between the form-shutter and the concrete. Consequently the overall defectiveness level was acceptable.

Table 3. Modified concrete mix design

Mix design (modified)	Weight (kg/m³)	Absolute volume (dm³/m³)
Air		25
Water	159	159
Cement (CEM I 42,5 R)	217	69
GGBS	217	76
Micro-Silica	0	0
Fly-Ash	0	0
FA 1 (natural sand)	410	155
FA 2 (crushed sand)	402	149
CA 1 (10 mm)	529	196
CA 1 (22 mm)	475	176
Admixture (Optima 100)	5,200	4,529
Admixture (Air + retarder)	0	0
Tot (kg/m³)	2414,2	

Furthermore, the compressive tests on the core samples showed excellent results. The results showed a very high strength of the material, concrete being classified in class C57/70, a class higher than the designed one's.

During the start phase, also the repair method for the surface defects was defined:

- Below 0.2 mm: no repairs
- From 0.2 mm to 1 mm: epoxy resin injection
- From 1 mm to 6 mm: V cut and grouting
- Above 6 mm: chipping and grouting

The material used for repairs is a non-shrinking concrete mix with small-size aggregates (10 mm max) and charged with textile fibres.

4.2   Geometry Control

A constant monitoring of the towers' geometry was carried out during the whole erection phases. For each leg four aiming points have been taken in consideration, for both the erection systems (Fig. 15).

Fig. 15. Aiming points

All the survey activities have been carried out in early morning, before sunrise, in order to minimize the effects of heat deformation.

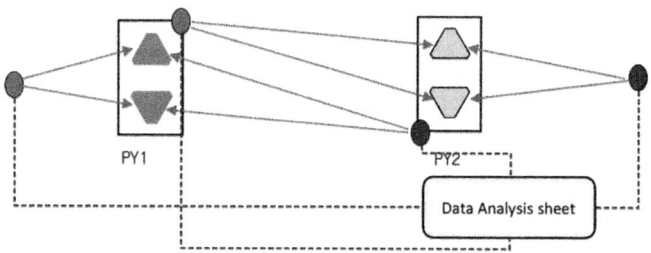

The 3BB survey activities have been carried out by 4 teams, 2 teams based on the European side and 2 teams based on the Asian side, the teams placed the survey instruments in the locations shown in the above scheme.

At the beginning stage, the geometry control was showing bad results. Up to the first 30 m the geometry was out of the imposed tolerances, reaching a mismatch value of almost 200 mm from the target value. Therefore corrective actions became necessary.

It has been decided to install an 18-meter-tall guide mast equipped with a horizontally controllable joint system and perform, during erection, slight adjustments of the slip-form position. Furthermore, it has been decided to limit the geometry error within a range of 50 mm between the target value and surveyed one.

When the geometry control exceeded the admissible range, the form-shutter position or the formwork position had to be adjusted by operating on the mast's horizontal joint system before continuing the casting process. The fixed tolerance has been restored within the range of 50 mm.

The following chart shows the transversal error monitored during the Europe-North leg erection when using the slip-forming system (Fig. 16).

The straight lines identify the envelope of the maximum/minimum transversal error, while the two lines into the envelop report the maximum/minimum transversal error of the whole section, along the leg height. Thanks to the corrective actions, from 30 m in height, the precision has been significantly improved reaching good results till +208 m.

The bridge's overall structural behaviour has been re-verified after completion considering the geometry gaps with respect to target values. This study confirmed that the current geometrical error has no influence on the bridge's global behaviour (Fig. 17).

During the slip-forming erection phase, thanks to the constant geometry monitoring, an additional problem became apparent: the legs' torsion.

This problem arises due to the complex legs' geometry and especially due to the legs' inclination. During the lifting of the form-shutter actions were generated causing the cross-section to rotate around its own axis.

Generally the section rotation can be prevented by the installation of a truss connecting the slip-forms, but due to the great distance between the two legs (70 m at the

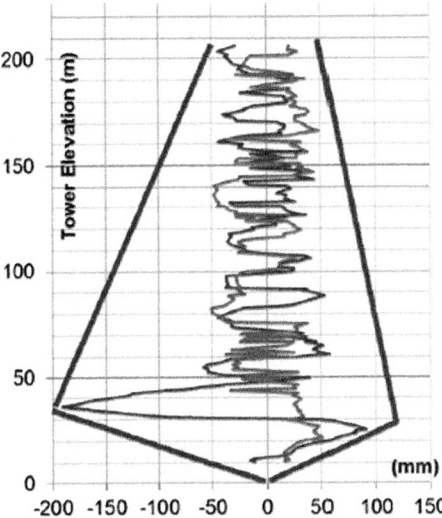

Fig. 16. Geometry control measurements

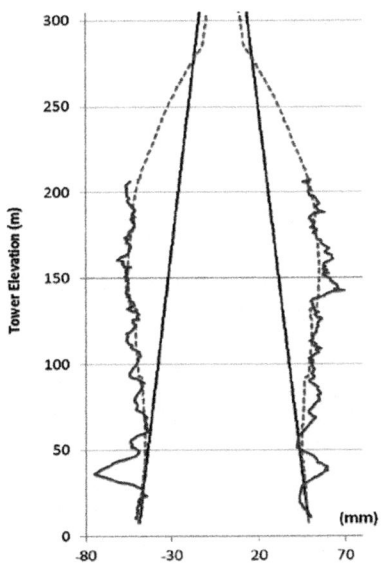

Fig. 17. Geometrical control and legs' precamber

shaft level) it wasn't technically possible. Therefore, the guide mast resulted useful also to mitigate the slip-form rotation.

At the beginning stage, when the guide mast wasn't installed yet, the slip-form rotation affected the thickness of the concrete cover. 40 mm is the minimum measured value and since it is lower than the minimum value for the splash zone (50 mm), it has been requested to verify whether that thickness could guarantee a long durability and appropriate protection from carbonation and chloride diffusion. The tests reported positive results.

5   Conclusions

At the beginning construction phase, several difficulties and quality-related problems became apparent from the moke-up test results.

Hence, proper full scale tests had to be carried out well in advance, with the purpose of verifying and determining all the parameters before the beginning of the works. Changes in mix design and optimisation during the construction phases represent a very critical aspect, due to their long development and long verification time.

In term of time of execution, the use of the slip-form system for this specific project turned out to be not effective, considering also the need of changing the formwork system at 208 m.

Taking in account the towers' construction timeline, it may be affirmed that if the ACS had been adopted since the beginning, the towers' construction duration would have been similar, may be slightly faster.

The below chart shows the comparison between the actual case (continuous line) and the assumption in which only the automatic climbing system is used (dashed line) (Fig. 18).

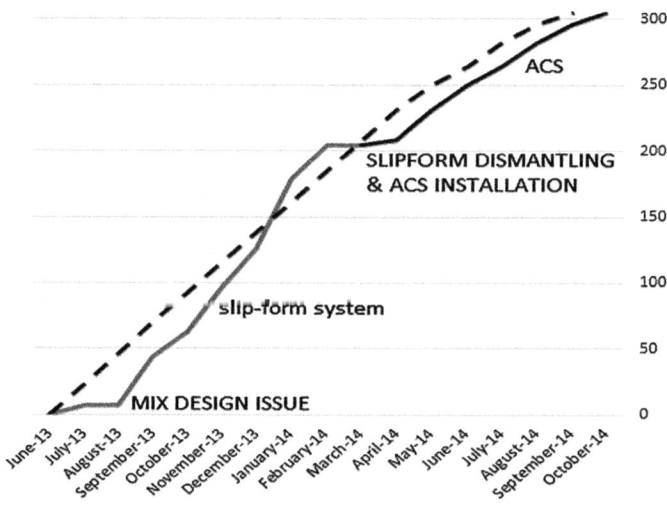

Fig. 18. Actual timeline vs. ACS assumption

However, we can assert that the erection of the four legs took a short time, just 17 months, despite all the optimisations needed. Undoubtedly, it was one of the most challenging activities within the construction of the 3rd Bosphorus Bridge (Figs. 19, 20 and 21).

Fig. 19. The towers at the initial stage

Fig. 20. 3rd Bosphorus Bridge tower

Fig. 21. 3rd Bosphorus Bridge tower

References

1. BB3 Concrete Works – Specifications
2. BB3 Technical Specifications steel works
3. BB3 Design Basis
4. BB3 Monitoring specifications
5. BB3 Method Statement for Construction of the Towers (until at EL+208.0 m)
6. BB3 Method Statement for ACS Operations
7. BB3 Material Approval – Concrete Mix Design
8. BB3 Precamber and Target Coordinate Report
9. BB3 Method Statement for Mock-up Test
10. BB3 T-ENG Design Report Concrete Pylon Sections
11. BB3 T-ENG Technical report – Concreting Technology
12. T-ENG BB3 Slipform difficulties – Acceptance of repairs on the damaged bottom part
13. T-ENG BB3 Slipform difficulties – Designer remarks on work plan for recommencing and conclusion report
14. BB3 Surface Treatment and Repair Method for Tower Concrete
15. BB3 Acceptance Criteria for Tower Concrete Reparations

Validation of NLFEA of Reinforced Concrete Walls Under Bidirectional Loading

B. Belletti[1(✉)], M. Scolari[1], J. Almeida[2], and K. Beyer[2]

[1] DICATeA, University of Parma, Parma, Italy
beatrice.belletti@unipr.it, scolari.matteo87@gmail.com
[2] EESD, École Polytecnique Fédérale de Lausanne (EPFL), Lausanne, Switzerland
{joao.almeida,katrin.beyer}@epfl.ch

Abstract. Nonlinear Finite Element Analysis (NLFEA) of the inelastic behaviour of RC walls are often carried out for uni-directional (in-plane) horizontal cyclic loading. In this paper the behaviour of RC walls with different cross-sections (T-shaped and U-shaped) subjected to bi-directional (in-plane and out-of-plane) loading is simulated by means of NLFEA. They are carried out with the software DIANA, using curved shell elements and a total strain crack model for concrete and embedded truss elements adopting Monti-Nuti model for the reinforcement. The aim of this paper is to validate this type of analysis by comparing the obtained results with experimental outcomes of two different RC slender walls, a T-shaped wall and a U-shaped wall, tested under quasi-static bidirectional cyclic load. In particular, the focus is on the comparison between different crack models (Fixed and Rotating crack models) and on the calibration of the Monti-Nuti model parameters for steel. NLFEA is found to acceptably simulate both the in-plane and out-of-plane behaviour observed during the experimental tests. The present work is the starting point for future research in which parametric studies on the influence of reinforcement content and detailing will be performed, assessing their influence on the bidirectional response of RC walls and namely on other less known deformation modes such as out-of-plane instability.

Keywords: Reinforced concrete walls · Thin walls · Out-of-plane instability
Nonlinear finite element analysis · Cyclic load

1 Introduction

Reinforced concrete (RC) walls are commonly used as the primary lateral force-resisting system for medium to high-rise buildings. Nevertheless, despite many years of research and subsequent evolution of code provisions, several RC walls still underperform when subjected to seismic actions, as demonstrated by the recent earthquakes in Chile (February 2010) and New Zealand (February 2011).

The observation of the damage occurred in these later seismic events (Wallace 2012) highlighted that many walls, in their failed configuration, were characterized by large out-of-plane displacements. The failure of these structural walls was hence caused or highly influenced by the out-of-plane buckling (also defined as out-of-plane

© Springer International Publishing AG, part of Springer Nature 2018
M. di Prisco and M. Menegotto (Eds.): ICD 2016, LNCE 10, pp. 32–48, 2018.
https://doi.org/10.1007/978-3-319-78936-1_3

instability), which is triggered in the end region of the wall. It is noted that this deformation mode may be the result of the application of pure in-plane cyclic loading, see Fig. 1.

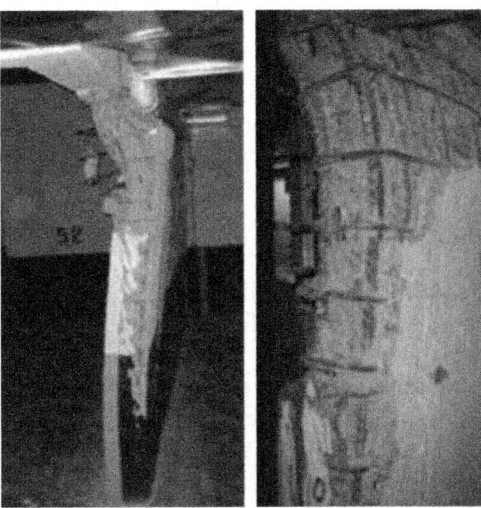

Fig. 1. Out-of-plane buckling of walls after 2010 Chile earthquake (Wallace 2012)

These seeming faults in the design of structural walls are emphasized also by analysing international code provisions. Indeed, the majority of standards treats the out-of-plane instability in a simplified way by imposing limits on the height to thickness ratio of the wall. Only the New Zealand code includes more sophisticated models based on the studies by Goodsir, Paulay and Priestley (Goodsir 1985; Paulay and Priestley 1993).

Prior to the Chile and New Zealand earthquakes this failure mechanism had only been observed in laboratory tests (Oesterle 1979; Vallenas et al. 1979; Thomsen IV and Wallace 2004). In the recent years the research on this topic has been continued by joining experimental test campaigns with advanced nonlinear finite element analysis (NLFEA).

NLFEA are a widely validated numerical tool to simulate the behavior of RC walls, both considering pushover (Belletti et al. 2013; Damoni et al. 2014) and cyclic analyses (Belletti et al. 2016a, b). Recently, this numerical tool was also applied to the study of out-of-plane instability. In particular, Dashti et al. (2014) and Dashti et al. (2015) studied by means of NLFEA the behavior of rectangular RC walls subjected to uni-directional (in-plane) horizontal cyclic loading.

The present paper focus on the behavior of walls subjected to bidirectional (in-plane and out-of-plane) horizontal loading, addressing in particular the influence of out-of-plane instability that can occur for thin members.

For this purpose, the experimental response of two test units was compared with the results of NLFEA. The latter are carried out using two different crack models, a "Fixed crack model" and a "Rotating crack model", in order to highlight the differences between them.

2   NLFE Model Description

The analyses herein shown were performed with DIANA 10 (Manie 2015). According to (Belletti et al. 2014), concrete was modelled using 4 nodes curved shell elements (named Q20SH in DIANA) with 4 Gauss integration points over the element area and 5 Simpson integration points over the element thickness. Reinforcing bars were modelled using embedded reinforcement considering perfect bond between reinforcement and concrete. Each bar element is characterized by 2 Gauss integration points along the truss axis.

2.1   Concrete Model

The Total Strain Crack Model, available in DIANA (Manie 2015), was used to model the concrete behaviour. Moreover, two different approaches were analysed in this paper: the "Fixed crack model" approach and the "Rotating crack model" approach.

For concrete in tension an exponential behaviour based on the definition of the fracture energy in tension, G_F, and of the crack bandwidth, h, have been adopted. G_F was calculated according to the proposal in fib-Model Code 2010 while h was assumed equal to the square root of the area of each element, as suggested by Guidelines for Non-linear Finite Element Analyses (2012).

For concrete in compression two different stress-strain relationships were used. The behaviour of unconfined concrete has been modelled with a parabolic relationship, according to (Feenstra 1993), based on the definition of the crack bandwidth, h, and of the fracture energy in compression, G_C, assumed equal to 250 times G_F (Nakamura and Higai 2001). The confinement effect due to the presence of stirrups in the boundary elements of some of the analysed walls was taken into account adopting the model proposed by Mander et al. (1988), as shown in Fig. 2.

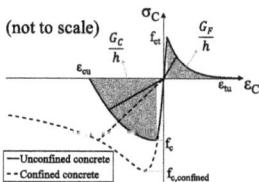

Fig. 2. Constitutive model adopted for confined and unconfined concrete

The reduction of the compressive strength of concrete, due to tensile strains perpendicular to the principal compressive direction, was taken into account according to Vecchio and Collins (1993). These authors fix the lower bound of this reduction curve in 0.6.

For what regards the fixed crack model, due to the fact that after cracking the reference coordinate system is fixed and determined by the crack direction, shear stresses and strains developed along the crack. In this paper the shear stiffness after cracking is reduced using a constant shear retention factor, equal to 0.03, that multiplies the elastic shear modulus of concrete.

2.2 Steel Model

The cyclic behaviour of steel was considered using the Monti-Nuti model (Monti and Nuti 1992) available in DIANA, Fig. 3.

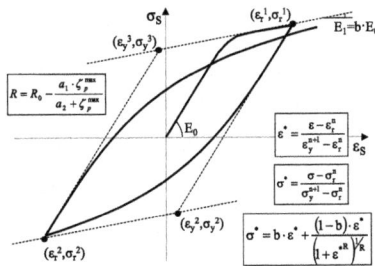

Fig. 3. Monti-Nuti model adopted for steel modeling

The model proposed by Monti and Nuti is expressed in terms of a dimensionless stress σ^* and strain ε^*. The curvature parameter R is a function of the initial curvature R_0, of the maximum plastic excursion developed ζ_p^{max} and of two material parameters a_1 and a_2. In order to investigate the influence of these parameters on the stress-strain relationship a preliminary parametric study was carried out.

The parametric study was based on the experimental campaign carried out on $\phi12$ bars tested at the ETH Zurich by Thiele et al. (2001), Fig. 4.

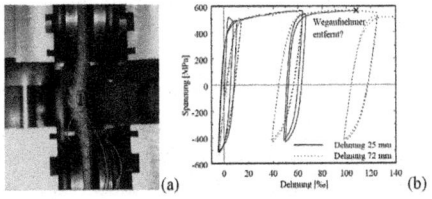

Fig. 4. Reference experimental campaign for parametric study on Monti-Nuti parameters: (a) experimental set-up, (b) experimental results (Thiele et al. 2001)

The reference parameters, according to Fragiadakis et al. (2007), were set equal to R0 = 20, a_1 = 18.45 and a_2 = 0.15. In the following the results obtained by changing the parameter R0 (Fig. 5-a), the parameter a_1 (Fig. 5-b), and the parameter a_2 (Fig. 5-c) are reported.

From Fig. 5-a it can be seen that the increasing of the parameter R0 leads to an increasing of the curvature of the stress-strain relationship, even if the scatter in NLFEA results is rather small and all the curves obtained by means of NLFEA do not quite match the experimental results.

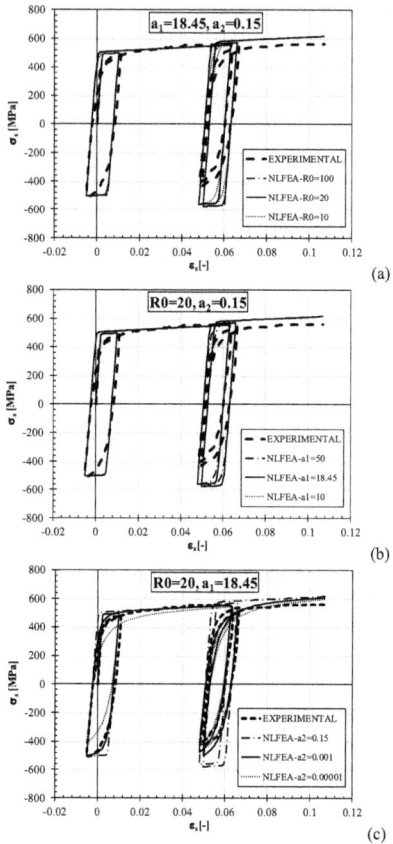

Fig. 5. Parametric studies on Monti-Nuti model: (a) influence of R0, (b) influence of a1, (c) influence of a2

The results reported in Fig. 5-b highlight how increasing the a_1 parameter the curvature decrease. Also in this case the scatter between the experimental and the NLFEA results is relatively large for all the different values of the a_1 parameter.

Finally, the curves reported in Fig. 5-c shown that the slope of the stress-strain relationship strongly depends on the a_2 parameter. In particular assuming $a_2 = 0.001$ the NLFEA curve fits well the experimental results. For this reason, the following final set of parameters (R0 = 20, $a_1 = 18.45$ and $a_2 = 0.001$) will be used for the analyses of the walls.

3   Case Studies

Two different case studies were analysed in this paper: a thin T-shaped RC wall, TW4 (Rosso et al. 2015), and a U-shaped RC wall, TUC (Constantin and Beyer 2016).

3.1   TW4

3.1.1   Experimental Set-up

TW4 is a T-shaped RC wall that was tested at EPFL (Rosso et al. 2015). TW4 is a 2:3 scale wall subjected to a combination of in-plane and out-of-plane quasi-static cyclic loading. The wall was 2000 mm tall, 80 mm thick and 2700 mm long. At the north end (see Fig. 6) the wall presented a flange 80 mm thick and 440 mm long. The foundation was 3600 mm long, 700 mm thick and 400 mm tall, while the top RC beam was 3160 mm long, 440 mm thick and 420 mm tall, Fig. 6.

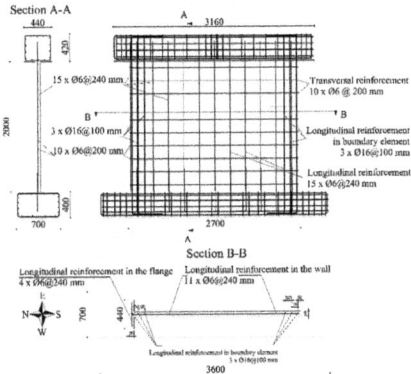

Fig. 6. Geometrical characterization and detailing of TW4 wall (measure in cm)

Figure 6 also reports the reinforcement details, which follow current design practices in Colombia. According to these detailing practices the reinforcement consisted of a single layer of grid reinforcement, characterized by a slight eccentricity with respect to the centerline of the section. The detail of the eccentricity of the rebar layer is shown in Fig. 7.

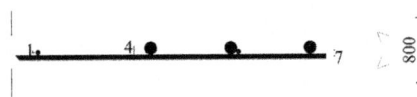

Fig. 7. Detail of the eccentricity of the single layer of reinforcement for specimen TW4 (measure in mm)

The longitudinal reinforcement is characterized by φ6 bars spaced of 240 mm, moreover both the extremities of the wall (the so-called boundary elements) were over reinforced with 3 longitudinal rebars of 16 mm. The transversal reinforcement is characterized by φ6 bars spaced of 200 mm.

During the test the axial load ratio was maintained constant and equal to 5% (equivalent to an applied constant axial force of 330 kN).

The test unit was subjected to horizontal bi-directional loading (in-plane and out-of-plane). The in-plane load was applied with a shear span of 10 m, achieved by coupling an horizontal actuator to impose the in-plane loading and two vertical actuators, acting on the wall ends, to impose the required bending moment. Two additional horizontal actuators were connected to the top beam to impose the horizontal out-of-plane displacement. The applied loading history is reported in Fig. 8.

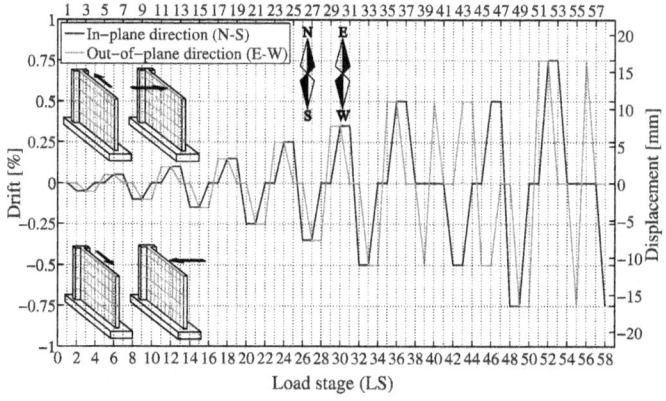

Fig. 8. Loading history applied for test unit TW4 (Rosso et al. 2015).

The material properties, derived from material tests, as well as the model parameters, for both concrete and steel are listed in Table 1.

Table 1. Main mechanical properties derived from material tests and model parameters for test unit TW4

	From material tests			Calculated and used for NLFEA			
	f_y [Mpa]	f_u [Mpa]	E_s [Gpa]	b [-]	R0 [-]	a_1 [-]	a_2 [-]
STEEL φ6	460	625	183.5	0.009	20	18.45	0.001
STEEL φ16	565	650	208.1	0.005	20	18.45	0.001
	f_c [Mpa]	E_c [Gpa]	f_{ct} [Mpa]	G_F [N/mm]	G_c [N/mm]	shear retention factor [-]	
CONCRETE	31.2	29.2	1.46	0.1356	33.9	0.03	

3.1.2 NLFE Model

A sketch of the model used for NLFEA is depicted in Fig. 9.

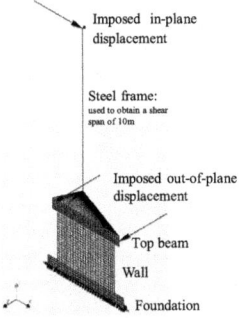

Fig. 9. TW4, NLFE model

The foundation and the top beam were modelled with concrete adopting an elastic behaviour while the wall was modelled with nonlinear behaviour for concrete and steel, according to the stress-strain relationships shown in Fig. 10, which are computed with the material properties listed in Table 1.

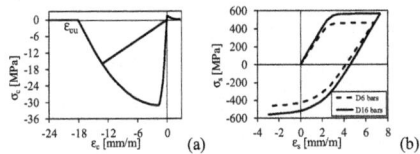

Fig. 10. TW4, constitutive models for (a) concrete and (b) steel

The steel frame indicated in Fig. 9 was adopted in the NLFE to simulate the applied experimental shear span of 10 m. This frame is obtained using 2-node beam elements, with 2 Gauss integration points along the beam axis each, connected to the upper part of the top RC beam with a hinged connection. The legitimacy of this frame modelling approach was validated by verifying the ratio between shear and bending moment at the base of the wall.

3.1.3 NLFE Results

Figure 11 describes the comparison between experimental results and NLFEA results for both "Fixed crack model" and "Rotating crack model" in terms of in-plane force *vs* displacement response. The points in which NLFEA reached the crushing of concrete are also signaled. Crushing of concrete is attained when the compressive strain of concrete reaches the ultimate value, ε_{cu}, Fig. 10-a.

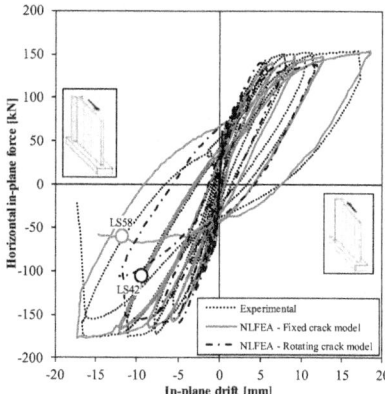

Fig. 11. TW4, in-plane force *vs* displacement response: comparison between experimental and NLFEA results

The results reported in Fig. 11 show that both the "Fixed crack model" and the "Rotating crack model" are able to predict the experimental response. Moreover, the "Rotating crack model" failed at LS42 (corresponding to −0.5% in-plane drift) as a result of the crushing of concrete at the base of the boundary element (Fig. 12-a), while the "Fixed crack model" failed at LS58 (corresponding to −0.75% in-plane drift) as a result of crushing of concrete triggered by damage induced by out-of-plane deformations (Fig. 12-c), which is the same failure mode that was experimentally observed (Fig. 12-b).

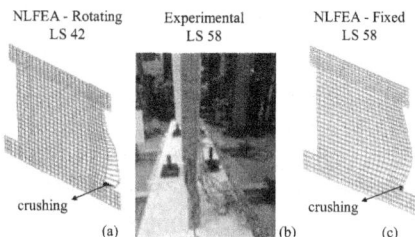

Fig. 12. TW4, failure modes: (a) NLFEA Rotating crack model, (b) Experimentally observed, and (c) NLFEA Fixed crack model (the contour in NLFEA results represent the compressive strain)

This fact can be explained considering that for the "Rotating crack model" the stress-strain relationship is evaluated in the principal direction also upon cracking, while for the "Fixed crack model" the reference system in which the stress-strain relationship is evaluated remains fixed upon cracking and thus a shear strain occurs that leads to the reduction of the compressive strain.

In order to highlight this a further analysis was carried out using the "Rotating crack model" by increasing the fracture energy in compression, G_C, from 250 times G_F (33.9 N/mm) to 500 times G_F (67.8 N/mm), where G_F represents the fracture energy in tension. The results are reported in Fig. 13.

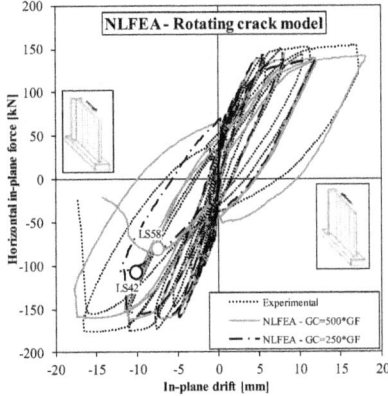

Fig. 13. TW4, influence of fracture energy in compression on "Rotating crack model"

Figure 13 shows that by increasing the fracture energy in compression the results obtained using the "Rotating crack model" are quite similar to the results obtained using the "Fixed crack model" and also the experimental results. Assuming the fracture energy in compression equal to 500 times G_F, failure occurs at LS58 (corresponding to −0.75% in-plane drift) due to the crushing of concrete in the compressive edge of the wall, which is in accordance with the experimental outcomes.

The mechanism of out-of-plane instability, as analytically described by Goodsir (1985) and experimentally confirmed by Rosso et al. (2015), occurred after the development of large tensile strains in the boundary element; due to the cyclic load applied, when the boundary element is reloaded in compression and before cracks close, the compression force is supported only by the vertical reinforcement potentially leading to out-of-plane instability. For this reason, the comparison between NLFEA and experimental results is extended to the evaluation of the out-of-plane displacement along the height of the wall in two different significant load steps, namely LS38 and LS47-48 when the load is applied from the flange to the free edge of the wall and in correspondence of an in-plane drift of 0 mm, when the out-of-plane was maximum.

The results reported in Fig. 14 are obtained using for the "Fixed crack model" $G_C = 250 \cdot G_F$ (33.9 N/mm) and for the "Rotating crack model" $G_C = 500 \cdot G_F$ (67.8 N/mm).

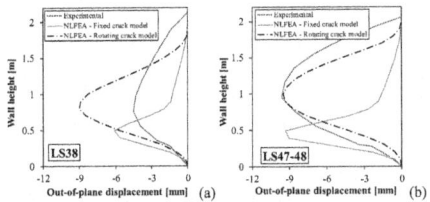

Fig. 14. TW4, out-of-plane displacement along the height of the wall: (a) LS38 and (b) LS47-48

From Fig. 14 it can be noted that both the "Fixed crack model" and the "Rotating crack model" are able to satisfactorily simulate the overall out-of-plane behaviour of the wall even if for the "Fixed crack model" the maximum out-of-plane displacement was detected at a lower height when compared with the experimental results and the "Rotating crack model" results.

3.2   TUC

3.2.1   Experimental Set-up

TUC is a U-shaped RC wall tested at EPFL (Constantin and Beyer 2016) and represents an half-scale model of the lower two stories of a prototype elevator shaft, Fig. 15.

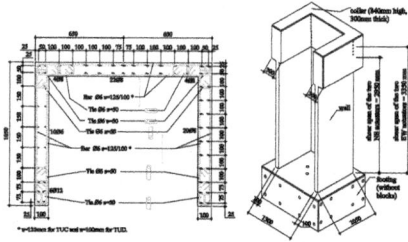

Fig. 15. Geometrical characterization and detailing of TUC wall (Constantin and Beyer 2016)

The transversal reinforcement is characterized by stirrups with a diameter of 6 mm spaced of 125 mm. Moreover, the extremities of the flanges, defined later as boundary elements, and the corners between the web and the flanges, are over reinforced with additional stirrups with a diameter of 6 mm spaced of 50 mm.

The longitudinal reinforcement was set different for the two flanges. One flange was detailed with vertical reinforcement mainly concentrated in the boundary elements while in the other flange and in the web the vertical reinforcement was uniformly distributed. The total reinforcement percentage for the flanges is approximatively the same, as listed in Table 2.

Table 2. TUC, vertical reinforcement percentage

	ρ_v total[a]	ρ_v conf.	ρ_v unconf.
Flange with distributed reinf. (East)	1.06%	1.34%	0.91%
Flange with concentrated reinf. (West)	1.01%	2.45%	0.31%
Web	1.16%	1.25/0.9%[b]	1.00%
Entire wall	1.09%	–	–

[a]ρ_v was computed by counting the corners towards the web.

[b]due to differences in flange reinforcement layouts reinforcement contents of confined corner regions differ slightly between the two corners.

During the test the specimen was subjected to an axial load ratio of 0.06 (equivalent to an axial force of 806 kN).

The horizontal load is applied by means of three actuators: the EW actuator loaded the web at a height of $h_{EW} = 3.35$ m while the NS actuators loaded the flanges at a height of $h_{NS} = 2.95$ m.

As the key objective of the experimental investigation was to understand the behaviour of the wall under diagonal loading, the main cycles were applied along the two diagonals. The detail of the applied loading history is reported in Fig. 16 with the definition of the loading directions adopted.

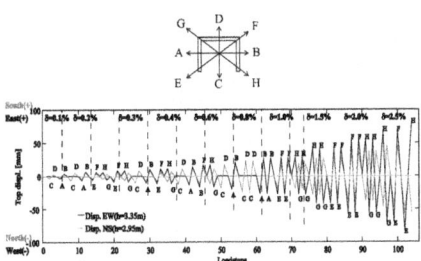

Fig. 16. TUC, loading history (Constantin and Beyer 2016)

The material properties and model parameters for both concrete and steel are listed in Table 3.

Table 3. TUC, material properties

	From material tests			Calculated and used for NLFEA			
	f_v [Mpa]	f_u [Mpa]	E_s [Gpa]	b [-]	R0 [-]	a_1 [-]	a_2 [-]
STEEL φ6	462	623	200	0.009	20	18.45	0.001
STEEL φ8	563	663	200	0.007	20	18.45	0.001
STEEL φ12	529	633	200	0.006	20	18.45	0.001
	f_c [Mpa]	E_c [Gpa]	f_{ct} [Mpa]	$f_{c,conf}$ [Mpa]	G_F [N/mm]	G_C [N/mm]	shear retention [-]
CONCRETE	42	31.6	3.2	50.7	0.14	35	0.03

3.2.2 NLFE Model

The model used for the NLFEA is represented in Fig. 17. Three different regions are considered in the model: the upper collar, made of elastic material, the confined concrete, in the zones characterized by the presence of additional stirrups, and the unconfined concrete, in the remaining zones. The influence of the foundation is neglected and hence the model restraints are located at the base nodes of the wall.

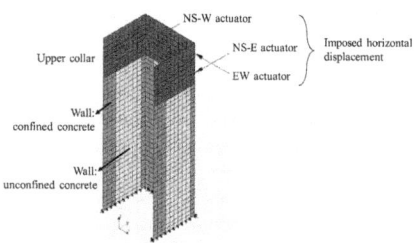

Fig. 17. TUC, sketch of NLFE model

The constitutive relationships adopted in NLFEA are based on the model parameters listed in Table 3.

They were defined following the formulations proposed above in Sect. 2 and are illustrated in Fig. 18.

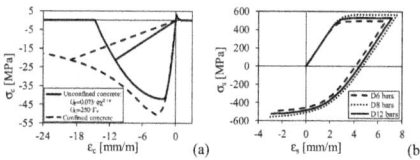

Fig. 18. TUC, constitutive models for (a) concrete and (b) steel

3.2.3 NLFE Results

Figures 19 and 20 describe the comparison between experimental results and NLFEA results for both "Fixed crack model" and "Rotating crack model" in terms of SRSS force *vs* displacement response when load is applied along the E-F diagonal and H-G diagonal respectively; the points in which the NLFEA failed due to crushing of compressive concrete are also indicated. The SRSS values were multiplied by the sign of the NS displacement for plotting the hysteresis loop consistently.

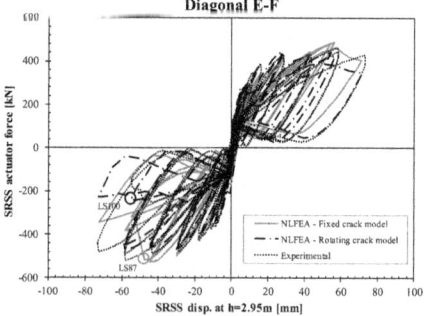

Fig. 19. TUC, SRSS force *vs* displacement along diagonal E-F

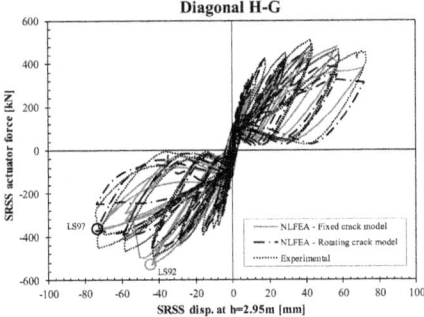

Fig. 20. TUC, SRSS force *vs* displacement along diagonal H-G

The results plotted in Figs. 19 and 20 show that in both directions the "Fixed crack model" failed before the "Rotating crack model", which is opposite to what had been observed for the previous test unit TW4. In particular, from Fig. 19 the "Fixed crack model" reached concrete crushing in the unconfined zone of the East flange at LS87 ($\delta = 2\%$) while the "Rotating crack model" reached the crushing of concrete at LS100 ($\delta = 2.5\%$). The same remark can be done analysing the specimen loaded along the diagonal H-G, Fig. 20. This fact can be explained considering that for the "Rotating crack model" the compressive strain in the flanges are mainly concentrated in the confined zone while, for the "Fixed crack model", the shear behaviour defined once the first cracking is formed tend to distribute the compressive strain along the whole flange.

In Fig. 21 the vertical strains on the inner and outer side of the flange ends, along the height of the wall, are investigated. These vertical strains, obtained for diagonal loading at drift of 1%, were derived from NLFEA and compared with experimental outcomes.

Figure 21 shows that both the "Rotating crack model" and the "Fixed crack model" are in good agreement with the experimental results. In particular, from Fig. 21 it can be observed that when the flange ends are in compression (Pos. E and Pos. H) the compressive strains at the base of the wall on the outer face are ~2 times larger than on the inner face, Fig. 21-a and c. On the other end, when the flange ends are in tension (Pos. F and Pos. G) the tensile strains in the outer face ends are close to the tensile trains in the inner face ends, Fig. 21-b and d. The good agreement obtained comparing NLFEA and experimental results in terms of vertical strains in the inner and outer sides of the flange is encouraging for the application of the current model in future parametric studies. Note that the large strain gradient under diagonal loading, exhibited when the flange ends are in compression, promotes the occurrence of local out-of-plane buckling of the wall, which was observed in the experimental test.

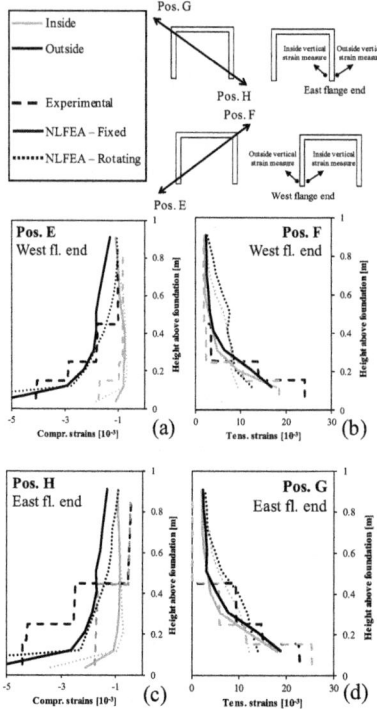

Fig. 21. TUC, comparison between experimental and NLFEA results: vertical strains on the inner and outer side of the flange computed along the height

4   Conclusions

In this paper the bidirectional response of reinforced concrete walls was investigated by means of NLFEA (nonlinear finite element analyses). The need for this research arises from the observed lack of numerical and experimental studies on this topic. The interaction of bidirectional loading with out-of-plane instability modes of thin walls was also considered.

The results demonstrate that NLFEA using curved shell elements and appropriately calibrated material constitutive models are able to predict not only the global behaviour of RC walls but also their local response. It thus seems logical to use this advanced numerical tool to further study wall configurations that are able to optimize the response under bidirectional loading. In particular, the influence of the content and detailing of longitudinal reinforcing bars, the confining effect due to the increase of stirrups, and the width of confined zone in the boundary elements should be addressed. The analyses of different boundary and loading conditions and the evaluation of the effect of slabs and torsional loading will also be carried out.

References

Belletti B, Damoni C, Gasperi A (2013) Modeling approaches suitable for pushover analyses of RC structural wall buildings. Eng Struct 57(12):327–338

Damoni C, Belletti B, Esposito R (2014) Numerical prediction of the response of a squat shear wall subjected to monotonic loading. Eur J Environ Civil Eng 18(7):754–769

Belletti B, Damoni C, Hendriks MAN, De Boer A (2014) Analytical and numerical evaluation of the design shear resistance of reinforced concrete slabs. Struct Concr 15(3):317–330

Belletti B, Scolari M, Vecchi F (2016a) Nonlinear static and dynamic finite element analyses of reinforced concrete shear walls using PARC_CL crack model. FraMCoS-9

Belletti B, Stocchi A, Scolari M, (2016b) Shell modelling of a 1/13 scaled RC containment vessel under cyclic actions with PARC_CL crack model. In: 8th international CONSEC, Lecco, Italy, 12–14 September

CEN (2004) Eurocode 8: Design provisions for earthquake resistance structures – Part 1. Bruxelles, Belgium

Constantin R, Beyer K (2016) Behaviour of U-shaped RC walls under quasi-static cyclic diagonal loading. Eng Struct 106:36–52

Dashti F, Dhakal RP, Pampanin S (2014) Simulation of out-of-plane instability in rectangular RC structural walls. In: Second European Conference on Earthquake Engineering and Seismology, Istanbul, Turkey

Dashti F, Dhakal RP, Pampanin S (2015) Development of out-of-plane instability in rectangular RC structural walls. In: 2015 NZSEE Conference

Feenstra PH (1993) Computational Aspects of Biaxial Stress in Plain and Reinforced Concrete. Ph.D. thesis, Delft University of Technology

fib – International Federation for Structural Concrete: fib Model Code for Concrete Structures 2010. Ernst & Sohn, Berlin (2013)

Goodsir WJ (1985) The design of coupled frame-wall structures for seismic actions. University of Canterbury, Christchurch

Guidelines for Non-linear Finite Element Analyses of Concrete Structures (2012) Rijkswaterstaat Technisch Document RTD:1016:2012, Rijkswaterstaat Centre for Infrastructure, Utrecht, 2012

Mander J, Priestley M, Park R (1988) Theoretical stress-strain model for confined concrete. J Struct Eng, 1804–1826. https://doi.org/10.1061/(ASCE)0733-9445(1988)114:8(1804)

Manie J (2015) DIANA User's manual, Release 10, TNO DIANA

Fragiadakis M, Pinho R, Antoniou S (2007) Modelling inelastic buckling of reinforcing bars under earthquake loading. In: ECCOMAS thematic conference on computational methods in structural dynamics and earthquake engineering, Rethymno, Crete, Greece, 13–16 June 2007

Monti G, Nuti C (1992) Nonlinear cyclic behaviour of reinforcing bars including buckling. J Struct Eng, ASCE 118(12):3268–3284

Nakamura H, Higai T (2001) Compressive fracture energy and fracture zone length of concrete. In: Shing, BP (ed.) ASCE 2001, pp 471–487. J Str Eng

Oesterle R (1979) Earthquake resistant structural walls: tests of isolated walls: phase II, Construction Technology Laboratories, Portland Cement Association

Paulay T, Priestley MJN (1993) Stability of ductile structural walls. ACI Struct J 90:385–392

Rosso A, Almeida JP, Beyer K (2015) Stability of thin reinforced concrete walls under cyclic loads: state-of-art and new experimental findings. Bull Earth Eng 14:455–484 (published online)

Thiele K, Dazio A, Bachmann H, (2001) Bewehrungsstahl unter zyklischer Beanspruchung. Institut für Baustatik und Konstruktion Eidgenössische Technische Hochschule Zürich, Mai 2001

Thomsen JH IV, Wallace JW (2004) Displacement-based design of slender reinforced concrete structural walls-experimental verification. J Struct Eng 130(4):618–630

Vallenas JM, Bertero VV, Popov EP (1979) Hysteretic behaviour of reinforced concrete structural walls. Report no. UCB/EERC-79/20, Earthquake Engineering Research Center, University of California, Berkeley

Vecchio FJ, Collins MP (1993) Compression response of cracked reinforced concrete. J Str Eng ASCE 119(12):3590–3610

Wallace J (2012) Behavior, design, and modeling of structural walls and coupling beams— Lessons from recent laboratory tests and earthquakes. Int J Concr Struct Mater 6(1):3–18

Shell Modelling Strategies for the Assessment of Punching Shear Resistance of Continuous Slabs

B. Belletti[1(✉)], R. Cantone[2], and A. Muttoni[2]

[1] DICATeA, University of Parma, Parma, Italy
beatrice.belletti@unipr.it
[2] Ibeton, École Polytecnique Fédérale de Lausanne (EPFL),
Lausanne, Switzerland
{raffaele.cantone, aurelio.muttoni}@epfl.ch

Abstract. The punching shear resistance formulation provided by Model Code 2010 is calibrated on the basis of experimental tests on isolated slabs supported on columns. According to Level of Approximation approach, several quantities are required for the design punching shear resistance assessment, like the resisting moment and the radius of the line of moment contraflexure. In this paper specific formulations are provided to adjust these quantities in order to take into account for moment redistribution and compressive membrane action effects. The punching shear resistance, mostly investigated for axisymmetric cases, in terms of loading and boundary conditions, will be analysed referring to actual rectangular RC continuous floors with orthogonal reinforcement layouts, largely adopted in practice. The results of nonlinear finite element analyses, carried out using PARC_CL crack model, are post-processed according to the Critical Shear Crack Theory to predict the punching shear strength of the continuous slab.

Keywords: Shear · Punching shear · Concrete slabs
Critical Shear Crack Theory

1 Introduction

Conventional design methods provide punching shear resistance of flat slabs according to empirical formulas based on datasets including isolated, circular or octagonal specimens. This setup means to model the slab region inside the line of moment contraflexure r_s (approximately $0.22L$ from the column axis where L is the span between adjacent columns). These testing procedures are, generally, quite straightforward even though they are not able to simulate redistribution phenomena occurring between hogging and sagging moment as well as compressive membrane effects. Hence, shear and punching resistance predictions may not consider several relevant aspects of the actual structural behaviour, namely, forces redistributions or non-axisymmetric situations typical of actual design cases. In addition, boundary conditions govern the distribution of shear forces ranging from parallel development for linearly supported slabs to radial development for slabs supported on columns,

© Springer International Publishing AG, part of Springer Nature 2018
M. di Prisco and M. Menegotto (Eds.): ICD 2016, LNCE 10, pp. 49–57, 2018.
https://doi.org/10.1007/978-3-319-78936-1_4

Fig. 1a. These main aspects may greatly influence the overall behavior of reinforced concrete slabs, structural solution commonly adopted in practice, Muttoni et al. (2013), MC2010, triggering possible inaccurate design predictions. Accordingly, the next sections will present new strategies for the modelling of RC floors taking into account the governing mechanisms and their contribution on the global response of these structural members commonly used for practical purposes, Fig. 1b–c. The approach of the Critical Shear Crack Theory (CSCT), Muttoni (2008) coupled with PARC_CL Crack Model, Belletti et al. (2013), (2015), will be adopted, performing Nonlinear Finite Element Analyses (NLFEA). The investigation of redistribution effects and membrane forces will be analysed by means of a multi-span test in the literature, carried out by Ladner et al. (1977) at EMPA.

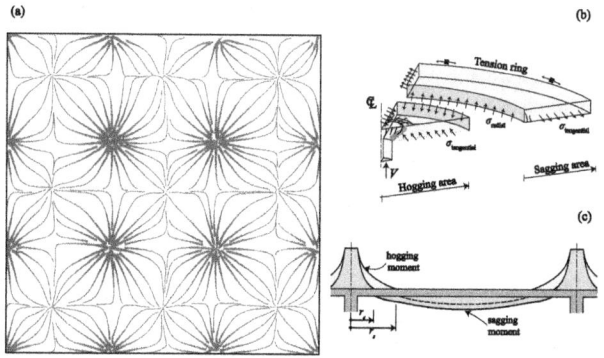

Fig. 1. (a) Shear fields for a continuous slab supported on 16 columns, Ladner et al. (1977) (b) Effect of the tension ring in the hogging moment area due to dilation of cracked concrete (c) Moment redistribution between hogging and sagging moment in a continuous slab

Afterwards, a parametric analysis will be performed in order to study the influence of several parameters on the punching strength of continuous slabs, namely, the hogging reinforcement ratio, the geometry of the specimen, contribution of sagging moment on the change of hogging moment area, Choi and Kim (2012), Hewitt and Batchelor (1975).

2   Mechanical Model Based on the Critical Shear Crack Theory and Multi-layered Numerical Modelling According to PARC_CL Crack Model

Experimental outcomes and theoretical considerations have shown how the punching shear capacity depends on the flexural state of deformations of the concrete member. Based on these assumptions, Muttoni (2008) and Fernandez et al. (2009), presented a theory that allows determining the punching shear strength V_R of two-way slabs as a function of the opening and roughness of a critical shear crack developing in the specimen. For two-way slabs, the punching shear strength V_R is correlated to the opening of a critical shear crack and proportional to the product between the rotation ψ and the effective depth of the member d, Fig. 3a–b. In particular, the punching strength is checked in the control perimeter at the control section, located at 0.5d from the point of maximum moment. Based on these hypotheses, the following failure criterion was proposed:

$$V_R = \frac{0.75 b_0 d \sqrt{f_c}}{1 + 15 \frac{\psi d}{16 + d_g}} \quad \text{MPa, mm} \tag{1}$$

where f_c is the concrete compressive strength and d_g the maximum aggregate size.

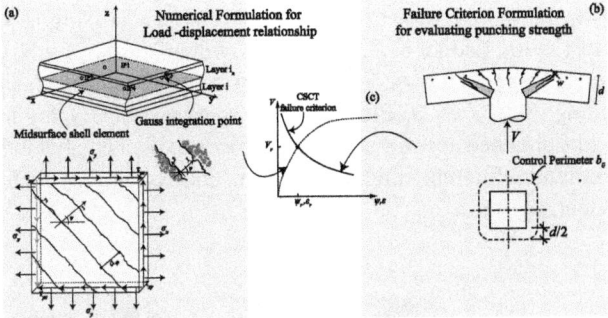

Fig. 2. (a) Shell modeling based on plane stress approach, Belletti et al. (2013) (b) Punching strength acc. Muttoni (2008) (c) Punching shear resistance evaluation

In order to take into account mechanical nonlinearities as well as shear and moment redistributions, multi-layered shell analyses can be performed achieving a suitable load-displacement relationship for concrete members. The constitutive model PARC_CL, Belletti et al. (2013), (2015), aimed at analysing the overall behaviour of reinforced concrete members subjected to plane stresses up to failure, Fig. 2a–b. This numerical model is based on a fixed crack approach and smeared reinforcement assumption. Regarding concrete and reinforcement mechanical behaviours, multiaxial state of stress and aggregate interlocking are considered as well as doweling action and tension stiffening contributions.

3 Experimental Campaign Under Investigation, Ladner et al. (1977)

The experimental programme consisted of several tests carried out on a reinforced concrete continuous slab 7.2 m × 7.2 m, with regular span of 2.4 m and supported on 16 columns of different sizes ranging from 100 mm to 320 mm (an inner column, two edge columns and a corner column for each size), Fig. 3. The nominal thickness of the slab was 110 mm with a constant effective depth $d = 80$ mm. The tested slab has been modelled

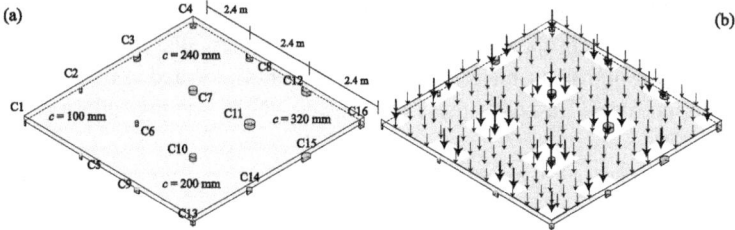

Fig. 3. View of the tested specimen under investigation

with multi-layered 8 nodes shell elements. Columns are simulated considering two different assumptions regarding boundary conditions at the supports. The first case considers spring elements working both in tension and compression according to stiffnesses evaluated from the test setup, Ladner et al. (1977), (providing $V_{NLFEA,\ CT}$ NLFEA punching shear resistance). As regards the second case, no stiffness in tension was provided at the columns (providing $V_{NLFEA,\ NT}$ NLFEA punching shear resistance). Figure 4a shows the punching strengths obtained according to the approach explained above. With respect to the type of the column, different control perimeters and eccentricity factor k_e have been adopted, Sageseta (2013).

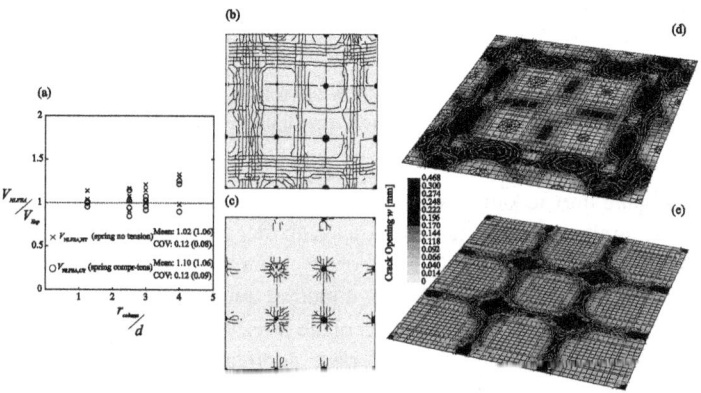

Fig. 4. (a) Comparison of test results to NLFEA predictions; Experimental and predicted cracking pattern, respectively, at the bottom (b)–(d) and top (c)–(e) sides evaluated at the end of the test campaign (crack openings w in mm), Cantone et al. (2016a)

It has to be highlighted that NLFEA analyses with CSCT failure criterion seem to be a very robust and consistent approach with a mean value close to 1.0 and a sound scatter of the results. It can be highlighted that the agreement between experimental and numerical results remains still valid for corner and edge columns where the axisymmetric condition is not satisfied anymore. Finally, Fig. 4b–e shows the experimental and numerically obtained cracking patterns at bottom and top side showing good agreement. For columns C12 and C11, NLFEA predictions seem not to be in good agreement.

4 Parametric Analysis for Studying the Influence of CMA and Moment Redistribution

In the next sections, parametric analyses have been performed through nine case studies accounting for reinforced concrete membrane nonlinearities by means of PARC_CL Crack Model and out of plane shear failures via CSCT failure criterion. Unlike standard axisymmetric models, this modeling approach will be able to consider orthogonal reinforcement layouts. Case a and Case b represent isolated slabs having a typical geometry adopted for specimens tested in laboratory (L = 3000 mm and L = 3950, respectively), while Case c represents the geometry of the corresponding continuous flat slabs (L = 6800 mm). Case c is modelled with no additional boundary conditions at the edges apart for symmetrical restrains in order to guarantee zero rotations (self - confined case). Three hogging reinforcement ratio in the column region are studied (ρ_{hog} = 1.5%/0.75%/0.375%). The choice of the reinforcement layouts aims at studying actual hogging and sagging reinforcement ratios common in practice. For this reason, regarding the case study c, a ratio ρ_{hog}/ρ_{sag} equal to 3 has been adopted over the columns (0.4L × 0.4L, where L is the span between adjacent columns) while a ratio equal to 2 in the slab strip connecting adjacent columns and a ratio equal to 1 in the remaining parts of the slab.

Table 1. Main assumptions of the parametric analysis, Cantone et al. (2016b)

	B [mm]	r_q [mm]	d_{nom} [mm]	c [mm]	f_c [MPa]	f_y [MPa]	Edge conditions
Case a	3000						Free edge
Case b	3950	1505	210	260	35	520	Free edge
Case c	6800	-					Self - confined

Marker for each case investigated

A uniformly distributed pressure is applied at the top of the slab simulating permanent and variable loads. The column size c is 260 mm for all case studies as the effective width d_{nom} is equal to 210 mm. The concrete compressive strength was fixed to 35 MPa whereas the yield strength of hogging and sagging reinforcement set up to

520 MPa, Table 1. The common design assumption, MC2010, consists in a constant radius of the line of moment contraflexure r_s corresponding to the hogging moment area. As shown in Fig. 5, the above assumption persists valid for an isolated specimen in which the value of r_s is constant at $0.22L$ up to failure. Nevertheless, for self - confined or fully - confined continuous slabs, moment redistribution occurs due to elastic activation of sagging reinforcement after cracking of concrete in the hogging area. This loss of stiffness in the hogging area triggers a reduction of the radius of the line of moment contraflexure r_s (and, consequently, of shear slenderness) shifting closer to the column, Einpaul et al. (2015).

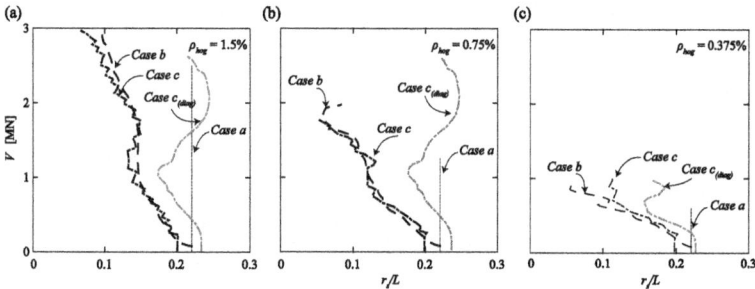

Fig. 5. Radius of the line of moment contraflexure r_s along the axis and the diagonal for ρ_{hog} under investigation

Focusing on the continuous slab of Case c, Fig. 5a and c highlight how moment redistributions are highly related to the hogging reinforcement ratio but still by the external confinement restraining the horizontal dilation. Figure 6 shows load – rotation curves for all case studies investigated in the current parametric analyses. Regarding Fig. 6a ($\rho_{hog} = 1.5\%$), it may be pointed out how no significant difference is observed between the behavior of an isolated specimen or an actual continuous slab in case of members without shear reinforcement. This shift becomes more visible by decreasing whether the hogging reinforcement ratio, Fig. 6b–c, or for members with shear

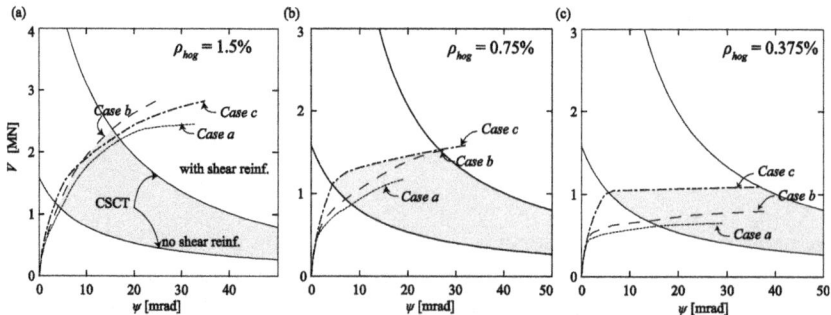

Fig. 6. V - ψ curves and intersection with CSCT failure criterion for ρ_{hog} under investigation

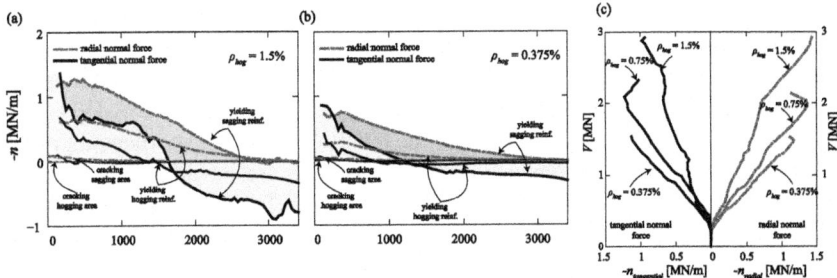

Fig. 7. Radial and tangential normal forces in the axis (a–b) for four significant loading step, namely, hogging cracking moment, sagging cracking moment, yielding of hogging reinforcement and yielding of sagging reinforcement; (c) Tangential and radial membrane forces in the critical section for Case c.

reinforcement. In these cases, CMA effects, moment redistributions and effect of dilation appear more pronounced, leading to different shear resistances and deformation capacities among different modeling approaches.

In addition, Fig. 7 points out the profiles of radial and tangential normal forces of the axial direction at four significant loading steps with respect to the self-confined continuous slab: namely, cracking in the hogging area, cracking in the sagging area, yielding of hogging reinforcement and yielding of sagging reinforcement. As already discussed above, the restrained dilation triggers tensile tangential normal forces in the outer part of the specimen.

Moreover, the radius of the tension ring shifts with increasing load due to the fact that the relative stiffness between hogging and sagging area is changing. In addition, Fig. 7c shows the profiles of compressive normal and tangential forces at the critical

Table 2. Design code provisions for members with and without shear reinforcement according MC2010 Level of Approximation;

MC2010	ρ_{hog} [%]	Level II [kN]		Level III [kN]		Level IV [kN]			
		UR	SR	UR	SR	UR	SR		
Case *a*	1.5	604.3	1167.7	673.9	1250.8	744.1	1376.0	1.23	1.18
	0.75	502.8	864.1	535.3	931.6	630.0	885.0	1.26	1.03
	0.375	376.7	620.7	404.5	672.2	460.0	520.0[**]	1.22	-[**]
Case *b*	1.5	604.3	1167.7	673.9	1250.8	760.0	1416.0	1.26	1.21
	0.75	502.8	864.1	535.3	931.6	649.0	995.0	1.29	1.15
	0.375	376.7	620.7	404.5	672.2	499.0	584.0[**]	1.33	-[**]
Case *c*	1.5	604.3	1167.7	673.9	1250.8	792.0	1399.0	1.31	1.19
	0.75	502.8	864.1	535.3	931.6	705.0	1072.0[**]	1.40	1.24
	0.375	376.7	620.7	404.5	672.2	649.0	774.0	1.72	1.25

[*]*UR*: member without shear-reinforcement; *SR*: shear-reinforced member; [**]Flexural plateau

section in the axial direction for Case *c*. It can remarked that, the membrane effects increase with decreasing reinforcement ratios.

Afterwards, design punching predictions according Level of Approximation by Model Code 2010 have been carried out in order to highlight how LoA IV is able to detect design resistances higher than Loa II and III (Table 2). The benefits of using LoA IV in terms of design punching shear resistance are much higher when compressive membrane actions play an important role (Case c) than in case of isolated slabs (Case a). The discrepancy between isolated and actual continuous members attains higher values when shear reinforcement is provided.

This capacity reserve is related to the activation of sagging moment up to failure guiding to an ultimate flexural mechanism different from the one of an isolated specimen.

5   Conclusions

An actual continuous slab supported on 16 columns, tested by Ladner et al. (1977), has been analysed with the PARC_CL Crack Model and the Critical Shear Crack Theory failure criterion. This investigation aims at studying the consistency of the adopted numerical approach and the contribution of compressive actions on the overall load – deformation response and ultimate punching strengths. Then, a parametric analysis has been carried out by comparing different hogging reinforcement ratio and modeling approaches in order to evaluate the contribution of moment redistributions and compressive membrane action on the overall load - deformation response. The investigation leads to the following conclusions:

- The numerical approach adopted with the mechanical model of CSCT shows sound agreement and robustness in terms of scatter of the results.
- Multi-layered shell modelling allows the investigation of non-axisymmetric situations as it is, for example, the case of edge or corner columns.
- Moment redistributions shift the line of moment contraflexure triggering a change in the hogging moment area (not constant anymore at 0.22L). The elastic activation of sagging moment up to failure is not considered in a common isolated specimen experiment. This contribution yields to flexural shear capacities clearly higher.
- CMA effects increase the resistant moment at the slab -column connection
- PARC_CL Crack model is able to capture compressive membrane forces in the critical region accounting for additional punching/shear carrying resistance capacities.
- Actual designing approaches, did not take into account the onset of membrane forces at different load levels.
- The contribution of the normal forces may be considered in a simplified manner in order to account for an increase of the hogging resistant moment, thus leading to a higher ultimate punching strength.

References

Belletti B, Walraven JC, Trapani F (2015) Evaluation of compressive membrane action effects on punching shear resistance of reinforced concrete slabs. Eng Struct 95:25–39

Belletti B, Esposito R, Walraven JC (2013) Shear capacity of normal, lightweight, and high-strength concrete beams according to model code 2010. II: experimental results versus nonlinear finite element program results. J Struct Eng 139(9):1600–1607

Cantone R, Belletti B, Muttoni A, Fernandez MF (2016) Approaches for suitable modelling and strength prediction of reinforced concrete slabs, fib Symposium, Cape Town, South-Africa, 21–23 November

Cantone R, Belletti B, Manelli L, Muttoni A (2016) Compressive membrane action effects on punching strength of flat RC slabs. In: CONSEC, 12–14 September, Lecco, Italy

Choi JW, Kim J-H (2012) Experimental investigations on moment redistribution and punching shear of flat plates. ACI Struct J 109(3):329–338

Einpaul J, Fernández Ruiz M, Muttoni A (2015) Influence of moment redistribution and compressive membrane action on punching strength of flat slabs. Eng Struct 86:43–57

Fernández Ruiz M, Muttoni A (2009) Applications of critical shear crack theory to punching of reinforced concrete slabs with transverse reinforcement. ACI Struct J 106(4):485–494

Fib Model Code for Concrete Structures 2010. Fédération internationale du béton. Ernst & Sohn, Germany (2013)

Hewitt BE, Batchelor B (1975) Punching shear strength of restrained slabs. J Struct Div ASCE 101(9):1837–1853

Ladner M, Schaeidt W, Gut S (1977) Experimentelle Untersuchungen an Stahlbeton-Flachdecke. EMPA Bericht no. 205, Switzerland. 96p (in German)

Muttoni A (2008) Punching shear strength of reinforced concrete slabs without transverse reinforcement. ACI Struct J 105:440–450

Muttoni A, Fernández Ruiz M, Bentz EC, Foster SJ, Sigrist V (2013) Background to the Model Code 2010 shear provisions – part II punching shear. Struct Concr 14(3):195–203

Sagaseta J, Tassinari L, Fernández Ruiz M, Muttoni A (2014) Punching of flat slabs supported on rectangular columns. Eng Struct 77:17–33

Inverse Identification of the Bond-Slip Law for Sisal Fibers in High-Performance Cementitious Matrices

S. R. Ferreira[1], M. Pepe[2], E. Martinelli[2(✉)], F. A. Silva[3],
and R. D. Toledo Filho[1]

[1] Civil Engineering Department, COPPE, Federal University of Rio de Janeiro,
Rio de Janeiro, Brazil
[2] Department of Civil Engineering, University of Salerno,
Fisciano, Salerno, Italy
e.martinelli@unisa.it
[3] Civil Engineering Department, Pontifical Catholic University
of Rio de Janeiro, Rio de Janeiro, Brazil

Abstract. The use of Natural Fibers (NFs) in Fiber-Reinforced Cementitious Composites (FRCCs) is an innovative technical solution, which has been recently employed also in High-Performance FRCCs. However, NFs are generally characterized by complex microstructure and significant heterogeneity, which influence their interaction with cementitious matrices, whose identification requires further advances in the current state of knowledge. This paper presents the results of pull-out tests carried out on sisal fibers embedded in a cementitious mortar. These results are considered for identifying the bond-slip law that describes the interaction between the sisal fibers and the cementitious matrix. A theoretical model, capable of simulating the various stages of a pull-out test, is employed as part of an inverse identification procedure of the bond-slip law. The accuracy of the resulting simulations demonstrates the soundness of the proposed theoretical model for sisal fibers embedded in a cementitious matrix.

Keywords: Sisal fibers · Fiber Reinforced Cementitious Composites
Bond-slip law · Pull-out test · Inverse identification

1 Introduction

In the last decade many researches have been developed with the aim of exploring the potential of composite materials, based on either polymeric or cementitious matrices, internally reinforced by fibers derived by plant leaves or branches (Netravali and Chabba 2003). These fibers, generally referred to as either "vegetal" or "natural", highlighted promising properties in terms of both strength (Silva et al. 2010) and durability (Melo Filho et al. 2013, Ferrara et al. 2014; Ferrara et al. 2015). Moreover, they have an apparent potential for enhanced sustainability with respect to similar materials reinforced with industrial fibers made of either steel (Toledo Filho 1997) or plastic fibers (Sreekumar et al. 2009).

© Springer International Publishing AG, part of Springer Nature 2018
M. di Prisco and M. Menegotto (Eds.): ICD 2016, LNCE 10, pp. 58–70, 2018.
https://doi.org/10.1007/978-3-319-78936-1_5

The aforementioned researches unveiled the multifold potential of Natural-Fiber-Reinforced Cementitious Composites in reducing the demand of raw materials, as they are mainly based on using renewable resources, and the supply and production costs, as the original plants are widely available, especially in tropical and sub-tropical zones.

Among the various Natural Fibers (NFs) investigated so far, those made from sisal (*agave sisalana*) leaves (Silva et al. 2009) have attracted a great interest in both material scientists and concrete technologists (Netravali and Chabba 2003). More specifically, their excellent properties in terms of tensile strength is the main motivation for using them as a reinforcement in composite materials (Silva et al. 2008). However, it should be recognized that the weak chemical bonds that can be established with matrices based on Portland cement results in low mechanical bond between these fibers and matrices, whose maximum strength might be estimated in the range of 0.32 and 0.72 MPa, according to experimental results reported in the scientific literature (Silva et al. 2011). Moreover, the high water absorption capacity of sisal fibers results in a volume expansion when they are added to the fresh cementitious matrix and, conversely, it produces a contraction when the matrix dries and, hence, a partial detachment between fibers and matrix at the hardened state.

Different procedures have been proposed for reducing water absorption capacity in natural fibers and improving fiber-matrix bond interaction: they are based on applying chemical and physical treatments of both matrix and fibers (Ferraz et al. 2011). For instance, the partial replacement of cement with *microsilica* have led to increasing the pullout resistance by about 24% (Toledo Filho 1997). This increase is related to the fineness of microsilica, which is capable to reduce porosity in the transition zone, hence enhancing the fiber-matrix bond. Furthermore, the use of alkaline solutions (Saha et al. 2010; Kundu et al. 2012) removes most of the surface non-cellulosic substances and increases roughness of their surface, hence enhancing the fiber-matrix bond. Simple treatments, such as soaking the fibers in distilled water followed by a drying process, also result in improving the fiber-matrix bond (Li et al. 2008). Moreover, a reduction in the fiber hydrophilicity can be achieved by means of wetting and drying cycles promoting *hornification* (namely, stiffening of the polymeric structure present in lignocellulosic materials, as defined by Claramunt et al. 2010). This treatment promotes a reduction in volumetric changes of pulps and fibers of natural origin, as well as a significant alteration in their mechanical properties, while acting also as a bridge between fibers and cementitious matrices and, hence, strengthening the interfacial bond (Claramunt et al. 2010).

This paper summarizes the results of pull-out tests carried out on *hornified* sisal fibers embedded in a cementitious mortar. Then, these results are employed in identifying the bond-slip law that describes the interaction between sisal fibers and cement-based matrix: an inverse identification procedure, based on a theoretical model capable of simulating the various stages of a pull-out test, is applied for this task (Ferreira et al. 2016).

2   Experimental Program

2.1   Materials and Processing

The sisal fibers used in the present study were obtained from sisal plants growing in farms located in the Bahia state, Brazil. They were extracted from the sisal plant leaves in the form of long fiber bundles: this process was executed by means of semi-automatic scrapers.

As regards microstructure, it consists of numerous individual fibers (fiber-cells), which are about 6-30 μm in diameter. The individual fiber-cells are linked together by means of the middle lamella. The chemical composition of sisal fibers includes approximately 60.5% cellulose, 25.7% hemicellulose, 12.1% lignin, 1% pectin and 1.6% ash (Fidelis 2012; Silva et al. 2009; Silva et al. 2010).

The cement-based matrix presented a mix design of 1:0.5:0.4 (binder: sand: water/binder ratio) by weight.

The binder was composed by 30% of Portland cement CP-32 F II, 30% of meta-kaolin and 40% of fly ash. This ratio of metakaolin and fly ash was aimed to guarantee the durability of the fiber once a matrix free of calcium hydroxide is obtained (Toledo Filho et al. 2009; Melo Filho et al. 2013).

The fly ash also ensured higher workability to the matrix that, within the context of high-performance composites, is a desirable property, as it provided a better homogenization of the natural fibers (Ferreira et al. 2015).

The sand was processed to obtain a maximum diameter of 840 μm and the superplasticizer was the Glenium 51 (type PA) with solids content of 31%. In addition, a viscosity modifier Rheomac UW 410, (manufactured by BASF), at a dosage of 0.8 kg/m^3 was also used in order to avoid segregation and bleeding during molding.

The matrix showed a flow table spread value of 450 mm according to the Brazilian standard NBR 13276 (2005) and a compressive strength at 28 days of 31 MPa, according to NBR 7215 (1996).

The mixtures were produced in a room with controlled temperature (21 $\pm$ 1 °C) using a mixer with capacity of 5 l. The mixing procedure consisted of the following stages:

– all dry components were homogenized in the mixer;
– the water and superplasticizer were added and mixed for 2 min at a speed of 125 RPM;
– the process was stopped during 30 s to remove the material retained in the mixer;
– the mixing procedure continued for 2 min at 220 RPM and, finally, for a further 5 min at 450 RPM.

A special mold was developed for preparing the specimens. After filling the mold with the matrix, the top cap was fixed and the fiber stretched slightly for alignment. The mortar was placed in plastic bags before being placed in the mold as to facilitate the casting process. Embedment length of 25 was analyzed. After 24 h, the specimens were demolded and placed in a fog room (HR% $\geq$ 95%) to moist curing for 7 days for the pullout test.

2.2 Hornification Process

The sisal fibers were placed in a container with water (T = 22 °C) during three hours to reach its maximum water absorption capacity. The drying process was carried out in a furnace at a temperature of 80 °C. The furnace used was equipped with an electronic temperature control and connected to a scale, with a tolerance of 0.01 g to record the loss of water.

The furnace was programmed to reach 80 °C at a heating rate of 1 °C/min and to maintain this temperature for 16 h. After 16 h of drying, the furnace was cooled down to the temperature of 22 °C in order to avoid possible thermal shock to the fibers. This procedure was repeated ten times. More details can be obtained elsewhere (Ferreira et al. 2015).

2.3 Testing

The sisal fiber's microstructure was investigated using a Hitachi TM3000 Scanning Electron Microscope (SEM). The microscope was operated under an accelerating voltage of 15 kV. A pre-coating with a thin layer of approximately 20 nm of gold was done to make the fiber conductive and suitable for analysis. In order to measure the fiber's cross-sectional area, for each single fiber used in the pullout and tensile test, an adjacent piece of the fiber (immediately next to the one tested) was kept for future measurement and morphology characterization using the SEM. Fiber-matrix interface zone was also investigated. The obtained images were post-processed using ImageJ, a Java-based image processing program.

The tensile tests were performed in an electromechanical testing machine Shimadzu AG-X with a load cell of 1 kN. The tests were performed on 15 sisal fibers using a displacement rate of 0.1 mm/min. The fibers with a gage length of 50 mm were glued to a paper template for better alignment in the machine and for a better griping with the upper and lower jaws in accordance with ASTM C1557 (2013). To calculate the tensile strength of the fibers, their diameters were measured by image analysis from micrographs obtained in a scanning electron microscope.

The pullout tests were performed in an electromechanical testing machine Shimadzu AG-X with a load cell of 1 kN. The tests were carried out using a displacement rate of 0.1 mm/min. The samples were fixed in the machine grips through a system with hinged-fixed boundary conditions. Fifteen tests were performed (embedded length of 25 mm).

3 Anlytical Modeling

An analytical model has been formulated for simulating the interaction between fiber and matrix in a pull-out process (Ferreira et al. 2016).

It assumes that:

– the fiber behaves in a linear elastic way;
– the matrix is supposed to be perfectly stiff;

- the interaction between fiber and matrix is based on a bond-slip law τ-s, invariant throughout the fiber length.

As for the last point, the model assumes the following bilinear bond-slip law (Fig. 1):

$$\tau(s) = \begin{cases} k_{el} \cdot s & if & |s| \leq s_{el} \\ \tau_r - k_{in} \cdot (s - s_{el}) & if & s_{el} < |s| \leq s_u \\ 0 & if & |s| > s_u \end{cases} \tag{1}$$

where $k_{el} = \tau_{max}/s_{el}$ is the slip modulus of the elastic branch ending at a slip value s_{el} with a corresponding bond stress τ_{max}. Moreover, τ_r is the residual bond stress, and k_{in} is the post-peak slip modulus, strictly positive in Eq. (1), resulting in a linear variation of stresses from τ_r to τ_u, the latter being achieved for a slip s_u.

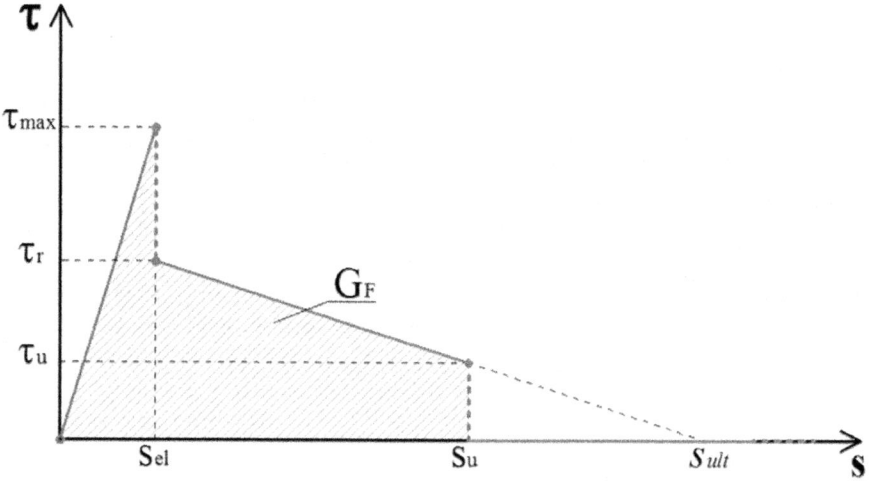

Fig. 1. Bilinear bond-slip law

The equilibrium equation of an infinitesimal element of fiber can be easily determined as already presented in several models already available in the literature (Caggiano et al. 2012), a general equilibrium condition can be determined (Fig. 2):

$$\frac{d\sigma f}{dz} - \frac{P_f}{A_f} \cdot \tau = 0 \tag{2}$$

Moreover, the axial strain ε_f developing in the fiber is strictly related to the interface slip s:

$$\varepsilon_f = \frac{ds}{dz} \tag{3}$$

Since the fiber is elastic $\sigma_f = E_f \varepsilon_f$ and, hence, Eq. (3) can be introduced in Eq. (2), in order to obtain a well-known differential relationship between the second derivative of the interface slips and the corresponding bond stress τ (Caggiano et al. 2012):

$$\frac{d^2s}{dz^2} - \frac{P_f}{E_f A_f} \cdot \tau(s) = 0 \tag{4}$$

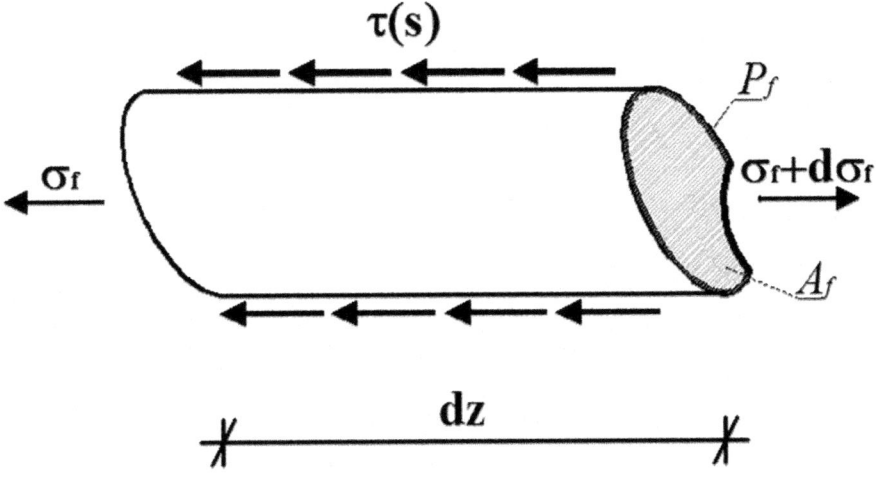

Fig. 2. Local equilibrium conditions on a free infinitesimal element of fiber.

A step-wise analytical solution can be obtained by considering the various states of stresses of the fiber-to-matrix interface resulting from the bilinear bond-slip law represented in Fig. 1. Further details on both the mathematical derivation and the numerical implementation of this solution are available in Ferreira et al. (2016).

Based on the above assumptions, the full range analytical expression of the applied pull-out force F_0 can be, as a function of the displacement s_0 at the end of the fiber embedment, on the loaded side.

Moreover, F_0 also depends upon the actual bond-slip law assumed for describing the interface behavior. Hence, the following conceptual relationship can be written:

$$F_0 = F_0(s_0; \boldsymbol{q}) \tag{5}$$

where q is a vector that collects the five parameters describing the interface law de-scribed in Fig. 1:

$$\boldsymbol{q} = \begin{bmatrix} s_{el} & s_u & \tau_{max} & \tau_r & \tau_u \end{bmatrix} \tag{6}$$

The aforementioned analytical model can be employed for determining the parameters describing the actual bond-slip interaction by means of an "inverse identification" procedure (Martinelli et al. 2012). More specifically, the following optimization problem has to be solved:

$$\bar{q} = \arg\min_{q} \Delta(q) \tag{7}$$

being

$$\Delta(q) = \sum_{i=1}^{n} \left[F_0\left(s_{0,i}; q\right) - F_{0,i}^{\text{exp}} \right]^2 \tag{8}$$

where $s_{0,i}$ is the displacement imposed on the free end of the fiber at the i-th increment of the experimental procedure, $F_{0,iexp}$ is the corresponding force and n is the number of displacement increments either in the experimental process or the current numerical analysis.

4   Results and Analysis

4.1   Morphological Characterization of Sisal Fiber

Table 1 summarizes the results derived from the SEM analysis by presenting the measured value in terms of area, perimeter and fiber's shape.

Table 1. Sisal fibers geometry

#	Area	Perimeter	Shape
	mm^2	mm	
1	0.01478	0.7900	Twisted
2	0.01739	0.8200	Twisted
3	0.03750	1.2300	Twisted
4	0.01753	0.6114	Horse shoe
5	0.01115	0.5619	Twisted
6	0.02270	0.8739	Twisted
7	0.01650	0.5330	Horse shoe
8	0.01641	1.0300	Twisted
9	0.02970	0.8281	Horse shoe
10	0.01571	0.5683	Arch
11	0.02391	1.0600	Twisted
mean	0.02030	0.8097	–

Based on the values of parameters reported in Table 1, the hornified sisal fiber presents high scattering in terms of morphological characterization. The area and perimeter values range from 0.001 to 0.003 mm^2 and 0.53 to 1.3 mm, respectively. This significant variation might be directly related to the significant variability in the shape of fibers' cross-sections.

This clue is confirmed by Fig. 3, which shows the three typical cross-section shapes exhibited by sisal fibers, each one presenting a different relationship between cross-section area and perimeter.

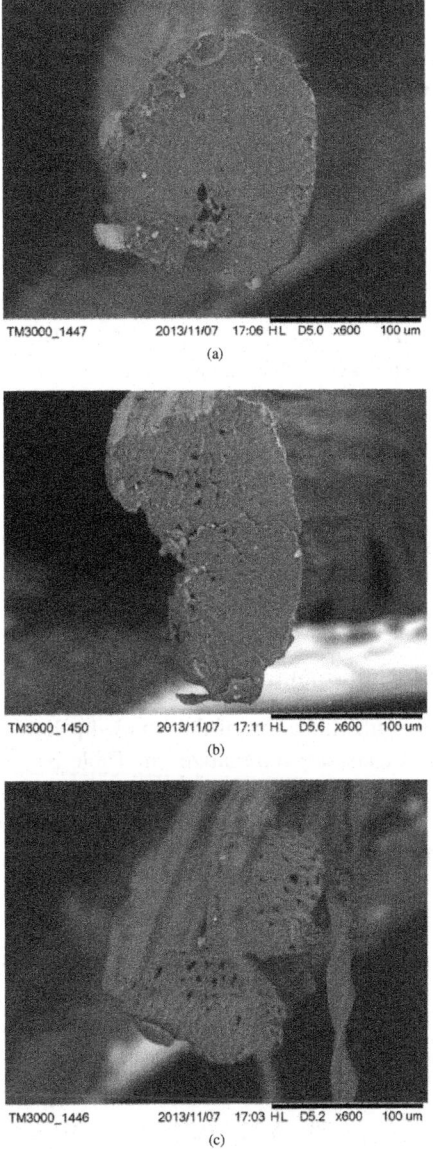

Fig. 3. Sisal fiber shape: horse shoe (a), arch (b), twisted (c).

As a matter of fact, within the leaf, there are three basic types of fibers, generally referred to as structural, arch and xylem fibers (Silva et al. 2011). Structural fibers have a *horse shoe* shape and a rough surface (Fig. 3a); *arch* shape fibers (Fig. 3b) grow in association with the conducting tissues of the plant (usually found in the middle of the leaf); *xylem* fibers grow opposite to the arch fibers, presenting a *twisted* shape (Fig. 3c), similar to double-helical DNA model.

4.2   Mechanical Properties

The results in terms of tensile strength and the corresponding elastic modulus for both untreated and hornified sisal fibers are summarized in Table 2. They show that hornification induce to a slight increase in tensile strength and strain at failure (about 5%), while it reduces the elastic modulus.

Table 2. Tensile tests results.

Treatment	Tensile strength		Elastic modulus	
	MPa	CoV (%)	GPa	CoV (%)
Natural	453.23	66.49	20.16	3.52
Hornification	474.86	26.46	18.05	2.75

In fact, wetting and drying cycles change the microstructure in natural fibers, which, in turn, modifies the polymeric structure of the fiber-cells resulting in higher tensile strength and strain. Moreover, it is worth highlighting that hornification reduces the variability affecting the mechanical response of sisal fibers, both in terms of tensile strength (coefficient of variation slumps from 67 to 26%) and elastic modulus (CoV reduces from 3.5 to 2.7%).

4.3   Bond-Slip Law Identification

The results of the inverse identification of the bond-slip laws performed on hornified sisal single fiber pull-out tests, are summarized in Table 3.

A total number of 15 tests were performed (embedded length equal to 25 mm), but 4 samples exhibited fiber fracture and for this reason their results are not considered in the present analysis for the bond-slip law identification.

Table 3. Parameters of the identified bond-slip law results.

#	s_e	s_u	τ_{max}	τ_r	τ_u	G_F
	mm	Mm	MPa	MPa	MPa	Nmm
1	0.08	2.96	0.31	0.28	0.21	0.36
2	0.11	4.19	0.27	0.22	0.13	0.36
3	0.50	8.05	0.14	0.13	0.09	0.41
4	0.28	5.22	0.35	0.27	0.25	0.86
5	0.30	5.21	0.44	0.30	0.29	1.12
6	1.57	5.76	0.27	0.13	0.02	0.53
7	1.31	5.20	0.62	0.34	0.13	1.20
8	0.20	7.90	0.39	0.19	0.15	1.40
9	1.00	5.55	0.15	0.12	0.08	0.30
10	0.36	3.35	0.31	0.26	0.12	0.31
11	0.62	3.95	0.20	0.20	0.19	0.38
mean	0.58	5.21	0.31	0.22	0.15	0.66
CoV (%)	86.7	31.5	44.5	34.0	52.4	62.1

The quality of the approximation that can be achieved by means of the proposed model is demonstrated by comparing experimental results and analytical simulations (in terms of F_0-S_0 curves) for each test (Fig. 4). The curves clearly demonstrate that the analytical solution proposed in this study is capable of accurately reproducing the bond behavior of hornified sisal fibers embedded in a cement matrix.

However, the results reported in Table 3 highlight that the bond-slip laws identified for the various tests are fairly scattered. In other words, defining a unique bond-slip law capable to describe the bond behavior of natural fibers in cement matrix does not seem realistic. This can be justified by considering that sisal fibers are heterogeneous, in terms of both mechanical and geometric properties, and this heterogeneity leads to a significant variability in the resulting adhesion with mortar. In fact, the resulting bond behavior in cement matrix is influenced by several parameters, such as fiber shape, fiber cross section area variation along the longitudinal axis and surface texture and roughness of the filaments.

Nevertheless, the identified bond-slip laws may be analyzed and compared in terms of some key mechanical parameters, such as the specific fracture energy G_F, which represents the area under the local bond-slip law in Fig. 1.

Before commenting into details the results in Table 3, it should be mentioned that the bond behavior of fibers in cement matrix can be mainly divided in three phases: adhesion, mechanical bond and frictional bond (Naaman, 1999). The first two mainly depends on the chemical compatibility between the fiber and the matrix and the surface texture of the fiber while the latter is, mainly, governed by the geometrical characteristic of the fiber.

Then, the values of G_F reported in Table 3 can be interpreted in the light of the above considerations.

As regards fracture energy, it should be noticed that the high dispersion, in terms of coefficient of variation CoV (around 60% as highlighted in Table 3), presented by the G_F can be mainly attributed to the morphological characteristics of sisal fibers. In fact, these fibers are characterized by a high variability in terms of transverse section properties (P_f and A_f) as well as in terms of overall straightness and this heterogeneity

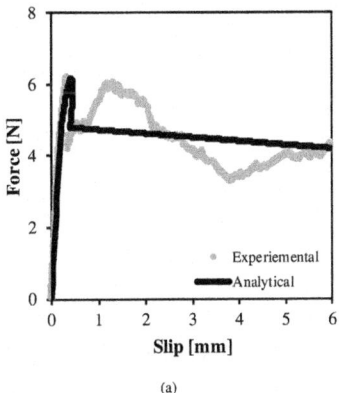

(a)

Fig. 4. Simulation examples of the force-slip response

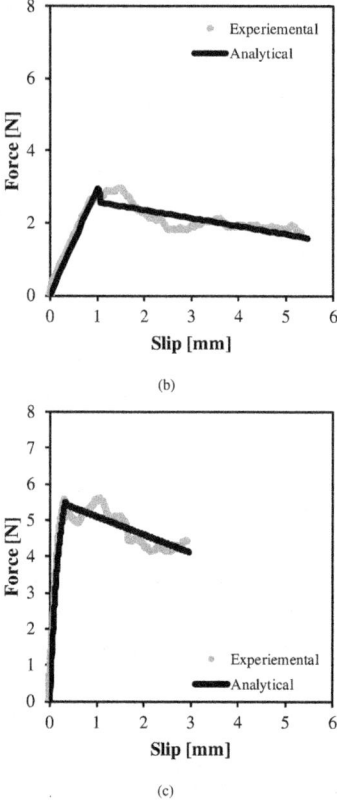

Fig. 4. (*continued*)

influence the frictional bond mechanisms that play a fundamental role on the definition of the fracture energy.

Conversely, the values of τ_{max} presents a lower value in terms of CoV. This can be explained by considering that the elastic branch is mainly controlled by the adhesion and mechanical bond mechanism occurring between fiber and matrix, both being more related to the chemical adhesion between sisal fiber and cement based matrix, rather than their geometric properties.

5   Conclusions

This paper was intended at investigating the bond behavior of hornified sisal fibers embedded in cement-based matrix. On the one hand, the reported experimental results demonstrate the potential of using sisal fibers as spear reinforcement in cementitious composites. On the other hand, the proposed theoretical model leads to scrutinizing the local bond-slip relationship characterizing the interaction between fiber and matrix.

The following main points can be remarked:

- the pull-out tests highlighted good bond properties of hornified sisal fibers: since the majority of specimens failed in debonding, one can reasonably recognize that their transfer length is longer than 25 mm and, hence, the majority of the tested specimens can be considered for charactering the bond behavior of sisal fibers;
- an analytical model, already formulated by the Authors for similar problems, was extended to the case of hornified fibers by introducing a more general bi-linear bond-slip law, whose characteristic features (e.g. discontinuities and non-zero ultimate stresses) resulted essential in simulating the observed behavior of the fibers under consideration;
- in spite of the significant variability affecting the geometry of fibers, reasonably stable values of the key parameters of the identified bond-slip laws were determined for the various tested specimens. Particularly, the variability of two main parameters, such as bond strength and fracture energy was analyzed.

The results of this study will pave the way towards a comprehensive understanding of the mechanical behavior of cementitious composites reinforced by natural fibers. Particularly, the research will move toward the structural scale by analyzing the response of structural members made of this the materials considered herein. Moreover, the proposed analytical model will be available for identifying the bond-slip laws resulting for other kinds of natural fibers (i.e. jute and curauá) and compare them with the one determined for hornified fibers.

Acknowledgements. The study is part of SUPERCONCRETE Project (H2020-MSCA-RISE-2014, n. 645704): the Authors wish to acknowledge the financial contribution of the EU-funded Horizon 2020 Programme. More specifically, it was partly developed during the mobilities of both Prof. Romildo D. Toledo Filho at the University of Salerno (Italy), and Dr. Marco Pepe at the Federal University of Rio de Janeiro (Brazil).

References

ASTM standard C1557 (2013) Standard test method for tensile strength and young's modulus of fibers, ASTM International, West Conshohocken, PA

Caggiano A, Martinelli E, Faella C (2012) A fully-analytical approach for modelling the response of FRP plates bonded to a brittle substrate. Int J Solids Struct 49(17):2291–2300

Claramunt J, Ardanuy M, Garcia-Hortal JA (2010) Effect of drying and rewetting cycles on the structure and physicochemical characteristics of softwood fibres for reinforcement of cementitious composites. Carbohydr Polym 79:200–205

Ferraz JM, Menezzi CHS, Texeira DE, Martins SA (2011) Effects of treatment of coir fiber and cement/fiber ration on properties of cement-bonded composites. BioResources 3:3481–3492

Ferrara L, Ferreira SR, Krelani V, Silva F, Toledo Filho RD (2014) Effect of natural fibres on the self healing capacity of high performance fibre reinforced cementitious composites. In: Proceedings SHCC3, 3rd international RILEM conference on strain hardening cementitious composites, pp 9–16

Ferrara L, Ferreira SR, Krelani V, della Torre M, Silva F, Toledo Filho RD (2015) Natural fibres as promoters of autogeneous healing in HPFRCCS: results from on-going brazil-italy cooperation, special publication, vol. 305, pp 11.1–11.10

Ferreira SR, de Andrade Silva F, Lima PRL, Toledo Filho RD (2015) Effect of fiber treatments on the sisal fiber properties and fiber–matrix bond in cement based systems. Constr Build Mater 101:730–740

Ferreira SR, Martinelli E, Pepe M, de Andrade Silva F, Toledo Filho RD (2016) Inverse identification of the bond behavior for jute fibers in cementitious matrix. Compos Part B Eng 95:440–452

Fidelis MEA (2012) Development and mechanical characterization of jute textile reinforced concrete (Doctoral Thesis), Civil Engineering Department, Universidade Federal do Rio de Janeiro (COPPE/UFRJ). (in Portuguese)

Kundu SP, Chakraborty S, Roy A, Adhikari B, Majumber SB (2012) Chemically modified jute fibre reinforced non-pressure (NP) concrete pipes with improved mechanical properties. Construct Build Mater 37:841–850

Li Y, Hu C, Yu Y (2008) Interfacial studies of sisal fiber reinforced high density polyethylene (HDPE) composites. Compos Part A Appl Sci Manuf 4:570–578

Martinelli E, Napoli A, Nunziata B, Realfonzo R (2012) Inverse identification of a bearing-stress-interface-slip relationship in mechanically fastened FRP laminates. Compos Struct 94(8):2548–2560

Melo Filho JA, Silva FA, Toledo Filho RD (2013) Degradation kinetics and aging mechanisms on sisal fiber cement composite systems. Cem Concr Compos 40:30–39

Naaman AE (1999) Fibers with slip-hardening bond. In: PRO 6: 3rd international RILEM workshop on high performance fiber reinforced cement composites (HPFRCC 3), pp 371–385

NBR 13276 (2005) Mortars applied on walls and ceilings - preparation of mortar for unit masonry and rendering with standard consistence index. ABNT, Rio de Janeiro

NBR 7215 (1996) Portland cement - determination of compressive strength, ABNT, Rio de Janeiro

Netravali AN, Chabba S (2003) Composites get greener. Mater Today 6:22–29

Saha P, Manna S, Chowdhury SR, Sen R, Roy D, Adhikari B (2010) Enhancement of tensile strength of lignocellulosic jute fibres by alkali-steam treatment. Bioresour Technol 101:3182–3187

Silva FA, Chawla N, Toledo Filho RD (2008) Tensile behavior of high performance natural (sisal) fibers. Compos Sci Technol 68:3438–3443

Silva FA, Mobasher B, Soranakom C, Toledo Filho RD (2011) Effect of fiber shape and morphology on interfacial bond and cracking behaviors of sisal fiber cement based composites. Cem Concr Compos 33:814–823

Silva FA, Mobasher B, Toledo Filho RD (2009) Cracking mechanisms in durable sisal reinforced cement composites. Cem Concr Compos 31:721–730

Silva FA, Toledo Filho RD, Melo Filho FA, Fairbairn EMR (2010) Physical and mechanical properties of durable sisal fiber cement composites. Construct. Build. Mater 24:777–785

Sreekumar PA, Thomas SP, Saiter JM, Joseph K, Unnikrishnan G, Sabu T (2009) Effect of fiber surface modification on the mechanical and water absorption characteristics of sisal/polyester composites fabricated by resin transfer molding. Compos Part A Appl Sci Manuf 40:1777–1784

Toledo Filho RD (1997) Materiais compósitos reforçados com fibras naturais: caracterização experimental. Doctoral thesis, Civil Enginnering Department, PUC-Rio, Rio de Janeiro, Brazil. (in portuguese)

Toledo Filho RD, Silva FA, Melho Filho JA, Fairbairn EMR (2009) Durability of compression molded sisal fiber reinforced mortar laminates. Construct Build Mater 23:2409–2420

Application and Advantages of a Balcony Thermal Insulation Element

T. Heidolf[1], E. Nusiner[2], S. Terletti[2], and D. Carminati[2(✉)]

[1] Halfen GmbH, Langenfeld, Germany
[2] Halfen s.r.l., Bergamo, Italy
Diego.Carminati@halfen.it

Abstract. With improved insulation standards over the last 30 years, and with energy efficiency becoming an ever more important design consideration, is necessary to provide a solution to the architectural and engineering problem of creating a thermal break while still providing a structural connection. Thermal bridges occurs when a more conductive (or poorly insulating) material allows an easy pathway for heat flow across a thermal barrier. The most common form is probably within the balcony slab which is usually uninsulated. With an "ad hoc" system is possible to eliminate this kind of problem. Effective thermal insulation using special systems reduces the risk of high levels of condensation, mould formation and the associated damage caused along the ceiling slabs inside of balconies. Thermal outflow and energy loss through the balcony slab is minimized. This paper analyse a specific solution that can guarantee a load capacity connection for balcony slab and at the same time a thermal efficiency. Moreover will be described the design model and the thermal performance analysis in winter and summer situation.

Keywords: Balcony connection · Insulation · Thermal bridge
Energy saving · Design concept · Test results

1 Introduction

Thermal bridges are weaknesses within a building's structure where heat and/or cold is transferred at a substantially higher rate than through the surrounding envelope area. There are basically two types of this phenomenon:

- geometric thermal bridges where part of the structure projects through the building envelope
- material thermal bridges where materials with different conductivity are used in combination

In practice, these effects often combine. A classic example of this is the balcony slab, where problems occur if the connection is not given serious consideration (Fig. 1).

The thermograph photograph above shows that if thermal bridges at balconies are not taken care of, the balconies act as "cooling fins"; conducting the heat off the building and cooling the rooms adjacent to the balconies.

© Springer International Publishing AG, part of Springer Nature 2018
M. di Prisco and M. Menegotto (Eds.): ICD 2016, LNCE 10, pp. 71–80, 2018.
https://doi.org/10.1007/978-3-319-78936-1_6

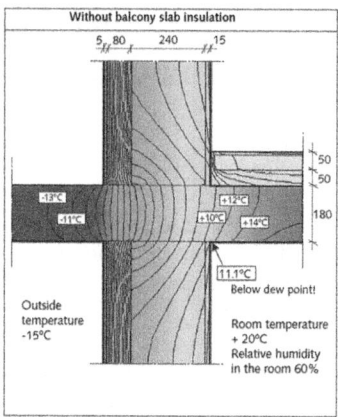

Fig. 1. Thermograph photograph of the thermal bridge at balcony

The effect of thermal bridges are:

- Higher energy consumption: due to the thermal outflow at the balcony connection, heat is drawn from every room resulting in a significant rise in heating costs and energy consumption.
- Mould formation: Interior temperatures of the adjacent rooms can drop well below the dew point. This leads to condensation, deteriorates plaster and paintwork and is an ideal condition for harmful mold formation! If there is sustained exposure to condensation, the building is subject to serious deterioration.
- Uncomfortable living space: Cold surface temperatures cause uncomfortable living space for occupant several reasons may lead to the necessity of strengthening of flat slabs.

2 System and Application

2.1 Balcony Insulation System

The HALFEN HIT-HP and HIT-SP is an element that guarantee the break thermal insulation of the balcony and, at the same time, a mechanical connection between the main slab and the balcony slab.

There are many types of HIT element depending on the load they have to transmit. Some elements has load capacity in terms of shear and bending moment, other only shear load bearing capacity. In general the element is composed by an upper part where are installed tension bars in stainless steel and a bottom part where patented compression shear bearings (CSB) take places. The special shape and the material properties of the fibre-reinforced high-performance mortar in the CSB, which is adapted to the application field, allows compressive and shear forces to be effectively transmitted.

All load components and the mineral wool, which serves as insulation and fire protection material, are firmly fixed in a patented, protective plastic box. The flexible combination of the units in the modular system significantly reduce transport and storage costs and therefore CO_2 emissions. This significantly improves flexibility and planning certainty, especially in the precast industry.

The innovative double-symmetric compression shear bearing (Fig. 2) eliminates the risk of installing in an incorrect orientation. This obviously reduces errors during installation and subsequently the risks of damage to property and injury. Additionally, effective transmission of positive and negative shear forces is possible without any additional load bearing elements and therefore without further reduction in thermal properties.

Fig. 2. Standard HALFEN HIT-MVX element.

2.2   Experimental Investigation

To characterize the behavior of the slab connection construction HIT-HP/SP, the load bearing behavior of the CSB as well as the slab connection construction HIT-HP/SP were analyzed with extensive FEM investigations performed by the company Nolasoft. Within the framework of the approval procedure, numerous building component tests on slabs with the HALFEN-Iso-Element HIT-HP/SP to dissipate bending moments and/or shear forces were performed at the TU Kaiserslautern [3]. Moreover, tests on the residual ultimate load after 120 min of fire exposure were conducted at the Leipzig Institute for Materials Research and Testing (MFPA Leipzig GmbH). The building materials investigations on the load bearing behavior of the fibre-reinforced high-performance mortar were made at the Bauhaus-University Weimar.

Based on the estimated design load capacity of the HIT elements, in the full-scale tests 10 load cycles were performed from (almost) zero to serviceability limit state and 3 load cycles from (almost) zero to ultimate limit state. Afterwards, the load was increased from (almost) zero up to maximum load. Subsequent to this peak load, the structure was unloaded and 3 additional load cycles from (almost) zero to ultimate limit state were applied.

In Fig. 3 the measured force deformation curve is shown for tests with concrete edge failure. The residual load capacity of the tested structure is very high, despite failure of the concrete edge or failure of the CSB. Therefore the behavior of a structure with a connection with HIT is ductile.

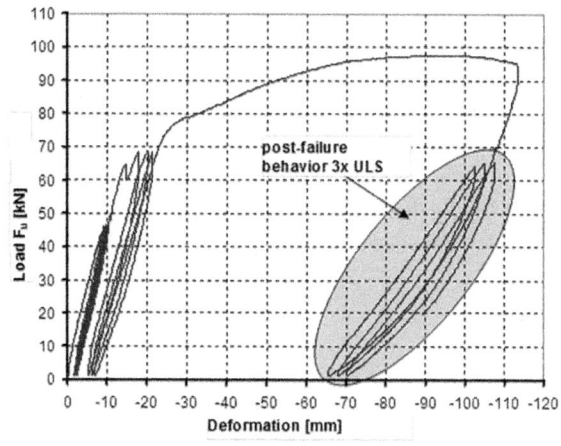

Fig. 3. Post-failure behaviour

2.3   Design Model

Based on the results of the FEM studies, the strut-and-tie model displayed in
ETA-13/0546 [1] was developed. The strut-and-tie model shall qualitatively show the
transfer of the internal forces of the balcony slab via the CSB to the floor slab (Fig. 4).

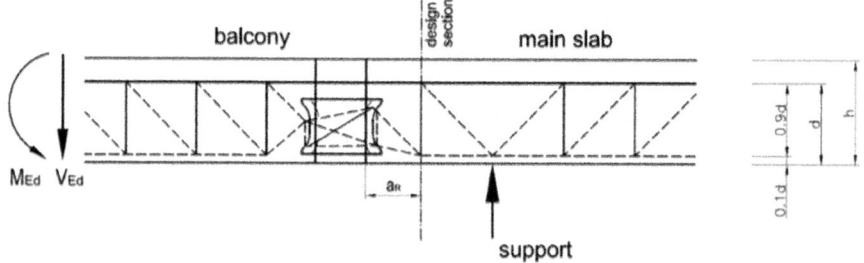

Fig. 4. Strut-and-tie model

Corresponding to the test results, the following models have to be considered for
the design of the slab connection HIT:

- Moment-shear interaction.
- Concrete edge failure of the floor slab.
- Shear force capacity of the adjacent components,

The moment-shear interaction model of the HIT-MVX (cantilevered slab connection) is used to determine the internal forces and the required number of tension bars.
Due to the interaction between moment and shear force, a separate determination of the
load bearing capacities for both loads is impossible. Therefore, both actions V_{Ed} and

M_{Ed} must be considered together. This is conducted by the determination of the maximum compression zone height x_c at the main slab side according to Eq. (1) (in [N] and [mm]):

$$x_c = \text{Max} \begin{cases} d_{CSB} - \sqrt{(d_{CSB})^2 - \dfrac{|M_{Ed}|}{\beta_{c1,M} \cdot n_{CSB}}} \\ \dfrac{h_{CSB}}{2} - \left(\left(\dfrac{h_{CSB}}{2}\right)^2 - \dfrac{V_{Ed}}{(\beta_{c2,V} \times n_{csb})} \right)^{0,5} \leq h_{CSB} \end{cases} \tag{1}$$

With:

$$d_{CSB} = h - c_{nom,s} - \frac{d_{s,1}}{2} - c_{CSB} $$

n_{CSB} number of CSB
h_{CSB} height of CSB (= 83 mm)
$\beta_{c1,M}$, $\beta_{c2,V}$ factors according to ETA-13/0546
c_{CSB} concrete cover of the CSB (= 15 mm)
$c_{nom,s}$ concrete cover of the tension bars
h element height
$d_{s,1}$ diameter of the tension bars
The concrete compressive, are calculated from the number of the compression shear bearings, the height of the concrete compression zone (assuming a constant stress) and the factor $\beta_{c1,M}$.

$$-F_{cd} = 2 \cdot x_c \cdot n_{CSB} \cdot \beta_{c1,M} \tag{2}$$

Maximum transferable shear force V_{Rd} in the design section is:

$$V_{Rd} = \text{Min} \begin{cases} n_{CSB} \cdot 16[kN] \\ |F_{cd}| \cdot \dfrac{(83\,mm - x_c)}{a_{CSB}} \end{cases} \tag{3}$$

With:
x_c = according to Eq. (1)
a_{CSB} distance between the shear forces
= 110 mm for HIT-HP
= 150 mm for HIT-SP.
The stainless steel bar sections of the tension bars are designed in a way that the welded common reinforcing steel becomes decisive. Therefore, only the cross section of the common reinforcing steel section of the tension bars has to be calculated according to Eq. (4).

$$A_s = \frac{F_{sd}}{f_{yd}} \tag{4}$$

With:

$F_{sd} = -F_{cd}$ according to Eq. (2)

$f_{yd} = f_{yk}/\gamma_s$

f_{yk} characteristic value of the yield stress of the reinforcing steel

$\gamma_s = 1,15$ (partial safety factor for reinforcing steel [6])

The other models are described in [1, 2, 5].

2.4   Comparison Between Test Result and Design Models

In Fig. 5, the experimental ultimate moment and shear loads, resp., and the calculated load bearing capacities are compared. It can be seen that the model to calculate the load bearing capacity of the balcony connection HIT-HP/SP is sufficiently precise. The calculation failure modes comply with the experimental failure and that the required safety level is adhered to.

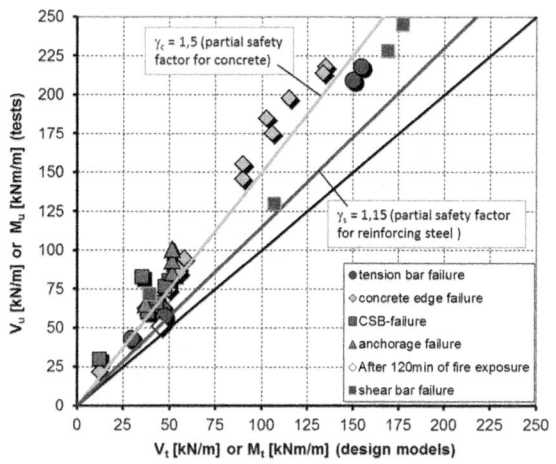

Fig. 5. Measured ultimate loads as function of the calculated values.

2.5   Thermal Performance Analysis

Structural thermal bridges such as balconies may lead to moisture problems resulting from lower temperatures on internal surfaces. Moreover, thermal bridges normally cause additional loss of heat.

Consequently, correct planning using thermally insulated balcony slab connections:

- prevent condensation and mould growth by fulfilling the minimum thermal insulation requirements according to DIN 4108-2,
- reduce the transmission heat losses in the area of the connections.

Depending on the temperature, the air can retain different amounts of moisture. With a rise in air temperature, the amount of storable moisture increases. In a room, the air is constantly moving (room air flow). The water content in a defined flowing air volume remains nearly constant.

However, the temperature of the air changes when the air flows along colder external components. The storage capacity of the air decreases as the air cools down, resulting in an increase in relative humidity. Condensation always occurs when the relative humidity reaches 100%. Assuming a room temperature of 20 °C and a relative humidity of 50% condensation would occur when the air cools down to approx. 9 °C (Fig. 6). If, under the given conditions, the temperature at the inner surface of an adjacent component, for instance the wall or the ceiling, is 9 °C or colder, then condensation will form on this surface.

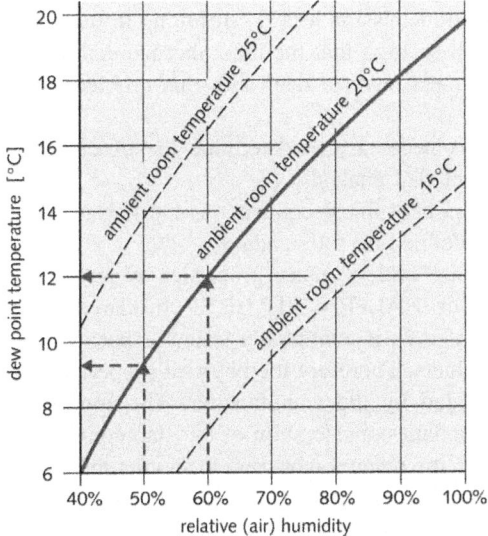

Fig. 6. Dew point diagram

Correct application of HALFEN HIT Insulated connections prevents the surface of the wall/ceiling from falling below the dew point and therefore prevents condensation. An increased relative humidity of approx. 80% above the surface of the component promotes mould growth.

In a standard scenario with an indoor temperature of 20 °C and a relative humidity of 50%, cooling down the air to approximately 13 °C raises the relative humidity to 80%.

HALFEN HIT Insulated connections prevent cooling of the adjacent components at the inside of the balcony below the critical temperatures for condensation and mould growth.

The criterion to prevent mould growth is the temperature factor f_{Rsi}. It is defined as the ratio of the lowest surface temperature minus outside temperature to the total temperature difference (inside temperature minus outside temperature).

DIN 4108-2 stipulates that the temperature factor f_{Rsi} must be higher than 0.7 for all component connections.

According to the National Technical Approvals Z-15.7-293, Z-15.7-309 and Z-15.7-312 the minimum thermal insulation requirement in accordance with DIN 4108-2 has already been proved and applies to the complete HALFEN HIT Insulated connections load range.

The Energy Saving Regulation (EnEV) specifies that the primary energy demand required to heat a building must be limited. To calculate this energy demand, thermal bridges through concrete balcony slabs must also be taken into account. Monolithic balcony systems without thermal separation have the same effect as cooling fins due to their geometry and therefore they cause substantial heat losses.

In most cases even when conforming to the specifications stipulated in DIN 4108, the calculated specific transmission loss H_T (resulting from standard cross-sections and thermal bridges) is still so high that the max. thermal ceiling set by the EnEV is not easy to maintain. Planners have to deal with this problem when they have to meet predefined criteria.

In these cases it is necessary to determine the exact transmission losses of all thermal bridges in a detailed analysis.

For structural component linear connections the linear thermal transmission coefficients (ψ-value) are defined by set standards.

All relevant thermal and acoustic properties were included in the European Technical Approval for HALFEN HIT-HP/SP Insulated connection. The thermal conductivity of the materials is continually tested as part of the internal and external monitoring of the products. Therefore the physical property values here are no longer just information provided by the manufacturer. The approved characteristic values allow a standards-compliant consideration of the slab connections.

In order to facilitate the planner work and to shorten the processing time, HALFEN has made an ψ-Calculator available. The required characteristics can be determined quickly and easily for a variety of installation situations. Since the HALFEN HIT has a standard fire resistance of 120 min, there are no different values for types with or without fire protection.

2.6 Example of Thermal Calculation

Figure 7 shows the temperature fields in the cross-section (shown as isotherms) for 6 different situations. The winter situation is calculated with a outdoor temperature of −15 °C and a room temperature of 20 °C. Without a facade insulation and without a thermal break between the balcony and the main slab the minimal surface temperature is 3,6 °C (Fig. 7a). A facade insulation of 22 cm improves the minimum surface temperature to 12.9 °C (Fig. 7b). According to Fig. 6 a relative humidity of approx. 65% above the surface of the component promotes mould growth. Figure 7c illustrates the advantages of the HALFEN HIT Insulated connection for the minimal surface temperature. With a surface temperature of 17,5 °C (HIT-HP) or 17,9 °C (HIT-SP) no condensation and mould growth will be arise. Furthermore the heat flow decreases form 201 W/m to 18 W/m and saves a lot of energy.

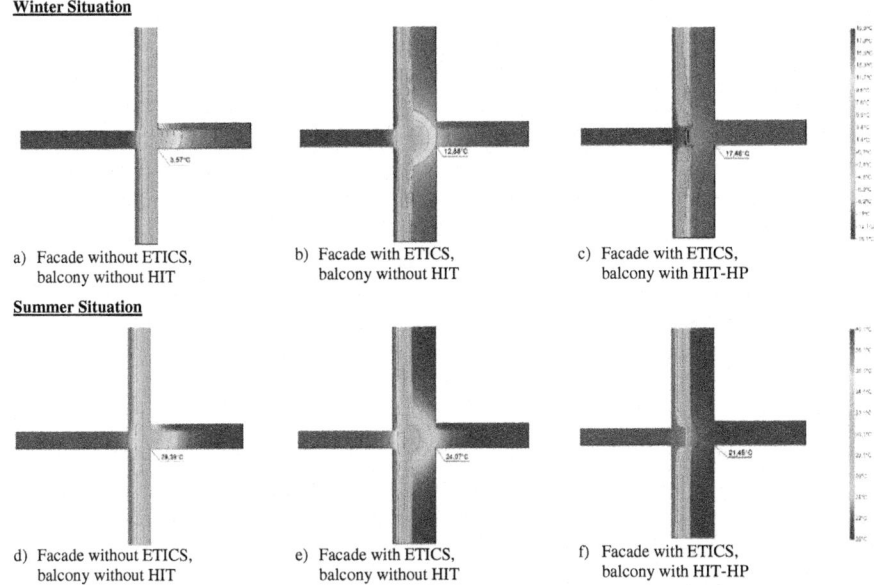

Fig. 7. Example of thermal calculation

Figure 7d–f shows the summer situation with a outdoor temperature of 40 °C and a room temperature of 20 °C. Although mold growth is not an issue, but a facade insulation and a insulated balcony connection reduced signified the energy costs for air conditioning (Table 1).

Table 1. Influence of facade insulation and balcony connection

	Facade without ETICS, balcony without HIT	Facade with ETICS, balcony without HIT	Facade with ETICS, balcony with HIT-HP	Facade with ETICS, balcony with HIT-SP
ψ-value [W/mK]	0,260	0,715	0,266	0,211
winter situation temperature θ_{min} [°C] heat flow Q [W/m]	3,6 200,9	12,9 36,8	17,5 17,7	17,9 15,8
summer situation temperature θ_{max} [°C] heat flow Q [W/m]	29,4 114,8	24,1 21,0	21,5 10,1	21,2 9,0

3   Conclusions

According to national and international standards reinforced concrete structures that would penetrate the exterior envelope of a building, for example, balcony slabs, should be thermally separated from the main floor slab. In practice, slab connection elements are used as standard for this purpose. Slab connection elements are also used to improve structural-physics requirements and therefore to prevent condensation and mould growth; they also prevent cracks caused by thermal expansion.

Conventional systems consist of an insulating layer (mostly EPS Expanded Polystyrene) and a static effective strut-and-tie system, which is formed from steel tension bars and steel shear bars as well as concrete or steel compression bearings. The static system for the HALFEN HIT Insulated connection HP/SP was developed further using patented compression shear bearings (CSB).

The flexible combination of the units in the modular system significantly reduce transport and storage costs and therefore CO_2 emissions can be reduced by up to 65%. This significantly improves flexibility and planning efficiency, especially in the precast industry. Furthermore, an optimal design of the elements for the intended application situation is possible. Because of the compact construction of the compression shear bearing boxes there are no protruding structural elements that may cause problems in transport or when storing semi precast balcony slabs. This factor also helps to keep transport costs down, avoids damage in transit and reduces the risk of installation errors.

This paper analysed a range of element nominated HIT from statically and thermal point of view. The testing investigation supported the statically method identify for such product.

The HALFEN HIT-HP/SP Insulated connection is the first and currently the only slab connection with a CE marking. The European Technical Approval ETA-13/0546, allows the elements, the design concept and the structural physical characteristics to be used in over 30 European countries. This increases safety in planning.

The new HALFEN HIT Insulated connections HP/SP, however, to a large extent use recycled materials. All other materials are recyclable. The HIT-HP/SP are therefore particularly environmentally friendly. They can be described as "green" slab connections.

References

1. European technical approval ETA-13/0546: Halfen Insulated Connection - HIT-HP MV, HIT-SP MV, HIT-HP ZV, and HIT-SP ZV; OIB Member of EOTA, Wien 29 June 2013
2. Heidolf T, Eligehausen R (2013) Design concept for load bearing thermal insulation elements with compression shear bearings. Beton- und Stahlbetonbau 108(3):179–187
3. Schnell J (unpublished) Tests on Halfen Iso-elements, various test reports for the period 2009 – 2015
4. EN 1990, 04.2002, Eurocode - Basis of structural design
5. Eligehausen R (unpublished) diverse Expert report on load capacity of slab connections using Halfen Iso-elements HIT-HP/SP
6. EN 1992-1-1, 12.2004, Eurocode 2: Design of concrete structures – Part 1-1: General rules and rules for buildings

Design Guidelines for Precast Structures with Cladding Panels in Seismic Zones

A. Colombo[1] and G. Toniolo[2(✉)]

[1] Assobeton, Italian Association of Precast Industry, Milan, Italy
[2] Department of Civil and Environmental Engineering,
Politecnico di Milano, Milan, Italy
giandomenico.t@gmail.com

Abstract. The paper presents a summary of the results of the experimental and analytical research on Improved fastening systems of cladding wall panels of precast buildings in seismic zones, performed within the scope of the European Project Safecladding (EU Programme FP7-SME-2012-2 – Grant Agreement n. 314122), as codified in the pertinent documents issued by the research Consortium in terms of Design Guidelines for practical applications. Further than the investigation on the current fastening systems and the related acceptability analysis for existing structures, the paper deals with the specific design criteria for the isostatic, integrated and dissipative fastening systems that should be the basis for the new precast constructions in seismic zones.

1 Introduction

The experience of recent earthquakes, like L'A-quila 2009, Lorca 2011 and Emilia 2012 (see Fig. 1), showed the inadequacy of the present design criteria of the fastenings of the cladding wall panels. In addition to this one, another widespread serious deficiency has been noticed in buildings not designed for seismic action: the loss of bearing of beams and floor elements in dry simple supports that entrust the transmission of horizontal forces only to friction without mechanical restrainers (see Fig. 2). Proper connection devices are needed able to transfer horizontal forces also in absence of the gravity action. And this is valid also with reference to the possible lateral overturning of beams. These connection devices should be dimensioned taking into account the stiffening effects of the cladding panels with respect to the response of the bare frame structure.

For what concerns the former deficiency, the basic principle shall be that the fall of cladding wall panels under earthquake action shall be prevented following the same requirement of no collapse stated for the main structure by Eurocode 8. To this end the fastening devices and the whole connections of wall panels shall be verified for the effects of the seismic action in terms of forces and/or displacements as calculated through the proper structural analysis.

Following this principle, the structural analysis of precast buildings under seismic action shall properly take into account the role of cladding panels in the seismic response of the overall construction assembly. The models used in the analysis shall represent as close as possible the real arrangement of the construction, including panels

© Springer International Publishing AG, part of Springer Nature 2018
M. di Prisco and M. Menegotto (Eds.): ICD 2016, LNCE 10, pp. 81–97, 2018.
https://doi.org/10.1007/978-3-319-78936-1_7

Fig. 1. Emilia 2012 – panel collapse

Fig. 2. Emilia 2012 – beam fall

and relative connections. Specific indications are given in the following clauses with reference to the systems of cladding connections discussed below.

The precast structures considered are frame systems made of columns and beams connected with horizontal floor diaphragms. In particular the roofs can provide rigid, deformable or null diaphragm action. To this frame system, for the peripheral cladding, a set of wall panels is added that, depending on the type of connections to the structure, may not interfere with the frame behaviour or may interfere leading to the interaction between the panels and the frame and to an increased stiffness of the system. In this dual wall-frame system a set of dissipative connections may be present able to attenuate the seismic response. All these structural systems have to be possibly analysed with proper specific calculation models.

2 General

As primary type of analysis for the current design practice the linear elastic analysis with response spectrum is proposed as regulated by Clause 4.3.3 of Eurocode 8, where the effects of energy dissipation at the ultimate limit state are represented by the behaviour factor q_p (see 5.11.1.4 of Eurocode 8). In particular from the analysis the forces and displacements on the connections of the panels are expected for the necessary pertinent verifications.

For new constructions the following suggestions refer to the modern production of precast concrete structural elements through processes under quality control as regulated by the present European harmonised standards.

With reference to the provisions of Chap. 5 of Eurocode 8, these structures, when provided with isostatic systems of wall panel connections, can have all the characteristics to be considered in ductility class high "DCH" (ductile reinforcement, proper detailing, overproportioned connections). However the use for frame structures of the high q_o factor given by Table 5.1 of Eurocode 8 to this class could lead to an excessive deformability incompatible with the requirements of the damage limitation state, with floor drifts larger than 1%. To fulfil this limit state a larger proportioning in size of the columns could be necessary (corresponding to a lower q_o factor), aware that in this way the frame structure will result over-dimensioned in strength. In any case it is highly recommended to apply the rules of DCH for reinforcement ductility, member detailing and connection over-proportioning (with $k_p = 1,0$). In particular, for a full exploitation of the ductility resources in compression of the longitudinal bars of columns, in the critical regions of the columns a spacing of stirrups $s \leq 3,5\phi$ should be adopted, with ϕ diameter of the longitudinal bars.

If an integrated system of wall panel connections is adopted, the resulting dual wall-frame structure could be classified in ductility class medium "DCM" with a lower behaviour factor.

For all connection systems, a ratio $\alpha_u/\alpha_1 = 1,0$ shall be taken, due to the null or low level of structural redundancy.

3 Existing Buildings

A structural analysis may be required for the verification of the seismic capacities of existing buildings, in terms of resistance and stability under the expected seismic action. Following the results of this verification, proper interventions of upgrading or retrofitting could be decided. These interventions should be performed according to the provisions of Eurocode 8 Part 3 and relevant National Annex. When it is technically and economically possible, a retrofitting should be made with the full fulfilment of the code requirements for new constructions. Otherwise the upgrading may correspond to an improvement of the capacity that doesn't reach the level required in the zone for the new constructions. In such a case lower return periods of the earthquake can be used leading to lower peak ground accelerations compared to those used for the new buildings in the zone. At present some countries allow the reduction factors as low as 0,6. In all cases the damage limitation requirement can be disregarded.

In the case of existing buildings originally not designed for seismic action, the analysis should be based on the assumption of a low ductility class for which a behaviour factor q = 1,5 shall be adopted. If the existing buildings were designed for seismic action, higher behaviour factors can be used, if justified by the calculations made and the structural details adopted at the time of construction.

While the elastic modal spectrum analysis is the first choice in the current design practice, more refined inelastic analysis might be required in some cases, due to the highly complex non-linear properties of the panel-to-structure interaction, in order to obtain more reliable and less onerous results.

For seismic loading, the deformation property of the current fastening devices, like the hammer-head straps of Fig. 3, has a double course. In the horizontal direction parallel to the plane of the panels, a low interaction between the panel and the bare frame may be expected until a certain deformation threshold is not overcome. Within this deformation limit, the panel connection system behaves essentially as isostatic and relatively simple structural models can be used, basically neglecting the structure-to-cladding interaction.

Fig. 3. Test on a hammer-head strap connection

When the quoted deformation limit is overcome, the free movement is stopped and more complex models should be used considering the interaction between the panels and the bare structure through the fastening devices. Figure 4 shows the elasto-plastic behaviour of the quoted hammer-head strap and a related possible model.

Anyway the consequent safety verifications will possibly demonstrate the inadequacy of the panel connection systems. In such a case proper upgrading or retrofitting interventions should be made. Any intervention of upgrading or retrofitting of panel connections on existing buildings should be made only when the adequacy of all the remaining parts of the structure has been verified to be compliant with the requirements of the chosen level of seismic resistance.

A little invasive technique of intervention on existing buildings to prevent the fall of wall panels under earthquake conditions consists of short slack cables connecting the panel to the main structural element as shown for vertical panels in Fig. 5.

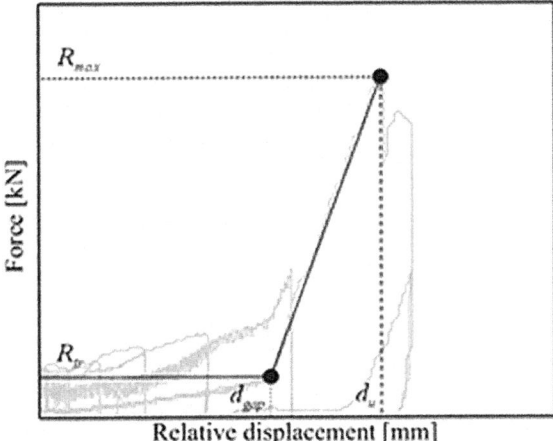

Fig. 4. Cyclic behaviour of hammer-head strap

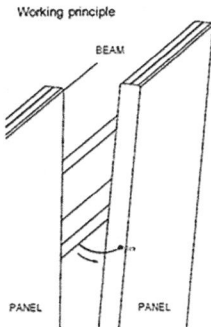

Fig. 5. Anti-fall restrainers

Considering that under earthquake conditions, after the possible failure of the original fastenings, the motion of panels is completely out of control, this solution can be used only for a quick upgrading of existing buildings in order to ensure their operativeness for a transitory period, waiting for a more reliable intervention.

Folded steel plates can be used, instead of the too rigid steel angles, in existing buildings to retrofit or upgrade the connections of horizontal panels to the structure (see Fig. 6). They ensure a good elastic and plastic deformability able to avoid the very high force response to the seismic action proper of the stiff wall systems. The installation of the steel folded plates doesn't require any special care for tolerances since they can be shaped in site on the existing situation and attached to column and panel with post-installed drilled fasteners.

Fig. 6. Folded plate angle

Folded steel plates can be effectively used also in new constructions. Figure 7 shows the use of folded steel plates for the connection of horizontal panels to the adjacent columns. Any panel is seated on a couple of steel brackets that ensure a vertical support independently from the other panels. Four folded steel plates are then fixed at the corners providing full translational restrains in all directions with a hyperstatic overall arrangement.

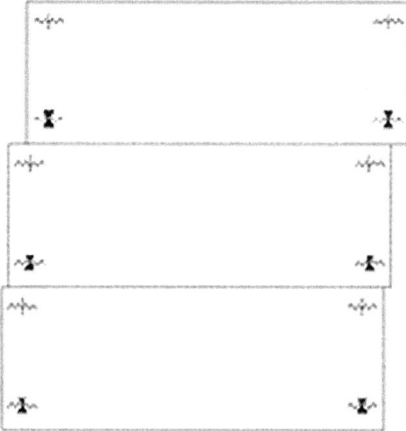

Fig. 7. Use of folded plates in horizontal panels

In the in-plane horizontal direction the folded plates have the elasto-plastic behaviour described in Fig. 8. In the vertical direction they can add a fixed bilateral restrain or leave a displacement freedom depending on the type of fastening on the column. The elastic deformability in the horizontal direction can attenuate the effects of thermal expansion.

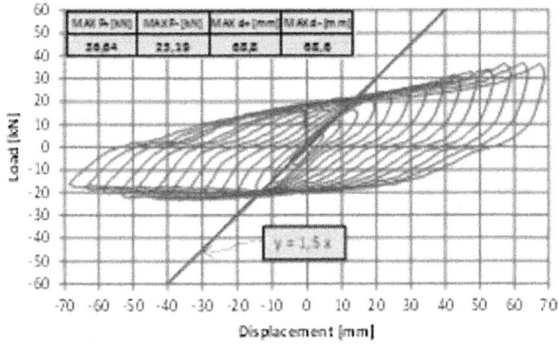

Fig. 8. Cyclic test on folded plate angle

4 Isostatic Connection System

In order to ensure the pure frame behaviour of the resisting structure, the connection system of the wall panels shall actually allow without reaction the large displacements of the frame structure under seismic action, except for possible minor unintended reaction effects due to friction or sealing. In this case one shall adopt sliding connection devices with adequate capacities (such as ±150 mm or greater) or pinned connectors for free rotations.

It should be stated beforehand that at present the isostatic systems, adapted for seismic purposes as described below, are of easy application with a sufficient reliability.

Considering that the vertical panels are placed over the foundation beam, to which they transmit their weight, and are supported horizontally by the roof beam with connections placed close to the top, the arrangements to obtain an isostatic connection system for these panels are those shown in Fig. 9. The first solution adopts hinged lower and upper supports so to have a *pendulum behaviour* for any single panel (see Fig. 9a). The second solution adopts fixed supports at the base of the panels and two sliding connections to the structure at the upper position so to have a *cantilever behaviour* uncoupled from the structure (see Fig. 9b). The third solution adopts a simple seating of the panels on bearings placed at the two edges of the base side together with a hinged connection to the structure at the upper position so to have a *rocking behaviour* for large displacements (see Fig. 9c). For all the three solutions, in the out-of-plane direction the panels are supported with an isostatic pendulum scheme (see Fig. 9d).

In the pendulum arrangement (Fig. 9a), for a given top displacement d the adjacent vertical sides of the panels display a relative slide of bd/h where h is the height of the upper support and b is the width of the panels. In the meantime the two adjacent sides get closer by a minor quantity that requires in any case a free spacing between the panels (few millimetres) closed by the sealant. The sealing between the adjacent panels may give a minor reaction to the motion that can be neglected. The rotating base supports of the panels may display a friction reaction to the motion that has very small (negligible) dissipative effects on the seismic response. Since only vertical compression

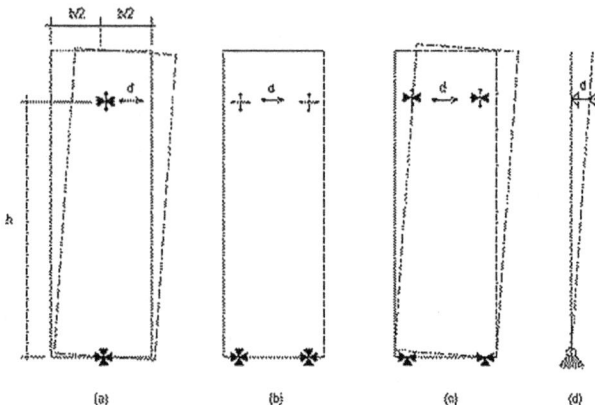

Fig. 9. Isostatic connection systems of vertical panels

forces are expected at the base supports, these can consist of simple seatings able to provide only an unilateral restraint in the vertical downwards direction. To allow for thermal expansion, the upper supports may be made of one or two lateral vertical slide channel bars able to transmit only horizontal forces.

The cantilever arrangement (Fig. 9b) keeps the panels still during the motion of the structure because of the horizontal slide channel bars placed at the upper position. To allow for thermal expansion vertical channel bars may be coupled to the horizontal ones (one in the beam and one in the panel). Sensible friction effects may arise due to the contemporary orthogonal forces caused by the biaxial vibratory motion. The base support of the panels can be provided with reinforcing bars protruding from the bottom and anchored by bond within corrugated sleeves inserted in the foundation and filled with no-shrinking mortar. Other types of dry mechanical connections may be adopted for the base support of the panels.

The rocking arrangement (Fig. 9) consists of the seating of the panels on the foundation through unilateral bearings that work only in compression. The two edges of the base side of the panel alternatively rise up during the rocking. Small horizontal actions applied to the upper connection of any single panel are equilibrated by its weight until they are $\leq Gb/(2h)$, where G is the weight of the panel, b is its width and h is the height of the applied horizontal action. In this condition the panels behave as integrated in the structure that becomes a dual wall-frame system with a much higher global stiffness. Under seismic action therefore the reacting system has an initial high horizontal stiffness that decreases when that limit is overcome and the panels begin to rock, behaving as isostatic. In the rotated position the panel, seated in its edge active bearing, provides a stabilizing constant horizontal force $H = Gb/(2h)$. At the reverse motion the panel seats back again on the two base lateral bearings restoring its initial stable equilibrium. The analysis of such vibration motion requires calculation codes with refined algorithms for the solution of the non linear equations.

The horizontal panels are connected externally to the adjacent columns to which they transmit their weight, being restrained horizontally by the same connections. The lowest panels can be seated directly with their weight on the foundation elements.

In this paper only the isostatic hanging (see Fig. 10a) and seated (see Fig. 10b) equivalent solutions are considered. The superimposed panels shall have a free spacing at the joint between the adjacent sides to allow the relative slide motion without friction. This joint is sealed with proper material (silicone) that may introduce a minor reaction effect. Any single panel is provided by two upper or lower vertical supports placed at the ends and fixed to the columns. One of them provides also the horizontal restrain in the plane of the panel. To allow for thermal expansion, the opposite one gives no horizontal reaction in the plane of the panel. At the opposite lower or upper side two couples of sliding connections are placed allowing the free horizontal and vertical displacements. All the four corner connections provide a fixed horizontal support orthogonal to the panel.

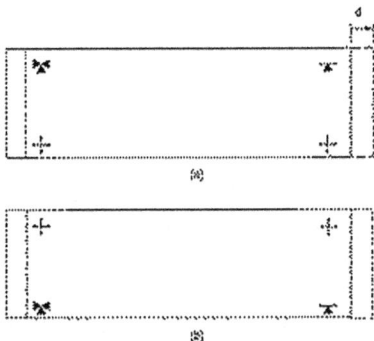

Fig. 10. Isostatic connection systems of horizontal panels

For buildings with isostatic arrangements of wall panel connections, the structural analysis under seismic action can refer to the frame system following the current design practice of such structures.

5   Integrated Connection Systems

In the integrated system the connections of each panel are arranged with a hyperstatic set of fixed supports. With this arrangement of connections the panels participate to the seismic response of the structure within a dual wall-frame system which has a much higher stiffness and a lower energy dissipation capacity compared to a pure frame and this leads to a structural seismic response with higher forces and lower displacements. The panel connections shall be proportioned by consequence, not with a local calcu-lation based on the mass of the single panel, but from the analysis of the overall structural assembly with its global mass.

The adoption of an integrated system has also some side effects such as those of a strong engagement of the floor diaphragm action necessary to take the inertia forces of the floors to the lateral resisting walls. This can lead to very high joint forces.

A typical hyperstatic arrangement of connections is shown in Fig. 11 for vertical panels. Four fastenings are used, one for each corner, the lower two attached to the bottom foundation beam with fixed connections, the upper two attached to the top beam with vertically sliding connections that allow the free thermal expansion of the panel. With this arrangement any panel acts as a vertical cantilever beam clamped at its bottom and pinned at its top.

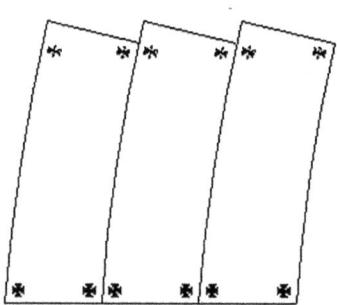

Fig. 11. Integrated connection system of vertical panels

Different types of base fixed connecting mechanisms can be adopted, like the connections with protruding bars, with bolted shoes and with bolted plates. Independently from the type of connection, a gap should be left during construction between the panel and the supporting beam, which is filled with high strength, non-shrinking grout after mounting the panels. The purpose of this bed of mortar is to form a uniform contact between the panel and the supporting beam, necessary to ensure friction and prevent sliding.

Due to the large in-plane stiffness of the panels in their integrated arrangement, significant internal forces can develop during strong earthquakes. Openings in the panels reduce their local strength and, therefore, appropriate resistance verifications should be performed.

Figure 12 shows a hyperstatic arrangement of connections for horizontal panels. Four fixed fastenings are used, one for each corner, attached to the contiguous columns. With this arrangement any panel acts as a horizontal beam clamped at its ends.

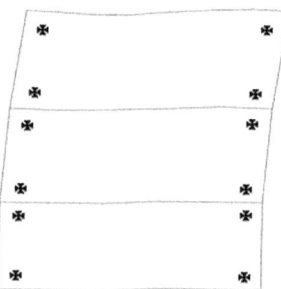

Fig. 12. Integrated connection system of horizontal panels

In case of horizontal panels, the fixed connections applied to the columns affect significantly their deformation during earthquakes, inducing the development of high local forces. Additionally, the insertion of adequate fastening devices in the reduced dimensions of the columns without endanger their resistance could be a difficult construction problem. For these reasons horizontal panel arrangement in integrated systems are not recommended.

6   Dissipative Connection Systems

Between the two extreme solutions of isostatic systems, with their large displacement demand, and integrated systems, with their high force demand, the dissipative systems of cladding connections offer an intermediate solution able to keep displacements and forces within lower predetermined limits.

In this section two structural arrangements are considered, one for vertical panels and one for horizontal panels. Actually in Sect. 3 a third arrangement has been presented, again for horizontal panels and the use of folded plates angles.

For vertical panels, starting from the isostatic pendulum arrangement of fastenings of Fig. 9a, a number of dissipative mutual connectors are added between the panels, opposing the relative slide at their vertical joints. Figure 13a shows this solution where the zig-zag lines indicate the position of the inter-panel dissipative devices. An equivalent solution, with an lower energy dissipation efficiency, can be obtained starting from the isostatic rocking system of Fig. 9c.

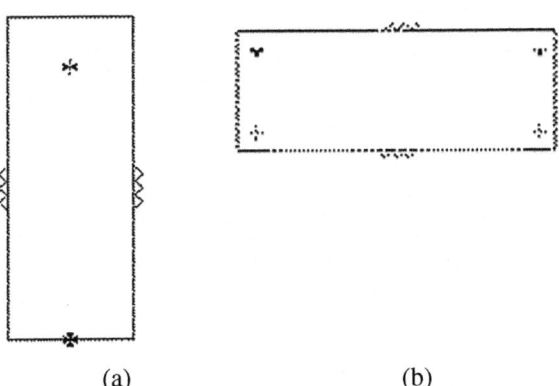

(a)   (b)

Fig. 13. Dissipative connection arrangements (a) for vertical, (b) for horizontal panels

For horizontal panels, starting from the isostatic hanging arrangement of fastenings of Fig. 10a, a number of dissipative mutual connectors are added between the panels, opposing the relative slide at their joints. An equivalent solution can be obtained starting from the isostatic seated arrangement of Fig. 10b.

The dissipative devices considered in this paper are the friction based connectors, the multi-slit plates and the steel cushions.

The friction devices are made of two steel T shape parts that are fixed with a symmetrical set of bolted fasteners to the adjacent panels in special recesses and coupled with two lateral bolted steel plates, as shown in Fig. 14. The length of the slotted holes made in the web of the T shape profiles gives the limit to the reciprocal slide between the parts. The tightening torque given to the bolts, controlled by dynamometric wrench, activates the friction between the plates determining the slip threshold shear force. Two brass sheets are interposed between the steel plates and the profiles to ensure the stability of the repeated slide cycles.

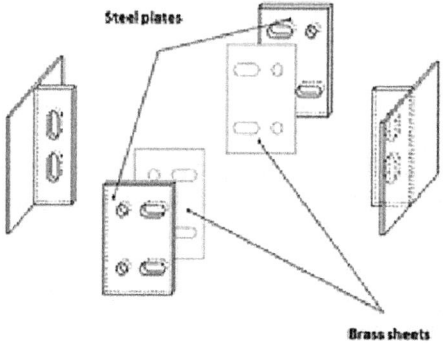

Fig. 14. Friction based dissipative devices

A centred longitudinal shear is transmitted through this device, limited to the threshold value. The cyclic dissipative behaviour is shown by the experimental diagram of Fig. 15; it can be represented with a good approximation by a rigid-pseudo plastic model. After a strong friction engagement the bolts shall be released for re-centring of the structure and re-tightened; the same steel plates and brass sheets can be re-used with the same potential behaviour.

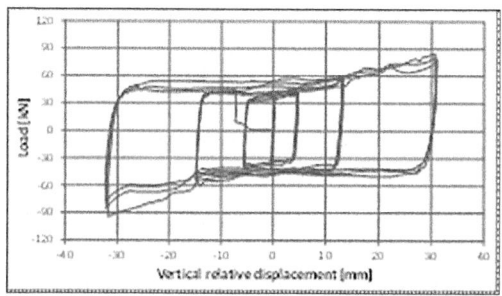

Fig. 15. Force-displacement cycles for friction devices

The multi-slit devices considered in this paper have a composition similar to the friction devices, where the two lateral solid steel plates used in the latter devices are replaced by as many lateral plates provided with slits of various shapes and sizes. The slits isolate a number of slender strips and in this way move the shear behaviour of the solid plates to a flexural diffused behaviour of the single strips which is suitable for energy dissipation (see Fig. 16).

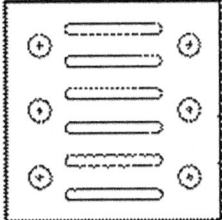

Fig. 16. A type of multi-slit dissipative device

The cyclic dissipative behaviour is shown by the experimental diagram of Fig. 17; it can be represented with a good approximation by a bi-linear hardening model. After significant plastic deformations, the multi-slit devices shall be removed for re-centring the structure and replaced with new ones.

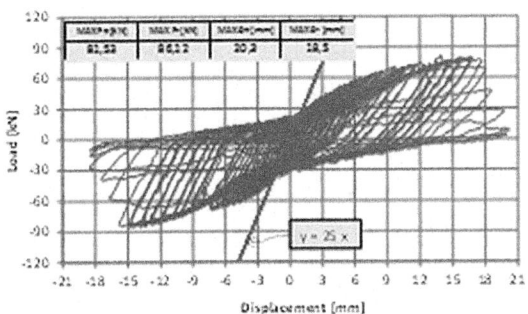

Fig. 17. Force-displacement cycles for multi-slit devices

The steel cushions considered in this document are made of a flattened ring-shape plate that is fixed with central fasteners to the adjacent panels in special recesses as shown in Fig. 18a. They can be utilized wherever multi-slit and friction devices are used. They can transmit longitudinal shear forces with the deformation mechanism described in Fig. 18b.

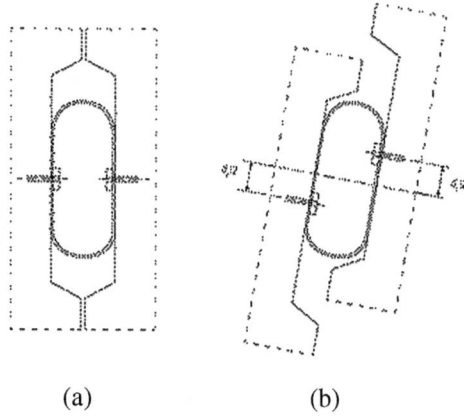

(a)   (b)

Fig. 18. Steel cushion and its deformation mechanism

The cyclic dissipative behaviour is shown by the experimental diagram of Fig. 19; it can be represented by an approximate bi-linear hardening model or by more accurate algebraic curves. After significant plastic deformations, the steel cushions shall be removed for re-centring the structure and replaced with new ones.

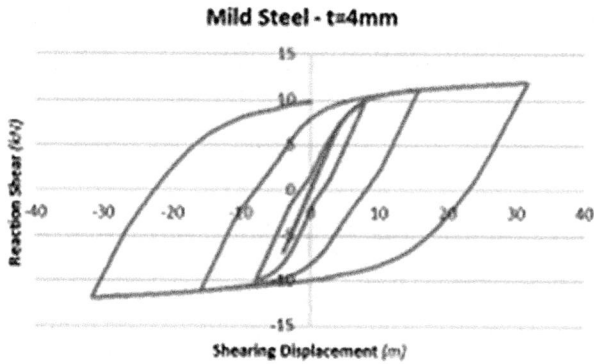

Fig. 19. Force-displacement cycles for steel cushions

With reference to vertical panels, in terms of roof drift d_x of an one-storey building, the relative slide play $\pm s_z$ between two adjacent panels (see Fig. 20) leads to $d_x = \pm s_z h/b$ that, for the common dimensions of the panels, corresponds to about three times its value.

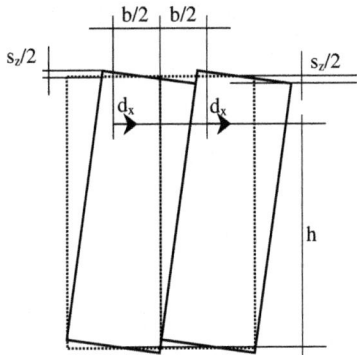

Fig. 20. Displacement mechanism of vertical panels

Figure 21 shows the forces transmitted between the panels and the frame structure, where (a) is the end panel that transmits a relevant vertical force to the foundation, (b) is an internal panel that exchanges vertical shear forces at both sides.

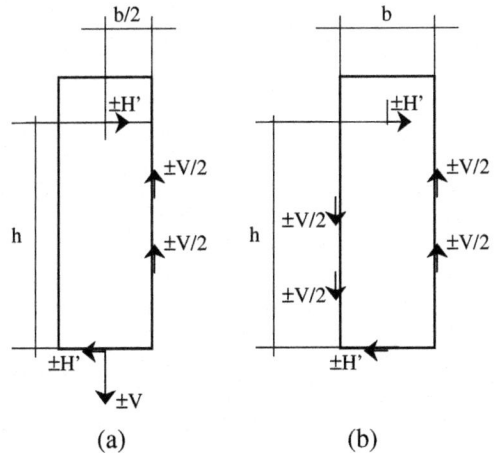

Fig. 21. Forces applied to (a) end panel, (b) internal panel

The horizontal resistance contributions can be deduced from the rotation equilibrium in terms of the shear V transmitted by the joint connections:

$$H' = V b/(2h) \qquad H = V b/h$$

If V is the threshold force of the rigid-pseudo plastic model of a friction dissipative device, the sum F of these horizontal forces corresponds to the maximum response of the structure. Compared to the maximum response F_{max} of the integrated dual wall-frame system, where the initial high stiffness is given by the large walls with fixed panel connections, this response leads to the required reduction (behaviour) factor

$$q = F_{max} / F$$

The calculation of F_{max} can be referred to the total vibrating mass of the building and to the maximum spectral response of the site. The equation above can be used to proportion the dissipative devices in number and strength assuming a proper (conservative) value of the q factor.

7 Conclusions

Recent earthquakes pointed out the often inadequate fastening systems of cladding panel to the main structure. For this residual weak point the last European research project Safecladding developed the proper design criteria, indicating the possible solutions of isostatic, integrated or dissipative connections. The first one is the simplest to be applied with the use of available products. The second solution draws very high seismic forces in the structural joints and need new improved connection devices. The third solution is at present out of the current construction experience, but it seems very promising for an effective application. The experience in the new constructions of these solutions will solve the questions still open for a generalised routine praxis.

Acknowledgments. In the present paper the authors summarized the outcome of the wide experimental and analytical research on Improved fastening systems of cladding wall panels of precast buildings in seismic zones, performed within the European Project Safecladding (EU Programme FP7-SME-2012-2 – Grant Agreement n. 314122). The work has been carried out by the research providers on the basis of the tasks committed to each of them. The results have been codified in the two detailed Design Guidelines for practical applications quoted in the References (Safecladding Project 2015a, b). The intellectual paternity belongs to the single research groups and to the consortium as a whole that assured an organic co-ordination of the different contributions.

With reference to the research of concern with its different activities, a mention shall be devoted to the research providers, with Dr Paolo Negro and Mr. Marco Lamperti for ELSA Lab. of the JRC (Ispra), Prof. Matej Fischinger and Prof. Tatjana Isakovic for Ljubljana group, Prof. Ioannis Psycharis for Athens group, Prof. Fabio Biondini and Dr. Bruno Dal Lago for Milan group, Prof. Faruk Karadogan and Prof. Ercan Yuksel for Istanbul group, in addition to Mr. Claudio Pagani and Antonello Gasperi for the Company BS Italia and the authors of this paper themselves.

The national industrial associations were represented by Mr. Bulent Tokman (TPCA, Turkey), Mr. Alejandro Lopez (ANDECE, Spain), Mr. Thomas Sippel (VBBF, Germany), in addition to Assobeton (Italy) and the Turkish Company YAPI. The general manager of the research project was Mr. Alessio Rimoldi of the international association of the precast industry BIBM.

References

Safecladding Project (2015a) Design guidelines for precast structures with cladding panels, Deliverable 6.1

Safecladding Project (2015b). Design guidelines for wall panel connections, Deliverable 6.2

Hybrid Industrial/Recycled SFRC: Experimental Analysis and Design

L. Vistos[1], D. Galladini[1], H. Xargay[1], A. Caggiano[1,2], P. Folino[1], and E. Martinelli[3(✉)]

[1] Facultad de Ingeniería (FIUBA), Instituto INTECIN (UBA-CONICET),
Universidad de Buenos Aires, Buenos Aires, Argentina
[2] Institut für Werkstoffe im Bauwesen, Technische Universität Darmstadt, Darmstadt, Germany
[3] Department of Civil Engineering, University of Salerno, Fisciano, Italy
e.martinelli@unisa.it

Abstract. This paper is intended as a practice-oriented contribution about the use of sustainable Fiber-Reinforced Concrete (FRC) in the design of structural members according to the provisions of the current codes and guidelines. More specifically, the work focusses on Hybrid Industrial/Recycled Steel Fiber-Reinforced Concrete (HIRSFRC) realised by combining tailored Industrial Steel Fibers (ISFs) with Recycled Steel Fibers (RSFs), the latter being obtained by recycling waste pneumatic tyres. First, the results of a series of experimental tests, carried out for characterising the behaviour of the aforementioned materials, are summarised. They are specifically considered for evaluating the parameters that are generally considered for describing the post-cracking response of FRC. Then, a parametric analysis on the sectional behaviour of beams made of the aforementioned HIRSFRCs is proposed: this is intended at highlighting the influence of the material behaviour on the ultimate bending moment and curvature of structural members.

Keywords: Hybrid-FRC · Fracture · Recycled Steel Fibers · Waste tyres
Constitutive Laws · Structural design

1 Introduction

In recent years the disposal of exhaust tyres has emerged as a big issue in waste management (Tchobanoglous and Kreith 2002) and the increasing amount of these waste actually constitutes a serious threat for both environment and human health (Sienkiewicz et al. 2012). Moreover, based on the "Council Directive 1999/31/EC" of the European Commission on the Landfill of Waste, as of 2003 post-consumer "whole tyres" could no longer be landfilled and, since July 2006, such regulations must be applied to both "whole" and "shredded" tyres (EU Directive 1999/31/EC).

Therefore, there are strong motivations for recycling these waste, as they can be easily turned into a eco-friendly source of secondary raw materials (Sharma et al. 2000). In fact, recycling processes of waste tyres mainly consist of separating the internal steel reinforcement from the rubber covering. Hence, rubber scraps and short steel fibers can be generally obtained by these processes and can be utilised in several valuable

© Springer International Publishing AG, part of Springer Nature 2018
M. di Prisco and M. Menegotto (Eds.): ICD 2016, LNCE 10, pp. 98–112, 2018.
https://doi.org/10.1007/978-3-319-78936-1_8

applications. Particularly, they can be used as concrete components in partial-to-total replacement of the ordinary constituents (e.g., natural aggregates and industrial fibers, respectively). On the one hand, rubber scraps find an interesting field of application as a partial replacement of ordinary stone aggregates for obtaining the so-called "rubber-ized concrete" (Centonze et al. 2012). On the other hand, recycled fibers can be poten-tially used in substitution of the industrial ones, commonly employed for producing Fiber Reinforced Cementitious Composite (FRCC) (Yung et al. 2013).

As a matter of fact, adding a small fraction (usually in the order of 0.5–1.0% in volume) of short fibers, during mixing, results in enhancing the toughness in the post-cracking response of cementitious materials, as those fibers induce a bridging effect across the opening cracks and, hence, a positive influence on their propagation (Naaman and Reinhardt 2006).

However, the fibers, employed in FRCC, need to have good mechanical properties, be easy to spread in concrete mixtures and durable when embedded into cementitious matrices (Qian et al. 2003). Many types of fibers (i.e., made of steel, glass, natural cellulose, carbon, nylon, polypropylene, etc.) have been used in FRCC and are widely available for commercial applications (ACI-544.1 1996). A total of 60 million tons of these kinds of fibers are currently employed every year around the world, and, then, their production requires a huge amount of raw materials (Bartl et al. 2005). Therefore, Recy-cled Steel Fibers (RSFs) obtained from waste tyres could contribute to reducing this demand. Particularly, they can directly be utilised as a dispersed reinforcement in concrete to obtain a material that could be designated as Recycled Steel-FRC. In this regards, some pioneer researches already demonstrated the feasibility of these applica-tions (Aiello et al. 2009).

Besides the research already available in the literature about the material-scale behaviour of Hybrid Industrial/Recycled Steel Fiber-Reinforced Concrete (HIRSFRC), this paper is intended as a practice-oriented contribution about the use of these cemen-titious composites in the design of structural members according to the provisions of the current codes and guidelines. First, the results of a series of experimental tests, carried out for characterising the behaviour of the aforementioned materials, are summarised. More specifically, starting from a FRC mixture with 0.5% (in volume) of Industrial Steel Fibers (ISFs), three more mixtures were prepared by replacing 25%, 50% and 100% in weight of such fibers with an equal amount of RSFs. Therefore, the mechanical behav-iour of conventional Steel (S)-FRC was observed in comparison with the one of both Hybrid Industrial/Recycled Steel Fiber-Reinforced Concrete (HIRSFRC) and RSFRC. The experimental work consisted of compression and four-point bending tests on notched beam specimens according to UNI-11039-1 (2003) and UNI-11039-2 (2003). They are specifically considered for evaluating the parameters that are generally consid-ered for describing the post-cracking response of FRC. Then, a parametric analysis on the sectional behaviour of beams made of the aforementioned HIRSFRCs is proposed: this analysis is intended at highlighting the influence of the material behaviour on the ultimate bending moment and curvature of structural members. Thus, this work prelimi-narily describes the key geometric and mechanical properties of the RSFs employed in this research (Sect. 2) and the mechanical behaviour of the analysed HIRSFRC (Sect. 3). After this, final considerations to briefly explain the main concepts behind the structural rules of FRCC in structural designs are outlined in Sect. 4 of this work. Final concluding remarks are reported in Sect. 5.

2   RSF from Waste Tyres

A quantity of 15 kg of RSFs was examined to obtain a comprehensive description of both their geometry and mechanical properties (Fig. 1). It is worth mentioning that, due to the possibility that fibers could derive from different recycling plants and/or countries, it is largely accepted in the literature that a specific identification is necessary to investigate the expected variability of both geometrical and mechanical properties for the employed RSF (Aiello et al. 2009).

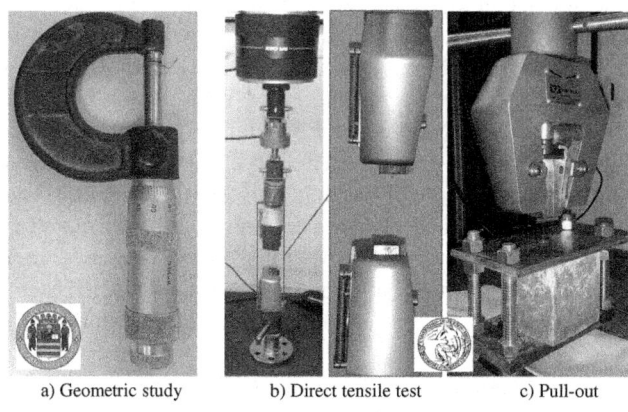

a) Geometric study   b) Direct tensile test   c) Pull-out

Fig. 1. Geometric and mechanical characterisation of RSF

As a result of the shredding and separation processes, RSFs under consideration have variable diameters and lengths, and often are characterised by irregular shapes with curls and twists. Therefore, the description of the main geometric parameters of these fibers deserved a dedicated investigation. Particularly, mechanical characterisation tests on RSF through tensile and pull-out tests were carried out at the Laboratory of Materials and Structures (LAME) of the University of Buenos Aires (Argentina). For the sake of brevity these results are omitted in this work, as they are available in Caggiano et al. (2015).

3   Experimental Tests

3.1   Materials and Methods

The results reported in this section were obtained from experimental tests performed according to UNI-11039-1&2 (2003).

FRC specimens tested in this study were prepared by adopting a unique mixture for the concrete matrix, which was also employed for preparing the plain concrete specimens considered as a reference (labelled as REF). This mixture was designed for a mean cubic compressive strength of 40 MPa target 28 days and prepared by using crushed limestone aggregates with a maximum aggregate size of 20 mm according to EN-12620 (2002), a constant cement content of 320 kg/m³ and a free water-to-cement ratio w/c of 0.51 (Martinelli et al. 2015).

Wirand Fibers (type FS7), hereafter referred to as ISFs, were considered in this study along with the RSFs. The relevant geometric and mechanical properties of ISFs are listed in the following: l_f = 33 mm (fiber length), d_f = 0.55 mm (nominal diameter), AR = 60 (aspect ratio), number of fibers per kg = 16100, f_t > 1200 MPa (failure strength in tension) and ε_u ≤ 2% (ultimate strain).

The concrete mixtures were prepared by mean of a laboratory mixer. Both coarse and fine aggregates were saturated and mixed; subsequently, cement, fibers and, finally, a super-plasticiser were added. The REF mixture was designed for a target slump value of 150–180 mm: a value of 175 mm was actually measured at fresh state. Moreover, the cementitious matrix composition of all FRC specimens was kept fairly unchanged; only the super-plasticiser quantity was slightly adjusted for controlling the influence of fibers on the resulting workability.

Three cube samples of 150 × 150 × 150 mm³ and beam specimens of 150 × 150 × 600 mm³ (Fig. 2) were cast in polyurethane moulds and duly vibrated. One of the cubic samples (labelled as "white") was extracted from each mixture before fiber mixing: it was tested in compression and compared with the corresponding FRC samples with the aim to observe the actual contribution of fibers on the compressive strength in each different mixture. After 36 h the concrete samples were demoulded. Then, the hardened beam samples were notched (through a 2.0 mm wide-slit) of 45 mm depth and starting from the bottom surface of the sample. Moreover, all concrete specimens were cured in a water bath (100% humidity), at a constant temperature of 22 °C, up to reach the 28 days of curing.

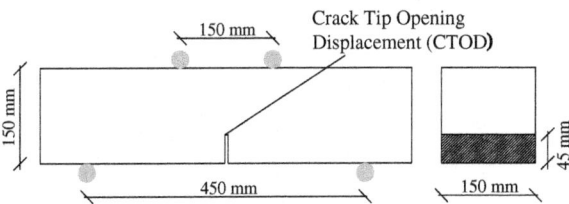

Fig. 2. 4-point bending test: geometry of the notched beam

Four FRC mixtures were prepared, always using 0.5% of fibers in volume of matrix, also combining the aforementioned ISFs and RSFs:

- RSFRC 0-05 with only ISFs (RSFs = 0%);
- RSFRC 25-05 with 25% of ISFs replaced by an equal amount of RSFs;
- RSFRC 50-05 with 50% of ISFs replaced by an equal amount of RSFs;
- RSFRC 100-05 with all RSFs.

Table 1 outlines the experimental programme reported in this paper. Four-point bending tests of notched beams were performed in displacement control (having displacement rate of 0.005 mm/s). Relevant load and displacement quantities were measured and recorded during all tests. Particularly, the crack-tip opening displacements were measured by means of dedicated transducers that monitored the relative displacements of the two sides of the notch tip. Furthermore, compressive tests were performed according to EN-12390-3 (2009) for measuring the cubic compressive strength of the SFRCs at the time of testing.

Table 1. Mixture and specimens.

Mix label	Compression (28 days)	Four-point bending test (28 days)
"REF"	3	3
RSFRC 0-05	3	3
RSFRC 25-05	3	3
RSFRC 50-05	3	3
RSFRC 100-05	3	3

3.2 Experimental Results

The results of compression tests are summarised in Table 2 reporting the average values of strengths obtained from cubic samples of the plain concrete and FRC mixtures considered in this study. As widely documented in the scientific literature, no significant difference was observed in terms of compressive strengths of both the so-called ''white'' and SFRC specimens. This means that, at least for the volume fraction considered in this study, the resulting compressive strength of FRC is mainly controlled by the matrix properties. Conversely, fibers only play a role in the post-cracking regime.

Table 2. Cubic compressive strengths in each mixture

Mix label	$f_{c,cube}$ at 28 days [MPa]	
	White	SFRC (mean of two)
REF	42.59 (mean of three)	
RSFRC 0-05	40.57	39.01
RSFRC 25-05	36.42	36.52
RSFRC 50-05	36.89	36.74
RSFRC 100-05	36.69	37.37

Four-point bending tests were performed with the aim of characterising the post-cracking behaviour of HIRSFRC samples: UNI-11039-1&2 (2003) provisions were taken into account for this purpose.

Figure 3 reports the experimental curves of the vertical load, P, versus the corresponding Crack Tip Opening Displacement ($CTOD_m$) curves, obtained in the tests:

CTOD$_m$ represents the mean of the two opposite CTOD measures. Based on the experimental evidence, the post-cracking response in bending of FRC specimens reinforced with only ISFs was characterised by a significant toughness (Fig. 3a).

The effect of replacing increasing amount of ISFs with an equal quantity of RSFs can be easily understood by analysing the curves depicted in Fig. 3. The post-cracking behaviour of FRC is generally characterised by a more pronounced softening range in specimens with a greater quantity of RSFs in substitution of ISFs. This is a result of the lower efficiency of the RSFs with respect to the I- ones, which are specifically designed to exhibit a good interaction with the concrete matrix. Particularly, recycled fibers are not straight, have no hooks and have (generally) lower aspect ratios: these are the main reasons explaining the (expected) decay resulting from replacing part (to total) of ISFs with an equal amount (in weight) of RSFs.

The steeper slope of the post-peak response observed for RSFRC 25-05 (Fig. 3b) is clearly due to the fact that the recycled fibers employed in those specimens need a wider crack opening for mobilising their bridging effect. The post-peak slope is even steeper for RSFRC 50-05 (Fig. 3c) and RSFRC 100-05 (Fig. 3d) where the actual volume fraction of RSF is even higher. Nevertheless, a significant increase in toughness can be observed for all FRC specimens with respect to the significantly brittle behaviour characterising the post-cracking response of the plain concrete.

3.3 Analysis of the Results

Three representative parameters, defined by UNI-11039-2 (2003), can be evaluated and compared for the FRC mixtures under investigation, with the aim of identifying and describing their post-cracking response. They are defined as the first crack strength (f_{If}) and two equivalent post-cracking strengths: (i) the first flexural strength ($f_{eq(0-0.6)}$) corresponds to a CTOD ranging between CTOD$_0$ and CTOD$_0$ + 0.6 mm which is supposed to be relevant for the Serviceability Limit State, whereas (ii) the second one ($f_{eq(0.6-3.0)}$) refers to a CTOD ranging between CTOD$_0$ + 0.6 and CTOD$_0$ + 3.0 mm which is rather significant for the Ultimate Limit State (di Prisco et al. 2009). CTOD$_0$ represents the CTOD corresponding to the peak load of the reference (plain concrete) specimen.

According to UNI-11039-2 (2003), the first crack strength values, f_{If}, defining the post-cracking response of HIRSFRC, was evaluated as:

$$f_{If} = \frac{P_{If} \cdot l}{b \cdot (h - a_0)^2} \tag{1}$$

where P_{If} represents the first crack load; b, h and l are the width, height and length of the beam, respectively, and a_0 represents the notch depth.

Figure 4 shows the comparisons of the mean values of first crack strength and the two equivalent crack resistances, defined in standard CTOD$_m$ ranges above defined. The following quantities, known as "equivalent crack strengths" are defined as follows:

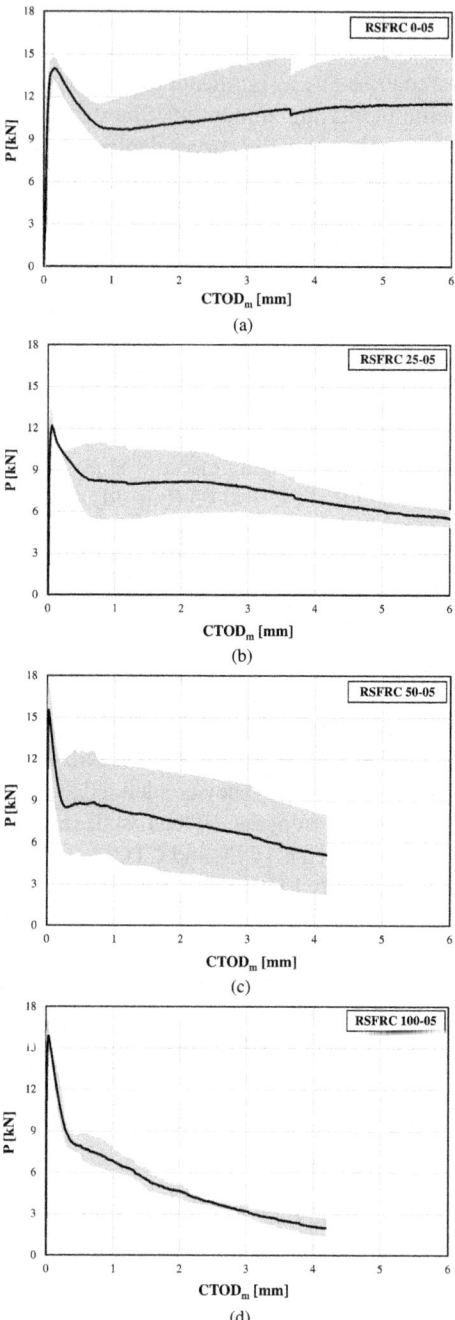

Fig. 3. Vertical force vs. $CTOD_m$ curves.

$$f_{eq(0-0.6)} = \frac{l}{b \cdot (h - a_0)^2} \cdot \frac{U_1}{0.6} \tag{2}$$

$$f_{eq(0.6-3.0)} = \frac{l}{b \cdot (h - a_0)^2} \cdot \frac{U_2}{2.4} \tag{3}$$

being U_1 and U_2 work capacity measures calculated according to UNI-11039-2 (2003).

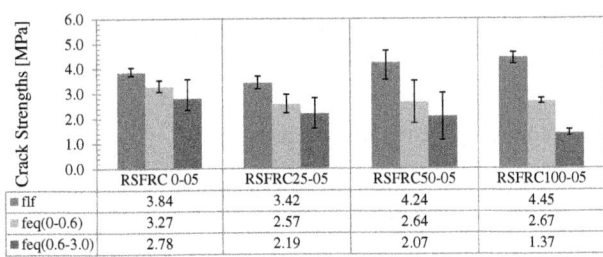

	RSFRC 0-05	RSFRC25-05	RSFRC50-05	RSFRC100-05
flf	3.84	3.42	4.24	4.45
feq(0-0.6)	3.27	2.57	2.64	2.67
feq(0.6-3.0)	2.78	2.19	2.07	1.37

Fig. 4. First crack, f_{lf}, and equivalent crack resistances, $f_{eq(0-0.6)}$ and $f_{eq(0.6-3.0)}$

As a matter of principle, the quantities U_1 and U_2 represent the area enclosed under the P-CTOD curves between the range [CTOD$_0$, CTOD$_0$ + 0.6 mm] and [CTOD$_0$ + 0.6, CTOD$_0$ + 3.0 mm], respectively. These results show that, as expected, all specimens, reinforced with a total amount of 40 kg/m^3 of steel fibers (equivalent to 0.5% in fiber volume fraction), mainly exhibit a crack-softening behaviour in the post-cracking regime.

Ductility indices can be considered as further objective measures of the fiber bridging actions and the following ductility measures were calculated:

$$D_0 = \frac{f_{eq(0-0.6)}}{f_{lf}} \quad \text{and} \quad D_1 = \frac{f_{eq(0.6-3.0)}}{f_{eq(0-0.6)}}. \tag{4}$$

Figure 5 reports the values of ductility indices (defined by Eq. 4) for the various tested beams. According to the classification of the UNI-11039-1 (2003), all the cementitious composites, tested in this experimental campaign, can be classified as "crack-softening" media, as both D_0 and $D_1 < 1$.

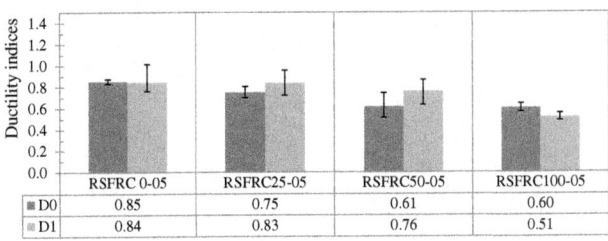

	RSFRC 0-05	RSFRC25-05	RSFRC50-05	RSFRC100-05
D0	0.85	0.75	0.61	0.60
D1	0.84	0.83	0.76	0.51

Fig. 5. Indices of the ductility D_0 and D_1

Finally, it is worth highlighting that UNI-11039-1 (2003) states that the D_0 index should not be lower than 0.5 for a FRC to be used in structural applications. Based on this criterion, Fig. 5 shows that all SFRC mixtures, even the one reinforced with only RSFs, can be considered as a structural fiber reinforced cementitious material.

4   Applications in Design

This section deals with outlining the design guidelines of FRC members under bending (with or without axial force) according to the provisions of the EN 1992-1-1 (2004) combined with the CNR DT 204 (2006) guidelines.

4.1   Constitutive Laws for Limit State Analyses

The following assumptions are considered for determining the ultimate moment resistance of fiber-reinforced concrete cross-sections:

1. Cross-sections in bending remain plane.
2. Strains in bonded rebars, whether in tension or compression, are the same as that of the surrounding FRCC (perfect adherence).
3. Stresses of FRCC in compression are derived from the design stress/strain relationship of ordinary (plain) concrete according to EN 1992-1-1 (2004).
4. Stresses of the reinforcing steel are derived from the design curves (EN 1992-1-1 2004).
5. The stress-crack opening law in uniaxial tension is defined for the post-cracking range of FRCC according to CNR DT 204 (2006).

Since points 1–4 of the above assumptions are largely employed in the calculation procedure of classical RC section, the discussion will be mainly focused at describing the stress-crack opening law of FRCC in tension (point 5). Particularly, two alternative stress-crack opening relationships, as proposed in Fig. 6, are alternatively suggested by the CNR DT 204 (2006) for the Ultimate Limit States (ULSs):

- a *rigid-plastic model*, based on a unique reference strength, f_{Ftu}:

$$f_{Ftu} = \frac{f_{eq2}}{3} \tag{5}$$

- and a *rigid-linear post-cracking model* (hardening, perfectly plastic or softening), formulated by means of two strength values: i.e., f_{Fts} and f_{Ftu}, respectively (Fig. 6):

$$f_{Fts} = 0.45 \cdot f_{eq1} \tag{6}$$

$$f_{Ftu} = k \cdot \left[f_{Fts} - \frac{w_u}{w_{i2}} \cdot \left(f_{Fts} - 0.5 \cdot f_{eq2} + 0.2 \cdot f_{eq1} \right) \right] \geq 0 \tag{7}$$

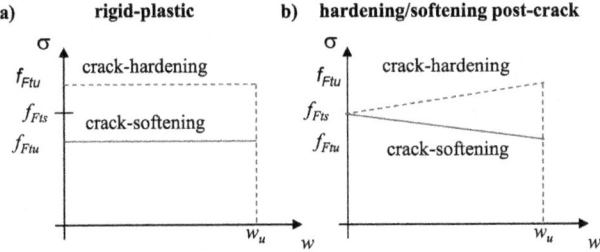

Fig. 6. Simplified constitutive stress-crack opening rules (CNR DT 204 2006)

where f_{eq1} and f_{eq2} are post-cracking strengths significant for the Serviceability (SLS) and ULS analyses, respectively.

For evaluating the f_{eq1} and f_{eq2} equivalent resistances, the following expressions must be used:

$$f_{eq1} = f_{eq(0-0.6)} \qquad (8)$$

$$f_{eq2} = f_{eq(0.6-3.0)} \qquad (9)$$

being $f_{eq(0-0.6)}$ and $f_{eq(0.6-3.0)}$ the equivalent crack resistances outlined in Sect. 3.

Further details about the post-cracking behaviour of FRCC, the partial safety factors (γ_F in Fig. 7), the k coefficient, the crack widths w_{i2} and w_u of Eq. (7) and the evaluation of the characteristic values (at 5% fractile) of f_{Fts} and f_{Ftu} can be found in the CNR DT 204 (2006), UNI-11039-1&2 (2003) and EN 1992-1-1 (2004) codes. For the sake of brevity these information are omitted herein.

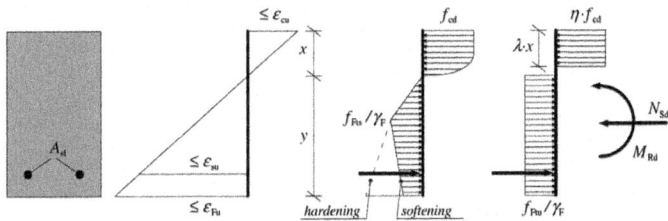

Fig. 7. ULS for bending with axial force.

4.2 Analysis of the Results

This section reports some examples of ULS calculations referring to a rectangular cross section in pure bending.

ULS, in the analysed examples, is reached when one of the following conditions is obtained (Fig. 7):

- Attainment of the maximum compressive strain at the FRCC, ε_{cu}: stress-block model was employed.

- Attainment of the maximum tensile strain at the steel rebars, ε_{su}. Two alternative models were considered for this purpose: (i) the strain-hardening model according to EN 1992-1-1 (2004), otherwise (ii) the elastic-perfectly plastic proposal considered in the old Italian code DM 9-1-1996 (1996);
- Attainment of the maximum tensile strain in the FRCC, ε_{Fu}.

The following geometric dimensions were considered for the transverse section: b = 20 cm and h = 60 cm. Moreover, two cases accounting for different longitudinal reinforcement contents were analysed: A_{sl} = 3.52 (low ratio) and 14.73 cm^2 (high ratio). Moreover, the following materials were considered:

(i) plain concrete labelled as "RC - white";
(ii) FRCC with 0.5% in volume fraction of industrial steel fibers ("RC + ISF100%");
(iii) FRCC with 0.5% in volume fraction of recycled steel fibers ("RC + RSF 100%");
(iv) FRCC with 0.5% in volume fraction of both ISFs and RSFs in equal weight (labelled as "RC + (I/R) SF 50–50%").

The design values of the FRCC compressive strength, f_{cd}, and the characteristic values of the equivalent post-cracking resistances, $f_{eq1,k}$ and $f_{eq2,k}$, were calculated following the experimental campaign reported in Sects. 2 and 3.

Starting from the design axial force N_{Sd} (which is null for the examples of pure bending analysed in this work), the ultimate bending moment M_{Rd} and curvatures were

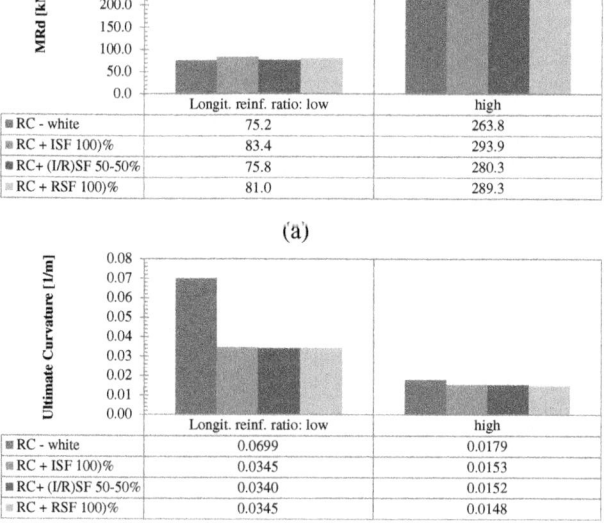

Fig. 8. (a) Ultimate bending moment M_{Rd} and (b) curvatures considering a rigid-linear post-cracking behaviour for the FRCC and ε_{ud} = +0.9 ε_{uk}

evaluated by imposing equilibrium conditions under the assumptions reported in the above sections.

Figures 8, 9, 10 and 11 report the results obtained in this parametric study. As a general comment, it can be highlighted that the steel rebars mainly control the value of design bending strength M_{Rd}, whereas the contribution of fibers is generally much smaller and, as expected, it is as marginal as the amount of rebars increases. A slight reduction in M_{Rd} is generally obtained for the RC + (I/R)SF 50–50% case. This is due to two combined effects: on the one hand, partial replacement of ISFs with RSFs leads to a reduction in the "average" post-cracking branch in the response of RC + (I/R)SF 50–50% with respect to RC + ISF100%; on the other hand, the higher scatter observed on RSFRC 50-05 (with respect to both RSFRC 0-05 and RSFRC 100-05) leads to characteristic values of the relevant parameters that are lower in RSFRC 50-05 than even than RSFRC 100-05. Consequently, in all the series of results the values of M_{Rd} obtained for the "hybrid" Industrial/Recycled-FRC are slightly lower than FRC having only ISFs and/or RSFs.

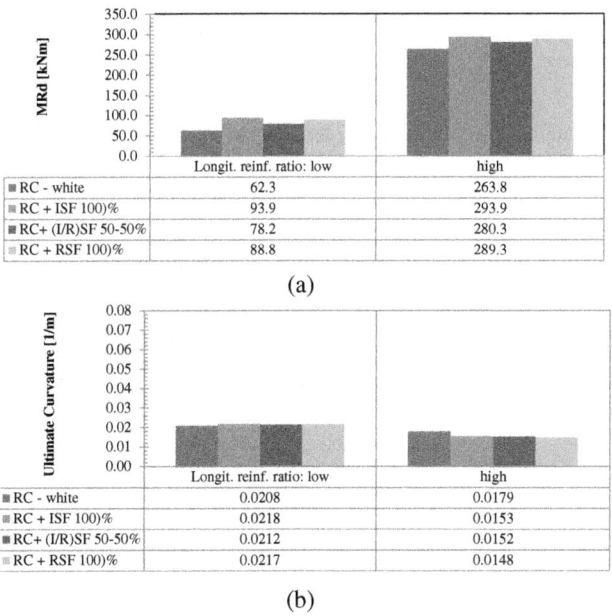

Fig. 9. (a) Ultimate bending moment M_{Rd} and (b) curvatures considering a rigid-linear post-cracking behaviour for the FRCC and $\varepsilon_{ud} = +1.0\%$

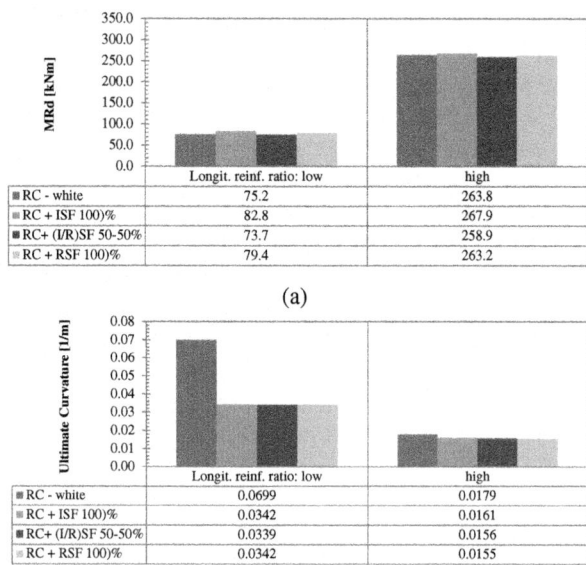

(a)

(b)

Fig. 10. (a) Ultimate bending moment M_{Rd} and (b) curvatures considering a rigid-plastic post-cracking behaviour for the FRCC and $\varepsilon_{ud} = +0.9\ \varepsilon_{uk}$

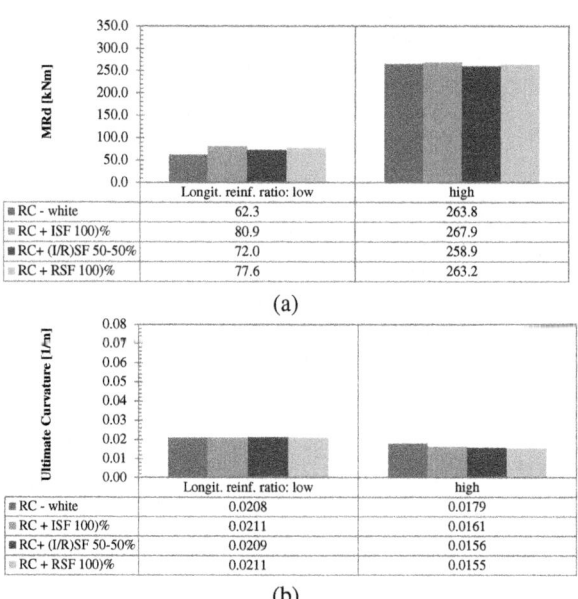

(a)

(b)

Fig. 11. (a) Ultimate bending moment M_{Rd} and (b) curvatures considering a rigid-plastic post-cracking behaviour for the FRCC and $\varepsilon_{ud} = +1.0\%$

Similar considerations can be done in terms of curvature, as no significant differences generally arise between the cases considered in this study. However, a significant difference only arises in the case of low amount of fibers, between the ordinary RC section and the other ones made of FRC. Figure 8 shows that the former is capable of developing an ultimate curvature significantly higher than the latter: this is a result of both the position of the neutral axis and the assumption of an ultimate axial strain $\varepsilon_{ud} = +0.9$ ε_{uk} in steel rebars (EN 1992-1-1 2004). However, as shown in Fig. 9, this result would not be achieved, if a lower ultimate axial value of axial strain (i.e., $\varepsilon_{ud} = +1.0\%$ by D.M. 9-1-1996) is assumed for steel rebars in tension.

Finally, the consequences of assuming rigid-linear or rigid-plastic post-cracking response can be highlighted by comparing Figs. 8, 9 and Figs. 10, 11, respectively. This comparison shows that assuming a rigid-plastic post-cracking response generally leads to slightly more restrictive predictions: however, this is a desirable feature for a simplified model.

5   Concluding Remarks

This paper addressed the mechanical behaviour of HIRSFRC at both material and structure scales. First of all, the results of a series of experimental tests on this hybrid material have been summarised with the aim to provide readers with both a general overview about the mechanical behaviour of HIRSFRC and an estimation of the main parameters that generally are considered in describing the post-cracking response of the materials under consideration. Secondly, a parametric analysis has been developed for highlighting the relationship between the material properties and design output, in the light of the current codes and guidelines about FRC. The material characterisation pointed out that HIRSFRC has potential for being used in structural applications. In fact, a somewhat limited reduction in the ductility parameters has been observed, even in the case of significant amount of RSFs employed in substitution of ISFs. The structure scale analyses, conversely, have shed a concerning light on both the actual influence of FRCs on the sectional response of RC beams. Hence, further researches are needed with the aim to reach a better control of the influence of RSFs on the mechanical behaviour of cementitious composites and a more consistent definition of Limit States in FRC beams.

Acknowledgements. The work proposed in this paper stems out of the activities of the SUPERCONCRETE Project (H2020-MSCA-RISE-2014, n. 645704): the Authors wish to acknowledge the financial contribution of the Europ. Union as part of the H2020 Programme.

References

ACI-544.1-96 (1996) State-of-the-art report on fiber reinforced concrete. Reported by ACI Committee 544, American Concrete Institute

Aiello MA, Leuzzi F, Centonze G, Maffezzoli A (2009) Use of steel fibres recovered from waste tyres as reinforcement in concrete: pull-out behaviour, compressive and flexural strength. Waste Manag 29(6):1960–1970

Bartl A, Hackl A, Mihalyi B, Wistuba M, Marini I (2005) Recycling of fibre materials. Process Saf Environ Prot 83(B4):351–358

Caggiano A, Xargay H, Folino P, Martinelli E (2015) Experimental and numerical characterization of the bond behavior of steel fibers recovered from waste tires embedded in cementitious matrices. Cement Concr Compos 62:146–155

Centonze G, Leone M, Aiello M (2012) Steel fibers from waste tires as reinforcement in concrete: a mechanical characterization. Constr Build Mater 36:46–57

CNR DT 204 (2006) Guidelines for design, construction and production control of fiber reinforced concrete structures. Nat Research Council of Italy

D.M. 9-1-1996 (1996) Norme tecniche per il calcolo, l'esecuzione ed il collaudo delle strutture in cemento armato, normale e precompresso e per le strutture metalliche

EN-12390-3 (2009) Testing hardened concrete. Part 3: compressive strength of test specimens. BSI

EN 1992-1-1 (2004) Design of concrete structures - Part 1-1: general rules and rules for buildings. Eurocode 2

di Prisco M, Plizzari G, Vandewalle L (2009) Fibre reinforced concrete: new design perspectives. Mater Struct 42(9):1261–1281

EU-Directive 1999/31/EC of the council of 26 April 1999 on the landfill of waste. Off J Eur Union L182:1–19

Martinelli E, Caggiano A, Xargay H (2015) An experimental study on the post-cracking behaviour of hybrid industrial/recycled steel fibre-reinforced concrete. Constr Build Mater 94:290–298

Naaman A, Reinhardt H (2006) Proposed classification of HPFRC composites based on their tensile response. Mater Struct 39:547–555

Qian X, Zhou X, Mu B, Li Z (2003) Fiber alignment and property direction dependency of FRC extrudate. Cem Concr Res 33(10):1575–1581

Sharma V, Fortuna F, Mincarini M, Berillo M, Cornacchia G (2000) Disposal of waste tyres for energy recovery and safe environment. Appl Energy 65(1):381–394

Sienkiewicz M, Kucinska-Lipka J, Janik H, Balas A (2012) Progress in used tyres management in the European Union: a review. Waste Manag 32(10):1742–1751

Tchobanoglous G, Kreith F (2002) Handbook of solid waste management. McGraw-Hill, New York

UNI-11039-1 (2003) Steel fibre reinforced concrete – definitions, classification and designation. UNI Editions, Milan, Italy

UNI-11039-2 (2003) Steel fibre reinforced concrete – test method to determine the first crack strength and ductility indexes. UNI Editions, Milan, Italy

UNI-EN-12620 (2002) Aggregates for concrete. Volume Ref. No. EN 12620:2002 E. European Committee for Standardization, Brussels

Yung WH, Yung LC, Hua LH (2013) A study of the durability properties of waste tire rubber applied to self-compacting concrete. Constr Build Mater 41:665–672

Comparison of Recent Code Provisions for Punching Shear Capacity of R/C Slabs Without Shear Reinforcement

M. Lapi, M. Orlando$^{(\boxtimes)}$, F. Angotti, and P. Spinelli

Dipartimento di Ingegneria Civile e Ambientale,
Università degli Studi di Firenze, Florence, Italy
`maurizio.orlando@unifi.it`

Abstract. In the last years the knowledge of the punching failure in R/C slabs increased thanks to several scientific studies. The progress obtained in this field is considerable, nevertheless achieved results are only taken into consideration by few Codes. The most updated code is the Model Code 2010, which adopted the Critical Shear Crack Theory (CSCT) for the punching shear capacity of R/C slab-column connections. At the same time, the EC2 formulation for punching is under revision, but the new formulation will not be available before three-four years. In this paper, the authors discuss main code provisions (ACI, current EC2, two proposals for revision of EC2, MC 2010, old Italian Recommendations) for punching shear capacity of R/C flat slabs without shear reinforcement. Through a parametric analysis, the authors investigate how each code takes into account the influence of main variables, which come into play in the punching phenomenon, on the evaluation of the punching capacity. Finally, results of each code formulation are compared with different literature experimental data.

Keywords: Punching failure · Flat-slab · CSCT · EC2 revision

1 Introduction

A flat-slab is a two-way structure that bears and transfers vertical loads to columns. This constructive system is often employed for multi-storey structures, used as offices and carparks, because it allows to increase the span between columns reducing the floor thickness. In other words it offers a greater flexibility in the choice of the internal layout allowing to reduce the building height.

Starting from fifties, for constructive reasons, the flat-slab deck are usually built without capitals. In this way the punching failure becomes predominant with respect to the flexural failure. The punching failure is due to a shear-stress concentration along the column's perimeter and it is characterized by a collapse surface with a truncated cone shape. This type of failure is rather brittle and it occurs without any warning sign. It is a local mechanism but it could bring to a progressive collapse of the entire building. For these reasons the punching issue is primary in the design of R/C flat-slab building. In the last years the knowledge of this failure mechanism increased thanks to several scientific studies. However, not all scientific results are adopted by international codes.

© Springer International Publishing AG, part of Springer Nature 2018
M. di Prisco and M. Menegotto (Eds.): ICD 2016, LNCE 10, pp. 113–132, 2018.
https://doi.org/10.1007/978-3-319-78936-1_9

The most updated code is the Model Code 2010 (fib 2010), which is grounded on the Critical Shear Crack Theory (CSCT) (Muttoni 2008). Furthermore, the EC2 formulation is under revision, but new provisions will not be available before three-four years.

2   Code Provisions

In this section the authors discuss main code provisions for the determination of the punching strength in R/C flat-slabs without shear reinforcement. In particular a parametric analysis of main variables that come into play in the punching failure, is presented. Models can be divided into two categories: empirical or mechanical. With regards to the first category, ACI 318 (ACI Committee 318 2014) and current version of EC2 (CEN 2004) are dealt with, while for the second category the Model Code 2010 and two proposals for revision of EC2 are dealt with. The first proposal is empirical and it has been developed at the Institute of Structural Concrete of RWTH Aachen University in Germany (Hegger et al. 2016). The second proposal is grounded on the CSCT (Critical Shear Crack Theory) and it has been developed at the EPFL in Switzerland (Muttoni et al. 2016). Furthermore, old Italian Recommendations (DM96 1996) are also analysed, as they could turn out to be a useful tool for preliminary design, although they can no longer be utilized for design purposes.

2.1   ACI 318 - 2014

ACI 318 formulation for punching of flat slabs is strictly empirical and its application is very easy. For slabs without shear reinforcement, the punching strength is the smallest of the three following values:

$$V_{ACI,a} = \frac{1}{6} \cdot \left(1 + \frac{2}{\beta}\right) \cdot \sqrt{f_{ck}} \cdot b_{0,ACI} \cdot d \tag{1}$$

$$V_{ACI,b} = \frac{1}{12} \cdot \left(\frac{\alpha_S \cdot d}{b_{0,ACI}} + 2\right) \cdot \sqrt{f_{ck}} \cdot b_{0,ACI} \cdot d \tag{2}$$

$$V_{ACI,c} = \frac{1}{3} \cdot \sqrt{f_{ck}} \cdot b_{0,ACI} \cdot d \tag{3}$$

where:

β is the ratio between long side and short side of the column;
f_{ck} is the characteristic compressive strength of concrete in MPa;
$b_{0,\,ACI}$ is the control perimeter set at $d/2$ from the border of the support region in mm;
d is the effective depth of the slab in mm;
α_s holds 40 for inner column, 30 for edge column and 20 for corner column;

Formulas (1) and (2) were developed to account for non-square columns and different positions of the column (inner, edge or corner), respectively.

2.2   EC2-2004

EC2 formulation is also strictly empiric, but unlike ACI 318 it takes into account the flexural reinforcement ratio and size effects. Thus, the punching strength of flat slabs without shear reinforcement is given as:

$$V_{EC2} = \max\left(V'_{EC2}; V''_{EC2}\right) \tag{4}$$

where:

$$V'_{EC2} = C_{Rd,c} \cdot \frac{b_{0,EC2}}{\beta} \cdot d \cdot k \cdot \left(100 \cdot \rho_l \cdot f_{ck}\right)^{\frac{1}{3}} \tag{5}$$

$$V''_{EC2} = v_{min} \cdot b_{0,EC2} \cdot d \tag{6}$$

$$C_{Rd,c} = \frac{0.18}{\gamma_c} \tag{7}$$

$b_{0,EC2}$ is the control perimeter set at *2d* from the border of the support with circular corners;

β is a coefficient that takes into account the eccentricity of the shear reaction; for structures where the lateral stability does not depend on the frame action between slabs and columns, and adjacent spans do not differ in length by more than 25%, following approximate values for β can be used:

- $\beta = 1.15$ for inner columns
- $\beta = 1.4$ for edge columns
- $\beta = 1.5$ for corner columns

d is the effective depth of the slab in mm;
k is a factor accounting for the size effect:

$$k = 1 + \sqrt{\frac{200}{d}} \leq 2 \tag{8}$$

ρ_l is the flexural reinforcement ratio; if ρ_l is greater than 2%, ρ_l is assumed equal to 0.02:

$$\rho_l = \sqrt{\rho_x \cdot \rho_y} \leq 0.02 \tag{9}$$

(ρ_x, ρ_y: reinforcement ratio in *x* and *y* direction)

f_{ck} is the characteristic compressive strength of concrete in MPa;
v_{min} is the minimum punching shear strength:

$$v_{min} = 0.035 \cdot k^{3/2} \cdot f_{ck}^{1/2} \tag{10}$$

2.3   MC 2010 (MC)

Model Code 2010, like SIA 262–2003 (Swiss Society of Engineers and Architects 2003), is grounded on the Critical Shear Crack Theory (CSCT). The punching failure depends on the slab rotation. For slabs without shear reinforcement, the punching strength is defined as:

$$V_{MC} = k_\psi \cdot \frac{\sqrt{f_{ck}}}{\gamma_c} \cdot k_e \cdot b_{0,MC} \cdot d_v \tag{11}$$

where:

k_ψ depends on the slab rotation:

$$k_\psi = \frac{1}{1.5 + 0.9 \cdot \psi \cdot d \cdot k_{dg}} \leq 0.6 \tag{12}$$

Ψ is the slab rotation, defined in the following, depending on the approximation level;
k_{dg} is the factor accounting for the influence of aggregate size, defined as:

$$k_{dg} = \frac{32}{16 + d_g} \geq 0.75 \tag{13}$$

d_g is the maximum aggregate size in mm;
γ_c is the partial safety factor for concrete material properties
k_e is a coefficient that takes into account the concentration of shear forces due to moment transfer between the slab and supported area. In cases where the lateral stability does not depend on frame action of slabs and columns, and adjacent spans do not differ in length by more than 25%, following approximated values may be adopted:

-   $k_e = 0.9$ for inner columns
-   $k_e = 0.7$ for edge columns
-   $k_e = 0.65$ for corner columns
-   $k_e = 0.75$ for corners of walls

$b_{0,MC}$ is the control perimeter set at a distance of d_v from the border of the support region with circular corners in mm;

d_v is the effective depth of the slab accounting for the effective level of the support region ($d_v \leq d$).

In this provision there are different levels of approximation. For each level a different expression of the slab rotation is defined. The rotation has to be calculated along the two main directions of the reinforcement.

Level I of approximation (LoA,I): "Fast pre-dimensioning"

For a regular flat slab designed according to an elastic analysis without significant redistribution of internal forces, a safe estimate of the rotation failure is:

$$\psi = 1.5 \cdot \frac{r_s}{d} \cdot \frac{f_{yd}}{E_s} \tag{14}$$

where:

r_s denotes the distance between the point where the radial bending moment is zero, and the support axis. For regular flat slabs where the ratio of spans is between 0.5 and 2, r_s can be calculated as the maximum of following values:

$$r_{s,x} \cong 0.22 \cdot L_x \tag{15}$$

$$r_{s,y} \cong 0.22 \cdot L_y \tag{16}$$

d is the effective depth of the slab;

f_{yd} is the design yield stress of the flexural reinforcement;

E_s is the Young modulus of the flexural reinforcement.

Level II of approximation (LoA,II): "Typical design of new structures"

In case where significant bending moment redistribution is considered in the design, the slab rotation can be calculated as:

$$\psi = 1.5 \cdot \frac{r_s}{d} \cdot \frac{f_{yd}}{E_s} \cdot \left(\frac{m_{Ed}}{m_{Rd}}\right)^{1.5} \tag{17}$$

where:

m_{Rd} is the average flexural strength per unit length in the support strip (for the considered direction);

m_{Ed} is the average moment per unit length for calculation of the flexural reinforcement in the support strip (for the considered direction);

– for inner columns:

$$m_{Ed} = V_{Ed} \cdot \left(\frac{1}{8} + \frac{|e_{u,i}|}{2 \cdot b_s} \right) \tag{18}$$

– for edge columns:

when calculations are made considering the tension reinforcement parallel to the edge:

$$m_{Ed} = V_{Ed} \cdot \left(\frac{1}{8} + \frac{|e_{u,i}|}{2 \cdot b_s} \right) \geq \frac{V_{Ed}}{4} \tag{19}$$

or perpendicular to the edge:

$$m_{Ed} = V_{Ed} \cdot \left(\frac{1}{8} + \frac{|e_{u,i}|}{2 \cdot b_s} \right) \tag{20}$$

– for corner columns:

$$m_{Ed} = V_{Ed} \cdot \left(\frac{1}{8} + \frac{|e_{u,i}|}{2 \cdot b_s} \right) \geq \frac{V_{Ed}}{2} \tag{21}$$

where:

$e_{u,i}$ is the eccentricity of the shear force resultant;
b_s is the width of the support strip for calculating m_{Ed}, defined as: m_{Ed}, defined as:

$$b_s = 1.5 \cdot \sqrt{r_{s,x} \cdot r_{s,y}} \leq L_{min} \tag{22}$$

Level III of approximation (LoA,III): "For special design cases or for analysis of existing structures"

This level of approximation is recommended for irregular slabs or for flat slabs where the ratio of span lengths is not included between 0.5 and 2.

The coefficient 1.5 used in previous equations can be replaced by 1.2 if:

– r_s is calculated using a linear elastic (un-cracked) model
– m_{Ed} is calculated from a linear elastic (un-cracked) model as the average value of the moment for design of flexural reinforcement over the width of the support strip b_s
– b_s can be calculated as in level II, taking $r_{s,x}$ and $r_{s,y}$ as the maximum value in the investigated direction.

Level IV of approximation (LoA,IV): "For special design cases or for more detailed assessment of existing structures"

The rotation ψ can be calculated on the basis of a non-linear analysis of the structure and accounting for cracking, tension-stiffening effects, yielding of the reinforcement and other non-linear effects relevant for providing an accurate assessment of the structure.

2.4   RWTH Proposal for Revision of EC2

In the proposal for revision of EC2 developed at RWTH Aachen (Hegger et al. 2016), the punching shear strength is calculated similarly to the current EC2 formulation. The only substantial difference is given by the presence of the coefficient k_λ accounting for the influence of column size and shear slenderness:

$$V_{Rd} = \max(V'_{RWTH}; V''_{RWTH})$$
(23)

$$V'_{RWTH} = C_{Rd,c} \cdot \frac{b_{0,RWTH.}}{\beta} \cdot d_v \cdot k_d \cdot k_\lambda \cdot (100 \cdot \rho_l \cdot f_{ck})^{1/3}$$
(24)

$$V''_{RWTH} = v_{\min} \cdot b_{0,RWTH} \cdot d_v$$
(25)

where:

$$C_{Rd,c} = \frac{1.8}{\gamma_c}$$
(26)

k_d is a coefficient accounting for the influence of size effects:

$$k_d = \left(1 + \frac{d_v}{200}\right)^{-\frac{1}{2}}$$
(27)

k_λ is a coefficient accounting for the influence of column size and shear slenderness:

$$k_\lambda = \left(\frac{a_\lambda}{d_v} \cdot \frac{u_0}{d_v}\right)^{-\frac{1}{3}}$$
(28)

a_λ/d_v is the shear span-depth ratio;

a_λ is the distance between the edge of the loaded area and the line of contraflexure; for non-symmetric cases a_λ can be calculated as:

$$a_\lambda = \sqrt{a_{\lambda,y} \cdot a_{\lambda,z}} \tag{29}$$

u_0/d_v is the specific column perimeter;
ρ_l is the flexural reinforcement ratio:

$$\rho_l = \sqrt{\rho_y \cdot \rho_z} \leq \min\left(0.02;\ 0.5 \cdot \frac{f_{cd}}{f_{yd}}\right) \tag{30}$$

f_{ck} (f_{cd}) is the characteristic (design) compressive cylinder strength of concrete in MPa;
f_{yd} is the design yield stress of steel in MPa;
d_v is the shear-resisting effective depth of the control section in mm;
$b_{0,RWTH}$ is the control perimeter set at 0.5d from the border of the support region with circular corners.

As regards v_{min} and β, they assume the same values as in EC2-2004.

2.5   EPFL Proposal for Revision of EC2

The proposal for revision of EC2 developed at EPFL in Switzerland (Muttoni et al. 2016) is a closed-form formulation based on the Critical Shear Crack Theory (Muttoni 2008).

The punching shear strength is calculated as:

$$V_{EPFL} = \frac{1}{\gamma_c} \cdot \frac{b_{0,EPFL}}{\beta} \cdot d_v \cdot k_u \cdot \left(100 \cdot \rho_l \cdot f_{ck} \cdot \frac{d_{dg}}{r_s}\right)^{1/3} \tag{31}$$

$$V_{EPFL} \leq \frac{0.55}{\gamma_c} \cdot \sqrt{f_{ck}} \cdot d_v \cdot b_{0,EPFL} \tag{32}$$

where:

$$k_u = 8 \cdot \sqrt{\beta \cdot \frac{d}{b_{0,EPFL}}} \geq 2.0 \tag{33}$$

β is a parameter accounting for concentrations of shear forces due to acting moment transfer between slab and supported area; in cases where the lateral stability does not depend on frame action of slabs and columns and where adjacent spans do not differ in length more than 25%, following approximated values may be adopted:

- $\beta = 1.15$ for inner columns
- $\beta = 1.4$ for edge columns
- $\beta = 1.5$ for corner columns
- $\beta = 1.35$ for corners of walls

f_{ck} is the characteristic compressive cylinder strength of concrete in MPa;

$b_{0,EPFL}$ is the control perimeter set at a distance of $d_v/2$ from the border of the support region with circular corners in mm;

d_v is the effective depth of the slab accounting for the effective level of the support region $(d_v \leq d)$;

ρ_l is the flexural reinforcement ratio limited to the maximum of 4%:

$$\rho_l = \sqrt{\rho_{l,x} \cdot \rho_{l,y}} \leq 0.04 \qquad (34)$$

$\rho_{l,x}$ and $\rho_{l,y}$ should be calculated as mean values over the width of the support strip b_s defined as:

$$b_s = 1.5 \cdot r_s \leq L_{min}[= \min(L_x, L_y)] \qquad (35)$$

r_s denotes the distance between the point, where the radial bending moment is zero, and the support axis. The value of r_s may be calculated using a linear elastic (un-cracked) model. Otherwise, for regular flat slabs where the lateral stability does not depend on frame action between the slabs and the columns, and where the adjacent spans do not differ in length by more than 25%, can be approximated to $0.22\,L_x$ or $0.22\,L_y$ for the x- and y- directions, respectively:

$$r_s = \sqrt{r_{s,x} \cdot r_{s,y}} \geq d \qquad (36)$$

d_{dg} is a coefficient taking account of concrete type and its aggregate properties:

- $d_{dg} = 32$ for normal weight concrete
- $d_{dg} = 16$ for light weight concrete

2.6 Old Italian Recommendations

Old Italian Recommendations (DM96 1996) for punching of flat slabs refers to a very simple mechanical model, where the punching capacity only depends on the concrete tensile strength. For slabs without shear reinforcement, the punching strength is given as:

$$V_{DM96} = 0.5 \cdot f_{ctk} \cdot b_{0,DM96} \cdot h \tag{37}$$

where:

$b_{0,\ DM96}$ is the control perimeter set at $d/2$ from the border of the support region;
h is the slab thickness;
f_{ctk} is the characteristic tensile strength of concrete.

3 Parametric Analysis

In this section the authors perform a parametric analysis to investigate the influence of different parameters on the punching strength predicted using different codes. In particular, the variation of the specific punching strength v_R is calculated at varying one of following parameters:

f_c concrete compressive strength
f_y steel yield strength
ρ flexural reinforcement ratio
b_0/d ratio between control perimeter and effective depth of the slab
d slab's effective depth
r_s/d shear span-depth ratio

Data chosen for the parametric analysis are the same used in (Muttoni 2008), which, for each investigated parameter, refers to specimens with different geometry and/or mechanical data. Results of the parametric analysis are summed up in following diagrams (Figs. 1, 2, 3, 4, 5 and 6) in terms of the specific punching strength v_R:

$$v_R = \frac{V_R}{\sqrt{f_{ck}} \cdot d \cdot b_0} \tag{38}$$

where V_R is the punching strength measured in kN and v_R is in $\sqrt{MPa}$. The coefficient β is taken equal to one, as no eccentricity is considered (null bending moment); furthermore, $\gamma_C = 1.0$ and mean values of material strengths are used instead of characteristic values: $f_c = f_{cm}$ and $f_y = f_{ym}$.

The parametric analysis highlights the limited capacity of some formulations to predict the punching capacity of R/C slabs without shear reinforcement. From previous graphs it results that ACI, MC-LoA,I and DM96 (1996) do not take into account the influence of some parameters on the punching strength.

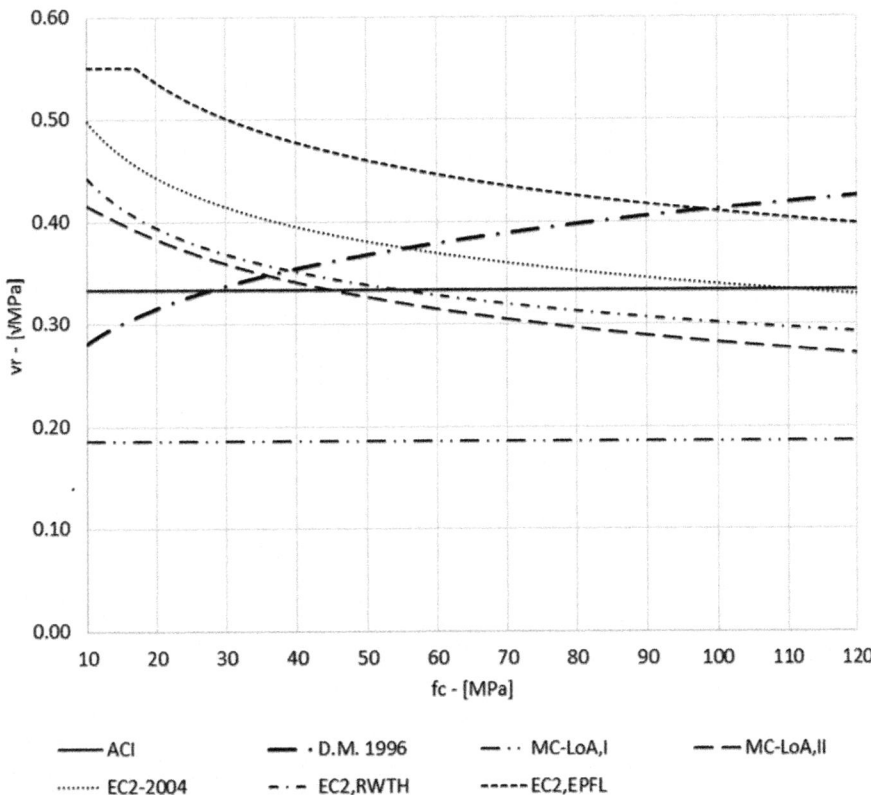

Fig. 1. Influence of concrete compressive strength on v_R (d = 98 mm, h = 125 mm, r_c = 75 mm, r_s = 850 mm, ρ = 0.8%, d_g = 10 mm, f_y = 550 MPa)

In particular, from all graphs it results that ACI provides a constant specific punching strength v_r equal to $0.33\ \sqrt{MPa}$, except for small values of f_y (Fig. 2) and ρ (Fig. 3) and high values of b_0/d (Fig. 4). ACI expression could give unsafe values of v_r for slender slabs and large columns ($r_s/d > 5$ and $b_0/d > 12$), in which cases other codes provide lower strength values. As regards the influence of concrete compressive strength, differently from other codes, DM96 (1996) gives increasing values of the specific punching strength at increasing the concrete strength (Fig. 1).

MC-LoA,I always provides a lower specific punching strength than other codes. This result is in agreement with the purpose of the level of approximation I, that is preliminary design based on safe hypotheses leading to quick and simple analyses. However it seems that MC-LoA,I is too much conservative.

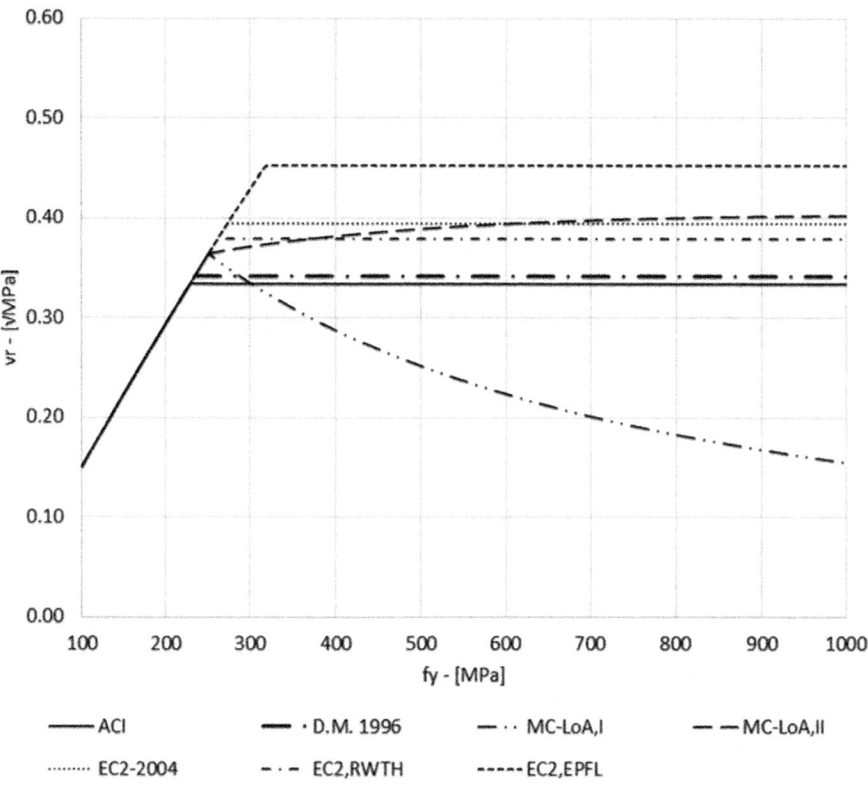

Fig. 2. Influence of steel yield strength on v_R (d = 114 mm, h = 152 mm, r_c = 162 mm, r_s = 982 mm, ρ = 1.15%, f_c = 24.6 MPa, d_g = 38.1 mm)

The current formulation of Eurocode 2 takes into consideration the influence of almost all variables on the punching strength, except for the slab slenderness (r_s/d). To overcome this lack, both proposals for revision of EC2 introduce the influence of the slab slenderness. In the RWTH proposal, the slenderness is taken into consideration through the coefficient $k_λ$, which includes the shear span-depth ratio $a_λ/d_v$ ($a_λ ≡ r_s$), in the EPFL proposal it is considered including r_s in the expression of the punching strength.

Results given by these two proposals at varying the slab slenderness are very similar (Fig. 6). Results given by the RWTH proposal and MC-LoA,II are also very similar, but v_R values are lower than the EPFL proposal, in particular at varying the concrete strength and the steel yield strength (Figs. 1 and 2). Nevertheless, the assessment of the reliability of different formulations requires the comparison with experimental data, which is presented in following Sect. 4.

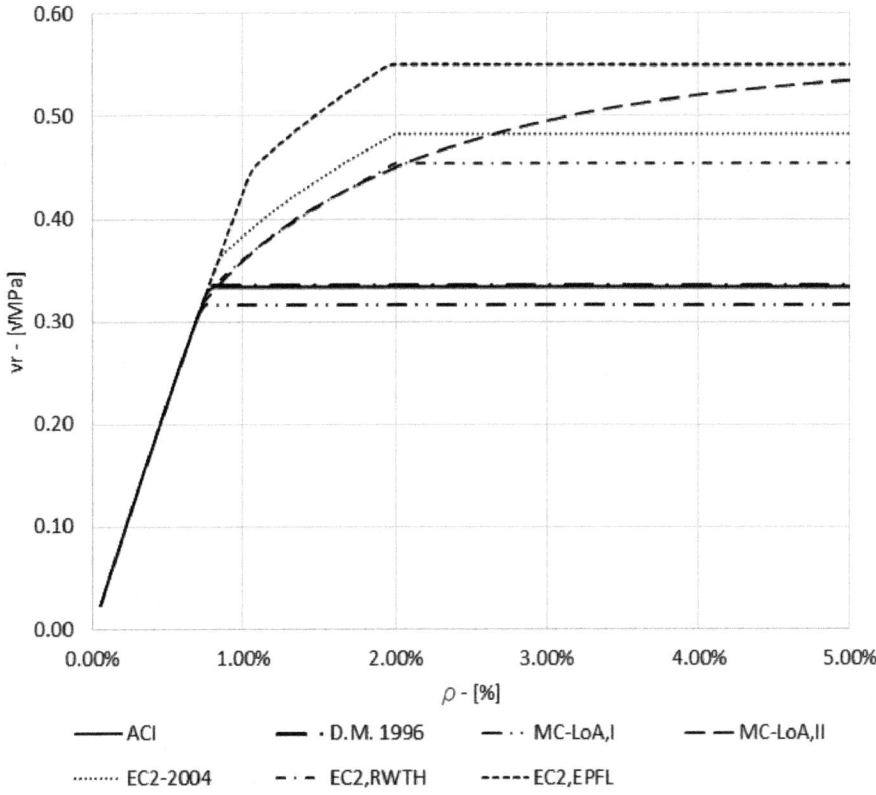

Fig. 3. Influence of flexural reinforcement ratio on v_R (d = 114 mm, h = 152 mm, r_c = 162 mm, r_s = 982 mm f_c = 22 MPa, d_g = 25.4 mm, f_y = 325 MPa)

4   Comparison with Experimental Data

In this section, values of the punching strength predicted using different code provisions are compared with literature experimental data. Following codes are taken into consideration for the comparison: EC2-2004, RWTH proposal and EPFL proposal for revision of EC2, and old Italian Recommendations (DM96 1996).

Several experimental campaigns are considered for a total of 173 slab specimens. Tests have been taken from a wider database, choosing those tests performed on specimens with similar geometry and reinforcement layout. All tests concern square isolated slabs equipped with uniformly distributed flexural reinforcement oriented along main axes x/y. The load is transmitted through line or points which react to the column load, along a circular or rectangular arrangement. Columns have square or circular cross-sections.

Results of the comparison are expressed in terms of the ratio V_{test}/V_{th} between the experimental (V_{test}) and the predicted value (V_{th}) of the punching strength, adopting all safety coefficients equal to one and using mean values of material strengths. The average

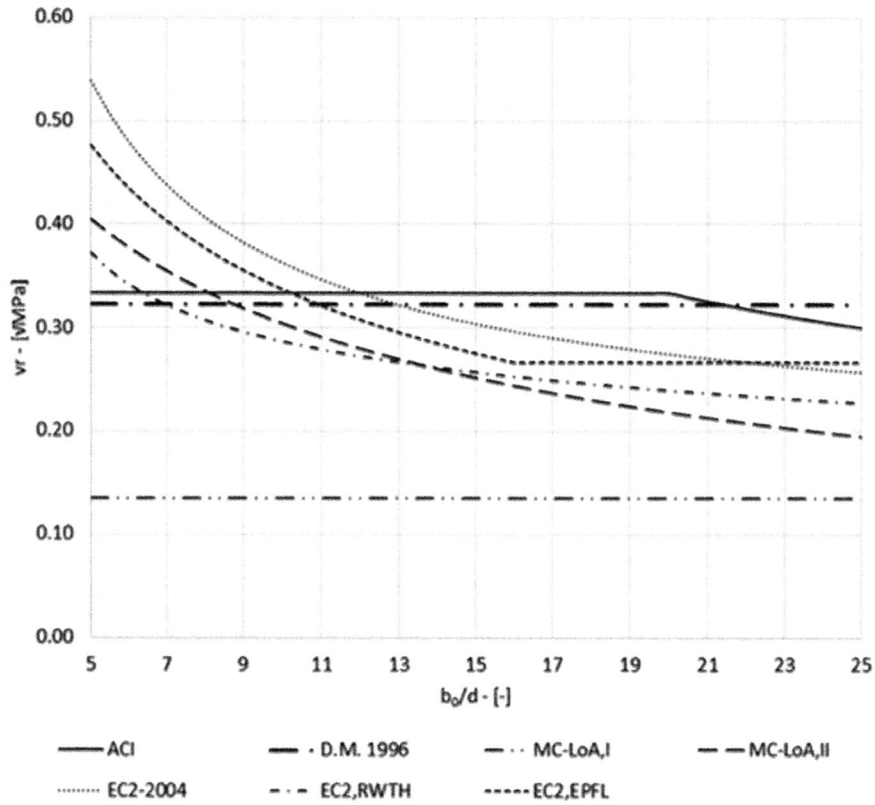

Fig. 4. Influence of punching shear control perimeter on v_R (d = 200 mm, h = 240 mm, r_s = 1270 mm, ρ = 0.8%, f_c = 33 MPa, d_g = 18 mm, f_y = 493 MPa)

value μ, the coefficient of variation CV and the 5%-quantile of the ratio V_{test}/V_{th} are listed in Table 1.

The current EC2 formulation gives an average value of the ratio V_{test}/V_{th} equal to 1.27, a CV equal to 0.33, and a 5%-quantile of 0.85. The unitary value of the ratio V_{test}/V_{th} corresponds to the 23%-quantile, meaning that in almost a quarter of all analysed cases the current EC2 formulation overestimates the experimental punching strength.

As regards the RWTH proposal for revision of EC2, the average of the ratio V_{test}/V_{th} is equal to 1.27 like EC2-2004, while the CV (=0.21) is the lowest among all formulations, and the 5%-quantile attains one of the highest values (0.93). As the unitary value of the ratio V_{test}/V_{th} corresponds to the 13%-quantile, in only 13% of analysed cases the RWTH proposal overestimates the experimental punching strength.

In summary, the RWTH proposal provides the same average strength of the current EC2 code (V_{test}/V_{th} = 1.27), but it gives a much lower CV value. Therefore, this proposal appears to be an improvement of the current EC2 code because, as results are less scattered.

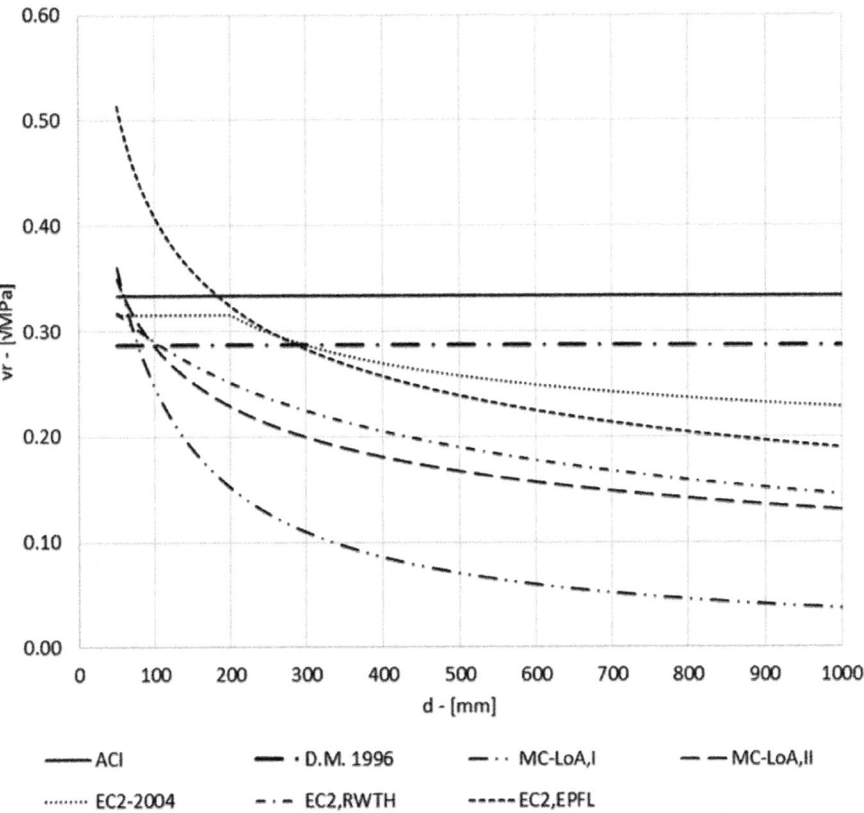

Fig. 5. Influence of effective depth on v_R (h = 1.08d, r_c = 0.71d, r_s = 6.9d, ρ = 0.33%, f_c = 30.5 MPa, d_g = 16 mm, f_y = 548 MPa)

The EPFL proposal for revision of EC2 provides the best estimate of the average ratio V_{test}/V_{th}, equal to $\cong 1.00$, and the *CV* value (=0.27) lies between RWTH proposal and current EC2 values. The 5%-quantile is quite low (0.65), leading to a higher probability of overestimation of the experimental strength than other codes. In fact, the unitary value of V_{test}/V_{th} corresponds to 56%-quantile of all 173 considered cases.

Finally, DM96 (1996) gives the worst results in terms of average value and *CV* of the ratio V_{test}/V_{th}, equal to 1.60 and 0.36, respectively, and the 5%-quantile is the highest (0.94). Strength predictions provided by this code seem too conservative, as the unitary ratio V_{test}/V_{th} corresponds to 9.8%-quantile, meaning that only in less than 10% of all studied cases DM96 (1996) underestimates the experimental punching strength. For these reasons, after comparisons with further experimental results, it could still be adopted in preliminary design since it is conservative and very easy to use, although it can no longer be used for accurate design of new structures or assessing existing ones.

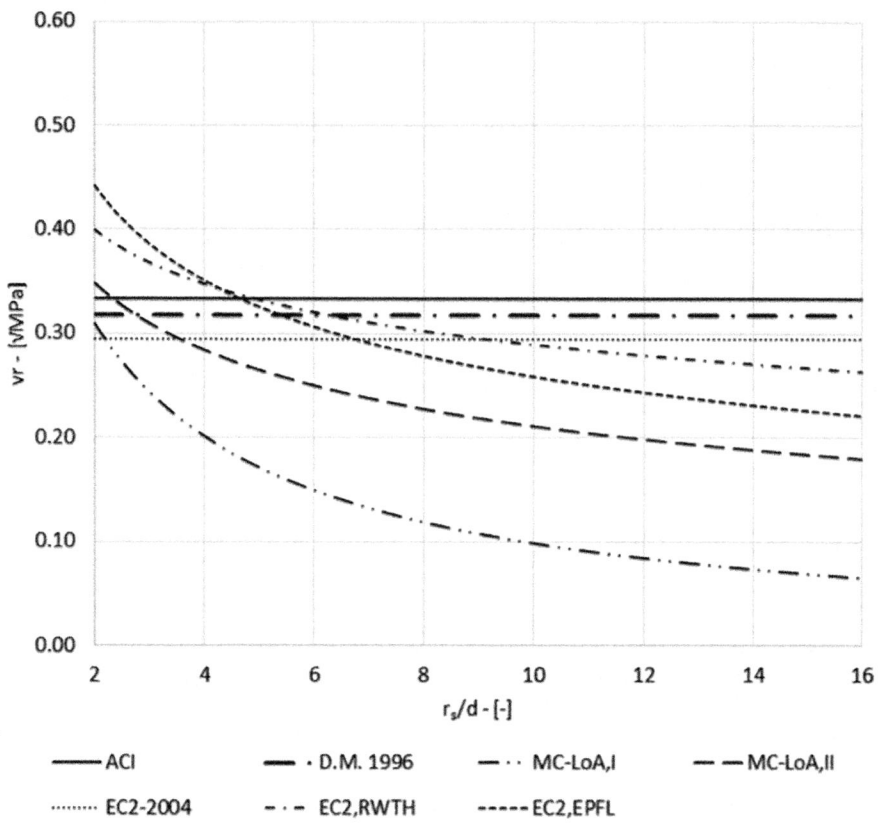

Fig. 6. Influence of slenderness on v_R (d = 300 mm, h = 360 mm, r_c = 300 mm, ρ = 0.5%, f_c = 30 MPa, d_g = 25 mm, f_y = 550 MPa)

Table 1. Comparison of literature test results with strengths predicted using code provisions: average μ, coefficient of variation *CV* and 5%-quantile of the ratio V_{test}/V_{th}.

Reference	Year	No. tests	EC2 2004 Current code			RWTH proposal for revision of EC2			EPFL proposal for revision of EC2			Old Italian Recomm. (DM96 1996)		
			μ	CV	5%-q	μ	CV	5%-q	μ	CV	5%-q	μ	CV	5%-q
Banthia et al.	1995	1	1.23	-	-	1.10	-	-	0.63	-	-	0.92	-	-
Broms	1990	1	1.00	-	-	1.12	-	-	0.89	-	-	1.34	-	-
Criswell	1974	4	1.02	0.10	0.93	1.08	0.13	0.96	0.72	0.24	0.57	1.12	0.25	0.86
Elstner & Hognestad	1956	22	1.04	0.11	0.94	1.08	0.11	0.94	0.94	0.15	0.75	1.50	0.20	1.01
Etter et al.	2009	1	1.02	-	-	1.30	-	-	1.03	-	-	1.38	-	-
Forssell & Holmberg	1946	7	1.32	0.06	1.22	1.39	0.06	1.30	1.14	0.06	1.05	2.38	0.06	2.22
Ghannoum	1998	3	1.03	0.10	0.96	1.13	0.10	1.05	0.92	0.10	0.86	0.93	0.05	0.88

(*continued*)

Table 1. (*continued*)

Reference	Year	No. tests	EC2 2004 Current code			RWTH proposal for revision of EC2			EPFL proposal for revision of EC2			Old Italian Recomm. (DM96 1996)		
			μ	CV	5%-q	μ	CV	5%-q	μ	CV	5%-q	μ	CV	5%-q
Graf	1938	2	0.97	-	-	1.03	-	-	0.82	-	-	1.77	-	-
Guandalini	2005	7	1.03	0.12	0.92	1.23	0.07	1.16	0.96	0.05	0.92	1.21	0.29	0.84
Lee et al.	2009	1	1.15	-	-	1.20	-	-	0.92	-	-	1.25	-	-
Li	2000	6	1.07	0.19	0.82	1.12	0.14	0.91	0.77	0.18	0.58	1.28	0.14	1.05
Lips	2012	5	0.98	0.09	0.88	1.18	0.04	1.13	0.94	0.08	0.86	1.43	0.10	1.24
Long & Masterson	1974	1	1.14	-	-	1.00	-	-	0.89	-	-	1.48	-	-
Manterola	1966	9	0.91	0.11	0.79	1.11	0.14	0.91	0.84	0.24	0.51	1.22	0.29	0.86
Marzouk & Jiang	1997	1	1.13	-	-	1.19	-	-	0.79	-	-	1.00	-	-
Marzouk & Hussein	1991	5	1.29	0.07	1.22	1.38	0.05	1.32	1.06	0.08	0.98	1.47	0.09	1.33
Matthys & Taerwe	2000	4	1.66	0.16	1.64	1.61	0.13	1.54	1.28	0.17	1.11	1.83	0.24	1.45
McHarg et al.	2000	1	1.07	-	-	1.17	-	-	0.96	-	-	1.13	-	-
Moe	1961	7	1.28	0.08	1.14	1.35	0.07	1.24	1.08	0.07	0.99	1.56	0.08	1.39
Mokhtar et al.	1985	1	1.07	-	-	1.13	-	-	0.78	-	-	1.22	-	-
Oliveira et al.	2000	2	1.18	-	-	1.30	-	-	0.94	-	-	1.27	-	-
Ospina et al	2003	1	1.06	-	-	1.12	-	-	0.90	-	-	1.03	-	-
Pilakoutas et al.	2003	1	1.34	-	-	1.47	-	-	1.11	-	-	1.60	-	-
Rankin & Long	1987	23	1.51	0.09	1.32	1.36	0.09	1.20	0.99	0.23	0.68	1.64	0.25	1.13
Regan	1984	29	1.73	0.44	1.09	1.54	0.27	1.13	1.27	0.33	0.90	2.07	0.47	1.33
Sistonen et al.	1997	10	1.28	0.06	1.18	1.39	0.07	1.28	0.85	0.08	0.78	1.59	0.08	1.46
Swamy & Ali	1983	2	1.13	-	-	1.23	-	-	0.93	-	-	1.07	-	-
Taylor and Hayes	1965	8	0.98	0.10	0.85	0.96	0.08	0.88	1.00	0.15	0.81	1.81	0.15	1.51
Timm	2003	3	1.05	0.01	0.97	1.03	0.09	0.95	0.99	0.11	0.892	1.72	0.09	1.57
Urban	1994	2	1.19	-	-	1.22	-	-	0.89	-	-	1.25	-	-
Widianto et al.	2010	1	0.84	-	-	0.90	-	-	0.81	-	-	0.68	-	-
Yamada et al.	1992	2	0.94	-	-	0.98	-	-	0.80	-	-	1.48	-	-
All tests		173	1.27	0.33	0.85	1.27	0.21	0.93	1.00	0.26	0.65	1.60	0.36	0.94

5 Conclusions

In this paper the authors presented and discussed formulations of main international codes (current EC2 version, ACI, MC2010, two proposals for revision of EC2), together with old Italian Recommendations, for the evaluation of the punching capacity of R/C slabs without transverse reinforcement. Firstly, formulations of different codes have been compared among them to evaluate if and how each of them takes into account the influence of different geometrical and mechanical parameters on the punching strength. Successively, values of the punching strength predicted using different codes have been compared with results of more than 170 literature experimental tests on specimens with similar geometry, reinforcement layout and load spatial distribution.

Results of the parametric analysis highlight that more recent formulations, that is to say Model Code 2010 and proposals for revision of EC2 are able to take into consideration all geometrical and mechanical variables which control the punching failure.

The current EC2 formulation does not take into account the influence of the slab's slenderness (r_s/d) on the punching capacity, giving unsafe results for high values of it ($r_s/d > 8$). For this reason, both proposals for revision of EC2 bridge this gap by introducing explicitly the dependence of the punching capacity from the parameter r_s/d.

Results of the comparison among different code formulations and literature experimental data allow for estimating the capability of each code in predicting the experimental strength, although with reference to a moderate number of cases. The current EC2 version overestimates the experimental punching capacity of less than 30%, with a significant data scattering around the average value. The RWTH proposal for revision of EC2 gives the same mean values of current EC2, nevertheless the data scattering is clearly lower and also the probability to overestimate the punching capacity. The EPFL proposal gives better results in terms of the mean value ($V_{test}/V_{th} \cong 1.00$), but, although the scattering is similar to the RWTH proposal, the probability of overestimating the punching capacity is evidently much higher. Nevertheless, to obtain results similar to the RWTH proposal, it would be enough to introduce a reductive coefficient of the theoretical strength.

The work presented in this paper is part of a wider and deeper study, which is currently in progress inside the task group CEN 250/SC 2/TG 4, in charge for the revision of EC2 sections referring to shear, torsion and punching of R/C structures. Results have preliminary nature, and they could be useful for the improvement of the two EC2 proposals.

References

Fédération Internationale du Béton (fib) (2012) Model Code 2010 - Final draft, Vol. 1-2, fédération internationale du béton, Bulletin 65, Lausanne, Switzerland

Muttoni A (2008) Punching shear strength of reinforced concrete slabs without transverse reinforcement. ACI Struct J 105(4):440–450

ACI Committee 318 (2014) Building Code Requirements for Structural Concrete (ACI 318-14) and Commentary. American Concrete Institute, Farmington Hills, MI

CEN (2004) Eurocode 2: design of concrete structures. Part 1-1: general rules and rules for buildings. EN 1992-1-1, December 2004

Hegger J, Sibuig C, Kueres D (2016) Proposal for punching shear design based on Eurocode 2. Institute of Structural Concrete RWTH Aachen University, Germany, 22 January 2016

Muttoni A, Fernández Ruiz M, Simões J, Cavagnis F (2016) Background document to provisions for Shear and Punching Shear Design – Closed Form solutions based on Model Code 2010 and Critical Shear Crack Theory. IBETON École Polytechnique Féderale de Lausanne, Lausanne, 3 June 2016

DM96 - Decreto 9 gennaio 1996 (1996) Norme tecniche per il calcolo, l'esecuzione ed il collaudo delle strutture in cemento armato, normale e precompresse per le strutture metalliche. Ministero dei Lavori Pubblici, Roma (in Italian)

SIA (2003) Code 262 for Concrete Structures, Swiss Society of Engineers and Architects, Zürich, 94 p

Banthia N, Al-Asaly M, Ma S (1995) Behavior of concrete slabs reinforced with fiber-reinforced plastic grid. ASCE J Mater Civil Eng 7(4):252–257

Broms CK (1990) Punching of flat plates – a question of concrete proprieties in biaxial compression and size effect. ACI Struct J 87(3):292–304

Criswell ME (1974) Static and dynamic response of reinforced concrete slab-column connections. ACI Spec Pubbl 42:721–746

Elstner RC, Hognestad E (1956) Shearing strength of reinforced concrete slabs. ACI J Am Concr Inst 28(53-2):29–58, July 1956

Etter S, Heinzmann D, Jaeger T, Marti P (2009) Versuche zum Durchstanzverhalten von Stahlbetonplatten (Tests on the Punching Behavior of Reinforced Concrete Slabs). Report No. 324, Institute of Structural Engineering (IBK), Swiss Federal Institute of Technology (ETH), Zurich, Switzerland, 64 p (in German)

Forssell C, Holmberg Å (1946) Stämpellast på Plattor av Betong. Betong (Stockholm) 31(2):95–123 (in Swedish)

Graf O (1938) Versuche über die Widerstandsfähigkeit von allseitig aufliegenden dicken Eisenbetonplatten unter Einzellasten. Deutscher Ausschuß für Eisenbeton, Heft 88 (in German)

Ghannoum CM (1998) Effect of high strength concrete on the performance of slab-column specimens. Master Degree thesis, Department of Civil Engineering and Applied Mechanics, McGill University, Montréal, Canada

Guandalini S (2005) Poinçonnement symétrique des dalles en béton armé. PhD thesis, École Polytechnique Fédérale de Lausanne, Lausanne, Switzerland

Lee JH, Yoon YS, Cook WD, Mitchell D (2009) Improving punching shear behavior of glass fiber-reinforced polymer reinforced slabs. ACI Struct J 106(4):427–434

Li K (2000) Influence of size on punching shear strength of concrete slabs. Master Degree thesis, Department of Civil Engineering and Applied Mechanics, McGill University, Montréal, Canada

Lips S, Fernández Ruiz M, Muttoni A (2012) Experimental investigation on punching strength and deformation capacity of shear-reinforced slabs. ACI Struct J 109:889–900

Long AE, Masterson DM (1974) Improved experimental procedure for determining the punching strength of reinforced concrete flat slab structures. ACI Spec Publ 42:921–938

Manterola M (1966) Poinçonnement de dalles sans armature d'effort tranchant. In: Comité Européen du Béton (Hrsg.): Dalles, Structures planes, CEB-Bull. d'Information No. 58, Paris (in French)

Marzouk H, Jiang D (1997) Experimental investigation on shear enhancement types for high-strength concrete plates. ACI Struct J 94(1):49–58

Marzouk H, Hussein A (1991) Experimental investigation on the behavior of high-strength concrete slabs. ACI Struct J 88(6):701–713

Matthys S, Taerwe L (2000) Concrete slabs reinforced with FRP grids II: punching resistance. ACI J Composites Constr 4(3):154–161

McHarg J, Cook WD, Mitchell D, Yoon Y (2000) Benefits of concentrated slab reinforced and steel fibers on performance of slab-column connections. ACI Struct J 97(2):225–234

Moe J (1961) Shearing strength of reinforced concrete slabs and footings under concentrated loads. Portland Cement Assoc D47:135

Mokhtar AS, Ghali A, Dilger W (1985) Stud shear reinforcement for flat concrete plates. ACI J 82(5):676–683

Oliveira DR, Melo GS, Regan PE (2000) Punching strengths of flat plates with vertical or inclined stirrups. ACI Struct J 97(3):485–491

Ospina CE, Alexander SDB, Cheng JJR (2003) Punching of two-way concrete slabs with fiber-reinforced polymer reinforcing bars or grids. ACI Struct J 100(5):589–598

Pilakoutas K, Li X (2003) Alternative shear reinforcement for reinforced concrete flat slabs. J Struct Eng 129(9):1164–1172

Rankin GIB, Long AE (1987) Predicting the punching strength of conventional slab-column specimens. Struct Eng Group, Part I, 82:1165–1186, April 1987

Regan PE (1984) The dependence of punching resistance upon the geometry of the failure surface. Mag Concr Res 36(126):3–8

Sistonen E, Lydman M, Huovinen S. (1997) The geometrical model of the calculation formula of the punching shear capacity of the reinforced concrete slab. Helsinki University of Technology, Laboratory of Structural Engineering and Building Physics, October 1997

Swamy RN, Ali SAR (1982) Punching shear behavior of reinforced slab-column connections made with steel fiber concrete. ACI J 79:392–406

Taylor R, Hayes B (1965) Some tests on the effect of the edge restrain on punching shear in reinforced concrete slabs. Mag Concr Res 17(50):39–44

Timm M (2003) Punching of foundation slabs under axisymmetric loading. PhD diss., Institute for Building Materials, Concrete Structures and Fire Protection of the Technical University Braunschweig

Urban T (1994) Nosnosc na przebicie w aspekcie proporcji bokow slupa), Badania Doswiadczalne Elementów I Konstrukcji Betonowych. Report 3. Lodz, Poland, 76 p (in Polish)

Widianto BO, Jirsa JO, Tian Y (2010) Seismic rehabilitation of slabcolumn connections. ACI Struct J 107(2):237–247

Yamada T, Nanni A, Endo K (1992) Punching shear resistance of flat slabs: influence of reinforcement type and ratio. Struct J 89(5):555–563

Mechanical Modelling of Friction Pendulum Isolation Devices

V. Bianco[1]([⊠]), G. Monti[1], and N. P. Belfiore[2]

[1] Department of Structural Engineering and Geotechnics,
Sapienza University of Rome, Rome, Italy
vincenzo.bianco@uniroma1.it
[2] Department of Mechanical and Aerospace Engineering,
Sapienza University of Rome, Rome, Italy

Abstract. Even though different versions of the Friction Pendulum Devices (FPD) can be found on the market and their effectiveness has been extensively proven by means of numerous experimental campaigns carried out worldwide, many aspects concerning their mechanical behaviour still need to be clarified. These aspects concern, among others: (1) the sequence of sliding on the several concave surfaces, (2) the influence of temperature on the frictional properties of the coupling surfaces, (3) the possibility of the *stick-slip* phenomenon, (4) the possibility of impact-induced failure of some components, (5) the geometric compatibility, and so on. These aspects are less clear the larger the number of concave surfaces of which the device is composed. This paper presents a new way of modelling the mechanical behaviour of the FPDs, by fulfilling: (1) geometric compatibility, (2) kinematical compatibility, (3) dynamical equilibrium, and (4) thermo-mechanical coupling.

Keywords: Base isolators · Friction pendulum devices · Kinematics
Dynamical equilibrium

1 Introduction

In recent years base isolation has become an increasingly applied structural design technique for both buildings and bridges located in highly seismic areas. Two basic types of base isolation can be identified (Kelly 1997; Taniwangsa and Kelly 1996): (1) by elastomeric bearings and (2) by a sliding system. According to the former approach, the building or structure is decoupled from the horizontal components of the earthquake ground motion by interposing a layer with low horizontal stiffness between the structure and the foundation. The latter approach works by limiting the transfer of shear across the isolation interface by allowing mutual sliding to occur. Many sliding systems have been proposed and some have already been implemented in practice. The Friction Pendulum (FP) system, firstly introduced by Zayas et al. (1987), is one of these sliding systems that has already been used for several projects (*e.g.* Mellon and Post 1998), both new and retrofit. It combines a sliding system with a restoring force. In fact it is composed of an articulated slider, whose surface is coated by a special interfacial material with the purpose to provide a suitable friction, sliding on a stainless steel

© Springer International Publishing AG, part of Springer Nature 2018
M. di Prisco and M. Menegotto (Eds.): ICD 2016, LNCE 10, pp. 133–146, 2018.
https://doi.org/10.1007/978-3-319-78936-1_10

concave surface (Fig. 1a). The concave surface geometrically provides the restoring force as the tangential projection of the applied gravity load. The FP system for seismic isolation has been recently manufactured as devices with multiple independent concave sliding surfaces, in order to provide adaptable behavior (*e.g.* Earthquake Protection Systems, Inc. 2003; Tsai et al. 2010). The Double concave Friction Pendulum (DFP) bearing (Constantinou 2004; Fenz and Constantinou 2006) is an adaptation of the traditional, well-proven single concave friction pendulum that allows for significantly larger displacements, given identical plan dimensions. The DFP bearing consists of two facing stainless steel concave surfaces (Fig. 1b). The upper and lower concave surfaces have radii of curvature R_1 and R_2 that might not be equal. While the coefficients of friction along the two kinematic pairs are μ_1 and μ_2, respectively.

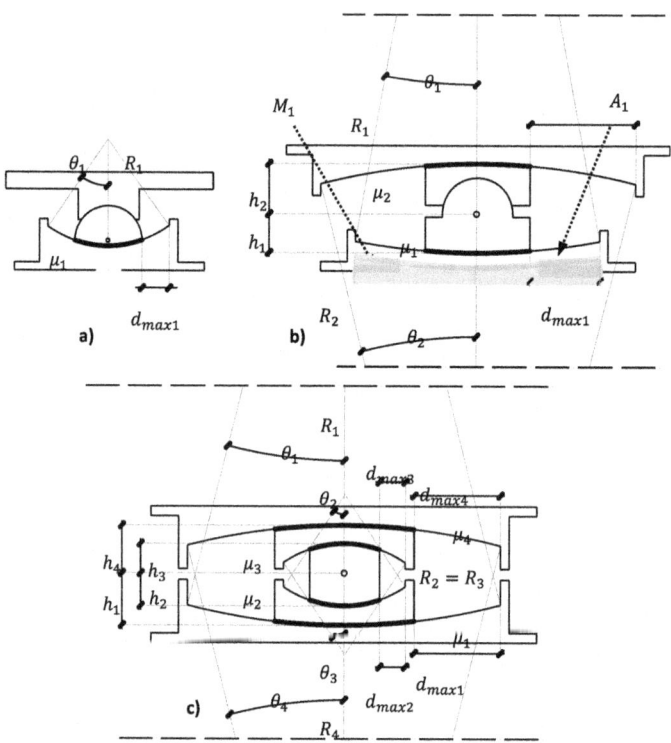

Fig. 1. Friction pendulum bearings: (a) Single Friction Pendulum (SFP), (b) Double Friction Pendulum (DFP), (c) Triple Friction Pendulum (TFP).

An articulated slider separates the two concave surfaces. The articulation is necessary for proper even distribution of pressure on the sliding surfaces and to accommodate differential movements along the top and bottom sliding surfaces when friction is unequal on these two latter. The Triple Friction pendulum bearing (TFP) is an even more advanced version of the original FP (Fig. 1c) that is composed of (a) two external concave plates, with radius of curvature and friction coefficient equal to $(R_1; \mu_1)$ and

$(R_4; \mu_4)$ respectively, (b) two inner sliding plates with radius of curvature and friction coefficient equal to $(R_2; \mu_2)$ and $(R_3; \mu_3)$ respectively, and (c) a sliding pad. The external convex surfaces of both the sliding plates and the sliding pad are coated with a lining material that has to provide the suitable friction coefficient. More recently even friction pendulum bearings presenting more than four sliding surfaces have been proposed (Tsai et al. 2010). The more sophisticated versions of the FP, which contemplate the presence of an increasing number of sliding concave surfaces, have the advantage to guarantee: (1) a certain adaptability to the given earthquake, despite being a passive system, as well as (2) a reduced footprint, with the same deformation capacity.

Friction plays a pivotal role in the functioning of the FP devices. However, it is a very complex phenomenon not completely understood yet: it is given by several concomitant physical phenomena whose relative importance varies as function of the involved parameters and contour conditions (e.g. Bowden and Tabor 1973; American Society for Metals 1992; Constantinou et al. 2007). The most frequently used interface in sliding bearings is made of PTFE or PTFE-like materials in contact with polished stainless steel. For these kind of interfaces, the dynamic coefficient of friction μ mainly depends on (Constantinou et al. 1999): (a) the sliding velocity, (b) pressure, (c) temperature and (d) time of loading.

Since the introduction of the friction pendulum devices for the seismic protection of structures, a lot of efforts have been made by the scientific community (e.g. Mokha et al. 1990; Nagarajaiah et al. 1991; Fenz and Constantinou 2008) in order to single out the most suitable analytical model of the horizontal force-displacement hysteretic curve characterizing the behavior of such devices. In fact, the hysteretic force-displacement curve needs to be implemented in the standard structural analysis softwares in order to assess the efficacy of the designed intervention to correctly protect the structure against the expected earthquake and in compliance with the code regulations. Even though more refined models of the overall hysteretic curve, each substantially based on the early work by Fenz and Constantinou (2008), have been recently proposed (e.g. Becker and Mahin 2012, 2013; Ray et al. 2013), they present common drawbacks, among which: (1) equilibrium conditions are fulfilled only in the horizontal direction, completely neglecting the rotational equilibrium, and (2) the influence of thermal effects are completely ignored. In fact, as to this latter aspect, it is well known that, when friction is involved in highly dynamic applications, heat is generated on the contact surfaces as a result of the transformation of mechanical energy into energy of thermal oscillations of molecules during friction. During an earthquake, with peak sliding velocity of the order of ≥ 400 mm/s, temperature could reach very high values and significantly affect the frictional properties of the PTFE-stainless steel sliding interface.

Moreover, such complex tribological systems are known to be susceptible to (1) the stick-slip and (2) the sprag-slip phenomena (e.g. Popov 2010). Where: the former consists of the likely alternation of time intervals of sliding to time intervals of no sliding, with consequent re-coupling of the ground shaking and superstructure movement, while the latter consists of vibrations that could be activated in the direction orthogonal to the sliding one.

In this scenario, many questions arise about the actual functioning of these devices. Do they always re-center? Otherwise, under which specific boundary conditions, either kinematic (displacements, velocities and accelerations) or tribological (friction) do they re-center? Similarly as what happens with a rolling wheel, must friction have an optimum value, for sliding to take place? In this view, a limit value of asymptotically null friction would allow those devices to work, or would it rather inhibit the triggering of the functioning of the devices? Is it only friction, at each phase of the device deformation, that dictates the triggering, rather than the inhibition, of sliding along the various concave surfaces? How does rotational equilibrium affect this aspect? Does the over-structure actually horizontally translate only, without any rotation, during an earthquake lacking the vertical component? Otherwise, under which specific conditions the over-structure does not undergo any rotation? Would it be necessary to outfit the isolation system with supplemental devices meant to suppress any possibility of rotation? Does the building actually move upward, as a response of the horizontal two dimensional earthquake? Or rather, as function of the soil deformability, it may move downwards, in order to accommodate vertical deformations? Does a complex soil-structure interaction take place during an earthquake?

It is evident that the topic is complex and needs to be faced from a multidisciplinary standpoint, involving also both mechanical and tribological engineering expertize (*e.g.* Scotto Lavina 1990; Belfiore et al. 2000).

With the aim to contribute to a better understanding of the mechanical behavior of the multiple FPs, thus attempting to answer the questions above, this work presents a new approach to model the thermo-mechanical behavior of such devices. It assumes, as a first approximation, that all the components the device is made of can be modelled as rigid bodies, thus neglecting, for the time being, any deformation. This modelling approach is aimed at fulfilling: (1) geometrical compatibility, (2) kinematic compatibility, (3) dynamic equilibrium, both translational and rotational, and (4) thermo-mechanical coupling.

2   Inspiring Physical Observation

Suppose to have a kind of flat double friction bearing (Fig. 2), composed of (a) two rigid horizontal steel plates, and (b) a rigid steel cylinder placed in between. Imagine that this device is loaded by a vertical force N and that friction at the two interfaces is governed by the Coulomb constitutive law (Fig. 2b), with a rigid-perfectly-plastic dependence of the friction force on the relative displacement $F(u)$, and threshold value $F_\mu = \mu \cdot N$. Imagine also that the friction coefficient μ at those interfaces, has the same value. If we impose a displacement on the lower plate, due to the rigid-plastic constitutive law of interfacial friction, we would immediately have, even for infinitesimally small values of imposed displacement, the mobilization of the friction threshold F_μ at each interface. Those two horizontal forces, equal in magnitude and opposite in direction, do generate a couple, to which corresponds a moment $M_S = F_\mu \cdot H$ that, in the initial configuration, is not balanced by any other couple. That moment would tend to overturn the pad by making it rotate around one of the corners.

In such tentative rotation, the points of application of the vertical force N would migrate to the pad corners still in contact with the relevant horizontal plates. In this way, even the vertical force would give rise to a couple $M_R = N \cdot d$, opposed to the overturning one and larger in absolute value, that would make the pad undergo both a

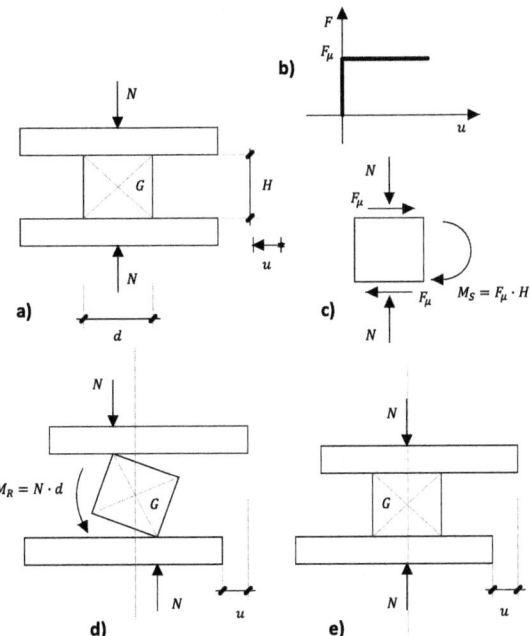

Fig. 2. Case of a flat double friction bearing: (a) undeformed initial configuration, (b) friction's Coulomb constitutive law, (c) pad free body diagram, (d) intermediate deformed configuration, and (e) final configuration of the whole device.

rigid body rotation around its centroid G and a simultaneous vertical translation. During such movement of the pad, the diagonally opposed corners, loaded by N, would also undergo sliding, up to the new, restored equilibrium configuration (Fig. 2e). It is the vertical weight force N that restores the geometric compatibility at each time step.

3 Geometric Compatibility

When it comes to multiple friction pendulum bearings, namely with concave spherical surfaces, the most advanced device currently available on the market is the Triple Friction Pendulum (TFP). It is composed of (Fig. 1c): (a) two external concave-surface-topped plates, both with radius $R = R_1 = R_4$, (b) two internal sliding plates, which are two straight cylinders whose bases, one convex and another concave, are two spherical caps belonging to two spherical surfaces with different radii, $R_1 \neq R_2$ and $R_3 \neq R_4$ respectively, and (c) an internal sliding pad, which is a straight cylinder with both bases composed of convex spherical caps belonging to surfaces with the same radius $R_2 = R_3$. The two external faces of the two larger plates are horizontal. The four sliding interfaces are composed by the superimposition, in pairs, of eight spherical caps, one constituted of plain stainless steel and the other coated by a particular liner that is generally PTFE or the like.

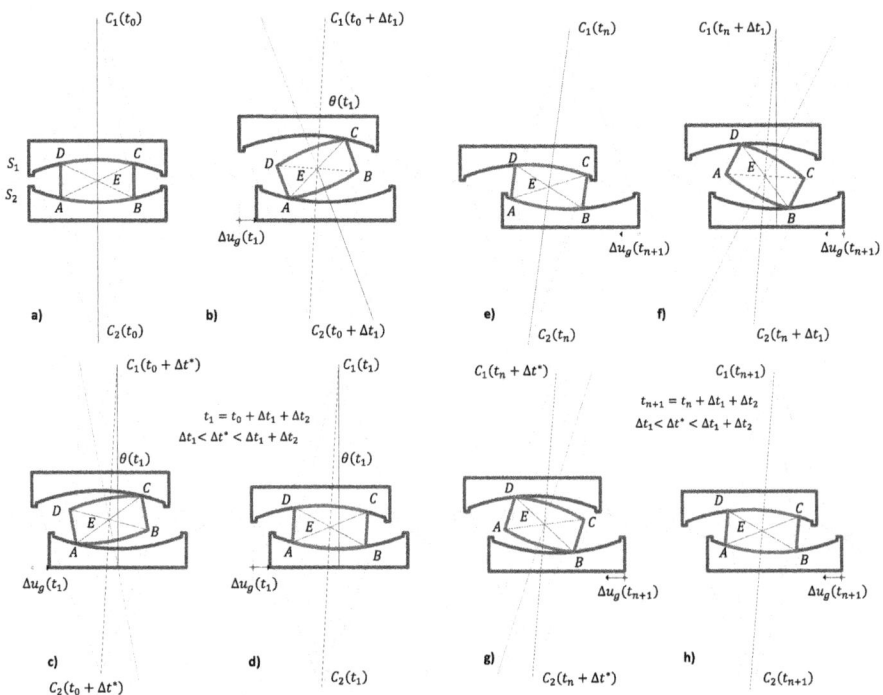

Fig. 3. Case of a DFP bearing subjected to an horizontal ground displacement contained in a radial plane: (a–d) rightward movement starting from a rest position, (e–h) leftward movement starting from a deformed configuration

The considerations herein presented are based on the following assumptions: the various parts constituting the *friction pendula* are considered as rigid bodies, thus neglecting, at least for the time being, any possible deformation. In this way, geometrical compatibility is fulfilled each time the two spherical caps constituting the two mating surfaces of a given interface are perfectly superimposed to each other.

The first and most important constraint is represented by the fact that the two outermost horizontal surfaces, lower and higher, can undergo a rigid body motion remaining horizontal, which means that they can only translate remaining parallel to themselves. This is due to the fact that, at the extrados of the so called *isolation plane*, the various isolators are connected by a rigid diaphragm (*e.g.* Italian Technical Regulations 2008) while, at the intrados, to the presence of the ground. In order for that condition to be fulfilled, it is necessary that either the two internal surfaces (S_2 and S_3) or the two external ones (S_1 and S_4) undergo sliding simultaneously while it is not preferable to allow mixed sliding, *e.g.* S_2 and either S_1 or S_4, since controlling the evolution of displacements along each concave surface would become extremely difficult. For this reason, the friction coefficient should be the same along the two surfaces of each of those two sliding pairs, *i.e.* $\mu_2 = \mu_3$ and $\mu_1 = \mu_4$.

It is preferable to start by analyzing the behavior of a double friction pendulum, composed of the internal pad and only two external plates, in a radial plane, which means in case of uni-directional imposed horizontal displacement (Fig. 3). At a generic time step t_n, the internal deformation undergone, as function of the imposed displacement $\Delta u_g(t_n)$, by the several members constituting the device, can be decomposed into two subsequent phases. When starting from the equilibrium condition (Fig. 3a–d), in the first phase, which means during the first part Δt_1 of the time increment Δt, the pad, due to the imposed displacement $\Delta u_g(t_1)$, rigidly rotates around one of the lowermost corners (A) and such rotation also yields a certain vertical displacement of the upper plate. During the second phase Δt_2 of the current time increment Δt, the pad undergoes a rigid rotation around its centroid (E) with this latter simultaneously rigidly translating along the straight line connecting the centers of the two spherical surfaces $\overline{C_1 C_2}$. This latter direction, resulting from the end of the previous sub-interval Δt_1, is inclined, with respect to the vertical direction, of an angle $\theta(t_1)$ that is the geometrical variable accounting for the final configuration assumed by the entire device (Fig. 3d). The same two-step deformation can be recognized if, at a generic time step t_n, the ground imposes a reversed displacement, and starting from a generic deformed configuration (Fig. 3e–h).

The only difference, in case of reversal, is that initial rigid rotation (Δt_1) involves the other diagonal of the pad (*i.e.* $\overline{DB}$ instead of $\overline{AC}$). During the second phase, sliding occurs along the corners of the involved diagonal of the pad, and friction is mobilized therein.

4   Mechanical Model

The analysis of the geometric compatibility of the friction pendulum devices described in the previous section suggests, in a consequential manner, the way in which it is possible to model the mechanical behavior of such devices in the three dimensional space. For the reasons already explained in the previous section, in the present work attention is focused on the dynamical behavior of a DFP subjected to a ground shaking, leaving the study of complete TFP to further future developments. Moreover, the study is herein limited to the case of unidirectional seismic attack. The dynamical behavior of a DFP at the instant t_n can therefore be limited to what happens in the generic plane π, singled out by (1) current position of the pad, singled out by polar $\theta(t_n)$ and azimuthal $\varphi(t_n)$ angles, and (2) direction, in the global reference system $oxyz$, of the current horizontal displacement $\Delta u_g(t_{n+1})$ imposed by the ground shaking, excluding any vertical earthquake, for the time being. Both phases at the current time step can be modelled by an articulated system of rigid bodies, different for each phase (Figs. 4 and 5). Even though the rigid bodies are expected to actually undergo some deformations, for the time being they are herein assumed to behave as perfectly rigid bodies. These latter form an open kinematic chain for both phases of motion.

During the first phase (Δt_1) (Fig. 4), the lower plate (Rigid Body 1) is assumed to move horizontally along a prismatic guide, without friction, and the motion is the one imposed by the earthquake and characterized by given displacement $x_g(t_{n+1})$, velocity $\dot{x}_g(t_{n+1})$, and acceleration $\ddot{x}_g(t_{n+1})$. The pad (Rigid Body 2) is kinematically constrained to both

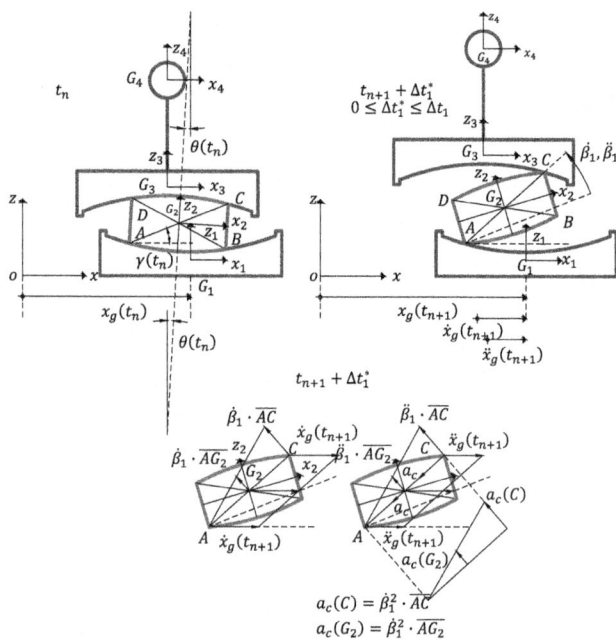

Fig. 4. Mechanical model, constituted of an Articulated System of Rigid Bodies (ASRB), to simulate the first phase of the behavior of the DFP at a given time step.

lower and upper plate (RB 3, 4) by two revolute joints in which the constraint force tangent to the concave surface is blocked to the maximum allowable Coulomb friction force. The upper sliding plate (RB 3) carries the weight of the over-structure that is herein assumed as a mass, rigidly connected to the plate by a massless rigid rod, and concentrated in rigid body 4. Substantially, rigid bodies 3 and 4 are assumed to form a single rigid body assemblage. Due to the motion imposed on rigid body 1, rigid body 2 undergoes a rigid rotation β_1 around one of its corners (point A in Fig. 4) characterized by angular velocity and acceleration $\dot{\beta}_1, \ddot{\beta}_1$. As a consequence of that (unknown) rotation, the assemblage of rigid bodies 3 and 4 undergoes a rigid planar translation, along axes x and z of the inertial reference system $oxyz$, maintaining contact with the pad in the pin-joint in point C of the pad. Note that the position assumed by the plates at the end of this first phase already singles out the final updated value of the inclination of the pad with respect to both the sliding plates $\theta(t_{n+1})$.

The second phase is modelled by the same assemblage of rigid bodies above, kinematically constrained between one another in a different way (Fig. 5). Rigid body 1 is blocked in its final position corresponding to the current time step t_{n+1}. Rigid body 2 undergoes a rigid roto-translation with its centroid G_2 translating along the straight line connecting the two centers of curvature of the two plates, along a linear prismatic guide without friction and contemporarily counter-rotates around its centroid with angular velocity and acceleration $\dot{\beta}_2, \ddot{\beta}_2$. Along the vertices of the diagonal involved ($\overline{AC}$ in Fig. 5), the pad is now constrained to the two plates (rigid bodies 1 and 3) by two

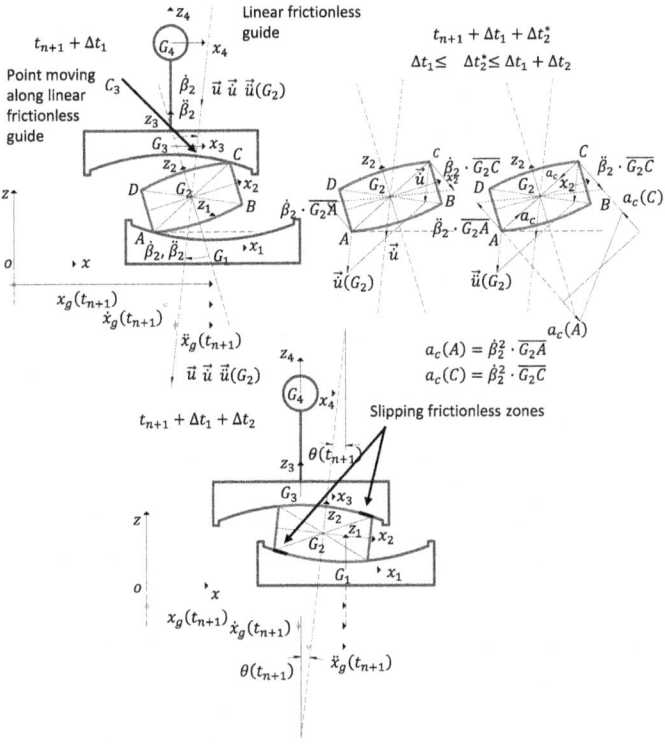

Fig. 5. Mechanical model, constituted of an Articulated System of Rigid Bodies (ASRB), to simulate the second phase of the behavior of the DFP at a given time step.

circular guides with Coulomb friction so that, during this phase, slipping takes place with consequent yielding of heat (Fig. 5). While kinematic quantities $\vec{\ddot{u}}(G_2)$ and $\ddot{\beta}_2$, which are the linear and angular accelerations characterizing the pad rigid roto-translation, are unknown since they depend on the dynamical characteristics of the system, the values of displacement $\vec{u}(G_2)$ and rotation β_2 are known since they are the quantities necessary to restore the geometrical compatibility by closing both kinematic couples implying that the pad axis superimposes to the straight line connecting the concave plates centers (compare previous sections). The assemblage of rigid bodies 3 and 4 translates with point C_3 moving (Fig. 5), without friction, along the straight line connecting the concave surfaces' centers $\overline{C_1 C_2}(t_n + \Delta t_1)$.

5   Kinematic Compatibility

A kinematic analysis, meant to study motion without considering the forces that produce it, is necessary. During the first phase, the velocity of point C, in which a revolute joint is placed, is given by (e.g. Scotto Lavina 1990; Belfiore et al. 2000):

$$\vec{v_C} = \dot{x}_g \cdot \vec{e_1} - \dot{\beta}_1 \cdot \vec{e_2} \times \vec{AC} \tag{1}$$

where $\vec{e_2}$ is the unit vector of axis y of the Inertial system $oxyz$ whose positive direction is towards the paper. While the acceleration is given by:

$$\vec{a_C} = \ddot{x}_g \cdot \vec{e_1} + \dot{\beta}_1^2 \cdot \vec{G_2A} - \ddot{\beta}_1 \cdot \vec{e_2} \times \vec{AC} \tag{2}$$

During the second phase, the velocity of point C, belonging to rigid body 2 (pad) is given by:

$$\vec{v_C} = \vec{\dot{u}}(G_2) + \dot{\beta}_2 \cdot \vec{e_2} \times \vec{G_2C} \tag{3}$$

while the velocity with which the rigid bodies 3 and 4 translate along the linear prismatic guide without friction, along the straight line connecting the position occupied by the two curvature centers at the end of step one, is given by the velocity of point C_3 (Fig. 5):

$$\vec{v_{C_3}} = \vec{\dot{u}}(G_2) \tag{4}$$

The acceleration of point C, belonging to rigid body 2, is given by Fig. 5:

$$\vec{a_C} = \vec{\ddot{u}}(G_2) + \dot{\beta}_2^2 \cdot \vec{G_2C} + \ddot{\beta}_2 \cdot \vec{e_2} \times \vec{G_2C} \tag{5}$$

and the acceleration of point C_3, is given by:

$$\vec{a_{C_3}} = \vec{\ddot{u}}(G_2) \tag{6}$$

6 Dynamical Equilibrium

Both mechanical models adopted to study the two phases of the planar motion of the DFP described in previous section are open kinematic chains (Shabana 2001). And each of them is characterized by one degree of freedom.

In fact, for the first phase, we have 3 rigid bodies, each characterized by 3 kinematic unknowns, for a total of $n = 3 \times 3 = 9$. The number of kinematic constraints is equal to $n_c = 7$ that are given by: (a) vertical translation and rotation (2) of the RB1, (b) two internal revolute joints ($2 \times 2 = 4$), and (c) rotation (1) of the assemblage of RB3 and RB4. The translational horizontal degree of freedom of the lowermost sliding plate (RB1) is kinematically controlled by the earthquake, while the rotational degree of freedom of the pad is not known a priori so that a force analysis is required and the system equations of motion must be formulated to obtain a number of equations equal to the number of unknown variables.

For the second phase, we have $n = 3 \times 3 = 9$ kinematic unknowns. The number of kinematic constraints is equal to $n_c = 8$ that are given by: (a) 3 for the lowermost plate that is assumed fixed, (b) 1 for the pad, since its centroid G_2 cannot move orthogonally to the straight line (prismatic frictionless guide) passing through the concave surfaces centers $\overline{C_1 C_2}(t_n + \Delta t_1)$, (c) 2 for the uppermost sliding plate assembled with the over-structure (RB3 + RB4), since they can only translate with the contact point at the end of phase 1 (*i.e.* point C_3 in Fig. 5) moving along the straight line above, thus neither rotating nor translating orthogonally to that line, and (d) 2 are given by the internal constraints constituted by the circular frictional prismatic guides, between the kinematic couples RB1-RB2 and RB2-RB3+RB4, since they do not allow mutual displacement at the contact point, along the relevant radius of curvature. The independent kinematic unknown is $\ddot{\beta}_2$ while the translational acceleration $\ddot{u}(G_2)$ can be expressed as function of $\ddot{\beta}_2$.

The first kinematic chain is both *kinematically* and *dynamically driven* (Shabana 2001) since one degree of freedom, that is the horizontal translation of the lowermost plate, is imposed by the ground movement and another, that is the rotation around one of the pad's corners, is governed by the forces involved, either inertia and external. On the other hand, the second kinematic chain is *dynamically driven*, since for the unknown, a force analysis is required. For both models, a relevant minimum number of differential equations can be written, in the ambit of the Embedding Technique (Shabana 2001), by making a suitable cut in one of the joints among the involved rigid bodies, and formulating the dynamic conditions of the resulting subsystem. The number of these obtained equations, which do not contain the joint reaction forces, is equal to the number of degrees of freedom of the system. Once the minimum number of differential equations are solved, the joint reaction forces can be obtained by the equations of motion obtained by a free-body analysis.

Proceeding in this way, for each kinematic chain, making the cut in correspondence of the revolute joint in point A, the following equilibrium equation, applying the D'Alambert principle, is obtained:

$$\vec{K}_A = \vec{K}_{G_2} + \overrightarrow{AG_2} \times \frac{d}{dt}\vec{Q}_2 + \overrightarrow{AG_3} \times \frac{d}{dt}\vec{Q}_3 + \overrightarrow{AG_4} \times \frac{d}{dt}\vec{Q}_4 = \vec{M}_A = \sum_{i=2}^{4} \overrightarrow{AG_i} \times \vec{P}_i \quad (7)$$

in which: $\vec{K}_A$ is the moment of the inertia forces with respect to point A, $\vec{K}_{G_2}$ is the moment of inertia of RB2 with respect to its centroid, $\vec{Q}_i$ is the momentum of the i-th RB, $\vec{M}_A$ is the moment of the external forces with respect to point A, and $\vec{P}_i$ is the self-weight force of the i-th RB. For each phase, Eq. (7) has to be specialized by substituting, in the expression of the i-th RB time-derivative of the momentum, the kinematically compatible expressions for the accelerations, as obtained in the previous section.

7 Thermo-Mechanical Coupling

During the second phase, slipping occurs along the two concave surfaces (Fig. 5) and heat is developed therein. It is possible both (1) to keep track of the temperature change of the plates and (2) to evaluate the change of the frictional threshold as function of the surface temperature. Thermo-mechanical coupling is obtained by defining a suitable law of variation of the frictional coefficient, as function of the temperature, as follows:

$$F_{\mu i}(t) = \mu_i[T_i(t)] \cdot N_i \tag{8}$$

where: $F_{\mu i}(t)$, μ_i, $T_i(t)$, N_i are (a) the friction strength, (b) the temperature dependent friction coefficient, (c) the temperature, and (d) the internal force orthogonal to the i-th concave surface.

At each time step, the increment of Temperature, yielded by the friction-induced heat, can be evaluated by the following thermal equilibrium equation:

$$c_i \cdot \dot{T}_i = f_i \cdot \dot{u}_i - \dot{Q}_i \tag{9}$$

where: $\dot{T}_i$ is the rate of change of temperature (*temperature/time*), c_i is the heat capacity (*energy/temperature*) of the i-th surface, given by the product of the steel specific heat c_s times the mass M_i of the concave surface involved in the heat exchange, and $\dot{Q}_i$ is the rate of heat exchanged with the surrounding environment (*energy/time*). For the sake of simplicity, M_i can be assumed coincident with the whole mass of the relevant steel plate (Fig. 1).

Moreover, the following assumptions may be made: (1) heat is generated by friction at each of the sliding interfaces, (2) heat conduction is assumed unidirectional, perpendicular to each concave surface, (3) heat loss due to radiation is assumed negligible, (4) bearing material (*PTFE* or the like) is assumed as a perfect thermal insulator so that heat generated at the sliding interface just flows towards each concave surface (in fact the thermal diffusivity of steel is ~ 20.0 W/m $\cdot^\circ$ C that is much larger than the one of PTFE ~ 0.24 W/m $\cdot^\circ$ C at 20 °C temperature).

The rate of heat exchange $\dot{Q}_i(T, t)$ by the i-th surface with the surrounding environment can be modelled by the Newton's law of cooling that is usually adopted to describe convective heat exchanges, that is:

$$\dot{Q}_i(T, t) = H_i \cdot (T_i - T_{air}) \tag{10}$$

where: H_i is the i-th concave surface heat transfer coefficient (*heat/temperature $\cdot$ time*), given by the product of the steel convectional heat exchange coefficient h_s times the area A_i of the portion of the concave surface effectively exchanging heat with the surrounding air whose temperature is T_{air}.

8   Conclusions

The so called Friction Pendulum Devices are complex tribological systems, composed of several kinematic couples, whose mechanical behavior is far from trivial. Many aspects, among which those related to the *stick-slip* and *sprag-slip* phenomena, still need to be clarified possibly involving both mechanical and tribological experts.

Moreover, most of the theoretical works available in the literature, aimed at reproducing the force-displacement hysteretic curve, have completely neglected both rotation and vertical translation of such devices during a horizontal ground motion. And they also seem to have not paid due attention to the geometrical compatibility.

In the present work, a new approach to model the mechanical behavior of such devices was proposed. It started from analyzing the likely geometrical compatibility of such devices during their functioning and ended up proposing a two-step deformation that restores geometric compatibility at each time step and also accounts for the possibility of stick-slip to occur. The main features of such approach, which is based on (1) geometrical compatibility, (2) kinematical compatibility, (3) dynamic equilibrium, and (4) thermo-mechanical coupling, were delineated.

For the time being, only the formal aspects of this approach were herein presented, and limited to the planar deformation of a DFP. As further developments, such approach will be extended to the tridimensional behavior of both DFP and TFP. This will be done paying due attention to the tribological issues of (a) wear, (b) thermal-induced effects, and (c) selection of the most suitable liner. Also questions related to the impact between the several components will be addressed.

References

American Society for Metals (1992) Friction, lubrication, and wear technology, vol. 18. ASM Handbook, Metals Park, Ohio

Becker TC, Mahin SA (2012) Experimental and analytical study of the bi-directional behavior of the triple friction pendulum isolator. Earthquake Eng Struct Dynam 41(3):355–373

Becker TC, Mahin SA (2013) Correct treatment of rotation of sliding surfaces in a kinematic model of the triple friction pendulum bearing. Earthquake Eng Struct Dynam 42(2):311–317

Belfiore NP, Di Benedetto A, Pennestrì E (2000) Fondamenti di meccanica applicata alle macchine. Casa Editrice Ambrosiana, Milano

Bowden FP, Tabor D (1973) Friction; an introduction to tribology. Heinemann, London

Constantinou MC, Tsopelas P, Kasalanati A, Wolff ED (1999) Property modification factors for seismic isolation bearings. Technical Report MCEER-99-0012, Multidisciplinary Center for Earthquake Engineering Research, State University of New York at Buffalo, Buffalo, NY

Constantinou MC (2004) Friction pendulum double concave bearing. NEES Report. http://nees.buffatlo.edu/docs/dec304/FP-DC%20Report-DEMO.pdf

Constantinou MC, Whittaker AS, Kalpakidis Y, Fenz DM, Warn GP (2007) Performance of seismic isolation hardware under service and seismic loading. Technical Report MCEER-07-0012, Multidisciplinary Center for Earthquake Engineering Research, State University of New York at Buffalo, Buffalo, NY

Earthquake Protection Systems (EPS), Inc. (2003) Technical characteristics of friction pendulum bearings. Technical Report, Vallejo, California

Fenz DM, Constantinou MC (2006) Behaviour of the double concave friction pendulum bearing. Earthquake Eng Struct Dynam 35:1403–1424

Fenz DM, Constantinou MC (2008) Spherical sliding isolation bearings with adaptive behavior: experimental verification. Earthquake Eng Struct Dynam 37:185–205

Italian Technical Regulations (2008) Norme Tecniche per le Costruzioni, NTC, Italy

Kelly JM (1997) Earthquake-resistant design with rubber, 2nd edn. Springer-Verlag, Berlin, New York

Mellon D, Post T (1998) Caltrans bridge research and applications of new technologies. In: Proceedings of U.S.-Italy workshop on seismic protective systems for bridges, Multidisciplinary Center for Earthquake Engineering Research, Buffalo, NY

Mokha A, Constantinou MC, Reinhorn AM (1990) Experimental study and analytical prediction of earthquake response of sliding isolation system with spherical surface. Technical Report NCEER-90-0020, National Center for Earthquake Engineering Research, University at Buffalo, SUNY, Buffalo

Nagarajaiah S, Reinhorn AM, Constantinou MC (1991) Nonlinear dynamic analysis of 3D base isolated structures. ASCE J Struct Eng 117(7):2035–2054

Popov V (2010) Contact mechanics and friction – physical principles and applications. Springer, Berlin

Ray T, Sarlis AA, Reinhorn AM, Constantinou MC (2013) Hysteretic models for sliding bearings with varying frictional force. Earthquake Eng Struct Dynam 42(15):2341–2360

Scotto Lavina G (1990) Riassunto delle Lezioni di Meccanica Applicata alle Macchine. Edizioni Scientifiche Siderea, Roma

Shabana AA (2001) Computational dynamics. John Wiley & Sons, New York

Taniwangsa W, Kelly JM (1996) Experimental and analytical studies of base isolation applications for low-cost housing, UCB/EERC-96/04. Earthquake Engineering Research Center, University of California, Berkeley, California

Tsai CS, Lin YC, Su HC (2010) Characterization and modeling of multiple friction pendulum system with numerous sliding interfaces. Earthquake Eng Struct Dynam 39(13):1463–1491

Zayas VA, Low SS, Mahin SA (1987) The FPS earthquake resisting system, Report No. 87-01, Earthquake Engineering Research Center, Berkley, CA

MID1.0: Masonry Infilled RC Frame Experimental Database

F. De Luca[1(✉)], E. Morciano[1,2], D. Perrone[2,3], and M. A. Aiello[2]

[1] Department of Civil Engineering, University of Bristol, Bristol, UK
flavia.deluca@bristol.ac.uk
[2] Department of Engineering for Innovation, University of Salento, Lecce, Italy
[3] Scuola Universitaria Superiore IUSS di Pavia, Pavia, Italy

Abstract. Experimental campaigns are a key tool for the evaluation of the behaviour of Masonry Infilled Reinforced Concrete (RC) frames. The case of masonry infills in RC frames is also a regional feature making the homogenous classification for a database a real challenge. This first attempt of a Masonry Infill Database (MID) includes a preliminary selection of experimental tests carried out on models of masonry-infilled RC frames under quasi-static or pseudo-dynamic loading. Each test is characterized in order to include, in a homogenous framework, all the relevant aspects of different experimental campaigns for easy access to the data for future applications. A Damage Classification is introduced, valid for both solid infill panels and for infill with openings. Finally, all monotonic backbones are fitted with a force-displacement piecewise linear approximation for future applications in Performance Based Earthquake Engineering.

Keywords: Reinforced Concrete · Masonry infills · Experimental Test Database

1 Introduction

The role played by masonry infills in reinforced concrete (RC) structures has been the subject of several studies in the last decades. Infills' presence is very common in the design practice of many seismic prone regions, especially in Mediterranean countries, and in general in Europe. A proper understanding of their seismic behaviour is required in order to reduce the risk of human and economic losses associated with seismic collapses.

Many studies have pointed out how significant is the influence of infills on the seismic response of structures. If, on one hand, this contribution increases the stiffness and the strength of structures, on the other, it may lead to some undesired consequences, such as the activation of failure modes associated with the interaction between the infill and the RC frame.

The interaction of the infill panel with the frame at global and local level can result in flexural or shear failures of the boundary RC structure. The above failures have often been observed during in-field campaigns after earthquakes and they have also been strongly evident in static and dynamic experimental tests (e.g., Mehrabi et al. 1996, FEMA306 1998; Fardis et al. 1999).

© Springer International Publishing AG, part of Springer Nature 2018
M. di Prisco and M. Menegotto (Eds.): ICD 2016, LNCE 10, pp. 147–160, 2018.
https://doi.org/10.1007/978-3-319-78936-1_11

Infill distribution in plan and elevation on the structure must be taken into account against the potential formation of soft storey mechanisms. Examples of progressive soft storey collapses likely caused by local interaction with infills are provided in the literature (e.g., Verderame et al. 2011, Negro and Colombo 1997). Last, but not least, one of the most important aspects related to this structural typology is the difficulty in performing quality controls of the materials and especially of the manufacturing process.

At the present time, there are several analytical and numerical studies on the behaviour of masonry infilled RC frames. Nevertheless, their "composite" nature imposes the necessity of ever newer and more sophisticated models and more reliable techniques to define their characteristics. Despite recent advances in computational power, a finite element micro-model (FEM) approach can be too intensive, especially for seismic risk assessments demanding numerous simulations. Instead, the equivalent strut macro-models seem to be a convenient approach. These simplified models replace the single panel with an equivalent strut having stiffness and width proportional to the characteristics of masonry, and the same thickness. Numerous equations have been proposed for the evaluation of strength and stiffness of the strut, but there is limited consensus on which equation is the most reliable; this is due to the limited systematic experimental experience available (Chrysostomou and Asteris 2012).

The assessment of the seismic performance and vulnerability is affected by several uncertainties, associated with the nature of the infill masonry and their interaction with the RC frame. Therefore, experimental campaigns turn out to be a key tool for the evaluation of the behaviour of the structural complex and to study the interaction between the infill panel and the frame. Unfortunately, it is quite rare to have a single, homogeneous experimental campaign fulfilling the wide variety of infills that can be found in practice. As a result, first, it is important to collect the experimental efforts to have a broader quantitative and controlled picture of the effect provided by infills to the structural behaviour of RC frames. This is the main framework of this study.

Based on the recent experience gained in structural and earthquake engineering, the idea of putting together a proper and, to some extent, fulfilling, database of experimental results on infilled RC structures is the main challenge herein.

For instance, in recent codes it is possible to find empirical formulations for the modelling of flexural and shear behaviour of RC elements, based on experimental databases previously collected by scientists and progressively updated over the years (e.g., Panagiotakos and Fardis 2001; Berry et al. 2004; Biskinis and Fardis 2010a, 2010b; Biskinis et al. 2004).

An experimental database of masonry infilled RC frames is a challenge far beyond the other experiences available in literature made for only the RC members. In fact, in this case the experimental campaign can vary significantly in many aspects (e.g., scale of the specimen, number of storeys, openings, etc.). Additionally, it is fundamental to account for the variety and variability of masonry, concrete, reinforcements, and finally their relative properties. Attempts to collect and homogenize experimental campaigns available in literature have been made recently (Cardone and Perrone 2015; Sassun et al. 2016) aimed at the use in loss estimation problems. The key issue is that a slight change in the above characteristics can change the relative ratios between relevant variables such as strength or stiffness, ending up in a significantly different behaviour (and mode of failure) for the masonry infilled RC unit.

The results of experimental campaigns have been fundamental not only to enlarge the experimental knowledge, but also for the further processes of calibration and comparison of the proposed models and predictive strategies. After an extensive analysis of all the experimental data available in literature and of the most effective techniques of modelling used for masonry-infilled RC frames, the significant properties and characteristics were selected in terms of:

- Materials used for the campaign;
- Geometric characteristics of the specimens;
- Characteristics, setup and scope of the test;
- Results of the campaign.

As it can be recognized by a first overview of literature sources on experimental tests on this topic, unfortunately not all the experimental campaigns are carried out or, at least documented, with the same level of accuracy. Therefore, after a preliminary selection of a large list of experimental references on the topic, it was necessary, at this preliminary stage, to select those sources that were significant, described in detail, and providing high quality data. In most cases, the refinement of the selection has ended up in the selection of recent studies (high quality of photos and artworks), even if some literature "milestones" are kept and considered first given their historical relevance and also given the significance they had for further experimental tests set up (e.g., Mehrabi et al. 1994).

The above approach was the best one for the selection of the proper entries for the database. Following up this preliminary step, it will be possible to include progressively all the references not included in this first step.

Four main objectives have been considered in this first version: (i) to include in the database a significant (in terms of number and variety) selection of the experimental tests carried out on scale models of masonry-infilled RC frames under quasi-static or pseudo-dynamic loading; (ii) to characterize each test in order to include all the significant aspects of each campaign for easy access to the data for future applications; (iii) to introduce a "Damage Classification" based on previous classifications and based on the analysis of the collected data, valid for both solid infill panels and for infill panels with openings; (iv) to model a force-displacement capacity backbone for the RC +masonry infill complex through the piecewise linear approximation by De Luca et al. (2013), aimed at different large scale and detailed earthquake and structural engineering applications. Even if the piece-wise linear fit cannot be of immediate use for calibration of infills analytical models, it can provide relevant results for large-scale approaches in which the full backbone of the complex RC+masonry infill is of interest.

Furthermore, the way the characteristics of each test are influencing the piece-wise linear fit can still provide useful insights for infill macro modelling efforts.

2 The Database

The Masonry Infill Database, in its current version, (MID1.0) includes the results of 16 experimental campaigns, collecting the results of 114 tests, in which 13 are bare tests (included for benchmarking purposes) and 101 are tests of masonry infills.

In Table 1, the campaign identification number (ID), the principal references, and the additional sources of information are listed for each experimental campaign included in the database. In many cases, in fact, it was necessary to retrieve complete information from other sources with respect to the principal one. Each test has a numerical ID based on the ID of the campaign and a progressive numbering for the specimen (e.g., 3_3, is the third specimen of the campaign by Mehrabi).

Table 1. References' list of the 16 experimental campaigns considered in MID1.0

ID	Reference	Additional sources
1	Kakaletsis and Karayannis (2009)	Kakaletsis and Karayannis (2008)
2	Haris and Hortobágyi (2015)	–
3	Mehrabi et al. (1994)	Mehrabi et al. (1996)
4	Crisafulli (1997)	Crisafulli et al. (2005)
5	Colangelo (2003)	Biondi et al. (2000) Colangelo (2005)
6	Colangelo (1996)	Biondi et al. (2000) Colangelo (2005)
7	Colangelo (2004)	Biondi et al. (2000) Colangelo (2005)
8	Al-Chaar et al. (2002)	–
9	Baran and Sevil (2010)	–
10	Calvi and Bolognini (2001)	Calvi et al. (2004)
11	Al-Nimry (2014)	–
12	Cavalier and Di Trapani (2014)	–
13	Basha and Kaushik (2012)	Basha and Kaushik (2016)
14	Zovkić et al. (2012)	Zovkic et al. (2013)
15	Pires and Carvalho (1992)	Skafida et al. (2014)
16	Kyriakides (2001)	Skafida et al. (2014)

The database has been collected in a Microsoft Excel© Spreadsheet with six separate sections. Section 1, SPECIMEN - summarizes specimen characteristics in terms of typologies of the tests, number of storeys, bays, test scale, type of vertical and horizontal loading applied, geometrical information about the presence of openings in the infill panel. Section 1 has 17 different entries. In Figs. 1, 2 and 3 the different distributions referred to the 113 specimens considered are provided for most relevant parameters of this section. Section 2, PANEL - summarizes all the characteristics of the infill panel and it has 23 entries including the characteristics of the masonry block, of the mortar and the masonry prism. Some entries in this section are often empty or not available (n.a.). In fact, the way the masonry infill is characterized geometrically and mechanically is really campaign-dependent and the high number of entries in this section allows MID1.0 being adaptable to the different choices that authors made.

Figures 4, 5 and 6 shows the distributions of some of the properties in Sect. 2 of the database. In this case the total number of specimens considered is 101, discarding the

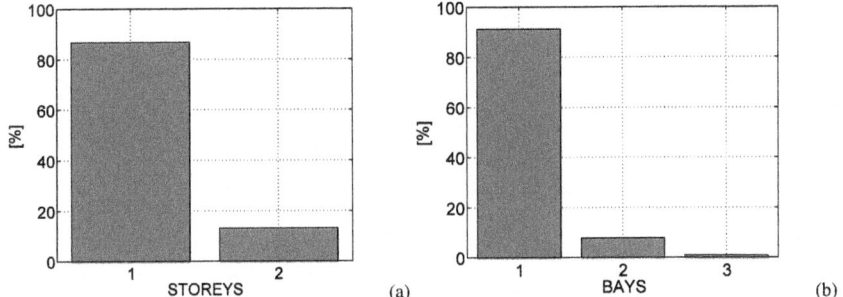

Fig. 1. (a) Number of storeys and (b) number of bays in MID1.0

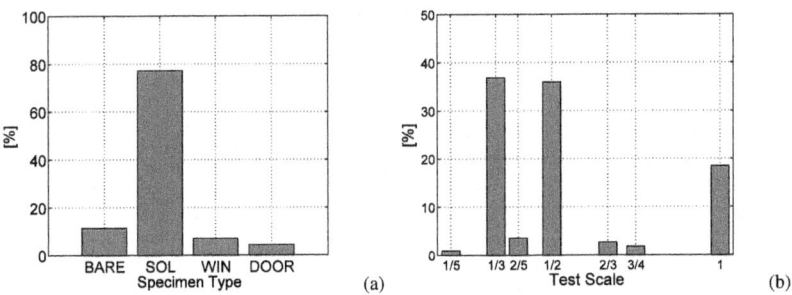

Fig. 2. (a) Specimen type BARE: bare, SOL: solid, WIN: window, DOOR: door (including information on the size of the opening), (b) test scale in MID1.0.

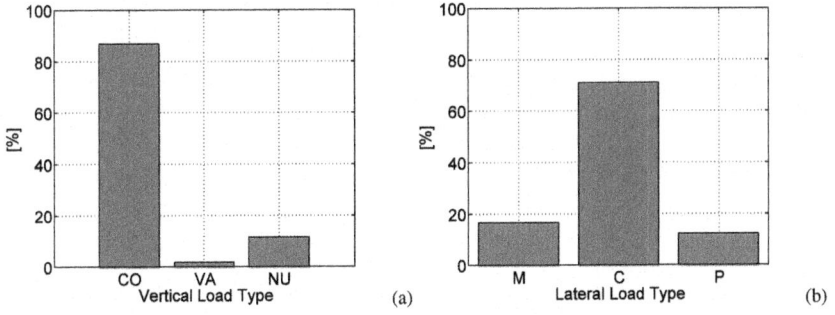

Fig. 3. (a) Vertical load type CO: constant, VA: variable, NU: null and (b) lateral load type M: monotonic, C: cylic (pseudo-static), and P(pseudo-dynamic) in MID1.0

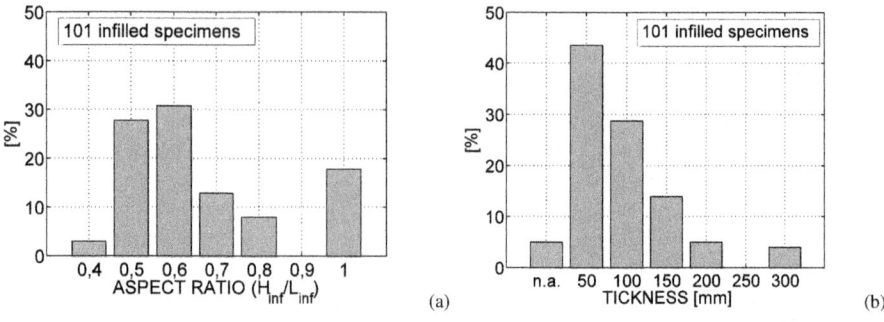

Fig. 4. (a) Aspect ratio, and (b) thickness of the infill panel in MID1.0

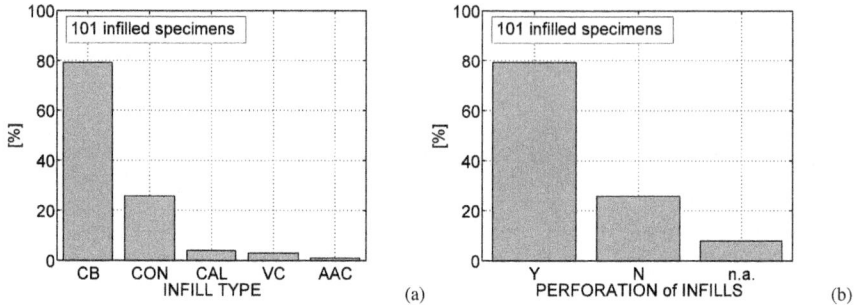

Fig. 5. (a) Infill type CB: clay bricks, CON: concrete, CAL: calcarenite, VC: vetro-ceramic, AAC: autoclaved aerated concrete, and (b) perforation of the infills Y: yes, N: no, n.a.: not available in MID1.0

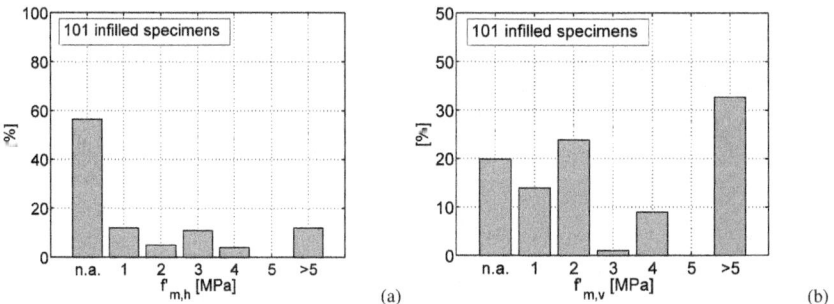

Fig. 6. (a) Horizontal and (b) vertical compressive strength of for the masonry infill prisms in MID1.0.

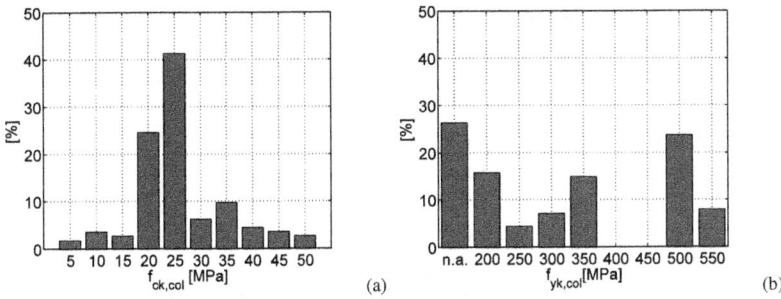

Fig. 7. (a) Concrete compressive strength ($f_{ck,col}$), and (b) steel yielding strength ($f_{yk,col}$) for RC columns in MID1.0

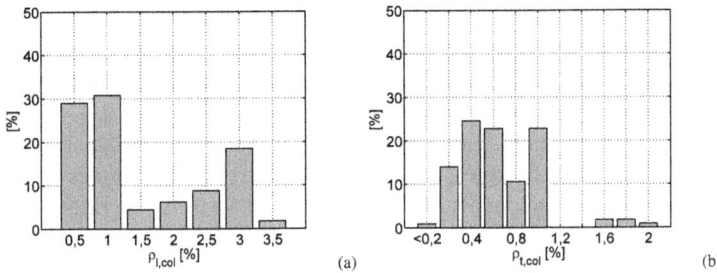

Fig. 8. (a) Longitudinal ($\rho_{l,col}$) and (b) transversal ($\rho_{t,col}$) reinforcement ratio of columns in MID1.0.

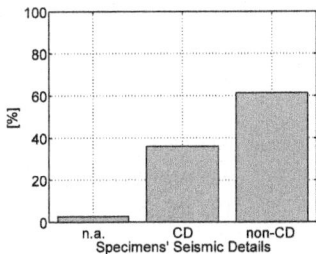

Fig. 9. Seismic detailing of the specimens CD: capacity-designed, non-CD: no capacity design, n.a.: not available in MID1.0.

13 bare tests included. Section 3, FRAME - summarizes all the characteristics of the reinforced concrete frame, including dimensions of columns, beams, base beam, and mechanical properties of concrete and reinforcement steel. This section has 25 entries. Figure 7 shows the distributions of two entries of this section; the compressive strength of concrete and yielding strength of steel in the columns.

Section 4, REINFORCEMENT - summarizes all the characteristics of longitudinal and transversal steel reinforcements in columns and beams of the RC frames. This section has 26 data entries. Columns' and beams' sections have been grouped in typologies in analogy to the approach employed in Berry et al. (2004). Figure 8 shows the geometric percentage of longitudinal and transversal reinforcement in columns. Figure 9 provides information on seismic detailing of the specimens. In this context "seismic detailing" should be interpreted as "capacity-designed"; i.e., conforming to general capacity design rules of modern codes.

2.1   Damage Classification and Failure Modes

Section 5, FAILURE MODE - includes information concerning the behaviour of the specimens to lateral loading and at collapse, it is referred only to the 101 infilled specimens. This section includes drifts at two different damage states (DS2 and DS3) and a novel classification of failure modes (it has three entries). This is an output section of the database; in fact, in many cases it was necessary to provide an interpretation of damage descriptions or, where possible, to carry out drift values from the force displacement behavior of the specimen.

The classification of failure modes for masonry infilled frames has been done in different studies such as Mehrabi et al. (1994), Asteris et al. (2011). In particular, specific attention was given to failure modes related to solid masonry infills, while only a few studies tried to classify the behaviour of masonry infilled frames with openings such as Kakaletsis and Karayannis (2008) or FEMA306 (1998). In order to generate a unique general arrangement, useful to characterize the behaviour of both solid infill specimens and those with openings, a novel failure mode classification is introduced in MID1.0. The classification is shown in Table 2 and visual representation of each failure mode is schematically shown in Fig. 10.

Table 2. Failure mode classification in MID1.0

ID	Failure mode	Component involved	Solid infill	Partial infilled
A	Corner crushing	Masonry panel	√	√
B	Diagonal cracking	Masonry panel	√	-
C	Sub-panel diagonal cracking	Masonry panel	-	√
D	Bed joints sliding	Masonry panel	√	√
E	RC frame failure	RC frame	√	√

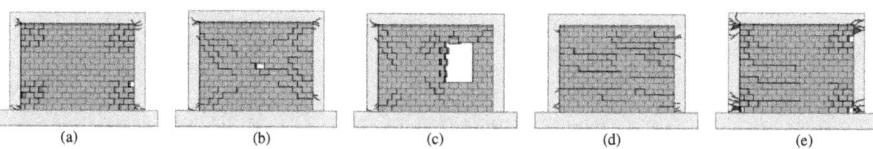

(a)   (b)   (c)   (d)   (e)

Fig. 10. Failure mode classification (a) corner crushing, (b) diagonal cracking, (c) sub-panel diagonal cracking, (d) bed joints sliding, (e) RC frame failure as classified in MID1.0

For each failure mode, a description is provided similarly to the approach of FEMA306 (1998). Description of *corner crushing* (A) is "complete loss of resistance of the blocks in the corners through spalling of face shells and diagonal cracking in the adjacent area. The damage can be spread into the whole panel but is more evident in the corners. This may lead to imminent formation of plastic hinges in the RC frame". *Diagonal cracking* (B) is "numerous diagonal cracks ranging from corner to corner of the infill panel, in both directions. Usually associated with slight crush of corners and/or with expulsion of blocks from the middle area of the panel". Description of *Subpanel Diagonal Cracking* (C) is "numerous diagonal cracks ranging from corner to corner of the sub-panel, usually from the edge of the opening to the corners. Usually associated with loss of stability of the opening, slight crush of corners in the sub-panel and/or with expulsion of blocks from the middle area".

Description of *bed joint sliding* (D) is "the main cracking pattern in the middle area of the infill is concentrated in a few (or a single) horizontal cracks in correspondence of bed joints, where the greatest displacement occurs. This failure mode could also happen through a more homogenous distribution of horizontal cracks". Finally, description of *RC frame failure* (E) is "the failure mode can happen by isolated collapse of a single element of the RC frame, typically related to a strength deficit, or after the formation of at least two out of the four previous failure modes. Almost uniform damage can be acknowledged in the infill, with complex cracking pattern, blocks spalling and large crushing".

In this first version of MID1.0, failure mode was defined only for the specimens in which the authors of the experimental campaigns were providing enough details to identify clearly the failure mode or in cases in which they were classifying the failure.

Failure mode classification is provided for 55 specimens only; Fig. 11 shows the distribution of failure modes excluding the cases in which this information was not available (i.e., 100% corresponds to 55 specimens). When the failure mode description of the specimen included characteristics of more than one failure mode and authors were mentioning more than one failure mode we considered hybrid categories (e.g., AB, AD, BE, CA, DA, DC, DE), in which the first letter identifies the dominant failure mode.

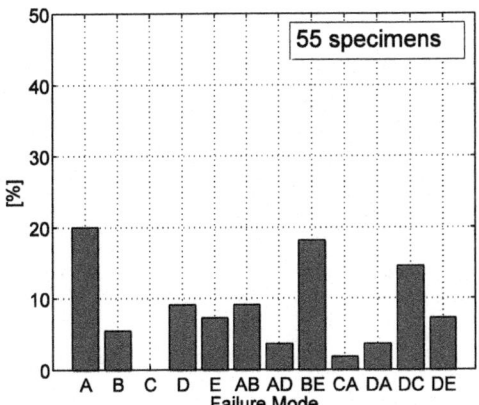

Fig. 11. Failure modes observed on 55 infilled specimens in MID1.0

Damage states (DS$_i$) were characterized and interpreted according to the EMS98 definition for damage to infills, see Grunthal et al. (1998) and De Luca et al. (2015) for further details. In particular, we defined as DS2, the so-called *moderate damage* (i.e., cracks in infill walls; fall of brittle cladding plaster; falling mortar from the joints of wall panels), and as DS3, the so-called *heavy damage* (i.e., large cracks in the infill wall). Damage state data were not available for all the specimens, DS2 was defined in 53 specimens, while DS3 was defined in 55 specimens over the total of 101 infilled specimens. In Figs. 12 and 13 are shown the distribution of DS2 and DS3 in MID1.0 over the total of specimens for which this information was provided.

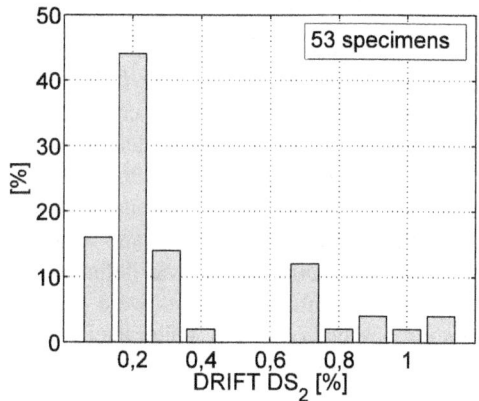

Fig. 12. DS2 drift of 53 infilled specimens in MID1.0

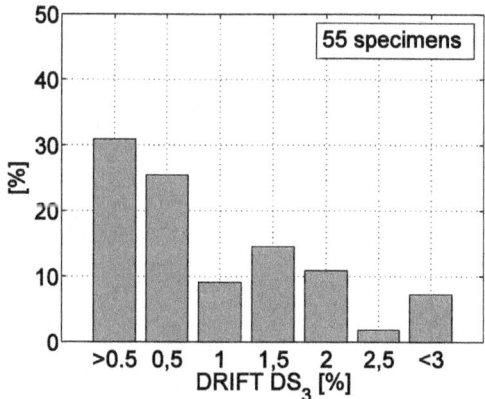

Fig. 13. DS3 drift of 55 infilled specimens in MID1

Section 6 of the database, PIECEWISE LINEAR MODEL includes the multi-linear fits of the monotonic envelope of each test included in the database. To do so, for each test, the monotonic envelope has been drawn. In the case of cyclic tests, positive and

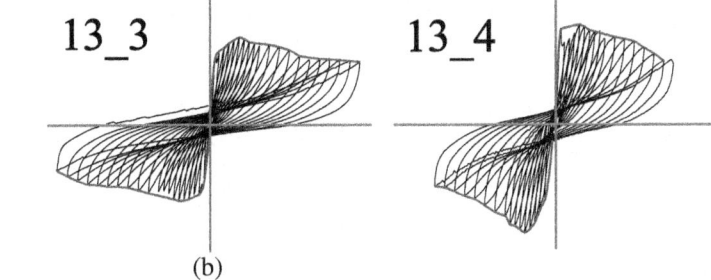

Fig. 14. Force-Displacement cyclic response of specimens 3 and 4 of campaign 13 (i.e., 13_3 and 13_4) and monotonic envelope produced in MID1.0 (adapted from Basha and Kaushik 2012 and 2016)

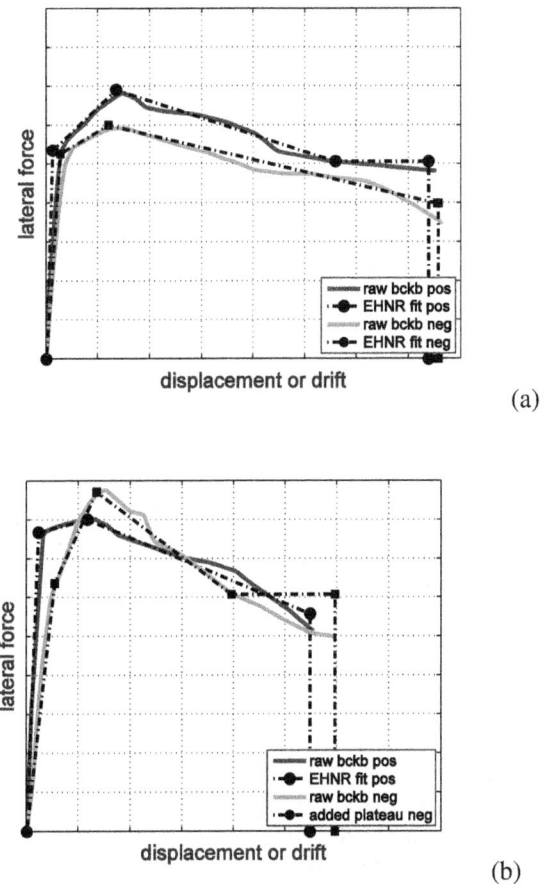

Fig. 15. Piecewise multi-linear fit of tests (a) 13_3 and (b) 13_4 in MID1.0

negative envelopes are considered. An example of the envelopes is shown in Fig. 14. This section of the database includes the coordinates of the five points of the fit for each monotonic branch of the test made according to the optimized algorithm by De Luca et al. (2013). All the 114 tests were fitted. For instance, Fig. 15 shows the multi-linear fits of positive and negative branches of the two tests in Fig. 14.

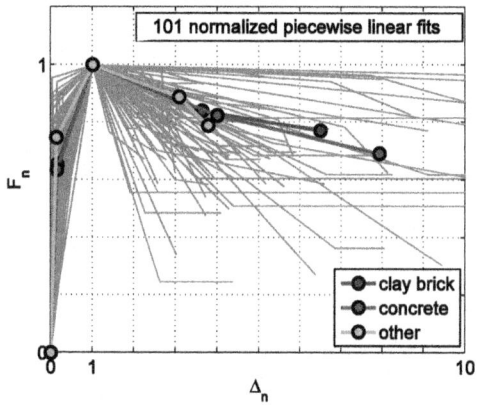

Fig. 16. Average normalized backbones per infill typology in MID1.0

The fits of the 101 infilled test (averaging positive and negative fits when both were available) were put together to have a preliminary normalized backbone shape for the test included in MID1.0. Figure 16 shows the average normalized shapes in MID1.0 for different infill typologies (i.e., clay brick, concrete, other) as classified in Fig. 5a.

3   Conclusions

MID1.0 is the first attempt of database of experimental tests on masonry infilled RC frames. It includes 101 test data. In this preliminary version, the database provides damage data, a failure mode classification and a normalized fit of all the tests for future analytical applications.

References

Al-Chaar G, Issa M, Sweeney S (2002) Behavior of masonry-infilled nonductile reinforced concrete frames. J Struct Eng 128(8):1055–1063

Al-Nimry HS (2014) Quasi-static testing of RC infilled frames and confined stone-concrete bearing walls. J Earthq Eng 18(1):1–23

Asteris PG, Kakaletsis DJ, Chrysostomou CZ, Smyrou EE (2011) Failure modes of in-filled frames. Electr J Struct Eng 11(1):11–20

Baran M, Sevil T (2010) Analytical and experimental studies on infilled RC frames. Int J Phys Sci 5(13):1981–1998

Basha SH, Kaushik HB (2012) Evaluation of shear demand on columns of masonry infileld reinforced concrete frames. In: Proceedings of the 15th world conference on earthquake engineering, Lisbon, Portugal, 24–28 September 2012

Basha SH, Kaushik HB (2016) Behavior and failure mechanisms of masonry-infilled RC frames (in low-rise buildings) subject to lateral loading. Eng Struct 111:233–245

Berry M, Parrish M, Eberhard M (2004) PEER structural performance database user's manual (version 1.0). University of California, Berkeley

Biondi S, Colangelo F, Nuti C (2000) La risposta sismica dei telai con tamponature murarie. CNR Gruppo Nazionale per la Difesa dai Terremoti (in Italian)

Biskinis DE, Roupakias GK, Fardis, MN (2004) Degradation of shear strength of reinforced concrete members with inelastic cyclic displacements. ACI Struct J, 101(6):773–783

Biskinis D, Fardis MN (2010a) Deformations at flexural yielding of members with continuous or lap-spliced bars. Struct Concr 11(3):127–138

Biskinis D, Fardis MN (2010b) Flexure-controlled ultimate deformations of members with continuous or lap-spliced bars. Struct Concr 11(2):93–108

Calvi GM, Bolognini D (2001) Seismic response of reinforced concrete frames infilled with weakly reinforced masonry panels. J Earthq Eng 5(02):153–185

Calvi GM, Bolognini D, Penna A (2004) Seismic performance of masonry-infilled RC frames: benefits of slight reinforcements. Invited lecture to sesto congresso nacional de sismologia e engenharia sismica, pp 254–276

Cardone D, Perrone G (2015) Developing fragility curves and loss functions for masonry infill walls. Earthq Struct 9(1):257–279

Cavaleri L, Di Trapani F (2014) Cyclic response of masonry infilled RC frames: experimental results and simplified modeling. Soil Dyn Earthq Eng 65:224–242

Chrysostomou CZ, Asteris PG (2012) On the in-plane properties and capacities of infilled frames. Eng Struct 41:385–402

Crisafulli FJ (1997) Seismic behaviour of reinforced concrete structures with masonry infills, Ph. D. thesis, Department of Civil Engineering, University of Canterbury, 404 p

Crisafulli FJ, Carr AJ, Park R (2005) Experimental response of framed masonry structures designed with new reinforcing details. Bull N Z Soc Earthq Eng 38(1):19–32

Colangelo F (1996) Pseudo-dynamic seismic response of infilled RC frames designed for gravity loading. In: Proceedings of the 11th world conference on earthquake engineering, Acapulco, Mexico, 23–28 June 1996

Colangelo F (2003) Experimental evaluation of member-by-member models and damage indices for infilled frames. J Earthq Eng 7(01):25–50

Colangelo F (2004) Pseudo-dynamic seismic response of infilled RC frames designed for gravity loading. In: Proceedings of the 13th world conference on earthquake engineering, Vancouver, British Columbia, Canada, 1–6 August 2004

Colangelo F (2005) Pseudo-dynamic seismic response of reinforced concrete frames infilled with non-structural brick masonry. Earthq Eng Struct Dyn 34(10):1219–1241

De Luca F, Vamvatsikos D, Iervolino I (2013) Near-optimal piecewise linear fits of static pushover capacity curves for equivalent SDOF analysis. Earthq Eng Struct Dyn 42(4):523–543

De Luca F, Verderame GM, Manfredi G (2015) Analytical versus observational fragilities: the case of Pettino (L'Aquila) damage data database. Bull Earthq Eng 13(4):1161–1181

Fardis MN, Bousias SN, Franchioni G, Panagiotakos TB (1999) Seismic response and design of RC structures with plan-eccentric masonry infills. Earthq Eng Struct Dyn 28(2):173–191

FEMA 306 (1998) Evaluation of earthquake damaged concrete and masonry wall buildings (FEMA 306). Applied Technology Council, Redwood City

Grüntal G (1998) European macroseismic scale EMS-98. European Seismological Commission, Sub-commission on Engineering Seismology, Working Group Macroseismic Scales, Luxembourg

Haris I, Hortobágyi Z (2015) Comparison of experimental and analytical results on masonry infilled RC frames for cyclic lateral load. Period Polytech Civil Eng 59(2):193

Kakaletsis DJ, Karayannis CG (2008) Influence of masonry strength and openings on infilled R/C frames under cycling loading. J Earthq Eng 12(2):197–221

Kakaletsis DJ, Karayannis CG (2009) Experimental investigation of infilled reinforced concrete frames with openings. ACI Struct J 106(2):132

Kyriakides MA (2001) Seismic retrofit of unreinforced masonry infills in non-ductile reinforec concrete frames using engineered cementitious composites. Dissertation, Department of Civil and Enviromental Engineering, Stanford University, Stanford

Mehrabi AB, Shing PB, Schuller MP, Noland JL (1994) Performance of masonry-infilled R/C frames under in-plane lateral loads. Rep. CU/SR-94, p 6

Mehrabi AB, Benson Shing P, Schuller MP, Noland JL (1996) Experimental evaluation of masory-infilled RC frames. J Struct Eng 122(3):228–237

Negro P, Colombo A (1997) Irregularities induced by nonstructural masonry panels in framed buildings. Eng Struct 19(7):576–585

Panagiotakos TB, Fardis MN (2001) Deformations of reinforced concrete members at yielding and ultimate. Struct J 98(2):135–148

Pires F, Carvalho EC (1992) The behaviour of infilled reinforced concrete frames under horizontal cyclic loading. In: Proceedings of the 10th world conference on earthquake engineering, Madrid, Spain, 9–24 July 1992

Sassun K, Sullivan TJ, Morandi P, Cardone D (2016) Characterising the in-plane seismic performance of infill masonry. Bull N Z Soc Earthq Eng 49(1):100–117

Skafida S, Koutas L, Bousias SN (2014) Analytical modeling of masonry infilled RC frames and verification with experimental data. J Struct 2014:1–17

Verderame GM, De Luca F, Ricci P, Manfredi G (2011) Preliminary analysis of a soft-storey mechanism after the 2009 L'Aquila earthquake. Earthq Eng Struct Dyn 40:925–944

Zovkić J, Sigmund V, Guljaš I (2012) Testing of R/C frames with masonry infill of various strength. In: Proceedings of the 15th world conference on earthquake engineering, Lisbon, Portugal, 24–28 September 2012

Zovkic J, Sigmund V, Guljas I (2013) Cyclic testing of a single bay reinforced concrete frames with various types of masonry infill. Earthq Eng Struct Dyn 42(8):1131–1149

The SonReb Method: Critical Review and Practical Aspects

G. Uva, F. Porco, and A. Fiore[(✉)]

DICATECh, Politecnico di Bari, Bari, Italy
andrea.fiore@poliba.it

Abstract. Current regulations require surveys on materials in order to identify one or more representative values of the in-situ concrete strength. In this context, the SonReb method is widespread. It correlates the in-situ concrete strength with the ultrasonic pulse velocity and rebound-hammer index. The method improves the reliability of both non-destructive methodologies that are less reliable if considered separately. However, the method neglects the numerical dispersion of the acquired resistances, making uncertain the reliability of every representative value identified. Uncertainty is inherent not only in the variability of the parameters that determine the values, but also in the use of literature formulations calibrated on "recurring concretes", i.e. concretes characterized by properties evenly variables (age, w/c ratio, …) that don't allow wider use. In this work, a critical review of the SonReb method is proposed, through a purely statistical approach with the aid of surveys on some school buildings in the province of Foggia built in the 60s and 80s.

Keywords: SonReb method · In-situ concrete strength · Material surveys
Safety assessment of existing buildings

1 Introduction

In the last years, the building development has slowly given way to the need to recover and preserve the large Italian existing heritage. This change of direction is confirmed by the current codes and national guidelines, which devote large sections to the issue of safety of existing buildings. With reference to the Reinforced Concrete structures (RC), a primary aspect of the cognitive process is certainly represented by the activities related to mechanical characterization of the materials. The characterization of materials is obtained by performing Destructive Tests (DT). These are to "take samples and run compression tests up to failure." The DT methods (among them the core drilling is the most widespread and used) are technically more delicate to perform. They can cause damage to the investigated structural portion, and have a greater economic burden. Consequently, with the core drilling is possible to take only a limited number of samples. The results could be unrepresentative of the overall characteristics of the structure in order to estimate the concrete strength. In addition, the values obtained on the same structural element can have a higher variability (Masi and Chiauzzi 2013). The sampling can be integrated with Non Destructive Testing (NDT) if the results are calibrated on those obtained by DT (Ministerial Circular 617 of 2009). The most

© Springer International Publishing AG, part of Springer Nature 2018
M. di Prisco and M. Menegotto (Eds.): ICD 2016, LNCE 10, pp. 161–171, 2018.
https://doi.org/10.1007/978-3-319-78936-1_12

widespread NDT in the field of surveys on materials are the rebound-hammer test and the ultrasonic pulse velocity test. They are based on the measurement of "related" physical quantities (the Rebound-hammer Index "Ir" and Ultrasonic Pulse Velocity "Vus") that allow to determine the quantity of interest (the concrete strength) "indirectly" through empirically derived relationships. For these reasons, most of the regulations (DM 2008, FEMA 2000; ACI 1998) and several studies proposed in the last forty years propose to correlate the results obtained by means NDT with the resistance values supplied by drilled cores in same points (Mikulic et al. 1992; Mahlotra 1976). The purpose of correlating DT and NDT is twofold; the first is to determine the concrete strength in the structural areas where only the results of the NDT are available, in order to extend the investigated points, the second aims is to expand the data sampling from which to identify one or more representative resistances (Porco et al. 2014). In addition, the use of NDT is also indicative to take into account, more broadly, the effects due to laying (Uva et al. 2013), even during the construction phases (Uva et al. 2014).

In order to reduce the uncertainties deriving by correlation between DT and NDT methods, it is common practice to use the "Combined Methods". These methods correlate the measurements acquired by two or more non-destructive investigations (Breysse et al. 2008). Currently there are no references UNI EN dedicated to combined methods but only details about are contained in RILEM NDT4 (1993) which is the reference code for this approach. In the case of RC structures, between the combined methods it is widely used the SonReb method that correlates, by means numerical regressions, the in-situ concrete strength "$Rc, situ$" with the variables that characterize the NDT. The underlying concept of the method is the assessment of the concrete strength in a point by balancing the factors that influence in a divergent way Ir and V_{US}, which are unreliable if they individually correlated to the resistance of the concrete core. Although there are numerous correlations proposed in the literature, with reference to "standard" concretes or directly obtained by substantial experimental investigations, there are no numerical relations of general validity. Because usually the tested concretes are different in nature from those under evaluation. A generalized application of method requires the determination of some corrective coefficients to take account of the above-mentioned differences (such as the content and type of cement, quality and size of the aggregates, the presence of additives …). Usually, in the case of existing buildings in reinforced concrete, these data are not available. Moreover, it should be noted, that the lack of a general correlation increases the possibility that the sampling is affected by a greater numerical dispersion. A significant statistical dispersion makes the average value unreliable to be hired as a "representative" of the in-situ resistance.

In accordance with the provisions of FEMA 356 (2000), several studies (Fiore et al. 2013; Pucinotti 2013), have shown the importance of checking the coefficient of variation "CV", i.e. the ratio between the standard deviation σ and the absolute value of the arithmetic average of the resistance values μ.

Given the uncertainties arising from the several correlations of the technical literature and their effects on the numerical dispersion of information acquired through DT and NDT investigations, this paper collects some critical considerations on the reliability of SonReb method for assessment of in-situ concrete strength. The considerations are based on the statistical treatment of the data collected from experimental surveys on the concrete carried out in the context of the seismic vulnerability assessment of some school buildings in the Province of Foggia (Italy). The comparative analysis between the numerical formulations normally used for the identification of the concrete strength and regressions on the examined concretes, will highlight the limitations of the method and the peculiarities of the analytical formulations more common.

2 SonReb: State of Art

The SonReb method (Sonic + Rebound) was born in the 60's from an idea of Facaoaru to combine the Rebound-Hammer Index I_r and Ultrasonic Pulse Velocity V_{US}. The underlying concept is to combine the results of two kinds of survey, which by the same factor in different way are often influenced, in order to improve the estimate of the in-situ concrete strength. For example, a high moisture content leads to underestimate I_r and, conversely, to overestimate the V_{US}, while by increasing the age of concrete (predominant feature for particularly dated existing buildings), there is the opposite tendency (Masi 2005).

The concrete strength is usually estimated by multi-regression analysis using as dependent variable the cubic resistance of RC concrete, obtained from tests on drilled cores ($f_{core}/0.83$), and as independent variables: I_r and V_{US}. Over the years, many authors have researched a universal law to be applied for any concrete. Some authors have combined the data samples of other authors thus obtaining consistent data set, without finding, however, the univocal law (Qasrawi 2000). Among various proposals, it is possible to identify the most frequently used expressions to correlate the above variables, which in their generical forms are shown below.

$$R_c = a \cdot V_{us}^b \cdot I_r^c \tag{1}$$

$$R_c = a + bV_{us} + cI_r \tag{2}$$

$$R_c = a \cdot e^{bV_{us}} \cdot e^{cI_r} \tag{3}$$

where, the Eq. 1 expresses the law in the form called "*power-power*" (Cianfrone and Facaoaru 1979), the Eq. 2 in the form "*bilinear*" (MacLeod 1971) and the Eq. 3 in "*double exponential*" form (Wiebenga 1968). a, b and c parameters represent the coefficients of the numerical regression. There are also those forms "polynomial" in three or four parameters (Bellander 1977) examined in this study for comparison purposes but which have not found widespread use.

All numerical formulations, in technical rules and reference guides, purely for concrete with "recurring features" were developed, i.e. concretes that are affected by variable factors roughly evenly (age, curing, w/c ratio, …). It is therefore well-established notion that none of the available expressions has a general validity. In fact, they provide resistance values with differences of the order of 30–40%. The choice of the value to be attributed to the concrete strength becomes very difficult. In addition, many formulations are specifically not valid for low-quality concrete (because of the low values of the ultrasonic pulse velocity in the material). Therefore, in the case of existing buildings, where the concretes often have different characteristics from those of new construction. Most frequently, the existing concretes have low resistance; consequently, the already uncertain reliability of some SonReb formulations can easily fail (Faella et al. 2011).

In Table 1, the formulations used for the purposes of the present work are shown. The most of them referable to the 1–3 general equations.

Table 1. Main SonReb formulations of technical literature.

Expression	Author/s
$R_C = 9.27 \cdot 10^{-11} V_{us}^{2,6} I_r^{1,4}$	RILEM – NDT4 (1993)
$R_C = 1.2 \cdot 10^{-9} V_{us}^{2,446} I_r^{1,058}$	Di Leo and Pascale (1994)
$R_C = 8.06 \cdot 10^{-8} V_{us}^{1,85} I_r^{1,246}$	Gasparik (1992)
$R_C = 8.925 \cdot 10^{-11} V_{us}^{2,6} I_r^{1,4}$	Dolce (2006)
$R_C = 2.756 \cdot 10^{-10} V_{us}^{2,487} I_r^{1,311}$	Bocca and Cianfrone (1983)
$R_C = 7.695 \cdot 10^{-11} V_{us}^{2,6} I_r^{1,4}$	Giochetti and Laquaniti (1980)
$R_C = -0.544 + 0.745 I_r + 0.951 V_{us}$	Tanigawa et al. (1984)
$R_C = 4.40 \cdot 10^{-7} \left(I_r^2 V_{us}^3 \right)^{0,5634}$	Del Monte and Lavacchini (2004)
$R_C = 1.453 \cdot 10^{-9} V_{us}^{2,6237} I_r^{0,5282}$	Brozovsky (2014)
$R_C = 1.6411 \cdot 10^{-9} V_{us}^{2,29366} I_r^{1,30768}$	Mulik et al. (2015)
$R_C = -34.51583 + 0.26511 I_r + 0.01385 V_{us}$	Faella et al. (2011)
$R_C = 0.00153 \left(I_r^3 V_{us}^4 \right)^{0,611}$	Arioglu and Koyluoglu (1996) richiamata da Erdal (2009)
$R_C = -25.568 + 0.000635 I_r^3 + 8.397 V_{us}$	Bellander (1979)
$R_C = -24.668 + 1.427 I_r + 0.0294 V_{us}^4$	Meynink and Samarin (1979)
$R_C = 0.0158 V_{us}^{0,4254} I_r^{1,1171}$	Kheder (1999)
$R_C = \frac{I_r}{(18.6 + 0.019 I_r + 0.515 V_{us})}$	Postacioglu (1985)

3   Case Study

With the OPCM 3274 (2003), the property owner is obliged to assess the seismic vulnerability of the own building. This obligation has sensitized researchers and engineers at the greater development of methods aimed at security checks. In this context, the Polytechnic of Bari and the Basin Authority of Puglia Region have signed an agreement aimed to carry out the seismic vulnerability assessments of the school buildings of the Province of Foggia. As part of the agreement, the University has drafted the "Guidelines for the assessment of the safety of public buildings with reinforced concrete or masonry structures". These to technicians in charge to carry out safety assessments were supplied, as a methodological guidance for the preparation of the required documentation.

The safety assessments have involved twenty school buildings with backbone in reinforced concrete and masonry. For the purposes of this study, the results of the investigation and not destructive of-structive collected on six school buildings (in blue

Fig. 1. Locations of the school buildings subject of surveys on materials. (Color figure online)

in Fig. 1) were considered. The buildings date back to the period from 1970 to 1980. They have mixes of the concrete presumably with similar properties because they are located within a small territorial perimeter.

The following Figs. 2, 3 and 4, the statistical size of the sampling through the distribution of the individual samplings of in-situ resistances, the rebound-hammer indexes and ultrasonic pulse velocities. Each population consists of #38 values.

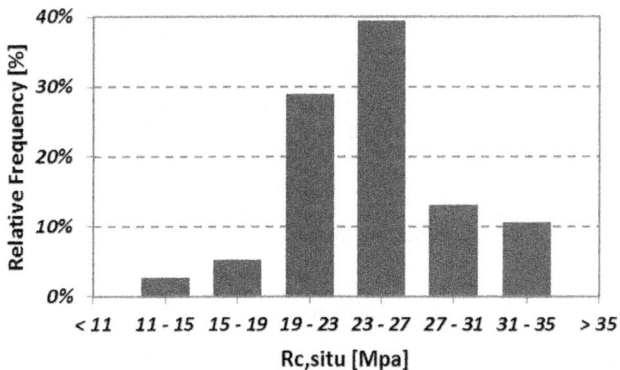

Fig. 2. In-situ concrete strengths distribution

By observing the distribution of velocities, the homogeneity of the concrete used in in-situ is apparent. In the range between 2800 and 4000 m/s (range that is attributable to a good quality concrete (Whitehurst 1951) fall within approximately 81% of the ultrasonic pulse velocities measured. The rebound-hammer indexes instead are, for more than 82%, greater than the value of 32, attributable according to technical references, to a good surface quality of concrete (Lenzi et al. 2010). For completeness, in Table 2, the main statistical parameters of the three samplings under examination.

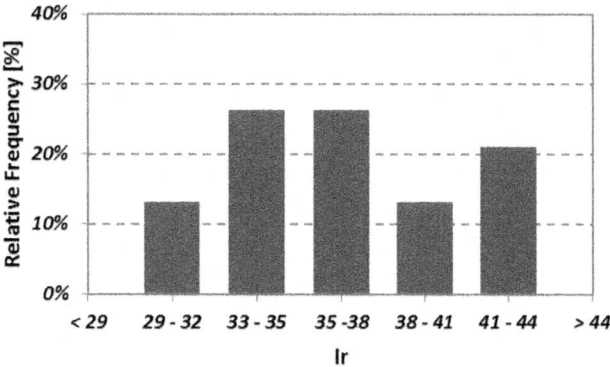

Fig. 3. Rebound-hammer indexes distribution

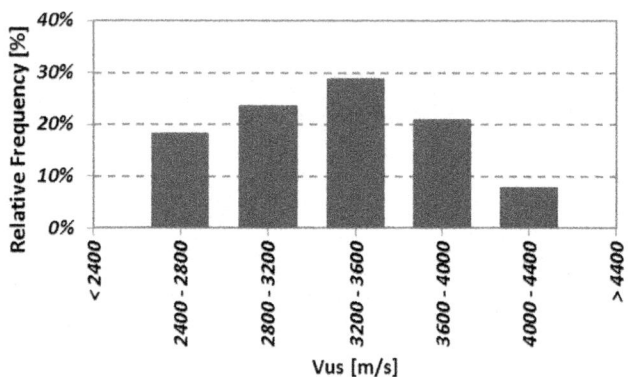

Fig. 4. Ultrasonic pulse velocities distribution

Table 2. Main statistical parameters of the considered samplings

	In-situ concrete resistances [MPa]	Rebound-hammer indexes	Ultrasonic pulse velocities [m/s]
μ	24.47	37.20	3300.65
σ	4.70	4.08	490.01
CV	19.2%	11.0%	14.8%

4   Results

4.1   Numerical Data Processing

On the basis of data populations described in the previous section, the SonReb method was applied by using the relationship between $R_{c,situ}$, I_r and V_{US} in according to generical analytic forms (see Eqs. 1–3). Furthermore, the method was applied by evaluating the concrete strengths with the multi-regression relationships collected in Table 1. The histogram of Fig. 5 shows the average values of the concrete strengths obtained according to the formulation determined by a specific multi-regression analysis on the sampling and expressions proposed in the technical literature. For comparison, within the histogram is reported the average value μ of the in-situ resistances "$Rc(carote)$" derived directly from the compression tests on cylindrical samples (drilled cores).

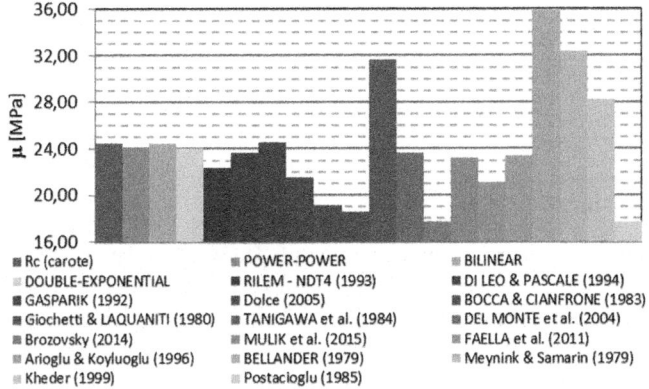

Fig. 5. Comparison between average concrete strengths

Some formulations provide representative average values very distant than those obtained with the Power-Power, Double Exponential or Bilinear numerical regressions. In particular, with the expression given by Bellander (1979) it is possible to obtain a resistance value equal to 35.99 MPa, while with the expression proposed by Brožovský (2014) a value of 17.60 MPa. Compared to the average value of the drilled cores equal to 24.47 MPa, percentage differences found with both relationships are, respectively, of +47% and −17%. In order to quantify the numerical distance from representative value measured by destructive tests, the estimate of standard deviation becomes important. The statistical parameter is a significant reference to quantify the actual degree of data dispersion with respect to a predetermined value. In this case, by assuming as the reference, the mean value of the resistances, the several formulations exhibit a considerable dispersion of the data with respect to the calibrated regressions on sampling (Fig. 6). In particular, the standard deviation is greater than 5 for all formulations collected in Table 1 with the exception of Postacioglu (1985), Kheder (1999) and Tanigawa et al. (1984).

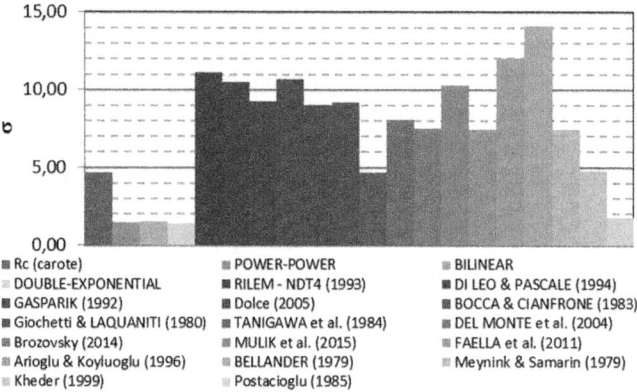

Fig. 6. Standard deviations

4.2 Some Remarks

The Italian and European rules identify three of Knowledge Levels (LC), in according to the number of tests performed. These represent the degree of knowledge of the building. Each level establishes the method of analysis to be used for security checks and the value of the "Safety Factor" to be applied to material resistances in conjunction with specific safety coefficients γ_m. This procedure allows to take into account the more or less detailed level as in the knowledge procedure (i.e. historical-critical analysis, geometrical survey, mechanical characterization of materials …). Therefore, the resistors used in static and seismic tests, are highly dependent on the number of data collected during the tests on the materials and the estimate of the average representative values expressed in terms of cylindrical compressive strength. This approach tends to neglect the numerical dispersion of the concrete elements. Dispersion is a recurring feature of the generic sampling data collected in surveys carried out on materials. Therefore, the estimated mean value is often non-unique and representative of the in-situ concrete, because it derives by a set of data having a high amplitude.

The FEMA 356 (2000) introduce a limitation on the use of the average value of the strength of the concrete in situ depending on the value assumed by the Coefficient of Variation CV. In particular, the use of the average value is allowed until CV $\leq$ 14%. The limit excludes the possibility that the value becomes representative of the whole data population under consideration Fig. 7 shows how the limitation indicated by FEMA 356 is frequently not satisfied.

In the case of SonReb method is possible to highlight how the use of known formulations involves of the data loss that are prohibitive in order to estimate the resistance value of the sampling. Only the estimate of formulations in the Power-Power, *Double Exponential* or *Bilinear* form calibrated directly on the population of resistances allows to identify a single representative value. Other important aspect lies in the advantage obtained by the use of NDT. The CV for the only drilled core resistances exceeds the limit of 14%, while using the SonReb method the

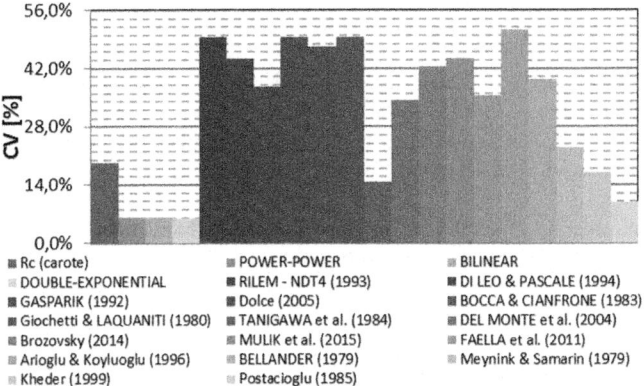

Fig. 7. Coefficients of variation

Coefficients are, respectively, equal to 6.0% (Power-Power), 6.1% (Double Exponential) and 5.8% (bilinear).

5 Conclusions

In this paper, with the aid of data collected from surveys carried on six school buildings in the Province of Foggia (Italy), some remarks about the reliability of the SonReb method were proposed. National and European standards push towards the identification of the average resistance value resulting from surveys on materials as parameter representative of the in-situ concrete. In this trend, there is the need to investigate if there are real benefits from the use of the combined method. The previously mentioned size, suitably remoulded in according to the Knowledge Level, is frequently, in the context of the safety assessments of existing buildings. However, based on homogeneous data was highlighted as the combined method is reliable only where it is directly calibrated on the same data sampling. Indeed, the application of literature formulations involves, as expected, coefficients of variation, and then, dispersions from the representative average value, very wide (even more than 50%).

By contrast, without resorting to "pre-packaged" formulations of literature, the use of the combined method is a valid aid for the structural checks. In fact, in the presence of a sufficient number of non-destructive tests, the adoption of the SonReb combined method reduces the numerical dispersion compared with the case constituted by the resistances arising from core samples (in the case study of about 10%).

Acknowledgements. The research presented in this article was partially funded by the Department of Civil Protection, Project ReLUIS-DPC 2014-2018.

References

American Concrete Institute (ACI) (1998) Non-destructive test methods for evaluation of concrete in structures, ACI 228.2R-98, Detroit, Michigan

Bellander U (1977) Concrete strength in finished structures: part 3. nondestructive testing methods. Investigation in laboratory in situ research. Sweden Cement Concrete Research Institute, 3:77

Bocca P, Cianfrone S (1983) Le prove non distruttive sulle costruzioni: una metodologia combinata. L'Ind Ital del Cem 6:429–436

Breysse D, Klysz G, Dérobert X, Sirieix C, Lataste JF (2008) How to combine several non-destructive techniques for a better assessment of concrete structures? Cem Concr Res 38:783–793

Brozovsky J (2014) Determine the compressive strength of calcium silicate bricks by combined nondestructive method. The Sci World J 2014:5 p. Article ID 829794, http://dx.doi.org/10.1155/2014/829794

Cianfrone F, Facaoaru I (1979) Study on the introduction into Italy on the combined non-destructive method for the determination of in situ strength. Mater Struct 12(5):413–424

Del Monte E, Lavacchini G, Vignoli A (2004) Modelli per la previsione della resistenza a compressione del calcestruzzo in opera. Ingegneria Sismica anno XXI no. 3

Di Leo A, Pascale G (1994) Prove non distruttive nelle costruzioni in c.a., II g. delle prove non distruttive (4)

Dolce M, Masi A, Ferrini M (2006) Estimation of the actual in-place concrete strength in assessing existing RC structures. In: Proceedings of 2nd international fib congress, Naples, Italy

Erdal M (2009) Prediction of the compressive strength of vacuum processed concretes using artificial neural network and regression techniques. Sci Res Essay 4(10):1057–1065

Faella C, Guadagnuolo M, Donadio A, Ferri L (2011) Calibrazione sperimentale del metodo SonReb per costruzioni della Provincia di Caserta degli anni '60–'80. In: Proceedings of 14th anidis conference, Bari, Italy

Federal Emergency Management Agency (2000) Prestandard for the seismic rehabilitation of buildings. FEMA 356, Reston Va

Fiore A, Porco F, Uva G, Mezzina M (2013) On the dispersion of data collected by in situ diagnostic of the existing concrete. Constr Build Mater 47:208–217

Gasparik J (1992) Prove non distruttive in edilizia, Quaderno didattico. AIPND, Brescia

Giochetti R, Laquaniti L (1980) Controlli non distruttivi su impalcati da ponte in calcestruzzo armato. nota tecnica 04. Università degli Studi di Ancona, Facoltà di Ingegneria, Istituto di Scienza e Tecnica delle Costruzioni, Ancona

Kheder GF (1999) A two stage procedure for assessment of in situ concrete strength using combined non-destructive testing. Mater Struct 32(6):410

Lenzi M, Versari D, Zambrini R (2010) Indagine Sperimentale di Calibrazione del Metodo Combinato SonReb. INARCOS, Bhavnagar

MacLeod G (1971) An assessment of two non-destructive techniques as a means of examining the quality and variability of concrete in structures. Rep. C1/sfB/ Eg/(A7q)UDC 666.972.017.620 179.142.454. Cement and Concrete Association, London

Malhotra VM (1974) Contract strength requirements - cores versus in situ evaluation. ACI J Proc 74(4):163–172

Masi A (2005) La stima del calcestruzzo in situ mediante prove distruttive e non distruttive. Il giornale delle prove non distruttive, N.1/2005

Masi A, Chiauzzi L (2013) An experimental study on the within-member variability of in situ concrete strength in RC building structures. Constr Build Mater 47:951–961

Meynink P, Samarin A (1979) Assessment of compressive strength of concrete by cylinders, cores and non-destructive tests. In: Proceedings on Quality Control of Concrete Structures RILEM Symposium, Session 2.1. Swedish Concrete Research Institute Stockholm, Sweden, pp 127– 134

Mikulic D, Pause Z, Ukraincik V (1992) Determination of concrete quality in a structure by combination of destructive and non-destructive methods. Mater Struct 25:65–69

Ministero delle Infrastrutture e dei Trasporti (2009) Circolare 2 febbraio 2009, n. 617: Istruzioni per l'applicazione delle "Nuove norme tecniche per le costruzioni" di cui al Decreto Ministeriale, 14 gennaio 2008

Ministero delle Infrastrutture e dei Trasporti (2008) DM 14/01/08: "Norme Tecniche per le Costruzioni", G.U. n. 29 del 4 febbraio 2008, Supplemento Ordinario n. 30

Mulik Nikhil V, Balki Minal R, Chhabria Deep S, Ghare Vijay D, Tele Vishal S (2015) The use of combined non-destructive testing in the concrete strength assessment from laboratory specimens and existing buildings. ISSN (PRINT): 2393-8374, (ONLINE): 2394-0697, vol 2, no. 5

Porco F, Uva G, Fiore A, Mezzina M (2014) Assessment of concrete degradation in existing structures: a practical procedure. Struct Eng Mech 52(4):701–721

Postacioglu B (1985) Nouvelle significations de l'indice sclerometrque schmidt et de la vitesse de propogation des ultra-sons. Mater. Struct. 447–451

Pucinotti R (2013) Assessment of in situ characteristic concrete strength. Constr Build Mater 44:63–73

Qasrawi HY (2000) Concrete strength by combined nondestructive methods simply and reliably predicted. Cem Concr Res 30:739–746

RILEM (1993) NDT 4 Recommendations for in situ concrete strength determination by combined non-destructive methods. Compendium of RILEM Technical Recommendations, E&FN Spon, London

Tanigawa Y, Baba K, Mori H (1984) Estimation of concrete strength by combined nondestructive testing method. ACI SP 82(1):57–65

Uva G, Porco F, Fiore A, Mezzina M (2013) Proposal of methodology for assessing the reliability of in-situ concrete tests and improving the estimate of the compressive strength. Constr Build Mater 38:72–83

Uva G, Porco F, Fiore A, Mezzina M (2014) The assessment of structural concretes during construction phase. Struct Surv 32(3):2–22

Wiebenga JG (1968) A comparison between various combined nondestructive methods to derive the compressive strength of concrete. In: Rep kB1-68- 61/1418. Inst TNO Veor Bouwmaterialen en Bouwconstructies. Delft

Whitehurst E (1951) Soniscope tests concrete structures. J Am Concr Inst 47:433–444

Innovative Structural Concretes with Phase Change Materials for Sustainable Constructions: Mechanical and Thermal Characterization

A. D'Alessandro[1](✉), A. L. Pisello[2,3], C. Fabiani[2], F. Ubertini[1],
L. F. Cabeza[4], F. Cotana[2,3], and A. L. Materazzi[1]

[1] Department of Civil and Environmental Engineering, University of Perugia,
Perugia, Italy
antonella.dalessandro@unipg.it
[2] Department of Engineering, University of Perugia, Perugia, Italy
[3] CIRIAF - Interuniversity Research Center, University of Perugia, Perugia, Italy
[4] GREA Innovació Concurrent, Universitat de Lleida, Lleida, Spain

Abstract. New phase change materials (PCMs) are promising fillers for the realization of multifunctional concretes, combining good mechanical properties with enhanced thermal storage capabilities within building envelope. These materials are currently receiving a growing interest in the scientific literature. Encapsulated PCMs result particularly suitable for applications in concrete. This paper presents a research on concretes doped with different contents of PCMs, up to the 5% of the total weight. Physical, mechanical and thermal experimental tests were carried out, in order to investigate the physical properties, the stress-strain behaviour, the ductility, the compressive strength, as well as the thermal conductivity, the diffusivity and the specific heat capacity of the novel concretes. The results of thermal tests demonstrated the effective enhancement of the thermal inertia of the materials, while mechanical tests showed performances compatible with structural applications. Overall, new multifunctional concretes with PCM inclusions appear promising for achieving sustainable and lightweight concrete structures.

Keywords: Phase change materials · Structural concretes
Cement-based composites · Stress-strain envelope · Thermal properties
Smart materials

1 Introduction

Novel nano- and micro-cementitious materials result a topic of growing interest in scientific literature (Shah et al. 2009; Ubertini et al. 2016). Cement-matrix materials result particularly suitable for a micro-structural modification, due to the presence of components and pores at different scales. The fillers can confer to the materials multifunctional properties, as new or enhanced capabilities (D'Alessandro et al. 2016a; 2016b; Laflamme et al. 2015; Yang and Che 2015). In particular, phase change

© Springer International Publishing AG, part of Springer Nature 2018
M. di Prisco and M. Menegotto (Eds.): ICD 2016, LNCE 10, pp. 172–183, 2018.
https://doi.org/10.1007/978-3-319-78936-1_13

materials (PCMs) possess high heats of fusion and, during the melting or solidifying transitions at a fixed temperature, are capable of storing or releasing energy (Sharma et al. 2009). Their use in building materials permits to optimize thermal-energy efficiency and construction sustainability of constructions (Kalnæs and Jelle 2015; Navarro et al. 2016; D'Alessandro et al. 2016a; 2016b). An optimal phase change material for concrete needs to exhibit appropriate mechanical performances, a proper phase change temperature and a great melting enthalpy. Also, the choice of such materials usually is also led considering other physical, technical and economic aspects, (Cabeza 2015; Fernandes et al. 2015). Research investigations present in literature mainly concern the capability of PCMs to improve the thermal performance of concrete, assessing whether the mechanical properties were severely flawed (Zhang et al. 2004; Lecompte et al. 2015). In order to ensure such acceptable properties, most of these studies focus on the integration of microencapsulated PCMs in cementitious admixtures, thus avoiding leakage problems which could negatively affect the compression resistance of the samples (Konuklu et al. 2015).

The present paper is aimed at investigating the mechanical and thermal performance of concretes with PCMs for structural applications in comparison to ordinary concrete. The paper is organized as follows. Firstly, a literature overview on the use of PCMs as multifunctional additives in concrete is presented in Sect. 1. Materials and samples developed within this work are presented in Sect. 2. The experimental methodology is described in detail in Sect. 3, while test results are presented in Sect. 4 and discussed in Sect. 5. Finally, the paper presents the main findings and conclusions.

2 Materials and Experimental Methods

2.1 Phase Change Materials

The PCMs utilized in the experimentation were Microtek Microencapsulated PCM. They appeared as a white powder (Fig. 1a) with a mean particle size between 17 and 20 μm. The core material inside the microcapsules, consisted of a inert polymer, was a paraffin-wax with heat absorption capabilities and a melting point at 18 °C. The PCM content was about 85–90% with respect to the total mass.

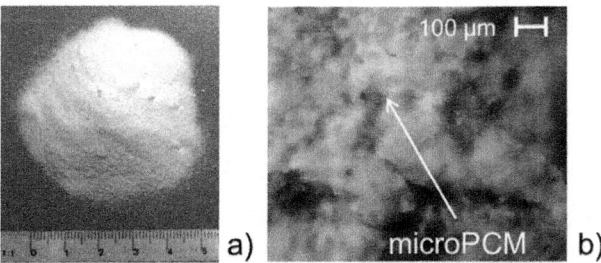

Fig. 1. (a) Appearance of MicroPCM; (b) Microscope enlargement of concrete with 5% of PCM

2.2   Concrete Preparation Procedures

Table 1 shows the components of the normal reference concrete and the concretes prepared with the addition of 1, 3 and 5% of PCMs with respect to the whole mass of the composite. The water/cement ratio was 0.45 for all the admixtures, except for 5% PCM-concrete, with a w/c ratio of 0.5, necessary to obtain similar workability. The cement was type 42.5, pozzolanic. The aggregates were constituted by sand and gravel with nominal dimensions from 0 to 4 mm and from 4 to 8 mm, respectively.

Table 1. Mix design of normal reference concrete and concretes with PCMs.

Components (kg/m^3)	Normal concrete	Concrete with PCM		
		1%	3%	5%
Concrete	524	511	486	447
Water	234	228	218	223
Sand (0–4 mm)	951	927	882	817
Gravel (4–8 mm)	638	622	592	548
PCM	-	24	71	102
Plasticizer	2.6	2.6	2.4	6.8
W/C ratio	0.45	0.45	0.45	0.5

The sample preparation process consisted in the preliminary mix of cement, sand, gravel and PCMs (Fig. 2a), and the subsequent addition of water and plasticizer (Fig. 2b). Then, the dough was mixed and carefully poured into oiled moulds, in order to avoid the separation of the fillers (Fig. 2c). The samples were unmolded for curing after some days (Fig. 2d). After mixing, microPCM capsules appeared intact, since no oil was observed in the composites.

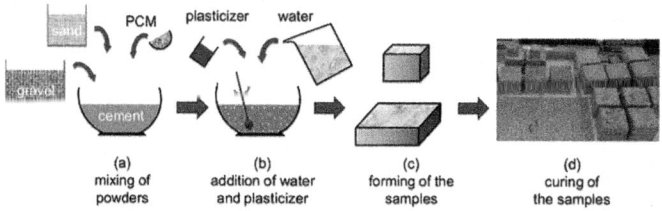

Fig. 2. Preparation process of concretes with PCMs

The samples were cubes with sides of 100 mm, for mechanical tests, and prisms with squared bases of 190 × 190 mm^2 and height of 50 mm, for thermal experiments. Figure 3 represents the structure of the normal and the PCM-concrete, with dispersed microcapsules. An enlarged image of a fragment of hardened concrete with PCMs can be observed in Fig. 1b. The microscope used for the elargements was type Bresser, Biolux NV, with magnifications up to 1280X.

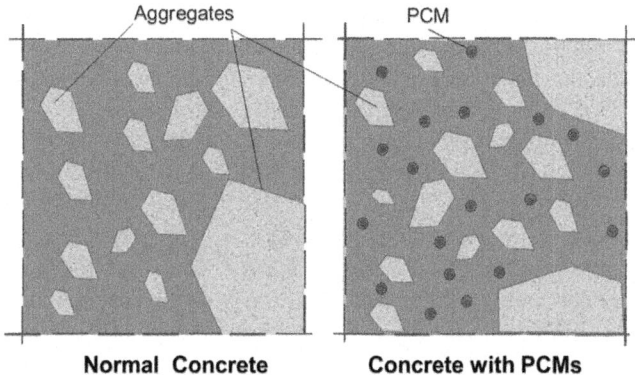

Fig. 3. Schematic representation of the structure of Normal Concrete and Concrete with microPCM

2.3 Experimental Setup

Mechanical Tests

After curing, the sample were measured and weight and then compression tests were carried out, using an Advantest machine, model Controls 50-C7600 (Fig. 4a) with a servo-hydraulic control unit model 50-C9842 (Fig. 4b), instrumented with three LVDT placed at 120° (Fig. 4c). The compressive loads were applied under displacement control, up to the collapse. Both resistance and ductility after peak have been investigated. The test speed was 2 µm/s, in agreement with the EN 12390-3 standard.

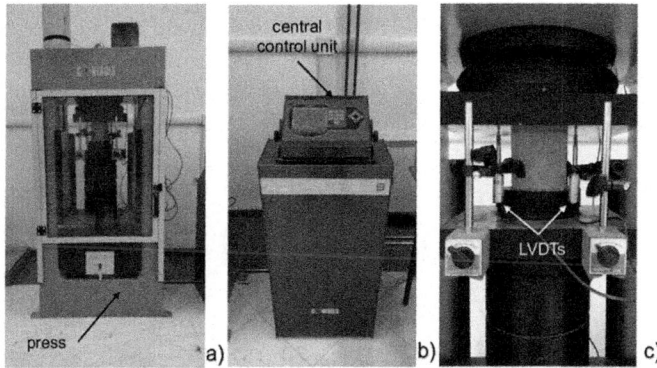

Fig. 4. Experimental setup for mechanical tests: (a) testing machine; (b) central control unit; (c) cubic concrete sample instrumented with LVDTs

Thermal Tests

The thermal investigations of the prismatic samples were carried out by means of Transient Plane Source (TPS) method, using a Hot Disk 2500 system, in accord with the ISO 22007-2 standard and an ATT DM340SR climatic chamber equipped with 12 PT100 thermocouples. In the environmental chamber the samples were insulated by XPS panels: just the upper side was exposed to the controlled environment of the chamber (Fig. 5a). Each sample was instrumented with four PT100: one on bottom surface, one on the upper surface and two probes on the lateral sides. The thermal program consisted of five subsequent segments, 8 h long, with linear varying temperature between 26 °C and 10 °C and a fixed RH value of 50%. The thermal tests in the climatic chamber were aimed at investigating the melting temperature of the PCMs within the concretes and the phase change enthalpy. The Hot Disk tests were carried out in a single sided configuration, installing a superinsulating material on the other side of the probe. These tests were aimed at assessing the thermal conductivity and the thermal diffusivity of the concretes, and to derive the specific heat.

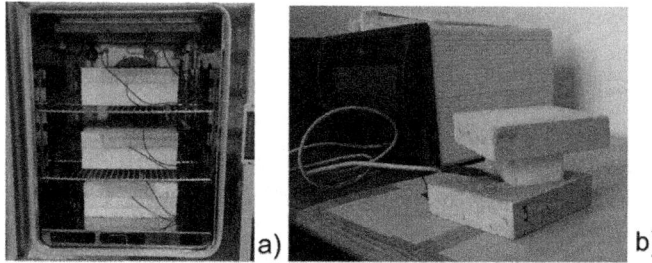

Fig. 5.   (a) Prismatic samples in the climatic chamber; (b) setup of the hot-disk

3   Mechanical Tests

The mass density of concretes with PCMs decreased with the increase of the filler content.

In particular, 5% PCM-concretes demonstrated a density reduction of about 11% with respect to normal concrete (Fig. 6a). So, they resulted suitable for the realization

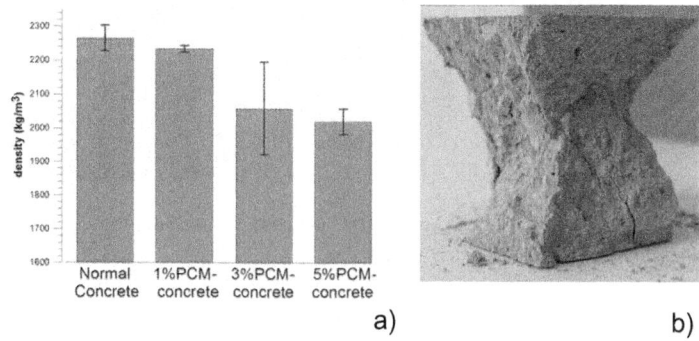

Fig. 6.   (a) Variation of density in concrete samples with different amounts of PCM; (b) sample of 5% PCM-concrete after compressive test

of lightweight concretes. Figure 6b represents the fracture pattern of a specimen of 5% PCM-concrete after compressive test: it exhibits a bi-pyramidal fracture, indicating a good behavior of the material.

The concrete samples were compared in terms of average compressive strength, R_m, and of characteristic compressive strength, R_{ck}, in order to evaluate their feasibility as structural materials. R_{ck} represents the strength value which the 5% of the tested samples don't achieve. According to literature, it is analytical defined as:

$$R_{ck} = R_m - k \cdot s \qquad (1)$$

where s is the standard deviation, and k depends on the number of tested samples. It was assumed equal to 3.40 and 2.75 for PCM-concretes (5 samples) and normal concrete (8 samples), respectively. Also, the coefficient of variation is particular relevant for structural applications because low values of this characteristic result in higher structural reliability:

$$CV = s/R_m \qquad (2)$$

All samples with and without PCMs exhibit the typical non-linear stress-strain curve, with decaying branch after peak.

Figure 7 shows the results of all the tested samples without PCM and with microPCM. As expected, the compressive strength of the concretes decreased with the increase of the content of PCMs because of the lower strength of PCM capsules with respect to aggregates.

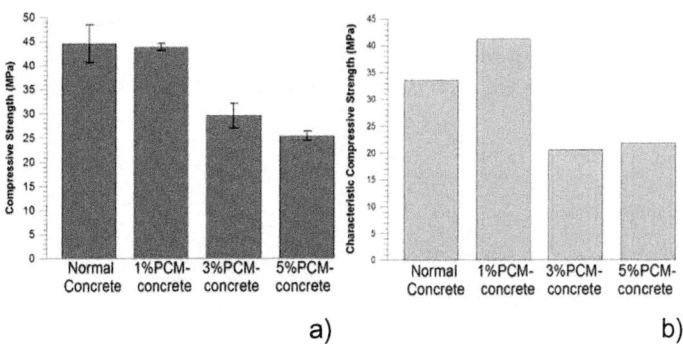

Fig. 7. (a) Average compressive strength with standard deviation, and (b) characteristic compressive strength of the tested concretes with different amounts of PCMs

The characteristic compressive strength shows instead a better performance of concretes with low amounts of PCMs. Microcapsules can provide higher specific area for nucleation sites in cement hydration, and could have a beneficial thermal effect in the composition of the hydration products.

Table 2. Compressive strengths, CV and densities of concrete composites normal and with PCMs

Type of concrete	R_m [MPa]	CV	R_{ck} [MPa]	Density [kg/m³]	Class
Normal	44.39	0.09	33.52	2265	C25/30
1% PCM-concrete	43.67	0.02	41.16	2233	C28/35
3% PCM-concrete	29.46	0.09	20.60	2057	C16/20
5% PCM-concrete	25.25	0.04	21.80	2018	C16/20

Table 2 reports the average compressive strength resulted from the axial compression tests. The table also shows the values of the coefficient of variation (CV) of the composites, their characteristic compressive strength and their average density. The CV of normal concrete tested in the experimentation is equal to 0.09, within the typical range of ordinary concretes. The CV values of PCM-concretes resulted smaller or equal than 0.09 and demonstrated the good reliability capabilities of the composites for structural applications. Moreover, 1% PCM-concrete possessed a coefficient of variation of 0.02, considerably less than the value of normal concrete. The low dispersion exhibited by resistances of 1% PCM-concrete resulted in an increased structural class from C25/30 of normal concrete to C28/35. Moreover, all concretes with PCMs demonstrated structural properties compatible with the use as structural concretes. Figures 8 represents the single stress-strain curves resulted from the compressive tests at displacement control, for all the normal concrete samples and samples with PCMs. The graphs show clearly the behavior of the materials with increased strain: the peak value represents the maximum compressive resistance, while the post peak branches are representative of the ductility properties. Each type of concrete exhibits a similar behavior even if a decrease of the maximum resistance can be observed. PCMs seem not to affect the ductile capabilities of the concretes. Figure 9 shows the average values of the stress-strain curves for each typology of concrete. The values were obtained by

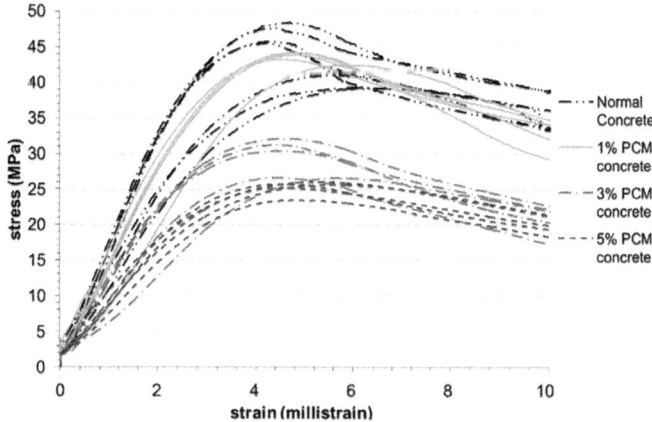

Fig. 8. Stress-strain curves of the complete uniaxial compression tests on concrete without and with PCM up to a strain of 10 µε.

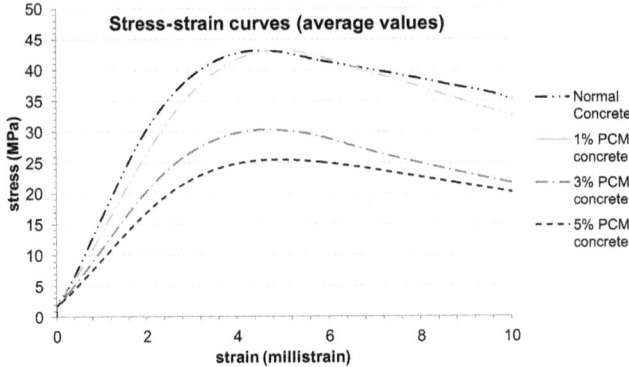

Fig. 9. Average values of stress-strain curves of the complete uniaxial compression tests on concretes without and with PCM up to a strain of 10 µε.

calculating the mean stress for fixed strains. The diagrams clearly show that the elastic modulus is greater for the concrete samples with higher compressive strength.

4   Thermo-Physical Tests

The thermo-physical tests were aimed at investigating the phase change materials in dynamic experimental conditions in order to assess the thermal benefits of the addition of PCMs in concrete.

The results coming from the thermal cycles show that all the samples followed the imposed temperature history. The differences between the temperature applied and measured on the concrete samples are due to the high thermal inertia of the cementitious material. Moreover, the samples with PCMs exhibit a peculiar behavior in the temperature range between 17 and 12°C, in the descending ramp, and in the range between 15 and 18 ° C in the increasing ramp (Fig. 10). The deflections resulted in the thermal profile of concrete with 5% of PCMs, with respect to the reference trend of the normal concrete sample, are associated to the phase change taking place in the fillers. Indeed, the paraffin inside the microcapsules has a nominal melting point of 18 °C. The first deflection referred to the solidification of the PCM, while the second one to the melting transition. During these phenomena, the PCM-concretes decrease their temperature at a lower level, since thermal energy is stored in the latent storage material, necessary to break or form molecular bonds in the PCMs. The effect of the presence of PCMs is visible up to about 2 h. Figure 11 represents the thermal behavior of concretes with 1% and 5% of PCMs, compared to the thermal behavior of the reference normal concrete, as recorded within the environmental simulation chamber. Phase change phenomena are clearly visible during cycles, closed to the transition phase temperature of the fillers. As shown in the diagrams, the concrete with the highest content of PCMs exhibited the greater deviation, resulting in a higher amount of energy absorbed and released by the concretes in phase change transition. Such behavior is highlighted also by the presence of a larger area included in the trends. Otherwise, the concretes with 1% of PCMs exhibit a thermal behavior almost linear, very similar to that of normal concrete.

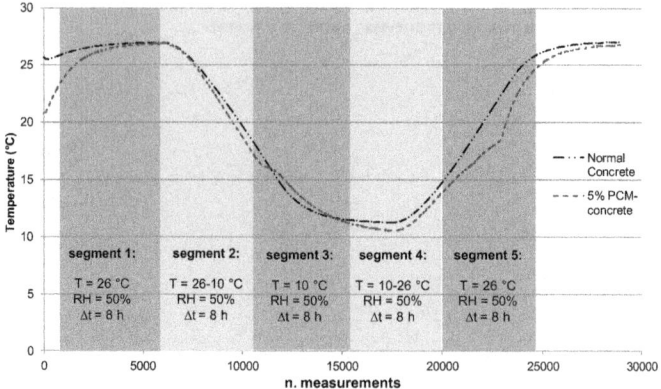

Fig. 10. Normal and 5% PCM-concretes' thermal profiles monitored by the PT100 probe at the bottom of the samples

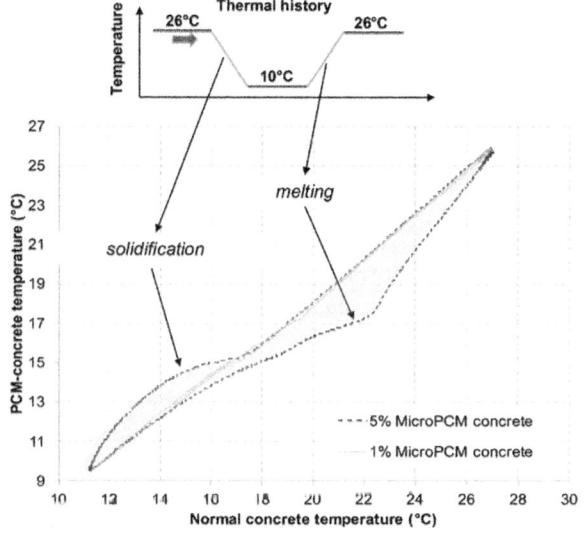

Fig. 11. Thermal profile of 1% and 5% PCM-concrete, as measured within the environmental simulation chamber

Figure 12 depicts the thermal conductivity and the diffusivity measured in the concrete samples with different amounts of PCMs, in comparison with the normal ones. The samples presented a slightly variation of both thermal conductivity and diffusivity in concretes with PCMs, with respect to normal concrete. In particular, the composite materials showed a reduction of the thermal conductivity, probably due to the small and

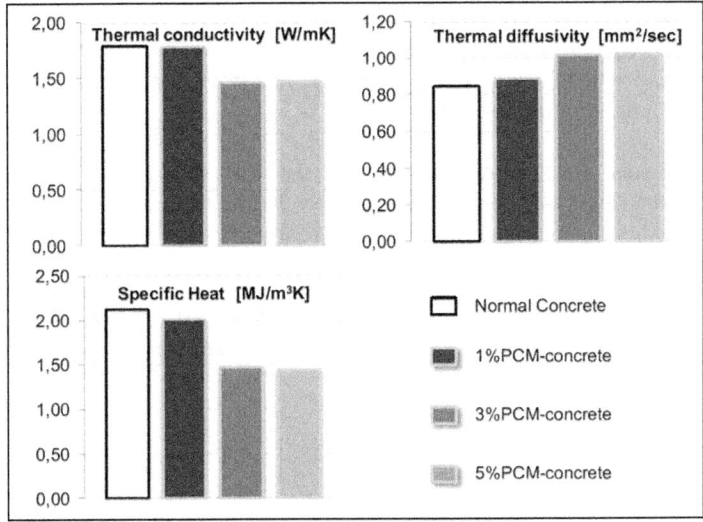

Fig. 12. Thermal conductivity, diffusivity and specific heat resulted by means of hot disk tests on normal concretes and with different contents of PCMs

closed pores created by the addition of the PCM microcapsules. Also, the capability to propagate the thermal wave in transient conditions resulted increased. The specific heat of the different concretes with and without PCMs was also tested, as key parameter influencing the thermal inertia of buildings and typically characterizing both sensible and latent thermal energy storage components. The presence of higher percentages of fillers resulted in a reduction of the specific heat superior to 20%. Indeed, microencapsulated materials seem to produce a noticeable effect on the thermal inertia of concretes.

5   Conclusions

This research was aimed at investigating the multifunctional performance of PCM-concretes for building applications.

Such concretes were made by adding increasing amounts of microencapsulated PCMs, from 1% to 5% with respect to the total mass of the material. The fillers had an internal core made of paraffin with melting temperature at 18 °C. Both mechanical and thermo-physical tests were carried out, in order to investigate the multifunctional properties of such materials and their applicability to structural civil engineering. The results of mechanical test demonstrated that PCM-concretes are compatible with structural applications as lightweight concretes. With respect to normal concretes, a reduction up to 11% in weight was observed. Moreover, small additions of PCMs resulted in an increase of the characteristic resistance. At the same time, the prototyped concrete manifested thermal capabilities suitable for building energy efficiency applications. Thermo-physical tests demonstrated that the phase transition is visible in PCM-concretes with higher amounts of fillers, resulting in a mitigation of the temperature fluctuations in the matrices.

Acknowledgements. Acknowledgments are due to the "CIRIAF program for UNESCO" in the framework of the UNESCO Chair "Water Resources Management and Culture". The research leading to these results has received funding from the European Union's Horizon 2020 research and innovation programme under grant agreement No. 657466 (INPATH-TES). The authors also thank the Microtek Laboratories, Inc. for providing the capsulated materials. The work is also partially funded by the Spanish government (ENE2015-64117-C5-1-R). Prof. Luisa F. Cabeza would like to thank the Catalan Government for the quality accreditation given to her research group (2014 SGR 123).

References

Cabeza LF (2015) Advances in thermal energy storage systems: methods and applications. Woodhead Publishing Series in Energy, No. 66

D'Alessandro A, Rallini M, Ubertini F, Materazzi AL, Kenny JM (2016a) Investigations on scalable fabrication procedures for self-sensing carbon nanotube cement-matrix composites for SHM applications. Cem Concr Compos 65:200–213

D'Alessandro A, Fabiani C, Pisello AL, Ubertini F, Materazzi AL, Cotana F (2016b) Innovative concretes for low carbon constructions: a review. Int J Low-Carbon Tech 12:289–309

Fernandes F, Manari S, Aguayo M, Santos K, Oey T, Wei Z, Falzone G, Neithalath N, Sant G (2014) On the feasibility of using phase change materials (PCMs) to mitigate thermal cracking in cementitious materials. Cem Concr Compos 51:14–26

Kalnæs SE, Jelle PB (2015) Review. Phase change materials and products for building applications: a state-of-the-art review and future research opportunities. Energ Build 94:150–176

Laflamme S, Ubertini F, Saleem H, D'Alessandro A, Downey A, Ceylan H, Materazzi AL (2015) Dynamic characterization of a soft elastomeric capacitor for structural health monitoring. J Struct Eng ASCE 141(8):04014186

Lecompte T, Le Bideau P, Glouanneca P, Nortershauser D, Le Massonb S (2015) Mechanical and thermo-physical behaviour of concretes and mortars containing phase change material. Energ Build 94:52–60

Navarro L, de Gracia A, Niall D, Castell A, Browne M, McCormack SJ, Griffiths P, Cabeza LF (2016) Thermal energy storage in building integrated thermal systems: a review. Part 2. Integr Passiv Syst Renew Energy 85:1334–1356

Shah SP, Konsta-Gdoutos MS, Metexa ZS, Mondal P (2009) Nanoscale modification of cementitious materials. In: Bittnar Z et al (ed) Nanotechnology in construction 3, pp 125–130. Springer

Sharma A, Tyagi VV, Chen CR, Buddhi D (2009) Review on thermal energy storage with phase change materials and applications. Renew Sustain Energy Rev 13(2):318–345

Ubertini F, Laflamme S, D'Alessandro A (2016) Smart cement paste with carbon nanotubes. In: Loh KJ, Nagarajaiah S (eds) Innovative developments of advanced multifunctional nanocomposites in civil and structural engineering, pp 97–120. Woodhead Publishing

Yang HS, Che YJ (2015) Influence of particle size distribution of fine and micro-aggregate on the microstructure of cement mortar and paste. Mater Res Innov 19:S1–S130

Zhang D, Li Z, Zhou J, Wu K (2004) Development of thermal energy storage concrete. Cem Concr Res 34(6):927–934

Konuklu Y, Ostryc M, Paksoy HO, Charvat P (2015) Review on using microencapsulated phase change materials (PCM) in building applications. Energ Build 106:134–155

Retrofitting RC Members with External Unbonded Rebars

A. Tinini, F. Minelli[✉], and G. A. Plizzari

Department of Civil, Environmental, Architectural Engineering and Mathematics,
University of Brescia, Brescia, Italy
fausto.minelli@unibs.it

Abstract. External unbonded rebars represents a suitable strengthening technique for the retrofitting of existing Reinforced Concrete (RC) members. Advantages regard the ease of installation, a minimum invasiveness and possibility of future inspections. Structurally, increment of flexural stiffness and bearing capacity and enhancement of shear-flexure behavior can be achieved. The presence of both bonded and unbonded bars introduces a change in the way the shear actions are resisted. Unbonded rebars develop an arch action component, with no bond present and constant force in the rebars, in addition to the beam action component, normally developing in presence of bonded bars. The present paper reports the results of four point loading tests on full-scale beam, with the aim of studying the influence of different bond condition. Moreover, the Double Harping Point technique, using external rebars and vertical deviators, is presented, with attention to the definition of the vertical equivalent stiffness of the deviators.

Keywords: Shear strength · Unbonded rebars · Beam and arch action
Retrofitting

1 Introduction

In recent years, an increasing attention has been devoted to the assessment and the structural rehabilitation of existing reinforced concrete (RC). This is mainly due to problem connected with structural deterioration, with more onerous requirements in Standards and Building Codes or with the change in use of the building, resulting in an increase of the loading.

The retrofitting of existing RC elements by means of external steel rebars was already studied in previous works (Cairns et al. 2005; Minelli et al. 2008). The technique is illustrated in Fig. 1. High yield threaded bars are applied to both sides of (or, in some application, under) a RC beam. The bars pass through yokes at the ends of the beam, where they are anchored by locknuts. On all but short spans there are benefits from use of deflectors to avoid a reduction in effective depth as the beam deflects. External bars can thus easily be installed by hand: no significant prestress is required and only sufficient force to firmly secure the bars or to obtain low post-tension action is applied.

© Springer International Publishing AG, part of Springer Nature 2018
M. di Prisco and M. Menegotto (Eds.): ICD 2016, LNCE 10, pp. 184–199, 2018.
https://doi.org/10.1007/978-3-319-78936-1_14

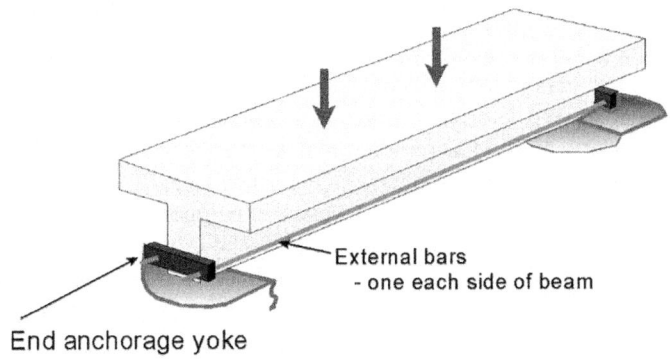

End anchorage yoke

Fig. 1. External unbonded bars reinforcement

The Authors highlighted some of the advantages of that reinforcement technique, resulting a more cost effective and less disruptive solution to the problem of strengthening simply supported RC elements. Moreover, the use of external unbonded reinforcement presents many of the merits of post-tensioning with unbonded tendons, dispensing at the same time with the need of specialist stressing operation. Less clearance is needed for the end anchorages, since there is no need to provide access for jacks.

The corrosion protection can be based on systems used for external prestressing. The technique is also compatible with principles of conservation, which require that a structure be returned to its original condition after any interventions.

Some disadvantages are also present: the system is limited to be used for the strengthening of simply supported beams, the end anchorages position represents a risk of anticipated failure (only in case it is not possible to install them over the support points) and space limitation may be present for systems applied under the element to be reinforced.

The concept of retrofitting by means of unbonded external reinforcement developed from observations of reinforced concrete beams when concrete around bars is broken out during repair actions. The study showed that the debonding of the longitudinal rebars might cause little reduction in strength (Cairns and Zhao 1993) or even an enhancement in ultimate strength of beams deficient in shear (Cairns 1995), thanks to development of an arch action component, which represent a more stable shear resistant mechanism. Without bond, the force in reinforcement must be constant along the length of the beam. The lever arm between reinforcement and concrete, and not bar force, must vary with bending moment. The neutral axis depth therefore also varies, in a manner consistent with arch rather than shear-flexure action. Concrete is, of course, better at resisting compressive stresses than shear stresses. It might thus be expected that this change from a purely flexural mode of failure towards a tied arch/flexure hybrid would be accompanied by an increase in shear capacity.

The potential of retrofitting using external bars anchored only at the ends of simply supported flexural elements (Fig. 1) has been demonstrated for flexural modes of failure (Cairns and Rafeeqi 2003), reaching a strength enhancements of the order of 100%.

In this paper, results concerning the behavior of member with embedded plain bars are also presented. Many existing structures, built after the second world war (up to the 70 s), present plain bars as longitudinal reinforcement. That type of rebars, not having lugs or any other surface deformation, are not able to transfer the bond stress by means of mechanical interlock. For this reason, the bond transfer mechanism is less efficient than in the case of members reinforced with deformed bars.

In the early Twentieth Century, an intensive experimental campaign on members with plain bars was performed (Abrams 1913). The Author highlighted two main resistant mechanisms: adhesive resistance and sliding resistance, the latter arising from inequalities of the bar surface and irregularity of its section and alignment together with the corresponding conformation in concrete.

Other researches further demonstrated that adhesive resistance can be attributed to the chemical adhesion mechanism and the micro-interlocking of concrete keys generated by the penetration of cement paste into the indentation of the bar surface (Stoker and Sozen 1970).

Tassios (1979) pointed out that a small slip is needed for activating micro-interlocking and developing the maximum adhesive resistance.

In recent years, an experimental research was carried out to investigate the effect of flexural cracking and bond loss on flexural behavior (Feldman and Bartlett 2008). The Authors tested two beams, critical in flexure, presenting the same contact area but different reinforcement ratios. The beam with the higher reinforcement ratio (0.98%) exhibited arch action because of the bond failure at about 60% of the maximum load. The other member, with a ratio of 0.33%, presented a predominant beam action up to the failure.

The present study was therefore undertaken with the aim of studying the influence of different bond condition (deformed, plain and unbonded rebars) on the shear-flexure behavior of RC members. Plain rebars, behaving similarly to deformed rebars for very low slip and to unbonded rebars for higher slip values (Tinini 2016), represent a type of reinforcement of great interest from the structural rehabilitation point of view. Therefore, they are included in the study in order to consider a more comprehensive case study.

2   Experimental Investigation

2.1   Specimen Geometry

The experimental program concerned seven full-scale beams tested under a four point loading system with a shear span-to-depth ratio a/d of 2.5, the most critical for shear resistance of RC members with no transverse reinfocement.

A total depth of 500 mm was chosen, with a gross cover of 40 mm. According to the a/d ratio considered, the shear span length was 1150 mm, while the total span length was of 4300 mm and the overall length 4500 mm. Four of the experimental specimens were made considering different type of embedded rebars (deformed (B), plain (S) and 2 unbonded (U), with deformed rebars placed through a corrugated plastic pipe that covered them), and three combining in pairs the different levels of bond condition considered (BS, BU and SU). Rebars having a diameter of 20 mm were used in case of bonded and unbonded rebars, while 22 mm diameter bars were used for plain bars. This resulted in a reinforcement ratio varying from 1.09% to 1.32%.

Figure 2 reports the main geometrical properties of the specimens.

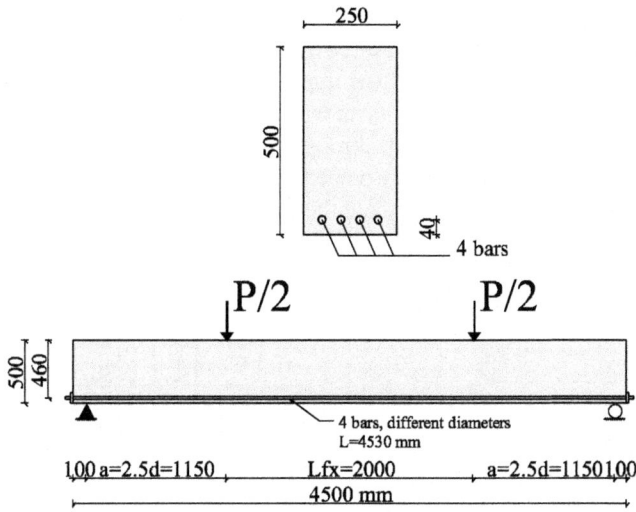

Fig. 2. Specimen geometry

2.2   Material Properties

All the beams were made with the same concrete mixture. The mix design consisted in 430 kg/m³ of Cement Portland II/A-LL 42.5R and 168 kg/m³ of water, resulting in a water/cement ratio (w/c) of 0.39. The maximum aggregate size was 14 mm. An amount of 3.85 kg/m³ of superplasticizer was added to the concrete in order to enhance the workability. According to European Standard EN 12350-2, the concrete showed a S4 consistency class.

In accordance with EN 12390-13 and EN 12390-3 five cylinders 80(Ø) × 210 mm and ten 150 mm cubes were used for the determination of the Modulus of Elasticity and the cubic compressive Strength of concrete, respectively. After 28 days of curing, the secant Young's modulus, E_{cm}, was 29.7 GPa, while the cubic compressive strength, R_{cm}, was 54.04 MPa. The cylinder compressive strength of concrete was analytically derived as $f_{cm} = 0.83R_{cm}$ and resulted 44.85 MPa.

Rebars properties were evaluated according to EN 15630-1; the yielding and ultimate tensile strength resulted respectively 522.0 MPa and 639.1 MPa for deformed rebars and 388.1 MPa and 567.6 MPa for plain bars.

2.3   Test Set-up and Instrumentation

A displacement-controlled test was adopted to allow a suitable test control during critical steps, such as in the case of abrupt cracking phenomena or load drops. This was obtained by adopting a worm electro-mechanical screw-jack having a loading capacity of 1000 kN and a stroke of 350 mm.

In the test loading frame (Fig. 3), the actuator was hanged at the laboratory strong floor (with a thickness of 1050 mm) and the load transferred to the top through transverse steel beams (two 2-UPN400) and 32 mm dywidag rebars, passing through holes in the strong floor. The applied load was measured by two load cells placed between the bearing steel plates of the dywidag bars and the upper 2-UPN400 beam. The top cross beam loaded the longitudinal steel beam at its midspan. The longitudinal beam, in turn, loaded the specimen in the two desired points through steel cylinders. These cylinders, welded to a 300 × 100 × 5 mm steel plate, were placed above a layer of neoprene with a thickness of 25 mm to better transfer and distribute the load between the frame and the beam.

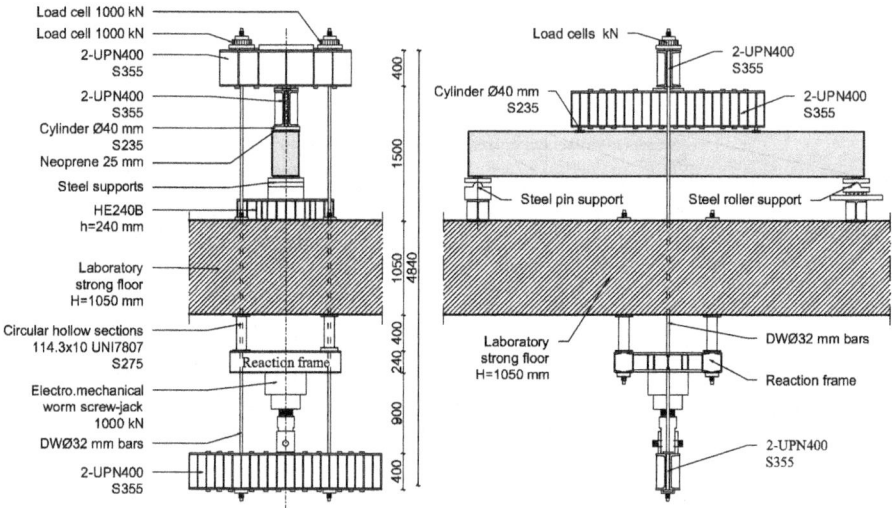

Fig. 3. Scheme of loading system adopted

Being the flexural span of specimens equal to 2000 mm, 2-UPN400 beam, identical to the lower transverse beam, were used as longitudinal beam.

A consistent number of instruments were used for monitoring the most important displacements and deformations of specimens, besides the applied loads. Linear Variable Differential Transformers (LVDTs) allowed the measurement of the vertical displacements of the beam and the supports. Potentiometric transducers were placed on both sides (back and front) of the specimens in the area of shear stresses (shear spans), to measure crack openings and strut deformations and, in the flexural span, to measure top compressive chord shortening and the elongation of the bottom chord at the bar level. The transducers for crack openings were placed with an inclination of 135° with respect to the horizontal.

The ones for the strut deformation were placed along the ideal connection line between the center of the support and the center of the loading point.

3   Experimental Results

3.1   Failure Mode

Figure 4 reports the comparison of the load-deflection (at midspan) curves for every specimen. Beam IDs report the number of bars with the related bond type, the shear slenderness and, in case of duplicate, an identification number. Firstly, it can be noticed that the influence of bond is very significant, determining a rather different structural behavior. For load levels significant for service conditions (around 100 kN), all specimens with bonded rebars exhibited a quite stable behavior with a crack onset and propagation highly controlled, and with overall deformations much smaller than those shown by the fully unbonded specimens.

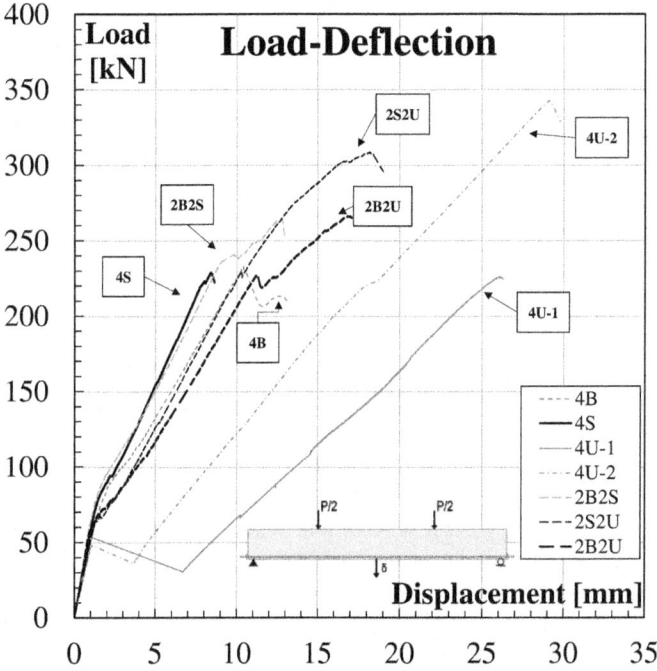

Fig. 4. Load-displacement chart

A noticeable tension stiffening effect was also reported. This behavior resulted more evident for the 4B specimen (4 deformed rebars), but became less effective with decreasing of bond level, resulting in a reduction of the specimen cracked stiffness.

In beams 4B and 4S, however, at a very low load level, a classical, sudden and brittle shear failure appeared through a wide crack running from the point load to the bottom reinforcement, toward the support, in the shear span. This brittle collapse occurred for crack widths, measured in the shear span, of about 0.3 mm, which represents a quite small value (i.e., one would think to be in a safe situation under such a structure), even though in accordance with classical literature on shear.

The failure load was almost the same, therefore no significant influence of the rebar type (deformed or plain) was shown for this specific cross-section. The reason is probably related to absence of debonding at the failure load level.

The beam with a mix of the two types of bars (2B2S) presented a higher bearing capacity than the other two members with only plain or only deformed rebars. Approaching the failure load of 4S and 4B specimens, a change in stiffness and a small load drop for the mixed beam can be observed, but no brittle failure occurred; indeed, the load further increased, reaching a value 16% higher than the one for bonded and plain specimens.

Also the other mixed specimen (2B2U and 2S2U) presented a greater bearing capacity, due to the presence of unbonded rebars and to the development, from the very beginning of the experimental test, of an arch action contribution.

The behavior exhibited by the two specimens with unbonded rebars was really different: after the first cracking load, a single, very wide flexure crack appeared in the flexural span between the loading points, in combination with considerable energy release and displacement. After that, only a few other flexure cracks, which were very wide, developed. For the specimens a flexure failure was attained through concrete crushing in the top chord, but well before the yielding of the reinforcing bars.

The need to perform an additional test on a copy of 4U-1 beam was related to a problem with the anchorage steel plates, due to concrete shrinkage, unrestrained by unbonded rebars. The shortening of the specimen resulted in a gap between the plate and the edge of the beam of about 1.5 mm, which did not allow the mutual interaction between concrete and longitudinal reinforcement (for unbonded elements, with no steel-to-concrete bond, the load can only be transferred by means of the anchorages). When the cracking load was reached for 4U-1 specimen, the gap was still open and the beam behaved like an unreinforced concrete member up to a deflection of 7 mm.

This problem was solved in 4U-AD2.5-2 specimen by inserting, a few days before the test, a layer of bedding mortar between the anchorage steel plates and the edges of the beam and welding the plates to the longitudinal bars. At the formation of the first flexural crack, the energy release and the displacement of the new member were significantly lower than in 4U-1 (Fig. 4). The vertical propagation of the crack was reduced and a top compressive strut thickness of 55 mm was measured. Because of that, the bearing capacity of the element was 50% greater but a ductile failure was not reached anyway.

The maximum deflection of unbonded members was 2–2.5 times greater than the one exhibited by the bonded specimens, while the bearing capacity of 4U-AD2.5-2 specimen turned out to be almost 1.5 times greater. No crack in the shear span, or inclined crack elsewhere, were reported.

The unbonded specimens, however, exhibited a cracking onset unacceptable under a design point of view. The first crack that arose in specimen 4U-1 was about 2.7 mm wide (1.5 mm for 4U-2 beam) and the midspan deflection suddenly dropped from 1.1 to 6.7 mm for 4U-1 (from 1.2 to 3.7 mm in the case of the 4U-2 beam). Moreover, the first vertical crack ran for almost 90% of the depth of the beam. Such a situation would have considered much more dangerous that the one mentioned for the bonded specimens. It is observed, on the contrary, that after this stage, flexural cracks develop in a fairly stable fashion.

Figure 5 illustrates the main shear crack width versus the load, for five of the specimens (the two unboned beam were obviously excluded, not presenting any crack in the shear span). The different bond condition influenced the cracking process; in fact, in bonded specimens (deformed and plain rebars), the collapse arose for very small crack width (~0.3 mm), as already explained. Decreasing the bond level, a more stable and wider crack propagation can be observed, reaching a crack width of 3 mm in 2S2U specimen. In specimen 2B2U, a sudden crack opening occurred for a width of 0.5 mm; afterwards, the beam was able to further increase the load and to reach a final crack width of 5.2 mm (with a more distributed crack pattern anyway). From that consideration, it can be concluded that the presence of both bonded and unbonded bars, introducing a mixed arch/beam action behavior, is able to increase the shear bearing capacity of reinforced concrete members and to allow stable shear cracks up to widths ten times greater than deformed rebars only.

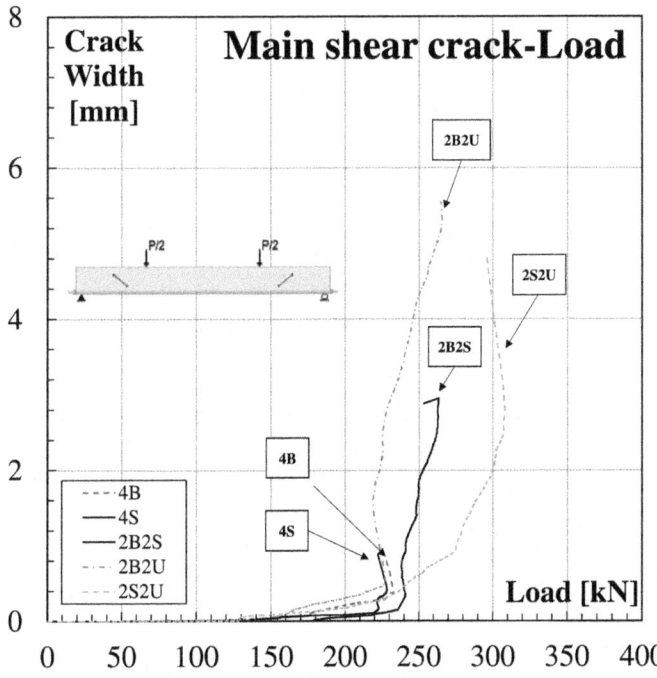

Fig. 5. Main shear crack-load chart

3.2 Crack Pattern

Figure 6 presents the crack patterns at failure for six of the specimen.

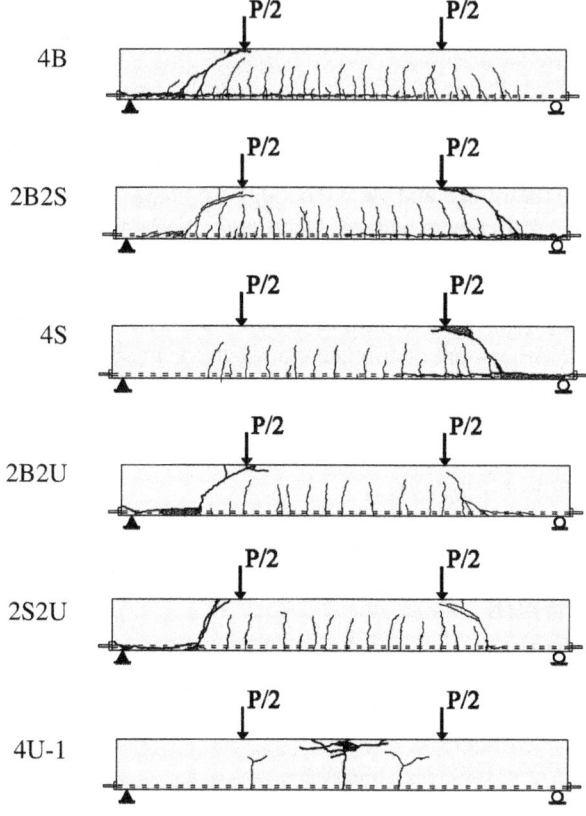

Fig. 6. Crack patterns at failure

Concerning 4B, flexure-shear crack appeared at a load intensity of 180 kN at a distance from the load point of almost 500 mm (similar to the beam depth), developing in an inclined crack towards the same loading point. This crack resulted the critical one along which the typical shear failure (block mechanism) took place. A considerable splitting along the longitudinal reinforcement could also be observed.

A similar behavior can be observed in most of the other members (2B2S, 4S, 2B2U and 2S2U), with the main difference that the critical shear crack resulted closer to the loading point in rebars having a lower bond resistance. Accordingly, the critical crack (failure crack) became steeper.

Looking at the flexural span it can be noted that, with rebars having a lower bond resistance, the flexural crack pattern resulted more spaced, with fewer and wider vertical cracks. In the evaluation process for assessment and rehabilitation of existing buildings, this could result in possible durability issues.

Concerning specimen 4U-1, a 2.7 mm wide crack (vertical) arose at a load level of 54 kN (first cracking point) with a propagation in 90% of the beam depth; a second shorter flexural crack developed for a load intensity of around 67 kN; another macro crack formed later on. All of the cracks tended to bifurcate and to become horizontal, clearly showing the arch shape. The final collapse occurred at the top chord of the member with concrete crushing and without cracks in the shear span.

4   Double Harping Point System

4.1   Introduction

This strengthening system presents many similarities with the unbonded tendons post-tensioning technique: however the use for retrofitting of RC elements is derived by a very similar technique quite common for the retrofitting of timber roof beam (Giuriani 2012).

The posts m (Fig. 7), due to the presence of the steel bar system, can be considered as two new vertical supports elastically unrestrained, with an equivalent stiffness k_V.

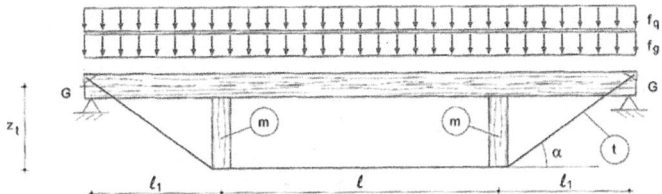

Fig. 7. Double harping point reinforcement for a timber roof beam (Giuriani 2011)

Giuriani suggested to consider these support as fixed ($k_V = \infty$) for normal dimension timber beams and for the area of reinforcement usually adopted ($A_s \geq 5$ cm^2). In this case, the ratio between the displacement of the reinforced member, w_r, and the displacement of the unreinforced, w, is less than 1/10.

These considerations are generally questionable in case of retrofitting of reinforced concrete elements. The material and geometrical properties in that case do not allow this simplification. Because of that, a proper evaluation of the equivalent stiffness offered by the reinforcement system is very important for considering the real redistribution of the load and avoiding underestimation of the internal forces on the member.

4.2   Evaluation of Actual Equivalent Stiffness

In a double harping point system for a reinforced concrete beam (Fig. 8), the distance between the anchorage point in the proximity of the supports and the centroid G of the transformed concrete section (e_1), the distance between the horizontal bars and the centroid G (e_2), the position of the posts (l_1) are all very important design aspects, influencing the efficiency of the reinforcement.

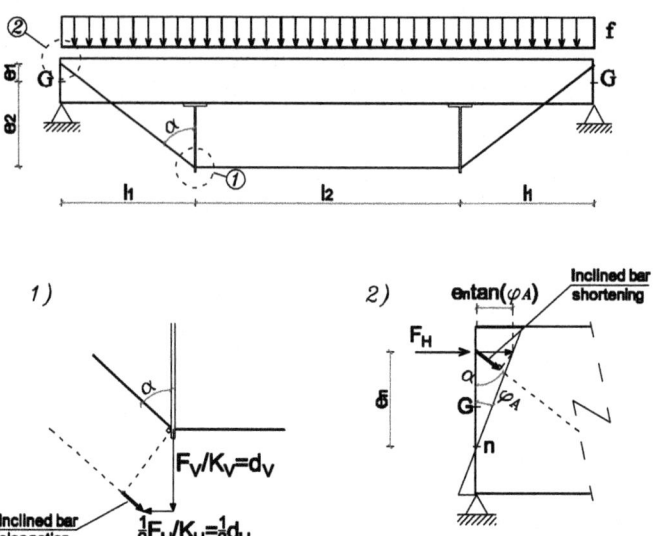

Fig. 8. Main geometric parameters and inclined bar compatibility (particulars 1 and 2)

The inclination of the bars in the external spans is a function of those parameters:

$$\alpha = \tan^{-1}\left(\frac{l_1}{e_1 + e_2}\right) \tag{1}$$

It is possible to observe that increasing the distance of the horizontal bars from the edge of the beam (increasing e_2), the value of α decreases, resulting in a higher vertical force in the posts. Assuming the effect of the post on the beam similar to that of a vertical axial spring of stiffness k_V, that results in an increase of equivalent stiffness.

However, due to maximum distance requirements for the reinforcement, the value e_2 cannot be increased freely.

If $e_1 \neq 0$, the anchorage point at the support level is not coincident with the centroid of the cross section of the beam. Because of that eccentricity, the element is not only subjected to a horizontal axial force but also to a concentrate bending moment at support.

The vertical equivalent stiffness of the reinforcement, k_V, does not depend only on the geometrical and material properties of the reinforcement, but also on the characteristic of the RC beam (flexural stiffness in the first time). The properties of the beam (neutral axis position, moment of inertia, moment of areas) for an element subjected to both bending moment and axial force are a function of the values of the internal forces. On their side, the internal force distribution depends on the equivalent stiffness of the support. That result in a circular reference, in which it is possible to consider three main parameters (Fig. 9):

- The equivalent vertical stiffness of the posts $[k_V]$;
- The neutral axis position $[c]$;
- The eccentricity of load to centroid $[e = u + y_{cen}]$.

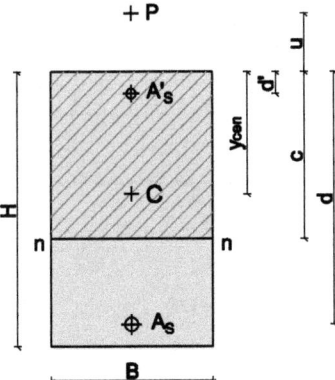

Fig. 9. Geometrical properties of the RC element (rectangular cross-section)

Only two of these unknowns are independent from the others and so, to solve the system, two equations are needed:

- Rotational equilibrium with respect to the point of eccentricity of load (*P*);
- A compatibility equation in term of elongation of the inclined reinforcement bars. The compatibility equation can be expressed as reported in Eq. 2:

$$\frac{F}{K} = \frac{F_V}{K_V} \cos \alpha - \frac{1}{2}\frac{F_H}{K_H} \sin \alpha - \left(e_n \tan \varphi_A\right) \sin \alpha \tag{2}$$

$$K = \frac{E_s A_{r,1}}{l_1} \sin \alpha \tag{3}$$

$$K_H = \frac{E_s A_{r,2}}{l_2} \tag{4}$$

Where K, K_H (Eqs. 3–4) are the axial stiffness of the inclined and horizontal bars and $A_{r,1}$, $A_{r,2}$ their cross sectional area respectively.

The first member of the equation represents the total elongation of the inclined bars (F/K). The second reports the elongation due to the vertical displacement of the system at the deviation point (F_V/K_V), the shortening due to the elongation of the horizontal rebars (F_H/K_H) and the shortening related to the rotation of the beam at the support (φ_A). The vertical displacement and the rotation can be evaluated considering the effect of three different load condition and using the Elastic Line Method for the concrete beam. The presence of the external reinforcement is taken into account including in the analysis a vertical axial spring with stiffness K_v, an Axial Force (N) and a Bending Moment (M_A) at the support location (Fig. 10a).

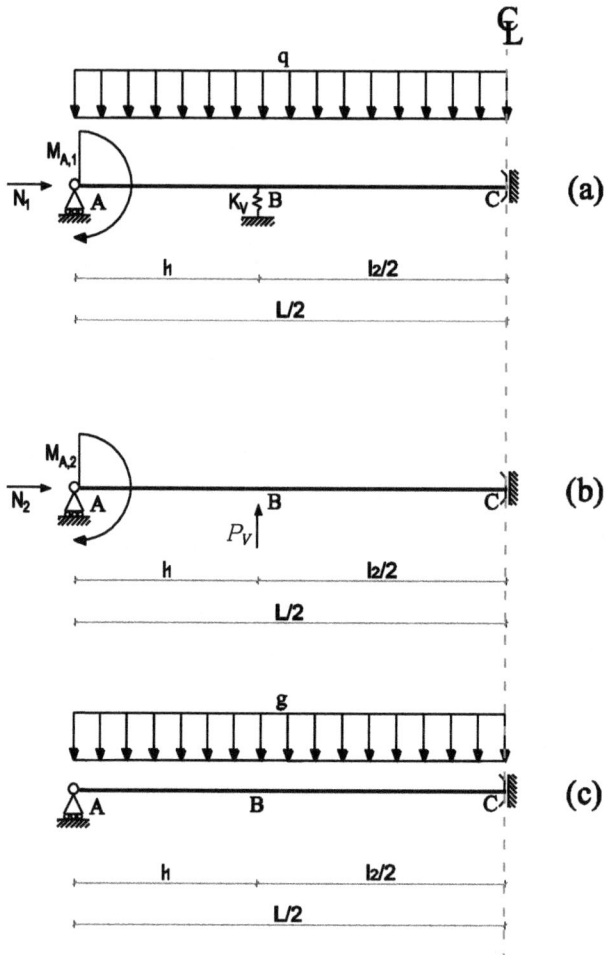

Fig. 10. Elastic line method: load cases

The first load (Fig. 10a) condition consider the effect of the load applied on the structure after the installation of the reinforcement system (q). The second condition (Fig. 10b) consider only the effect of the prestress action (P_V) that can be apply to the external reinforcement. Finally, the last load condition (Fig. 10c) considers the effect of the part of load already applied before the installation of the reinforcement (g).

The rotation in A and the internal force along the length of the member can be evaluated from the combination of the three load cases. It is worth noting that only rotation $\varphi_{A,1}$ (from load case in Fig. 10a) must be considered in Eq. 2, being the only one contributing to the deformation of the inclined bars after the application of external loads.

Figures 11 and 12 report the shear and bending moment diagram in case of a retrofitted one-way slab. The load applied comes from a typical SLS combination that can be considered in the rehabilitation of an existing building. A pretension of about 89 MPa is applied to the external bars.

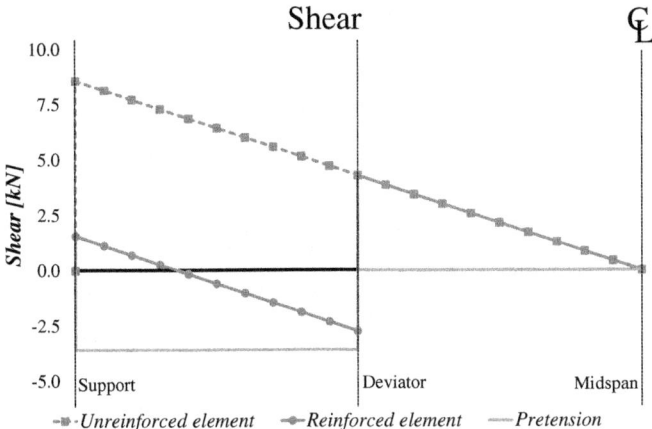

Fig. 11. Shear force over half of the span for reinforced and unreinforced one-way slab

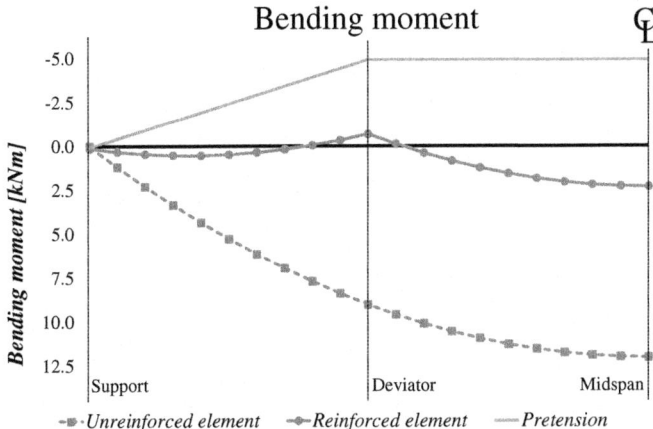

Fig. 12. Bending moment over half of the span for reinforced and unreinforced one-way slab

The external reinforcement system leads to a change of the static configuration of member. The effect is more evident in the external part of the span, between the support and the deviator; in the internal span, instead, the shape of the diagrams is similar to that for the unreinforced structure. Due to the presence of the reaction from the post, a discontinuity is present in the shear and a peak in the bending moment diagram.

The shear (Fig. 11) in the central span is exactly the same for the reinforced and the unreinforced element. The concentrated force at the deviation point leads to a drop in the diagram and to an important reduction in term of shear action in the external part of the smember. Therefore, the improvement in the behavior of the element is mainly related to a reduction of the shear demand. Looking at the bending moment diagram (Fig. 12), at the deviation point the maximum negative moment is reached; in the central span, the shape of the diagram is similar to the one for the unreinforced beam but the reaction coming from the post translates that portion of diagram, reducing the maximum positive moment at the midspan. Also in flexure, the reinforcement effect leads to a reduction in term of forces acting on the element

5   Concluding Remarks

In this paper, the results of seven tests on flexure-shear critical beams and a model for the evaluation of the equivalent stiffness of double harping point system have been presented. Based on the results, the following conclusions can be drawn:

1. Good steel-to-concrete bond (deformed rebars) tends to anticipate the collapse by determining a critical shear failure at very low load and deflection levels; this is mainly due to the cracking process governed by the compatibility between steel and concrete. The lack of bond, on the other hand, determines a flexural behavior, even though under considerable deflections and localizations of cracks.
2. Intermediate levels, in which both beams and arch action can develop together, demonstrate the possibility to increase the shear bearing capacity and to reach wider shear critical cracks.
3. The double harping point system results an interesting technique for the retrofitting of existing RC elements experiencing shear-flexure deficiency. The presence of the deviators as elastically restrained supports lead to a reduction of the shear and bending moment action. An improvement in term of stiffness, due to the presence of the external unbonded longitudinal reinforcement can also be achieved.

Acknowledgements. The Authors gratefully acknowledge Engineers Esti Federica, Caldera Silvia and Migliarini Mauro for assistance in carrying out laboratory testing and data processing.
     The fundamental contribution of the rebar provider (Priuli s.a.s. and Alfa Acciai S.P.A.) and the concrete provider (Italcementi Group) is also gratefully acknowledged.

References

Abrams DA (1913) Test of Bond Between Concrete and Steel, University of Illinois Bulletin No. 71, University of Illinois, Urbana, IL, p 240
Cairns J (1995) Strength in shear of concrete beams with exposed reinforcement. In: Proceedings of ICE structures & buildings, vol 110, May 1995, pp 176–185
Cairns J, Minelli F, Plizzari GA (2005) Strengthening RC beams by external reinforcement. In: Proceedings of the international conference on concrete repair, rehabilitation and retrofitting, Cape Town, South Africa, 21–23 November 2005
Cairns J, Rafeeqi SFA (2003) Strengthening of reinforced concrete beams by external unbonded bars: experimental investigation: theoretical investigation. Proc Instit Civil Eng 156(1):27–48
Cairns J, Zhao Z (1993) Structural behaviour of concrete beams with reinforcement exposed. Proc Instit Civil Eng Struct Build 99:141–154
EN 12350-2 (2009) Testing fresh concrete – Part 2: Slump Test

EN 12390-13 (2013) Testing hardened concrete. Determination of secant modulus of elasticity in compression

EN 12390-3 (2009) Testing hardened concrete. Compressive strength of test specimens

EN15630-1 (2004) Steel for the reinforcement and prestressing concrete – Part 1: Test methods

Feldman LR, Bartlett FM (2008) Bond in flexural members with plain steel reinforcement. ACI Struct J 105(5):552–560

Giuriani E (2012) Consolidamento degli Edifici Storici. Utet Scienze Tecniche, Torino (in Italian)

Minelli F, Plizzari GA, Cairns J (2008) Flexure and Shear behavior of RC beams strengthened by external reinforcement. In: Proceedings of the international conference on concrete repair, rehabilitation and retrofitting II, Cape Town, South Africa, 24–26 November 2008, pp 377–378

Stoker MF, Sozen MA (1970) Investigation of Prestressed Reinforced Concrete for Highway Bridges. Part V: Bon Characteristic of Prestressing Strends, University of Illinois Bullettin No. 503, University of Illinois, Urbana, IL, pp 116–119

Tassios TP (1979) Properties of Bond between Concrete and Steel under Load Cycles Idealizing Seismic Actions, Comité Euro-International du Béton, Bulletin No. 131, Paris, France, pp 67–121

Tinini A (2016) Investigating Existing Beams and Slabs Experiencing Flexure and Shear deficiency, PhD thesis, University of Brescia, Brescia

Balanced Lift Method – Building Bridges Without Formwork

J. Kollegger[✉] and S. Reichenbach

Institute for Structural Engineering, TU Wien, Vienna, Austria
johann.kollegger@tuwien.ac.at

Abstract. The balanced lift method is a building bridge method that was developed at the TU Wien. The most common methods to build bridges are the ones using falsework or the cantilever method, but a rather uncommon method, the lowering of arches is seen as the origin of the balanced lift method. The idea was to create a method which would allow a bridge to be built in a very fast manner without the usage of falsework, using prefabricated elements and mounting all parts together in a position – in this case vertically – that would simplify the construction process. In order to reach the final state of the bridge, the vertically assembled parts are rotated into their final horizontal position. This article contains a description of the development of the method, a large scale test will be portrayed and an already designed bridges using the balanced lift method will be shown.

Keywords: Precast concrete elements · Post-tensioning
Bridge construction method · Large-scale test

1 Introduction

When comparing construction methods for arch bridges and regular beam type bridges, it can be observed that in some cases the same construction techniques are used for the production of the arch and the bridge girder. For example in situations when the height above the ground and the size of the arch are not too large the most economical construction will be the one using falsework.

Figure 1 shows the falsework used for the support of the concrete during casting of the arch of the Egg-Graben Bridge. This bridge was designed at the Vienna University of Technology with the idea in mind to create a bridge without mild steel reinforcement in the bridge deck. Post-tensioning tendons encapsulated in plastic ducts and watertight anchorages were used to provide a sound deck slab with a thickness of 500 mm. It could be shown (Berger et al. 2011) that the serviceability and ultimate limit load states according to Eurocode could be fulfilled with this new design approach.

The Egg-Graben Bridge (Fig. 2) was a winning structure of the 2014 competition for the "fib Awards for Outstanding Concrete Structures". In the statement of the jury (fib 2014) it was mentioned that: "The jury highly appreciated the consistent application of durability philosophy. The bridge deck is intended to have a long service life with very little maintenance costs because the bridge deck is constructed exclusively with

© Springer International Publishing AG, part of Springer Nature 2018
M. di Prisco and M. Menegotto (Eds.): ICD 2016, LNCE 10, pp. 200–215, 2018.
https://doi.org/10.1007/978-3-319-78936-1_15

Fig. 1. Falsework of the arch of the Egg-Graben Bridge

encapsulated post-tensioned reinforcement and watertight anchorage. No other reinforcement is used. Therefore, the electrolytic corrosion in the deck is excluded. In this way water insulation and pavement were also saved. The concrete itself is meant to resist both physical and environmental loads. The bridge also fulfils high aesthetic expectations."

Fig. 2. Egg-Graben Bridge (© Pez Hejduk, Austria)

Another construction method which is frequently used for the erection of concrete arches, is the cantilever method. The bending moments due to the dead weight of concrete during construction are reduced by applying stay cables. The forces in the stay cables can be adjusted in the different construction stages in order to keep the bending moments in the cantilever small. A spectacular example for an application of the cantilever method for arch construction is the Hoover Dam Bypass (Fig. 3). The jury of the 2014 fib Awards for Outstanding Concrete Structures made the following statement about this bridge: "As a result of excellent engineering, the Hoover Dam Bypass bridges the Colorado River at 275 m above the water level. The Hoover Dam Bypass is a breathtaking example of civil engineering in the deep canyon of the Colorado River and its rocky cliffs."

Fig. 3. Hoover Dam Bypass (© Jamey Stillings Photography, USA)

Yet another method for the construction of concrete arches starts with a vertical production of the arch halves using climbform. At the very bottom of each vertical arch half a hinge is located, which enables a rotation of the arch half from the initial vertical to the final inclined position. Contrary to the cantilever construction method where usually several stay cables are placed on each arch half during construction, only one pair of stays can be attached at one location to an arch half when the method of lowering of arch halves is applied, in order to work with a statically determined structure during the delicate lowering operation. Therefore, larger bending moments, compared to the cantilever construction method, occur in the arch when the lowering of arch halves is applied.

The lowering of arch halves was first applied by Riccardo Morandi in the construction of the Lussia Bridge in 1955 in Italy as described by Troyano (2003). The advantage of this method is the accelerated construction of the arch which can be done quicker in the vertical position using climbform than in an inclined position using a formwork for cantilever construction. Nowadays the method of lowering of arches is regularly applied for bridges in Japan and Spain. Figure 4 shows a picture of the lowering process of an arch half during the construction of the Arnoia Bridge in the year 2012. Each of the two arch halves had a length of approximately 70 m and a weight of 11000 kN.

A tension force of 1100 kN was required in order to move the arch half from the vertical position to an inclined starting position for the lowering process. The cable elongation during the lowering of each arch half was equal to 33 m and the maximum force in each of the two tendons was equal to 2400 kN.

The comparison of construction methods for arches and bridge girders shown in Table 1 reveals that there are counterparts for the construction on falsework and the cantilever construction method. However, there was no method with a vertical production of the bridge girder proposed previously. The idea for the balanced lift method was conceived when a construction method was sought which starts with a vertical production of the bridge girders as a counterpart to the method of lowering of arch

Fig. 4. Arnoia Bridge in Spain – lowering of arches (© VSL Hravy Lifting, Switzerland)

halves. The balanced lift method is a new construction method. This statement is proven by the patents granted in Germany, USA, Russia, Canada, China, Japan, Australia and Europe.

Table 1. Comparison of construction methods for arches and bridge girders

Arches	Bridge girder
Erection on falsework	Erection on falsework
Cantilever construction method	Balanced cantilever method
Lowering of arches	Balanced lift method

2   Bridge Across the River Lafnitz

The design of bridges according to the balanced lift method will be described in this section for the example of two bridges in the south-eastern part of Austria. For the new S7 motorway "Fürstenfelder Schnellstraße" between Riegersdorf and the national border between Austria and Hungary, the rivers Lafnitz and Lahnbach must be crossed. The lengths of the Lafnitz Bridge and the Lahnbach Bridge are roughly 120 m and 100 m, respectively. The cross section of the S7 motorway (Fig. 5) in this line section is traced out for two separate directed lanes, therefore the bridges across the rivers should be erected separately each with a width of 14.5 m, regarding prospective reconstruction measures.

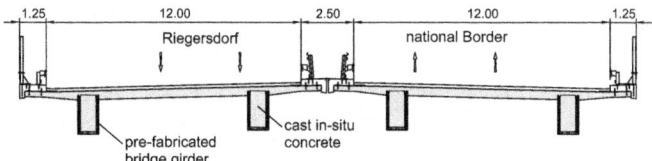

Fig. 5. Cross section of the bridge across the River Lafnitz

The construction areas where the two bridges for the S7 motorway are to be built are ecologically sensitive and part of the nature reserve "Natura 2000". The bridges are basically needed to cross the rivers and to provide options for a deer pass. To avoid encroachment into the natural habitat, an erection on falsework is not accepted by the highway management company ASFINAG. The construction site should be as small as possible and kept to the central pier and abutments. To meet all these requirements a construction of the bridges would only be possible by the balanced cantilever method, incremental launching or the balanced lift method.

Before the alternative design using the balanced lift method was introduced, the plan was to build the bridges by incremental launching of steel bridge girders (Fig. 6). The cross section was, in order to withstand the bending moments during the launching process, very high compared to the cross section height achieved with the balanced lift method. The big difference in heights, 4.6 m versus 2.0 m, can be achieved due to the compression struts which reduce the span lengths immensely (Fig. 7). The alternative design for the post-tensioned concrete bridges was based on a cross-section with a plate girder as shown in Fig. 5. It was proposed to build the central section of the webs by the balanced lift method as shown in Fig. 7, to install the end sections of the webs by mobile cranes placed behind the abutments, and to build the deck slab similarly to the original design by a formwork carriage. In the course of the preparation of the alternative design, the abutments and the locations of the central piers were rotated in plan by 30° with respect to the longitudinal axis of the bridge in order to react to the location of the

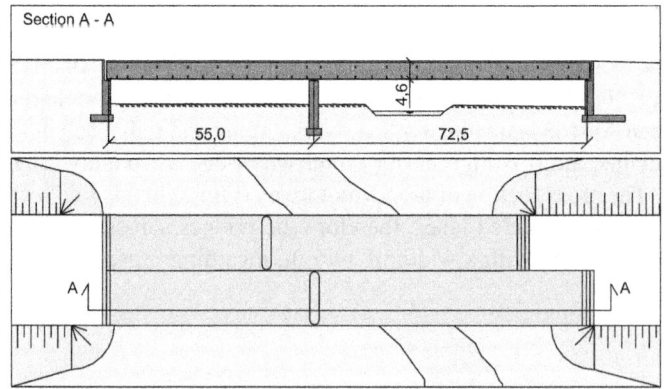

Fig. 6. Original design: steel-concrete-composite bridge

riverbed and to provide an improved design for the deer pass. These changes resulted in a bridge design with two equal spans.

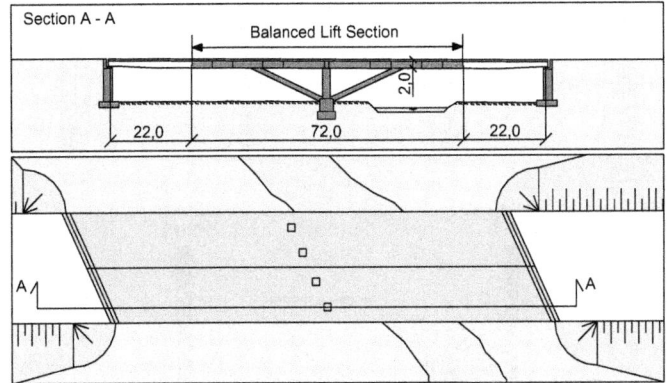

Fig. 7. Design based on the balanced lift method for a post-tensioned concrete bridge

It could be shown that the construction costs for the post-tensioned concrete bridges erected with the balanced lift method amounted to only 70% of the calculated costs of the composite bridges. When the Austrian highway management company ASFINAG became convinced of the financial benefits from a design based on the balanced lift method, a detailed design for the two bridges for crossing the rivers Lafnitz and Lahnbach was commissioned.

The first steps in the construction process of the S7 motorway bridges will not be any different from using conventional bridge construction methods. The foundations, the abutments and the piers must be cast. If a bridge with low piers is being built, an auxiliary pier, which is connected to the concrete pier, is needed. The auxiliary pier consists of two sections of a tower crane positioned on both sides of the pier and connected by a platform at the top. The compression struts, which are, in order to decrease weight, made out of hollow reinforced precast concrete elements with small element thickness, are assembled adjacent to the pier (Fig. 8 Construction phase 1). When using the balanced lift method, the weight of the bridge girders during the lifting (for bridges with high piers) or lowering (for bridges with low piers) operations is of utmost importance. This is why not only the compression struts but also the bridge girders are made out of thin precast elements.

To enable the construction of bridges of different sizes and span lengths a variety of bridge girders with different cross sections has been developed. The bridge girders with the chosen cross section are then assembled adjacent to the pier and the compression struts (Fig. 8 Construction phase 2). The bridge girders were designed as U-shaped thin walled prefabricated elements. The wall thickness was equal to 70 mm and the thickness of the bottom slab equal to 120 mm. The bridge girders with a total length of 35.5 m had to be divided into two parts of 19 m and 16.5 m respectively, because of the limited space available at the construction site. The joining of the two parts of the bridge girder

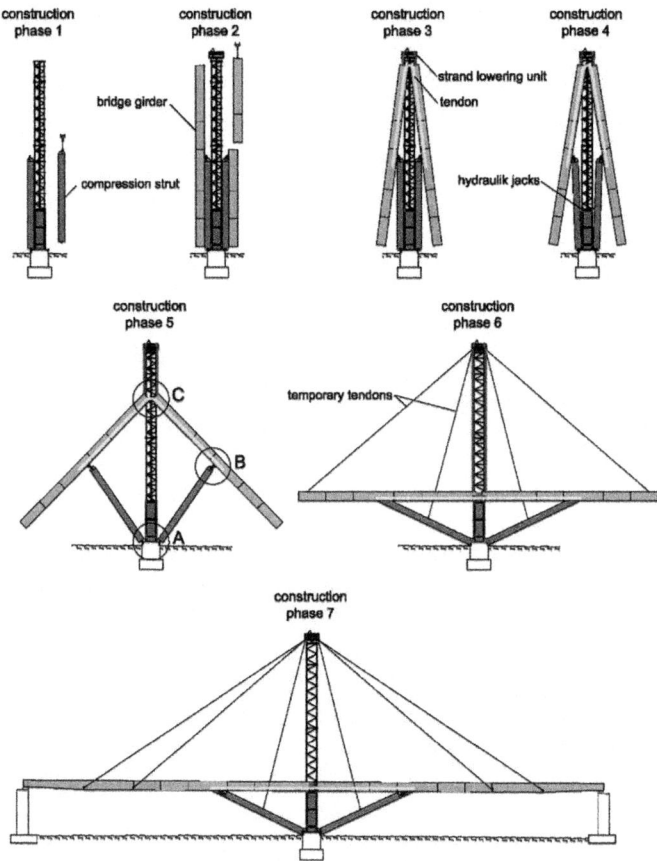

Fig. 8. Construction phases for the Lafnitz Bridge

is accomplished by a 20 mm joint with high-strength mortar and four monostrands which ensure a compressive stress at the joint in the subsequent construction phases.

In order to start the lifting or lowering operation, two bridge girders must be connected. This is achieved by tilting the bridge girders and by the installation of a tendon consisting of 16 monostrands (Fig. 8 Construction phase 3). This tendon, indicated in Fig. 8, has to ensure equilibrium during the lowering process by carrying the tensile forces. After the monostrands are slightly stressed, the lowering operation is started.

Similar to the method of lowering of arch halves, which requires that the arch halves are moved from the initial vertical position to a stable starting position for the lowering process, the bridge girders also have to be shifted to a secure starting position. The rotation of the bridge girders from the vertical position to the tilted position shown in construction phase 3 in Fig. 8, can be achieved easily by applying small horizontal forces to the top ends of the bridge girders. However, the small change in shape of the mechanism consisting of bridge girders, connecting tendon, and compression struts from construction phase 3 to construction phase 4, has to be carried out in a controlled symmetrical manner, either by pulling at the lower ends of the bridge girders or by pushing apart the compression struts. In order to restrict the works to the area of the central pier only, the second option was chosen. Two hydraulic jacks have to be fixed to the top of the pier. A horizontal force of 20 kN and a stroke of 275 mm for each jack is sufficient to achieve the transformation from construction phase 3 to construction phase 4 of Fig. 8.

In construction phase 5 the lowering process is carried out as is shown in Fig. 8. The relationship between the total force in the strand lowering units and the vertical movement of the top point of the bridge girders, designated as joint C in Fig. 8, is shown in Fig. 9. The whole mechanism is a statically determined system during the lowering process like in the method of lowering of arch halves.

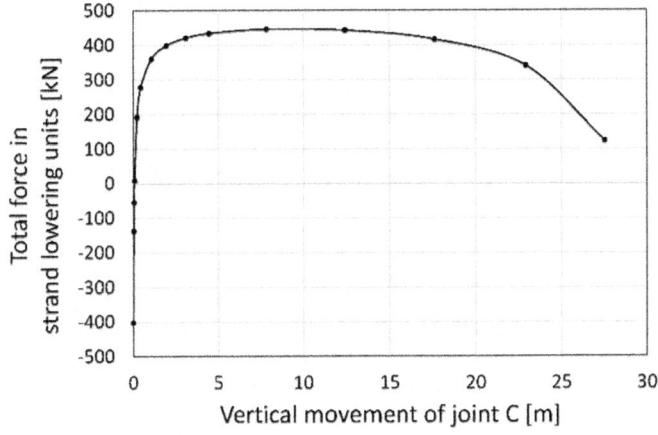

Fig. 9. Relationship between total force in the strand lowering units and the vertical movement in Joint C

An advantage over the method of lowering of arches is the fact that in the balanced lift method all forces are in equilibrium within the structural system consisting of bridge girders, connecting tendon and compression struts and therefore no forces have to be carried by stays and anchored to the ground. A closer examination of Fig. 9 shows that the relationship of force and vertical movement is strongly nonlinear and even starts with a negative force of 400 kN. A negative force indicates that in the beginning of the lowering process a vertical force pushing downward on top of the auxiliary pier would have to be applied. After a small movement of only 70 mm this force changes to a tensile force. Obviously, pushing downward at joint C to start the lowering process is an impractical and even dangerous option. Therefore, a design with two horizontal jacks at the top of the pier was prepared and a method statement was worked out for the interaction between horizontal jacks and strand lowering units. When the maximum stroke of the horizontal jacks equal to 500 mm is reached, the tensile force in the strand lowering units is already equal to 80 kN.

The bridge girders are then slowly rotated to the final horizontal position (Fig. 8 Construction phase 5). The maximum total force in the strand lowering units during this operation is equal to 650 kN and decreases to 150 kN in the final position. The compression struts are crucial to the rotating process. Once the bridge is in its final position, the importance of the compression struts does not diminish, since they become an integral part of the finished bridge. Due to the compression struts the span lengths are decreased therefore enabling a construction of a slimmer bridge in comparison to a bridge without compression struts.

Design drawings of the joints A, B, and C in the starting position and the final horizontal position are shown in Fig. 10. The rotation of approximately 75° and 165° at joints A and B, respectively, require careful detailing of the steel plates and the connection to the concrete, taking into account all eccentricities which occur during the change of geometry when the bridge girders are transformed from the vertical starting position to the final horizontal position. The rotations of the bridge girders of approximately 90° at joint C are enabled by a saddle which guides the connecting tendons (Fig. 10).

Once the balanced lift part has been completed the geometry of the bridge girder is corrected by releasing or stressing the connecting tendon. By filling the nodes A and C with in-situ concrete the geometry of the system is frozen. After the installation of temporary tendons (Fig. 8 Construction phase 6) the compression struts and the bridge girders at nodes B are filled with in-situ concrete.

The construction phase 7, shown in Fig. 8, consists in installing prefabricated bridge girder elements connecting the balanced lift part with the abutments. After adjusting the forces in the temporary tendons, the U-shaped prefabricated elements forming the bridge girder are filled with layers of cast in-situ concrete. Once the concrete has hardened, the temporary tendons are released and the construction phases 1 to 7 are repeated for the construction of the second web of the bridge. The installation of a bridge deck with the aid of a formwork carriage completes the bridge construction (Figs. 5 and 7).

3   Large Scale Test of the Balanced Lift Method

In the course of a research project exploring applications of thin-walled concrete elements in bridge construction, it became possible to carry out a large scale test of the balanced lift part of the bridges described in the previous section. The design of all elements (pier, bridge girders, compression struts) was based on a 70% scale of the designs of the S7 motorway bridges, therefore the test structure had a total length of 50.4 m.

As in the original designed structure the bridge parts were made of thin precast elements in order to keep the weight of the parts, which would be rotated during the balanced lift operation, to a minimum. As explained before, the 25 m long bridge girder walls consist of 70 mm thick concrete elements which are normally used as slab elements in buildings. The outside dimensions of the cross-section of the bridge girders corresponded exactly with being a 70% scaled down cross section of the S7 bridges. The girder had a height of 1.26 m and the width varied from 0.7 m to 1.4 m, with the larger width at the connection points with the compression struts (Fig. 12). However, the

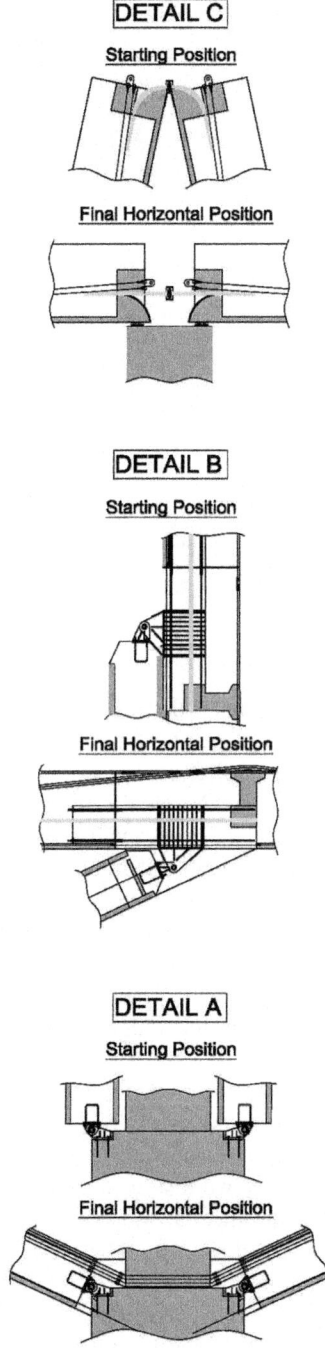

Fig. 10. Joints A, B and C before and after the lowering process

transverse
concrete beam

anchorage for
monostrands

lifting point

saddle for 12
monostrands

Fig. 11. View of end section of bridge girder

thickness of the elements with 70 mm for the side walls and 120 mm for the bottom slab
was identical to the design of the bridges for the S7 motorway.

Since the thin-walled bridge girder, consisting only of the U-shaped reinforced
concrete and with a weight of 208 kN, would have been too fragile for transport,
assembly or any lifting procedures, the cross section had to be enhanced. A truss made
of reinforcing bars (transverse bars with a diameter of 20 mm and diagonals with a
diameter of 12 mm) were welded to reinforcement bars protruding from the precast side
wall elements at the top side of the cross-section. With the help of the truss, the U-shaped
cross section was converted into a box section which turned out to be stable enough for
all further assembly and lifting operations. To enable future post-tensioning operations
and for the installation of monostrands, the bridge girders were equipped with transverse
concrete beams.

The transverse concrete beam and the saddle for the tendon at joint C is displayed
in the photograph of the end section of the bridge girder (Fig. 11), corresponding to
"Detail C" of Fig. 10. The transverse concrete beam carried the force of four monostrands. Two of these monostrands were stressed before the bridge girder was transported to the construction site. The other two monostrands were stressed when the bridge
girder was assembled in the vertical position. The lifting point is also anchored in the
transverse beam as can be seen in Fig. 11.

For the compression struts, a U-cross-section would have been impractical during
the later filling with concrete, therefore the 13 m long compression struts were pre-cast
as hollow reinforced concrete elements with a width of 1.4 m, a height of 0.875 m and
a weight of 15 tons each. The side walls were as thick as the bottom slab of the bridge

girders. The top and bottom slabs were cast with a thickness of 112 mm in order to integrate the anchor cones for the tie rods. The tie rods would have the job of carrying part of the concrete pressure during the filling of the hollow compression struts with cast in-situ concrete.

Due to the fact that the S7 bridges are bridges with low piers and this test structure is based on the design of the S7 bridge, an auxiliary pier was needed for the mounting process. For this reason, two 24 m long and 102 kN heavy sections of a tower crane with plan dimensions of 1.2 m by 1.2 m, which would serve as auxiliary piers, were fixed to the foundation slab. Two steel C-profiles were used in order to create vertical guiding rails between the two tower crane sections.

The top parts of the steel profiles were not connected with each other and the distance between the vertical guiding rails amounted to 800 mm.

The hinge connection in the bottom of the structure consisted of 30 mm thick steel plates at the ends of the compression struts and concrete filled steel tubes with an outer diameter of 150 mm placed on the foundation. The steel plates had to be positioned very accurately on the concrete filled tubes to provide a proper hinge connection. After the

Fig. 12. The large-scale test structure during the mounting

compression struts had been placed in a correct position they were temporarily fixed to the pier and the auxiliary piers.

The prefabricated bridge girders could then be transported from the prefabrication plant to the stockyard. A 100 ton mobile crane lifted one bridge girder after the other (Fig. 12) to be fixed to the top end of the compression struts using four 16 mm bolts. Even though the cylindrical openings in the joining steel plates only had a tolerance of 1 mm the assembly operation was carried out without any problems. A close-up view of the connection of the bridge girder to the compression strut, which corresponds to "Detail B" in Fig. 10, can be seen in Fig. 13.

Fig. 13. Connection detail of bridge girder and compression strut.

After only four days of assembly, the bridge was ready for the balanced lift part of the construction process. In order to start the rotating process, the bridge girders had to be tilted so that the top ends would touch each other. Subsequently, twelve monostrands were installed in the two bridge girders. The monostrands were anchored at the connecting points of each bridge girder with the compression struts. In the top section of the bridge girders, circular saddles with a radius of 544 mm had been formed during pre-casting (Fig. 11). After the installation of the twelve monostrands, the strands were slightly stressed to prevent sagging.

The lowering process was achieved by slowly and simultaneously lowering the two top points of the bridge girders with the aid of two mobile cranes (Fig. 14). The maximum lifting force, which had been calculated to 270 kN, corresponded well with the lifting force measured by the cranes.

Fig. 14. Lowering of the top points of the bridge girders with the aid of mobile cranes

Fig. 15. Large-scale model of the balanced lift part of the Lafnitz Bridge

Fig. 16. Bridge girder with 70 mm prefabricated wall elements

After the bridge had been rotated from the vertical into the horizontal position, the geometry of the structure was checked. By stressing or relaxing the installed mono-strands the geometry of the structure could be easily adjusted. This turned out to not be needed since the test structure's geometry was within an acceptable range (Figs. 15 and 16).

With the aid of the large scale test it could be demonstrated that the balanced lift method works and that the joint details can be built using simple connection details with steel plates and bolts (joint B) and a saddle with tendons (joint C). The radius of the tendons at joint C is much smaller than in usual post-tensioning applications. The reduction of the ultimate force as a function of the curvature of the saddle was determined experimentally (Kollegger et al. 2012). It could be shown that the monostrands do not lose load carrying capacity even for a small radius of only 0.5 m.

4   Conclusions

By employing the balanced lift method, the spans of the bridge girders are reduced by the compression struts, thus enabling considerable savings in construction materials. The proposed method will be especially advantageous for bridges with high piers and span lengths between 50 and 250 m. The usage of temporary piers enables an economic application of the balanced lift method for bridges with piers of modest height, as for example the two bridges on the S7 motorway.

Another advantage of the balanced lift method is the fact that all assembly and mounting operations are concentrated at the pier and that the rotation of the bridge girders can be carried out much faster than by horizontal launching of the bridge girders. The small space requirements and the high construction speed might be of advantage when an obstacle like a railway line or a busy motorway has to be spanned by a bridge and the interruption of the traffic routes has to be kept to a minimum.

Acknowledgements. The financial support of the large scale test by Österreichische Forschungsförderungsgesellschaft (FFG), ASFINAG, ÖBB Infrastruktur AG and Vereinigung Österreichischer Beton- und Fertigteilwerke (VÖB) is gratefully acknowledged. The good cooperation with Schimetta Consult GmbH during the detailed design of the bridges for the S7 motorway, and with Franz Oberndorfer GmbH & Co. KG during the construction of the large scale test structure in Gars am Kamp, is also gratefully acknowledged.

The authors are also thankful to Wilhelm Ernst & Sohn Verlag GmbH & Co. KG for the permission to use material in this paper, which was previously published in the journal "Structural Concrete" (Kollegger et al. 2014).

References

Berger J, Bruschetini-Ambro Z, Kollegger J (2011) An innovative design concept for improving the durability of concrete bridges. Struct. Concr. 12(3):155–163

Fédération international du béton (fib). Awards given out every 4 years to Outstanding Concrete Structures. 2014 fib Awards for Outstanding Concrete Structures – International Federation of Structural Concrete (2014)

Kollegger J, Gmainer S, Lehner K, Simader J (2012) Ultimate strength of curved tendons. Struct. Concr. 13(1):42–55

Kollegger J, Foremniak S, Suza D, Wimmer D, Gmainer S (2014) Building bridges using the balanced lift method. Struct. Concr. 15(3):281–291

Troyano LF (2003) Bridge Engineering – A Global Perspective. Telford, Madrid

Fire Resistance

Improved Fire Resistance by Using Different Types of Cements

Éva Lublóy[1], Katalin Kopecskó[2], and György L. Balázs[1(✉)]

[1] Department of Construction Materials and Technologies, Budapest University of Technology
and Economics, Budapest 1521, Hungary
balasz@vbt.bme.hu
[2] Department of Engineering Geology and Geotechnics, Budapest University of Technology
and Economics, Budapest 1521, Hungary

Abstract. Composition and microstructure of hardened cement paste have important influences on the properties of concrete exposed to high temperatures. An extensive experimental study was carried out to analyse the post-heating characteristics of concretes subjected to temperatures up to 800 °C. Major parameters of our study were the content of supplementary materials (slag, fly ash, trass) of cement (0, 16 or 25 m%) and the value of maximum temperature. Our results indicated that (i) the number and size of surface cracks as well as compressive strength decreased by the increasing content of supplementary materials of cements due to elevated temperature; (ii) the most intensive surface cracking was observed by using Portland cement without addition of supplementary materials. The increasing content of the supplementary material of cement increased the relative post-heating compressive strength. Tendencies of surface cracking and reduction of compressive strength were in agreement, i.e. the more surface cracks, the more strength reduction.

Keywords: Fire resistance · Concrete · Cement · Supplementary materials

1 Introduction

Effects of high temperatures on the mechanical properties of concrete were studied as early as the 1940s (Schneider 1988). In the 1960s and 1970s fire research was mainly directed to study the behaviour of concrete structural elements (Kordina 1997). There was relatively little information available about the concrete properties during and after fire (Waubke 1973).

Recent fire cases called again the attention to fire research. Different concrete types may also mean different possibilities for fire design *(EN 1991-1-2; ACI/TMS 216; Guideline on Verification Method for Fire Resistance and Exemplars)* Environmental protection requires the use of cements with low clinker content. Concrete is a composite material that consists mainly of mineral aggregates bound by a matrix of hardened cement paste. Composition and internal structure of hardened cement paste has an important influence on the properties of concrete exposed to high temperatures (Khoury 2001). Characteristics of concrete during and after the heating process depend on the

© Springer International Publishing AG, part of Springer Nature 2018
M. di Prisco and M. Menegotto (Eds.): ICD 2016, LNCE 10, pp. 219–228, 2018.
https://doi.org/10.1007/978-3-319-78936-1_16

type of cement, the type of aggregate and the interaction between them (Khoury 1995; Khoury et al. 1995; Budelman 1987).

Effect of cement type has been hardly studied. Early studies on concretes made of the two commonly used cement types, Portland cement or Portland cement with low amount of slag indicated almost the same residual compressive strength over 400 °C (Schneider and Lebeda 2000; Schneider 1986). The properties of cement paste itself are also influenced by the mixing proportions of the constituents (fib 2007; Barbara and Iwona 2015; Yun and Hung-Liang 2015). Behaviour at high temperatures depends on parameters like water/cement ratio, amount of CSH (calcium-silicate-hydrates), amount of $Ca(OH)_2$ and degree of hydration. Different cement pastes can perform differently in fire (Grainger 1980).

Grainger (1980) studied cements with or without pulverised fly ash (PFA) subjected to a series of temperatures ranging from 100 °C to 600 °C with an interval of 100 °C. In his research the dosage of PFA was 20%, 25%, 37.5% and 50% of the total mass of binders. It was found that the addition of PFA could improve the residual compressive strength of cement paste. The beneficial effect of PFA as part of the binder was in good agreement with the results of Dias et al. (1990) and Xu et al. (2001). Khoury et al. observed that 400 °C was a critical temperature for Portland cement concretes, above which concretes would disintegrate on subsequent post cooling exposure to ambient conditions (Khoury et al. 1990).

In fire, the heated surface region of concrete loses its moisture content by evaporation from the surface and by migration into the inner concrete mass driven by the temperature gradient (Schneider and Weiß 1997). Due to high temperature, the structure and mineral composition of concrete changes. The analysis was made by thermogravimetry (TG). Around 100 °C the mass-loss is caused by water evaporating from the micropores. The decomposition of ettringite ($3CaOAl_2O_3 \cdot 3CaSO_4 \cdot 32H_2O$) takes place between 50 °C and 110 °C. At 200 °C there is a further dehydration, which causes small mass loss. The mass loss of the test specimens with various moisture contents was different till all the pore water and chemically bound water were gone. Further mass loss was not perceptible around 250–300 °C (Hinrichsmeyer 1987; EN 1992, 2002). During heating the endothermic dehydration of $Ca(OH)_2$ takes place between temperatures of 450 °C and 550 °C (1). This endothermic reaction is accompanied by loss of mass (Thielen 1994).

$$Ca(OH)_2 \rightarrow CaO + H_2O \uparrow \tag{1}$$

In the case of concretes made of quartz gravel aggregate, other influencing factor is the change in crystal structure of quartz α formation into $\rightarrow \beta$ formation at 573 °C. This transformation followed by volumetric increase that influences the strength detrimentally (Hoj 2005). Dehydration of calcium-silicate-hydrates was found at the temperature of 700 °C (Gambarova 2004).

Based on this information, we carried out an extensive experimental study to analyse the post-heating characteristics of concretes subjected to temperatures up to 800 °C. Major parameters of our study were types of supplementary materials (ground granulated blastfurnace slag, fly ash, trass) and the substitutions of cement clinker (0, 16 or 25 m%) in addition to the value of maximum annealing temperature.

2 Experimental Program

2.1 Cements

Main purpose of our experimental study was to determine the influences of supplementary materials (slag, fly ash, trass) content of cements on post-heating characteristics of hardened cement paste as well as that of concrete. The following cements were involved in the comparative study: *ordinary Portland cements* (CEM I 52,5 N; CEM I 42,5 R); *sulphate resistant Portland cement* (CEM I 42,5 N-S; *Portland slag cement* (CEM II/A-S 42,5 N); *Portland trass cements* (CEM II/A-V 42,5 R; CEM II/B-V 32,5 R) and *Portland fly ash cement* (CEMII/A-P 42,5 N).

Cement clinkers and additives were ground together during the production of cement. The oxidative compositions of cements are given in Tables 1a and 1b.

Table 1a. Chemical composition of tested cements (m%) (data provided by the former Holcim Hungária Ltd.)

	CEM I 42,5 R	CEM I 42,5 N-S	CEM II/A-V 42,5 R	CEM II/B-V 32,5 R	CEM II/A-P 42,5 N
SiO_2	19.71	20.16	23.75	26.15	28.23
Al_2O_3	4.46	3.83	6.68	7.79	6.10
Fe_2O_3	2.97	6.03	4.73	5.33	3.42
CaO	64.59	62.9	55.31	50.48	54.54
MgO	1.0	1.88	2.66	2.64	1.0
K_2O	0.69	0.43	0.85	0.95	1.15
Na_2O	0.31	0.41	0.47	0.51	0.67
SO_3	2.63	2.6	2.79	2.81	2.84
Cl	0.02	0.009	0.027	0.026	0.01

Table 1b. Chemical composition of tested cements (m%) (data by Duna-Dráva Cement Heidelberg Cement Group)

	CEM I 52,5 N	CEM II/A S 42,5 N
SiO_2	20.59	22.77
Al_2O_3	5.55	5.83
Fe_2O_3	3.21	2.97
CaO	65.02	60.30
MgO	1.44	2.51
SO_3	2.88	3.00
K_2O	0.78	0.80
Na_2O	0.11	0.13
Cl	0.0055	0.0056

2.2 Test Specimens and Variables

The test variables were:

– types of cements: Portland cements (ordinary and sulphate resistant), Portland slag cement, Portland trass cements and Portland fly ash cement;
– maximum temperatures of heat loading: 50 °C, 150 °C, 300 °C, 500 °C and 800 °C.

The test constants were:

– water to cement ratio (w/c = 0.43);
– cement content. In this paper the studied specimens were cement paste cubes);
– testing of specimens started at 28 days and finished at 30 days.

The studied characteristics were:

– development of surface cracking (studied by macroscopic observation);
– change in compressive strengths (relative residual compressive strengths) due to heat loadings.

2.3 Test Methods

The studied specimens were cube cement paste specimens with dimensions of 30 mm. Cast specimens were removed after 24 h from the formwork, then specimens were stored in water for 7 days and kept at laboratory conditions (temperature 20 ± 2 °C, 65 ± 5% relative humidity) until testing in accordance with the standard (28 days). The experiments on specimens finished at the age of 30 days.

Our experimentally applied heating curve was similar to the standard fire curve (in accordance with EN 1991.1.2) up to 800 °C. Specimens were kept for two hours at the actual maximum temperature levels. Specimens were then slowly cooled down in laboratory conditions for further observations. During the heat load a program controlled electric furnace was used. The compressive strength was measured on the heat loaded and, than cooled down specimens and the average values of the measurements were analysed.

3 Results and Discussions

Results on surface cracking and residual compressive strength after exposing to high temperatures are presented and discussed herein.

3.1 Development of Surface Cracks

Development of surface cracks as a result of the elevated temperatures is presented in Fig. 1.

Type of cement	Temperature of heat load	
	500 °C	800 °C
CEM I 52,5 N		
CEM I 42,5 R		
CEM I 42,5 N-S		
CEM II/A-S 42,5 N		
CEM II/A-P 42,5 N		
CEM II/A-V 42,5 R		
CEM II/B-V 32,5 R		

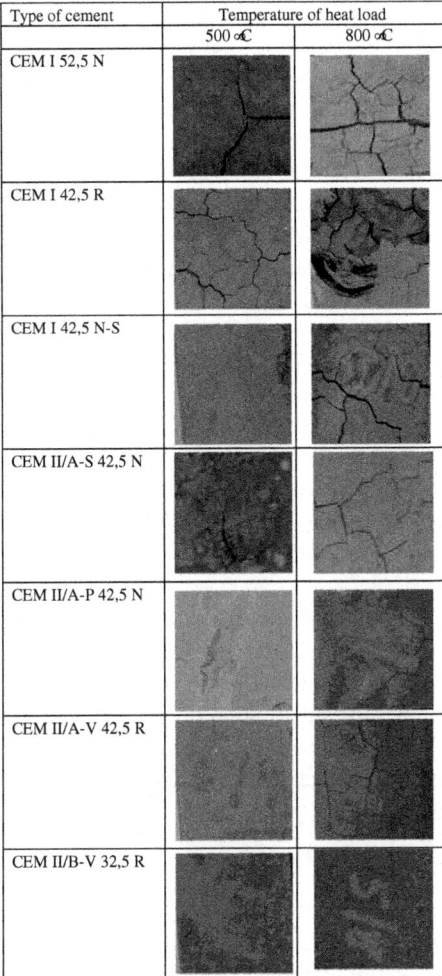

Fig. 1. Effect of cement type on the development of surface cracks as a result of the elevated temperature (hardened cement paste specimens, at the age of 30 days, w/c = 0.43)

Temperature loading on hardened cement paste specimens caused chemical and physical changes leading to surface cracks. There were no macroscopic observable changes on the surface of cubes due to heat loads up to maximum temperature 500 °C.

In the case of *Portland cement specimens* (CEM I 42,5 R, CEM I 52,5 N) cracks already formed by heating up to 500 °C, and the size and number of cracks considerably increased by heating up to 800 °C (see 1st and 2nd rows of Fig. 1). Crack development is explained by the chemical reactions in the hardened cement paste, i.e. dehydration of portlandite, $Ca(OH)_2$ at about 450 °C and decomposition of CSH at about 750 °C (see 1st and 2nd rows in Fig. 1).

In addition to these after cooling rehydration of CaO could considerably increase the extent of the crack development with further changes in volume. Rehydration of CaO takes place by the humidity of air and followed by expansion.

In the case of *sulphate resistant Portland cement* (CEM I 42,5 N-S) only small cracks appeared by heating up to 500 °C, and the amount and size of cracks increased in the case of specimens heating up to 800 °C (see 3rd row in Fig. 1). The hydration of aluminate phase (Brownmillerite) in *sulphate resistant Portland cement* is much slower than for tricalcium-aluminate (C_3A) in *ordinary Portland cement* (Kopecskó 2006). The unhydrated part of ferrite type aluminates could positively influence the resistance against high temperatures.

In the case of *Portland slag cements* (CEM II/A-S 42,5 N, CEM II/B-S 32,5 R), *Portland trass cements* (CEM II/A-V 42,5 R, CEM II/B-V 32,5 R) *and Portland fly ash cement* (CEM II/A-P 42,5 N) *specimens* only small cracks appeared by heating up to 500 °C, and the amount and size of cracks increased in the case of specimens heating up to 800 °C (see 4th row in Fig. 1).

With the increasing substitution of cement clinker by the supplementary material the production of portlandite during hydration in the hardened cement paste decreases; thus, less portlandite dehydrates at about 450 °C, which may give the explanation of decreased amount of cracks.

The supplementary materials are usually latent hydraulic (GGBS) or pozzolanic (trass, fly ash) materials. During the hydration process they consume part of the portlandite forming calcium-silicate-hydrates. The other cause of the decrease of portlandite is the diluting effect: the higher the substitution rate of the clinker, the smaller the amount of portlandite formed. The rate (speed) of pozzolanic reaction is influenced by many factors such as the specific surface/average grain size of the clinkers/supplementary materials, the type of SCMs, etc.

3.2 Compressive Strength

In Fig. 2 the compressive strengths of the hardened cement paste specimens are presented related to the compressive strength measured at 20 °C ($f_{c,T}/f_{c,20}$ called residual relative compressive strength) as functions of the maximum temperature and the cement type.

The relative residual compressive strength decreases up to 150 °C heat loading, then for some cement types slight increase is observable up to 300 °C. In the case of higher temperatures than 300 °C the residual relative compressive strength decreases again (Fig. 2). Specimens loaded up to 300 °C show higher residual strength comparing with the average strengths measured on specimens loaded up to 150 °C because the intensive dehydration in the temperature interval between 60 and 180 °C probably causes the hydration of the unhydrated cement grains in the microstructure.

In the case of *Portland cement specimens* (CEM I 42,5 R, CEM I 52,5 R, CEM I 42,5 N-S) the average of residual relative compressive strength of the test specimens was 28%, 35% and 45% heating up 500 °C and further 1%, 10% and 28% by heating up to 800 °C (Fig. 3).

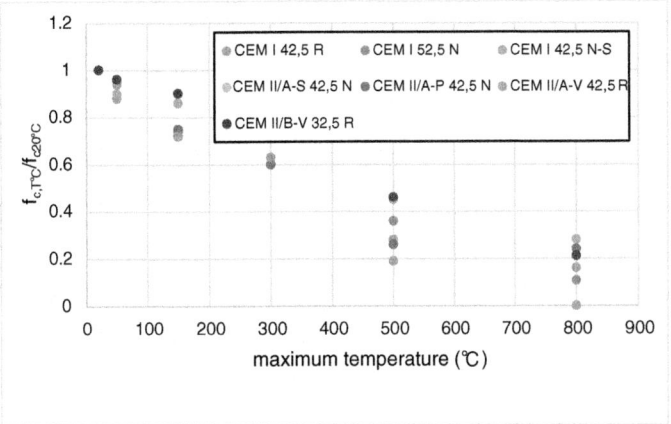

Fig. 2. Residual compressive strength of hardened cement paste with different cement types (strength values are related to strength values of 20 °C); Reference values (N mm^{-2}): CEM I 42,5 R, 76.36; CEM 52,5 N, 78.5; CEM I 42,5 N-S 57.17; CEM II/A-S 42,5 N, 98.5; CEM II/A-P 42,5 N 57.62, CEM II/A-V 42,5 R, 47.34; CEM II/B-V, 48.52.

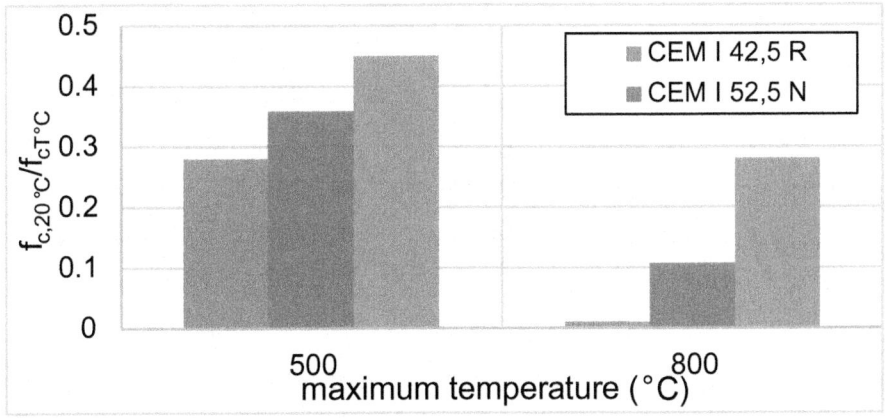

Fig. 3. Residual compressive strength of hardened cement paste with different cement types (strength values are related to strength values of 20 °C); Reference values (N mm^{-2}): CEM I 42,5 R, 76.36; CEM I 52,5 N, 78.5; CEM I 42,5 N-S 57.17.

For *Portland slag cement specimens* (CEM II/A-S 42,5 N) the average of the residual relative strength was 41% by heating up to 500 °C and further 17% heating up to 800 °C, respectively (Figs. 4 and 5).

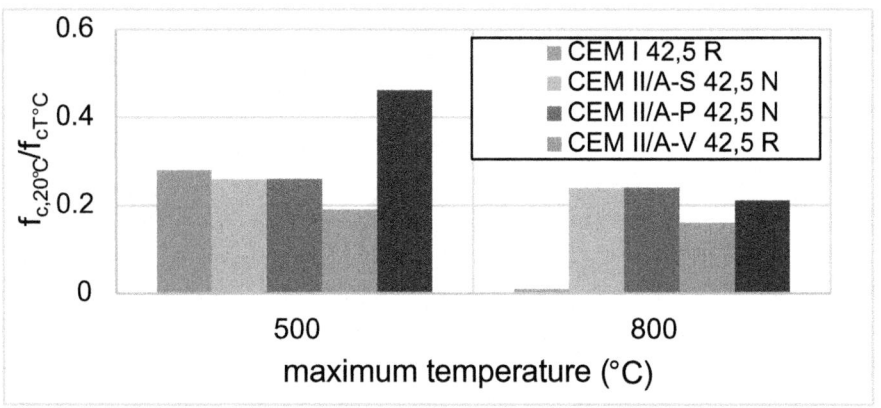

Fig. 4. Residual compressive strength of hardened cement paste with different cement types (strength values are related to strength values of 20 °C); Reference values (N mm^{-2}): CEM I 42,5 R, 76.36; CEM II/A-S 42,5 N, 98.5; CEM II/A-P 42,5 N 57,62, CEM II/A-V 42,5 R, 47.34; CEM II/B-V, 48.52.

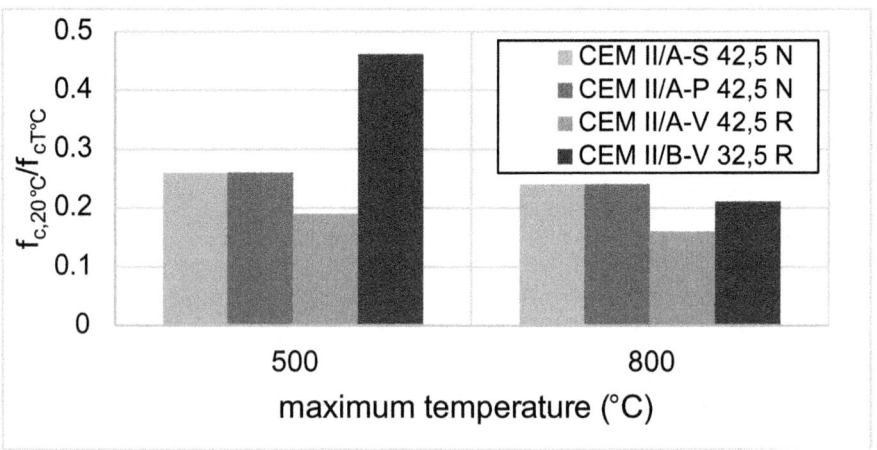

Fig. 5. Residual compressive strength of hardened cement paste with different cement types (strength values are related to strength values of 20 °C); Reference values (N mm^{-2}): CEM I 42,5 R, 76.36; CEM II/A-S 42,5 N, 98.5; CEM II/A-P 42,5 N 57,62, CEM II/A-V 42,5 R, 47.34; CEM II/B-V, 48.52.

For *Portland trass cement specimens* (CEM II/A-V 42,5 R) the average of the residual relative strength was 19% by heating up to 500 °C and further 16% heating up to 800 °C, respectively (Figs. 4 and 5).

For *Portland fly ash cement specimens* (CEM II/A-P 42,5 N; CEM II/B P 32,5 R) the average of the residual relative strength was 26% and 46% by heating up to 500 °C and further 24% and 21% heating up to 800 °C, respectively (Figs. 4 and 5).

The results of the compressive strength tests are in accordance with findings on crack development. The most significant cracks appeared on test specimens made of Portland cement, and the compressive strength loss was also the highest.

4 Conclusions

The purpose of the present study was to analyse the post-heating characteristics of hardened cement paste. Main experimental parameters were the cement types (7 different types: ordinary Portland cements, sulphate resistant Portland cement, Portland slag cement, Portland trass cements, Portland fly ash cement) and the maximal temperature of heat treatment up to 800 °C (20 °C, 50 °C, 150 °C, 300 °C, 400 °C, 500 °C, 600 °C, 800 °C), respectively.

Water to cement ratio was kept constant (w/c = 0.43). Present studies included analysis of surface cracking, compressive strength.

The following conclusions can be drawn from our test results:

1. The composition of cement has an important influence on the post-heating characteristics of cement paste.
2. Relative post-heating compressive strength increases with addition of the supplementary material to the cement.
3. The post-heating behaviour of hardened cement paste specimens were negatively influenced by rapid type cements both for surface cracking as well as for compressive strength.
4. The application of sulphate resistant Portland cement was found to be more favourable at high temperatures.
5. Amount of surface cracking (sum of lengths and widths) is reduced with addition of supplementary materials (here GGBS, fly ash and trass) to the cement.

References

American Concrete Institute (1997) ACI/TMS 216 standard method for determining fire resistance of concrete and masonry construction assemblies. Framington Hills, Oakland

Barbara P, Iwona W (2015) Comparative investigations of influence of chemical admixtures on pozzolanic and hydraulic activities of fly ash with the use of thermal analysis and infrared spectroscopy. J Therm Anal Calorim 120:119–127

Budelman H (1987) Strength of concrete with different moisture content after elevated temperature (Zum Einfluss erhöhter Temperatur auf Festigkeit und Verformung von Beton mit unterschiedlichen Feuchtegehalten), Heft 76, Braunschweig. ISBN 3-89288-016-6

Dias WPS, Khoury GA, Sulivan PJE (1990) Mechanical properties of hardened cement paste exposed to temperatures up to 700 °C. ACI Mater J 87:160–166

Eurocode 1 (1991) Basis of design and actions on concrete structures. Part 2-2: actions on structures exposed to fire, EN 1991-1-2:2002, November 2002

Eurocode 2 (1992) Design of concrete structures. Part 1 general rules - structural fire design EN 1992-1-2:2002, 25 February

fib bulletin 38 (2007) Fire design of concrete structures- materials, structures and modelling. ISBN 978-2-88394-078-9

Gambarova PG (2004) Opening addresses on some key issues concerning R/C fire design. In: Gambarova PG, Felicetti R, Meda A, Riva, P (eds) Proceedings for fire design of concrete structures: what now? What next? 2–3 December 2004

Grainger BN (1980) Concrete at high tempereatures. Central Electricity Research Laboratories, Leatherhead

Hinrichsmeyer K (1987) Analysis and modeling of concrete subjected to high temperature (Strukturorientierte Analyse und Modellbeschreibung der thermischen Schädigung von Beton) Heft 74 IBMB, Braunschweig

Hoj NH (2005) Fire design of concrete structures. In: Balázs L, Borosnyói A (eds) Proceedings of fib symposium on keep concrete attractive, Budapest, 23–25 May 2005, p 1097–1105

Khoury GA et al (2001) Fire design of concrete materials structures and modelling. In: 1st fib Congress, Osaka, Japan, October 2001

Khoury GA (1995) Effect of heat on concrete material. Imperial College report, p 73

Khoury GA, Grainger BN, Sullivan PJE (1995) Transient thermal strain of concrete: literature review, conditions within specimen and behaviour of individual constituents. Mag Concr Res 37:48–56

Khoury GA, Sarshar R, Sulivan PJE (1990) Factors affecting the compressive strength of unsealed cement paste and concrete at elevated temperatures up to 600 °C. In: Proceedings of 2nd international workshop on mechanical behaviour of concrete under extreme thermal and hygral conditions, Weimar, p 89–92, ISSN 0863-0720

Kopecskó K (2006) A gőzölés hatása a cement klinkerek és cementek kloridion megkötő képességére. In English: Chloride ion binding capacity of clinker minerals and cements influenced by steam curing. PhD thesis, p 100

Kordina K (1997) Fire resistance of reinforced concrete beams. (Über das Brandverhalten punktgeschützter Stahbetonbalken), Deutscher Ausschuss für Stahlbeton, Heft 479, Beuth Verlag GmbH, Berlin, ISSN 0171-7197

Ministry of Construction (2001) Taika Seinou Kenshouhou no Kaisetu Oyobi Keisanrei tosonso Kaisetu. Guideline on Verification Method for Fire Resistance and Exemplars. Inoue Shoin

Schneider U (1988) Concrete at high temperatures - a general review. Fire Saf J 13:55–68

Schneider U, Lebeda C (2000) Fire protection of buildlings. (Baulicher Brandschutz). W. Kohlhammer GmbH, Stuttgart, ISBN 3-17-015266-1

Schneider U (1986) Properties of materials at high temperatures, concrete, 2nd edn. RILEM Publ., Gesamthochschule Kassel, Universität Kassel

Schneider U, Weiß R (1997) Kinethical treatment of thermical deterioration of concretes and its mechanical influences. Cem Concr Res 11:22–29

Thielen KC (1994) Strength and deformation of concrete subjected to high temperature and biaxial stress-test and modeling, (Festigkeit und Verformung von Beton bei hoher Temperatur und biaxialer Beanspruchung Versuche und Modellbildung), Deutscher Ausschuss für Stahlbeton, Heft 437, Beuth Verlag GmbH, Berlin, ISSN 0171-7197

Xu Y, Wong YL, Poon CS, Anson M (2001) Impact of high temperature on PFA concrete. Cem Concr Res 31:1065–1073

Yun L, Hung-Liang C (2015) Thermal analysis and adiabatic calorimetry for early-age concrete members. J Therm Anal Calorim 122:937–945

Waubke NV (1973) Physical analysis of strength reduction of concrete up to 1000 °C. (Über einen physikalischen Gesichtspunkt der Festigkeitsverluste von Portlandzement-betonen bei Temperaturen bis 1000 °C Brandverhalten von Bauteilen), Dissertation, TU Braunschweig

Robustness Against Accidental Actions

Dynamics of Strongly Curved Concrete Beams by Isogeometric Finite Elements

Flavio Stochino[1]([✉]), Antonio Cazzani[1], Gian Felice Giaccu[2], and Emilio Turco[2]

[1] Department of Civil, Environmental Engineering and Architecture, University of Cagliari, Cagliari, Italy
fstochino@gmail.com
[2] Department of Architecture, Design and Urban Planning, University of Sassari, Alghero, Sassari, Italy

Abstract. The standard finite elements approach for the dynamics of curved beam is usually based on the same energy functional used for straight beam, in other words an energy form that is essentially derived from de Saint–Venant's theory. In case of strongly curved elements this approximation yields to not negligible errors, in particular for stress assessments. For this reason, in this work a different formulation, based on the Winkler's simple kinematic assumptions, is adopted. In this way a non diagonal constitutive matrix is obtained and the computational efficiency of NURBS (Non Uniform Rational B–Splines) shape functions is added to an accurate representation of the constitutive relations. In this paper the natural frequencies and mode shapes of plane curved concrete beams are obtained. Computational cost and results accuracy is assessed with respect to closed form solutions and literature results.

Keywords: Strongly curved beams · Isogeometric analysis · Beam dynamics

1 Introduction

Curved concrete bridge structures are widespread throughout the world. From the biggest one, the Mike O'Callaghan–Pat Tillman Memorial Bridge on the Colorado river (US) characterised by 323 m span, to small overpasses, see Fig. 1, the structural characteristics are the same. Indeed, this kind of structure converts part of its vertical load into horizontal forces that are transmitted to the soil by the foundations. The advantages of a mostly compressive behaviour are patent in case of concrete structure given the non-symmetric material stress-strain law.

The dynamics of this kind of structures have been a current research trend: see Sevim et al. (2016), Li et al. (2016), Zong et al. (2016), Kadkhodayan et al. (2016), Cazzani et al. (2016a). Indeed, the effects of earthquake, wind and impulsive loadings (see Stochino and Carta 2014; Stochino 2016) make the dynamic behaviour of paramount relevance.

Analytical solutions for this kind of problems are not always available; in this case the numerical solution is the only way to obtain a response. Often, complex structures

© Springer International Publishing AG, part of Springer Nature 2018
M. di Prisco and M. Menegotto (Eds.): ICD 2016, LNCE 10, pp. 231–247, 2018.
https://doi.org/10.1007/978-3-319-78936-1_17

Fig. 1. Concrete arch overpass in Germany.

like bridges are discretized by finite elements, see Buffa et al. (2015); Stochino et al. (2018) or boundary elements, see Aristodemo and Turco (1994).

The standard approach for the dynamics of curved beam is usually based on the same energy functional used for straight beam, in other words an energy form that is essentially derived from de Saint–Venant's theory. In case of strongly curved elements this approximation yields to not negligible errors, in particular for stress assessments. For this reason, in this work a different formulation, based on the Winkler's simple kinematic assumptions, is adopted. In this way a non diagonal constitutive matrix is obtained and the computational efficiency of NURBS (Non Uniform Rational B–Splines) shape functions, see Cazzani (2016b; c; d; e; f) and Bilotta et al. (2010), is added to an accurate representation of the constitutive relations.

In this paper the natural frequencies and mode shapes of plane curved concrete beams are obtained. Computational cost and results accuracy is assessed with respect to closed form solutions and literature results.

2 Problem Statement

2.1 Constitutive Laws and Modal Analysis

Vibration analysis can be developed starting from Hamilton's principle. In the case of zero external forces, it states:

$$\delta \int_{t_1}^{t_2} \int_0^l (\Xi - \mathrm{T})ds\, dt = 0 \tag{1}$$

where Ξ and T respectively are the elastic strain energy density and the kinetic energy density.

For a plane curved beam the elastic strain energy density can be written in a general form as:

$$\Xi = \frac{1}{2}E\left(c_{11}\varepsilon^2 + 2c_{12}\varepsilon\chi + c_{22}\chi^2 + \frac{c_{33}}{2(1+v)}\gamma^2\right), \tag{2}$$

where ε, χ and γ are the strain parameters, c_{ij} are parameters which depend on the geometry and E and v represent the elastic Young's modulus and the Poisson's coefficient of the material. Almost always the choice for this parameters derives from those derived from de Saint–Venant solution for a straight beam, *i.e.* $c_{11} = A$, $c_{12} = 0$, $c_{22} = I$ and $c_{33} = A_t$, A and I are, respectively, the area and the inertia of the cross–section, while A_t denotes the shear effective area. This choice produces acceptable results except for the strongly curved beams.

In this work the simple and effective Winkler's approach, see Winkler (1858), is also considered. It is based on two assumptions:

- cross sections of curved beam remain plane;
- the stress distribution on the cross section follows a hyperbolic law.

Following Winkler's assumption, the elastic coefficients are defined as:

$$c_{11} = \int \frac{R}{R-y} dA \tag{3}$$

$$c_{12} = \int \frac{Ry}{R-y} + \frac{y^2}{R-y} dA \tag{4}$$

$$c_{22} = \int \frac{Ry^2}{R-y} dA \tag{5}$$

$$c_{33} = \int \frac{R^2}{B^2(R-y)^2} \left(\frac{C_{11}}{C_{22}} - \frac{\Omega}{RA} \right)^2 dA, \tag{6}$$

where R is the curvature radius, A is the beam cross section and Ω is the portion of the cross- section area lying below the value of y.

Strain parameters depend on displacements u, w and φ, in formulae:

$$\varepsilon = u' - w/R$$

$$\gamma = w' + \frac{u}{R} + \varphi, \tag{7}$$

$$\chi = \varphi',$$

where $(\cdot)'$ denotes the derivative respect to the arc–length parameter s.

Kinetic energy density can be expressed as:

$$T = \frac{1}{2} \rho \left(A \left(\dot{u}^2 + \dot{w}^2 \right) + I\dot{\varphi}^2 \right), \tag{8}$$

where ρ is the mass density and the dot indicates the time derivative.

Using whatever finite element discretization, elastic strain energy and kinetic energy can be approximated as:

$$\epsilon = \int_0^l \varXi ds \approx \frac{1}{2}\mathbf{q}^{\mathrm{T}}\mathbf{Kq}, \tag{9}$$

$$\tau = \int_0^l T ds \approx \frac{1}{2}\dot{\mathbf{q}}^T\mathbf{M}\dot{\mathbf{q}} \tag{10}$$

where the vector $\mathbf{q}$ collects the parameters used to represent the displacement field, M and K are the mass and stiffness matrices respectively.

Assuming

$$\mathbf{q} = \mathbf{q}_m e^{-i\omega_m t}, \tag{11}$$

where $\mathbf{q}_m$ is the mode shape, ω_m the radian frequency and i the imaginary unit, imposing the stationarity of Hamilton's functional gives the eigenvalue problem:

$$\left(K - \omega_m^2 M\right)\mathbf{q}_m = \mathbf{0} \tag{12}$$

The definition of mass matrix M and those of stiffness matrix K descend from the particular finite element discretization used. In the next section we define specifically that deriving from isogeometric discretization.

2.2 Isogeometric Discretization

In this work the geometry of a curved beam is described using a NURBS (Non Uniform Rational B-Splines) interpolation; for a complete and more accurate description of this kind of interpolation please see Cottrell et al. (2009). A curve x = x (ξ) has a p-degree NURBS representation when there exist $n \in N$, control points $\mathbf{P}_i \in \mathbb{R}^2$, with the associated weights $g_i \in \mathbb{R}$, $i = 1...n$, and a *knot vector*, i.e. a set $\Pi = \left\{0 = \xi_1 \leqslant \xi_2 \leqslant ... \leqslant \xi_{n+p+1} = 1\right\}$ such that, for any $\xi \in [0,1]$:

$$x(\xi) = \sum_{i=1}^{n} R_{i,p}(\xi)\mathbf{P}_i \tag{13}$$

where the NURBS basis $\{R_{i,p}\}$ is expressed by:

$$R_{i,p}(\xi) = \frac{B_{i,p}(\xi)g_i}{\sum_{i=1}^{n} B_{i,p}(\xi)g_i} \tag{14}$$

The B-splines basis $\{B_{i,p}(\xi)\}$ of order p is expressed as a function of the basis corresponding to order $p - 1$ by the Cox-De Boor recursive formula:

$$B_{i,0}(\xi) = \begin{cases} 1 & \text{if } \xi_i < \xi < \xi_{i+1} \\ 0 & \text{otherwise} \end{cases} \tag{15}$$

$$B_{i,p}(\xi) = \frac{\xi - \xi_i}{\xi_{i+p} - \xi_i} B_{i,p-1}(\xi) + \frac{\xi_{i+p+1} - \xi}{\xi_{i+p+1} - \xi_{i+1}} B_{i+1,p-1}(\xi). \tag{16}$$

The main goal of the isogeometric approach is to exactly describe the geometry of the problem by means of NURBS interpolation and to adopt the same interpolating basis for representing the generalized displacements:

$$u(\xi) \approx \sum_{i=1}^{n} R_{i,p}(\xi)u_i, \tag{17}$$

$$w(\xi) \approx \sum_{i=1}^{n} R_{i,p}(\xi)w_i, \tag{18}$$

$$\varphi(\xi) \approx \sum_{i=1}^{n} R_{i,p}(\xi)\varphi_i, \tag{19}$$

by means of control points u_i, w_i, and φ_i.

Hence, denoting the Jacobian of the transformation by J, the generalized strains assume this form:

$$\varepsilon = \frac{\partial u}{\partial \xi}/J - w/R$$

$$\gamma = \frac{\frac{\partial w}{\partial \xi}}{J} + \frac{u}{R} + \varphi \tag{20}$$

$$\chi = \frac{\partial \varphi}{\partial \xi}/J,$$

The elastic strain energy for the eth element is given by:

$$\mathcal{E}_e = \frac{1}{2}\left(u_e^T K_e^{uu} u_e + w_e^T K_e^{ww} w_e + \varphi_e^T K_e^{\varphi\varphi} \varphi_e + u_e^T K_e^{uw} w_e + u_e^T K_e^{u\varphi} \varphi_e + w_e^T K_e^{w\varphi} \varphi_e\right), \tag{21}$$

where u_e, w_e and φ_e collect, respectively, the control points u_i, w_i and φ_i related to the e_{th} element. The stiffness matrixes $K_e^{uu}, K_e^{ww}, K_e^{\varphi\varphi}, K_e^{uw}, K_e^{u\varphi}, K_e^{w\varphi}$, take into account the elastic strain contribution of $\varepsilon, \gamma, \chi$ and also for the coupled term $\varepsilon\chi$.

Using the basis matrices R^e related to the eth element and a standard finite element assemblage procedure it is possible to obtain the global stiffness matrix K.

In the same way the kinetic energy for the e_{th} element can be computed:

$$\mathcal{T}_e = \frac{1}{2}\left(\dot{u}_e^T M_e^{uu} \dot{u}_e + \dot{w}_e^T M_e^{ww} \dot{w}_e + \dot{\varphi}_e^T K_e^{\varphi\varphi} \dot{\varphi}_e\right) \tag{22}$$

where $\dot{u}_e, \dot{w}_e, \dot{\varphi}_e$ collect the control points $\dot{u}_i, \dot{w}_i, \dot{\varphi}_i$ and $M_e^{uu}, M_e^{ww}, K_e^{\varphi\varphi}$ are the contributions to the mass matrixes of the three above mentioned displacements. Also in this case the global mass matrix M can be easily built.

The stiffness and mass matrixes were evaluated by means of Gauss' quadrature rule. In this paper the number of Gaussian points is always assumed equal to the degree p of the spline basis functions due to its efficiency as proved in Cazzani et al. (2016b).

3   Results and Discussion

3.1   Clamped-Clamped Circular Arch

In the following, non-dimensional frequency presented in Eq. (23) is used:

$$\widehat{\omega} = \omega L^2 \sqrt{\frac{\rho A}{EI}} \tag{23}$$

in which L is the span length of curved beam. The slenderness ratio is defined as L/r where r is the radius of gyration of the cross-sectional area $r = I/A$.

The first ten eigenfrequencies and eigenmodes have been calculated for the clamped clamped concrete circular arch presented in Fig. 2 considering both the De St Venant's and Winkler's constitutive relationships. Several meshes and slenderness ratio have been considered.

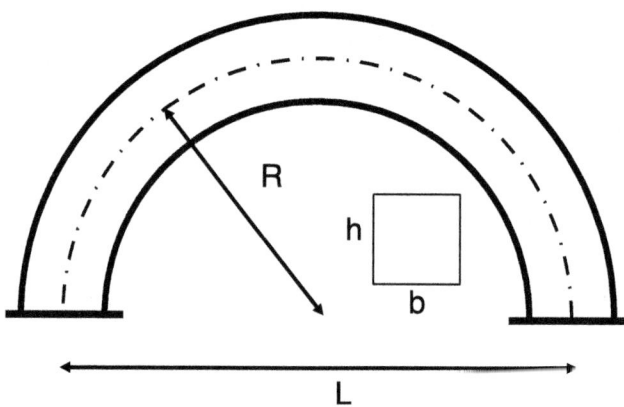

Fig. 2. Circular arch geometry and cross section.

Two main cases have been considered with $L/r = 300$ and $L/r = 20$ and the analytical reference solution has been obtained following the approach proposed in Tüfekçi and Arpaci (1998). The concrete material model is elastic linear and E_C is the concrete Young's modulus, k_s is the shear coefficient, G is the Shear modulus their ratio is fixed: $k_s G/E_C = 0.3$.

Tables 1 and 2 report the eigenfrequency for the thin circular arch ($L/r = 300$) following de Saint Venant's approach while Tables 3 and 4 refer to the same problem solved by Winkler's approach. As mentioned in Sect. 2.2 p represent the degree of the NURBS representation.

Table 1. Dimensionless values of first five natural frequencies, for clamped-clamped circular arch. (L/r = 300 and h/L = 0.011) using de Saint Venant's constitutive model.

p	DOF	Mode 1	2	3	4	5
2	48	19.308	45.497	92.579	158.42	261.10
	96	17.630	38.940	72.734	112.40	164.25
	192	17.526	38.542	71.549	109.70	158.60
3	51	17.538	38.667	72.412	113.29	170.94
	99	17.520	38.518	71.484	109.56	158.37
	195	17.519	38.516	71.476	109.53	158.28
4	54	17.520	38.521	71.527	109.87	160.18
	102	17.519	38.516	71.476	109.53	158.28
	198	17.519	38.516	71.475	109.53	158.27
5	33	17.545	38.973	76.136	131.14	237.52
	57	17.519	38.517	71.479	109.57	158.60
	105	17.519	38.516	71.475	109.53	158.27
	201	17.519	38.516	71.475	109.53	158.27
Ref.		17.519	38.516	71.475	109.53	158.27

Table 2. Dimensionless values of natural frequencies from 6th to 10th, for clamped-clamped circular arch. (L/r = 300 and h/L = 0.011) using de Saint Venant's constitutive model.

p	DOF	Mode 6	7	8	9	10
2	48	399.23	583.51	610.68	831.88	934.96
	96	222.90	295.78	375.55	475.31	552.56
	192	212.50	277.32	345.86	427.04	503.39
3	51	245.12	354.88	491.19	605.24	714.09
	99	212.18	276.98	345.74	427.74	505.43
	195	211.92	276.34	344.34	424.66	500.41
4	54	219.25	300.09	402.24	558.48	582.36
	102	211.93	276.35	344.39	424.84	500.78
	198	211.92	276.33	344.32	424.64	500.37
5	33	426.65	604.60	732.14	897.51	1303.7
	57	213.72	284.40	369.25	493.16	569.07
	105	211.92	276.33	344.33	424.65	500.41
	201	211.92	276.33	344.32	424.64	500.37
Ref.		211.92	276.33	344.32	424.64	500.37

The relative difference is reported in Table 9 and it is negligible in this case. Convergence of the numerical solution to the reference one is assessed in every case. It can be noted that lower eigenfrequencies require a lower number of degrees of freedom to reach an accurate result.

Table 3. Dimensionless values of first five natural frequencies, for clamped-clamped circular arch. ($L/r = 300$ and $h/L = 0.011$) using de Winkler's constitutive model.

p	DOF	Mode				
		1	2	3	4	5
2	48	17.753	40.050	78.513	130.96	213.54
	96	17.525	38.547	71.614	109.97	159.53
	192	17.521	38.521	71.485	109.55	158.31
3	51	17.535	38.643	72.293	112.92	170.05
	99	17.521	38.521	71.489	109.56	158.37
	195	17.521	38.520	71.483	109.54	158.29
4	54	17.521	38.524	71.533	109.88	160.18
	102	17.521	38.520	71.482	109.54	158.29
	198	17.521	38.520	71.482	109.54	158.29
5	33	17.546	38.977	76.143	131.16	237.55
	57	17.521	38.520	71.486	109.58	158.62
	105	17.521	38.520	71.482	109.54	158.29
	201	17.521	38.520	71.482	109.54	158.29
Ref.		17.521	38.520	71.482	109.54	158.29

Table 4. Dimensionless values of natural frequencies from 6th to 10th, for clamped-clamped circular arch. ($L/r = 300$ and $h/L = 0.011$) using Winkler's constitutive model.

p	DOF	Mode				
		6	7	8	9	10
2	48	329.98	513.19	574.99	805.62	822.01
	96	214.86	282.75	356.49	447.45	530.35
	192	212.00	276.48	344.59	425.11	501.04
3	51	243.39	351.82	486.92	604.05	706.73
	99	212.16	276.91	345.60	427.45	505.02
	195	211.94	276.36	344.37	424.70	500.45
4	54	219.24	300.06	402.18	558.36	582.35
	102	211.95	276.38	344.43	424.88	500.83
	198	211.94	276.36	344.36	424.68	500.42
5	33	426.69	604.62	732.30	897.62	1303.81
	57	213.75	284.43	369.29	493.22	569.10
	105	211.94	276.36	344.36	424.70	500.46
	201	211.94	276.36	344.36	424.68	500.42
Ref.		211.94	276.36	344.36	424.68	500.42

The situation is quite similar considering the thick arch ($L/r = 20$). Tables 5 and 6 present the de Saint Venant's eigenfrequencies while Tables 7 and 8 the Winkler's ones. The relative difference is reported, also in this case, in Table 9, and it is always under 2%.

Table 5. Dimensionless values of first five natural frequencies, for clamped-clamped circular arch. ($L/r = 20$ and $h/L = 0.173$) using de Saint Venant's constitutive model

p	DOF	Mode				
		1	2	3	4	5
2	48	14.437	25.269	42.241	43.652	61.055
	96	14.430	25.253	42.207	43.601	60.943
	192	14.430	25.252	42.205	43.598	60.938
3	51	14.430	25.252	42.206	43.600	60.943
	99	14.430	25.252	42.205	43.598	60.938
	195	14.430	25.252	42.205	43.598	60.938
4	54	14.430	25.252	42.205	43.598	60.938
	102	14.430	25.252	42.205	43.598	60.938
	198	14.430	25.252	42.205	43.598	60.938
5	33	14.430	25.254	42.224	43.635	61.168
	57	14.430	25.252	42.205	43.598	60.938
	105	14.430	25.252	42.205	43.598	60.938
	201	14.430	25.252	42.205	43.598	60.938
Ref.		14.430	25.252	42.205	43.598	60.938

Mode shapes of the thick circular arch are presented in Figs. 3 and 4, both symmetric and skew modes can be noted.

In order to prove the efficiency of the isogeometric approach a further test have been developed considering a quite thin arch *(L/r = 150)*. For this structure the complete eigenfrequency spectrum was calculated using different meshes with the same number of degrees of freedom, but with different NURBS degree *p*. The de Saint Venant's model was chosen for this test, but similar results (not reported here for the sake of synthesis) have been obtained for the Winkler's one.

In Fig. 5 the full eigenfrequency spectrum is plotted considering 300 degrees of freedom. The y-axis represents the ratio between the numerical solution and the analytical reference solution. It can be noted that for the first 2/3 of the spectrum the accuracy of the isogeometric finite element model is very high. Only after the 200th mode the errors are not negligible.

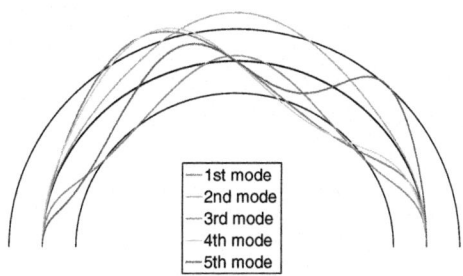

Fig. 3. Thick circular arch mode shapes from 1st one to 5th one.

Table 6. Dimensionless values of natural frequencies from 6th to 10th, for clamped-clamped circular arch. ($L/r = 20$ and $h/L = 0.173$) using de Saint Venant's constitutive model.

p	DOF	Mode				
		6	7	8	9	10
2	48	72.486	89.532	96.868	121.13	125.38
	96	72.197	89.393	96.022	119.23	125.13
	192	72.184	89.387	95.988	119.16	125.12
3	51	72.200	89.398	96.066	119.43	125.16
	99	72.183	89.387	95.987	119.16	125.12
	195	72.183	89.387	95.986	119.16	125.12
4	54	72.184	89.388	95.996	119.22	125.13
	102	72.183	89.387	95.986	119.16	125.12
	198	72.183	89.387	95.986	119.16	125.12
5	33	72.899	89.918	101.678	126.57	130.33
	57	72.183	89.387	95.988	119.17	125.12
	105	72.183	89.387	95.986	119.16	125.12
	201	72.183	89.387	95.986	119.16	125.12
Ref.		72.183	89.387	95.986	119.16	125.12

Table 7. Dimensionless values of first five natural frequencies, for clamped-clamped circular arch. ($L/r = 20$ and $h/L = 0.173$) using de Winkler's constitutive model

p	DOF	Mode				
		1	2	3	4	5
2	48	14.544	25.496	42.556	43.778	61.145
	96	14.541	25.485	42.528	43.737	61.039
	192	14.540	25.485	42.527	43.736	61.034
3	51	14.541	25.485	42.528	43.737	61.039
	99	14.540	25.485	42.527	43.735	61.034
	195	14.540	25.485	42.527	43.735	61.034
4	54	14.540	25.485	42.527	43.735	61.034
	102	14.540	25.485	42.527	43.735	61.034
	198	14.540	25.485	42.527	43.735	61.034
5	33	14.541	25.487	42.527	43.769	61.254
	57	14.540	25.485	42.527	43.735	61.034
	105	14.540	25.485	42.527	43.735	61.034
	201	14.540	25.485	42.527	43.735	61.034
Ref		14.540	25.485	42.527	43.735	61.034

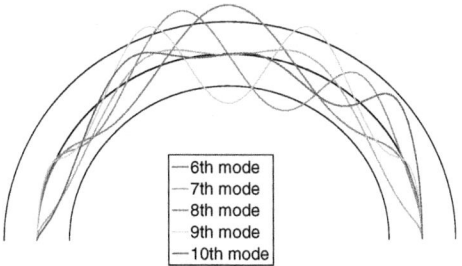

Fig. 4. Thick circular arch mode shapes from 6th one to 10th one.

Table 8. Dimensionless values of natural frequencies from 6th to 10th, for clamped-clamped circular arch. ($L/r = 20$ and $h/L = 0.173$) using Winkler's constitutive model.

p	DOF	Mode				
		6	7	8	9	10
2	48	72.808	88.786	97.303	121.60	123.96
	96	72.523	88.638	96.379	119.43	123.73
	192	72.510	88.632	96.340	119.35	123.72
3	51	72.526	88.643	96.418	119.61	123.75
	99	72.509	88.632	96.339	119.34	123.72
	195	72.509	88.632	96.338	119.34	123.72
4	54	72.511	88.633	96.347	119.39	123.72
	102	72.509	88.632	96.338	119.34	123.72
	198	72.509	88.632	96.338	119.34	123.72
5	33	73.187	89.127	101.919	124.88	130.15
	57	72.509	88.632	96.339	119.35	123.72
	105	72.509	88.632	96.338	119.34	123.72
	201	72.509	88.632	96.338	119.34	123.72
Ref.		72.509	88.632	96.338	119.34	123.72

3.2 Hinged-Hinged Parabolic Arch

The modal analysis of a parabolic arch hinged at both ends is presented here. The cross section is squared as reported in Fig. 6, while the ratio L/R' is 0.3.

The shear parameter is $k_sG/E_C = 0.3$ also in this case, while two slenderness values have been considered: $L/r = 150$ and $L/r = 10$.

Unfortunately it was not possible to find an analytic solution for every frequency, but an interesting work which reports similar numerical results is Luu et al. (2015).

Table 9. Percentage difference of natural frequencies evaluated by de Saint Venant's and Winkler's approach for clamped-clamped circular arch.

L/r = 300		L/r = 20	
Mode	Difference	Mode	Difference
1	0.01%	1	0.76%
2	0.01%	2	0.92%
3	0.01%	3	0.76%
4	0.01%	4	0.31%
5	0.01%	5	0.16%
6	0.01%	6	0.45%
7	0.01%	7	0.84%
8	0.01%	8	0.37%
9	0.01%	9	0.15%
10	0.01%	10	1.12%

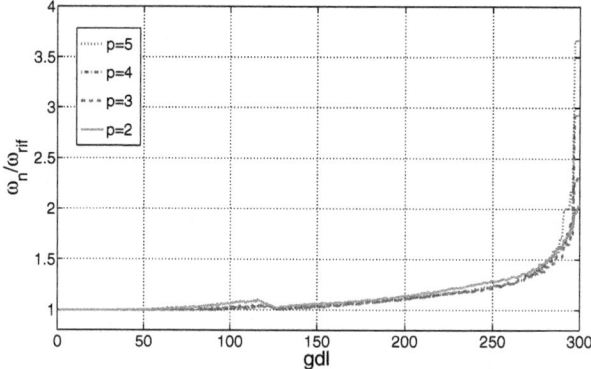

Fig. 5. Circular arch eigenfrequency spectrum for different meshes.

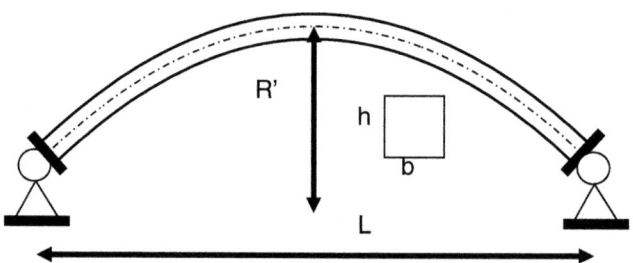

Fig. 6. Parabolic arch geometry and cross section.

Table 10. Dimensionless values of first five natural frequencies, for hinged-hinged parabolic arch. ($L/r = 10$) using de Saint Venant's constitutive model

p	DOF	Mode				
		1	2	3	4	5
2	50	14.164	15.745	29.904	30.671	46.634
	98	14.163	15.744	29.903	30.665	46.606
	194	14.163	15.744	29.903	30.664	46.604
3	29	14.164	15.744	29.903	30.670	46.653
	53	14.163	15.744	29.903	30.664	46.605
	101	14.163	15.744	29.903	30.664	46.604
4	32	14.163	15.744	29.903	30.685	46.609
	56	14.163	15.744	29.903	30.664	46.604
	104	14.163	15.744	29.903	30.664	46.604
5	35	14.163	15.744	29.903	30.664	46.605
	59	14.163	15.744	29.903	30.664	46.604
	107	14.163	15.744	29.903	30.664	46.604

Table 11. Dimensionless values of natural frequencies from 6th to 10th, for hinged-hinged parabolic arch. ($L/r = 10$) using de Saint Venant's constitutive model.

p	DOF	Mode				
		6	7	8	9	10
2	50	54.336	55.616	62.352	63.882	75.951
	98	54.334	55.616	62.265	63.881	75.945
	194	54.334	55.616	62.261	63.881	75.945
3	29	54.337	55.616	62.561	63.883	75.954
	53	54.334	55.616	62.263	63.881	75.945
	101	54.334	55.616	62.261	63.881	75.945
4	32	54.334	55.616	62.302	63.882	75.947
	56	54.334	55.616	62.261	63.881	75.945
	104	54.334	55.616	62.260	63.881	75.945
5	35	54.334	55.616	62.269	63.881	75.946
	59	54.334	55.616	62.261	63.881	75.945
	107	54.334	55.616	62.260	63.881	75.945

The first ten eigenmodes and eigenfrequencies have been calculated considering both the de Saint Venant's (see Tables 10 and 11) and Winkler's (see Tables 12 and 13) models. The relative differences are reported in Table 14 and only for the thick arch case the maximum difference reaches 4%, instead differences are negligible for the thin arch case.

Table 12. Dimensionless values of first five natural frequencies, for hinged-hinged parabolic arch. ($L/r = 10$) using Winkler's constitutive model

p	DOF	Mode				
		1	2	3	4	5
2	50	14.554	15.187	28.956	30.922	46.365
	98	14.553	15.187	28.955	30.913	46.325
	194	14.553	15.187	28.955	30.913	46.323
3	29	14.553	15.187	28.956	30.918	46.370
	53	14.553	15.187	28.955	30.913	46.324
	101	14.553	15.187	28.955	30.913	46.323
4	32	14.553	15.187	28.955	30.913	46.327
	56	14.553	15.187	28.955	30.913	46.323
	104	14.553	15.187	28.955	30.913	46.323
5	35	14.553	15.187	28.955	30.913	46.324
	59	14.553	15.187	28.955	30.913	46.323
	107	14.553	15.187	28.955	30.913	46.323

Table 13. Dimensionless values of natural frequencies from 6th to 10th, for hinged-hinged parabolic arch. ($L/r = 10$) using de Winkler's constitutive model.

p	DOF	Mode				
		6	7	8	9	10
2	50	52.253	53.884	61.608	64.639	72.799
	98	52.250	53.884	61.481	64.638	72.788
	194	52.250	53.884	61.474	64.638	72.787
3	29	52.252	53.884	61.758	64.638	72.797
	53	54.334	53.884	61.477	64.638	72.787
	101	52.250	53.884	61.474	64.638	72.787
4	32	52.250	53.884	61.507	64.638	72.790
	56	52.250	53.884	61.474	64.638	72.787
	104	52.250	53.884	61.474	64.638	72.787
5	35	52.250	53.884	61.480	64.638	72.787
	59	52.250	53.884	61.474	64.638	72.787
	107	52.250	53.884	61.474	64.638	72.787

Table 14. Percentage difference of natural frequencies evaluated by de Saint Venant's and Winkler's approach for hinged-hinged parabolic arch.

L/r = 150		L/r = 10	
Mode	Difference	Mode	Difference
1	0.03%	1	2.75%
2	0.03%	2	3.71%
3	0.03%	3	3.17%
4	0.04%	4	0.81%
5	0.03%	5	0.60%
6	0.03%	6	3.84%
7	0.02%	7	3.11%
8	0.04%	8	1.26%
9	0.04%	9	1.19%
10	0.02%	10	4.16%

The mode shapes for the former case are reported in Fig. 7.

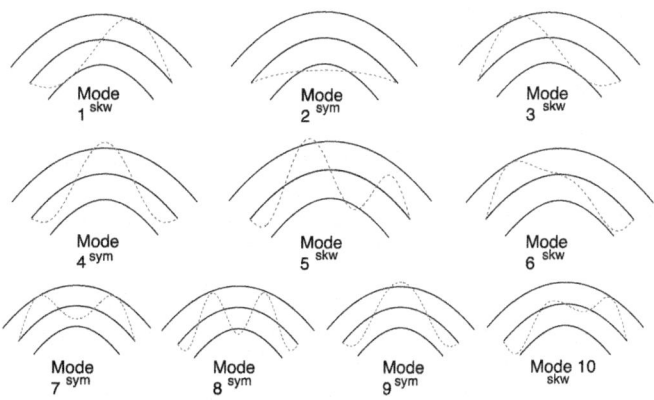

Fig. 7. Thick parabolic arch mode shapes from 1st one to 10th one.

4 Conclusions

The modal analysis of concrete strongly curved beams have been developed considering both a diagonal constitutive law (de Saint Venant's approach) or a not diagonal one (Winkler's approach). The structures were discretised using isogeometric finite elements based on NURBS shape functions.

Two case have been analysed: a fully clamped circular arch and a parabolic hinged arch considering different slenderness ratios. The numerical results have been validated considering literature benchmark study and available analytical results.

The relative eigenfrequencies obtained by the two approaches are very similar also in case of very thick arches. This result is quite different from what achieved in the static case, see Cazzani et al. (2016f). It seems that the accuracy of Winkler's model become more important for higher modes.

Isogeometric analysis produced accurate results also with low numbers of degrees of freedom as proved with the eigenfrequency spectrum presented in Fig. 5.

Acknowledgements. The financial support of MIUR, the Italian Ministry of Education, University and Research, under grant PRIN 2010–2011 (project 2010MBJK5B—*Dynamic, Stability and Control of Flexible Structures*) is gratefully acknowledged.

References

Aristodemo M, Turco E (1994) Boundary element discretization of plane elasticity and plate bending problems. Int J Numer Meth Eng 37(6):965–987

Buffa F, Causin A, Cazzani A, Poppi S, Sanna G, Solci M, Stochino F, Turco E (2015) The sardinia radio telescope: a comparison between close-range photogrammetry and finite element models. Math Mech Solids 22:1005–1026. https://doi.org/10.1177/1081286515616227

Bilotta A, Formica G, Turco E (2010) Performance of a high-continuity finite element in three-dimensional elasticity. Int J Numer Meth Biomed Eng 26(9):1155–1175

Cottrell JA, Hughes TJR, Bazilevs Y (2009) Isogeometric analysis: toward integration of CAD and FEA. Wiley, New York

Cazzani A, Wagner N, Ruge P, Stochino F (2016a) Continuous transition between traveling mass and traveling oscillator using mixed variables. Int J Non-Linear Mech 80:82–95

Cazzani A, Malagù M, Turco E (2016b) Isogeometric analysis of plane-curved beams. Math Mech Solids 21(5):562–577

Cazzani A, Stochino F, Turco E (2016c) On the whole spectrum of Timoshenko beams Part I: a theoretical revisitation. Zeitschrift für angewandte Mathematik und Physik (ZAMP) 67(24):1–30

Cazzani A, Stochino F, Turco E (2016d) On the whole spectrum of Timoshenko beams Part II: further applications. Zeitschrift für angewandte Mathematik und Physik (ZAMP) 67(25):1–21

Cazzani A, Stochino F, Turco E (2016e) An analytical assessment of finite element and isogeometric analyses of the whole spectrum of Timoshenko beams. J Appl Math Mech Zeitschrift für Angewandte Mathematik und Mechanik (ZAMM) 96(10):1220–1244. https://doi.org/10.1002/zamm.201500280

Cazzani A, Malagù M, Turco E, Stochino F (2016f) Constitutive models for strongly curved beams in the frame of isogeometric analysis. Math Mech Solids 21(2):182–209

Sevim B, Atamturktur S, Altunişik AC, Bayraktar A (2016) Ambient vibration testing and seismic behavior of historical arch bridges under near and far fault ground motions. Bull Earthq Eng 14(1):241–259

Kadkhodayan V, Aghajanzadeh M, Mirzabozorg H (2016) Seismic assessment of arch dams using fragility curves. Civ Eng J 1(2):14–20

Li Y, Cai CS, Liu Y, Chen Y, Liu J (2016) Dynamic analysis of a large span specially shaped hybrid girder bridge with concrete-filled steel tube arches. Eng Struct 106:243–260

Luu AT, Kim NI, Lee J (2015) Isogeometric vibration analysis of free-form Timoshenko curved beams. Meccanica 50(1):169–187

Stochino F, Cazzani A, Poppi S, Turco E (2018) Sardinia radio telescope finite element model updating by means of photogrammetric measurements. Math Mech Solids 22(4):885–901. https://doi.org/10.1177/1081286515616046

Stochino F, Carta G (2014) SDOF models for reinforced concrete beams under impulsive loads accounting for strain rate effects. Nucl Eng Des 276:74–86

Stochino F (2016) RC beams under blast load: reliability and sensitivity analysis. Eng Fail Anal 66:544–565

Tüfekçi E, Arpaci A (1998) Exact solution of in-plane vibrations of circular arches with account taken of axial extension, transverse shear and rotatory inertia effects. J Sound Vib 209 (5):845–856

Zong Z, Xia Z, Liu H, Li Y, Huang X (2016) Collapse failure of prestressed concrete continuous rigid-frame bridge under strong earthquake excitation: testing and simulation. J Bridge Eng 21:04016047

Winkler E (1858) Formänderung und festigkeit gekrümmter körper, insbesondere der ringe. Der Civilingenieur 4:232–246

Simplified Non-linear Analyses for Fast Seismic Assessment of RC Frame Structures: Review and Proposal

N. Caterino[1,3(✉)] and E. Cosenza[2,3]

[1] Department of Engineering, University of Naples Parthenope, Naples, Italy
nicola.caterino@uniparthenope.it
[2] Department of Structures for Engineering and Architecture,
University of Naples Federico II, Naples, Italy
[3] Institute for Technologies of Construction (ITC), National Research Council (CNR),
San Giuliano Milanese, Milan, Italy

Abstract. When dealing with large scale seismic vulnerability assessment, usual FEM model based software cannot be a suitable choice to capture the lateral behaviour and seismic capacity of each of the several buildings located in that area. Approximate methods based on the fundamental of structural analysis could be more appropriate. They are practical due to their ease of use and capability to drive to results, even approximated, starting from few input data available. Therefore, they are especially suitable when the full knowledge of structural details cannot be available and the computational effort has to be strongly limited. This study presents the review of some simplified pushover methods available in the literature. Then a combined use of two approximate procedures, each addressed to a different aim, is proposed. These methods have been applied to RC frame structures, to preliminarily compare their effectiveness in simulating the "exact" results would come from a professional FEM structural software.

Keywords: Pushover analysis · Seismic capacity · Approximate methods

1 Introduction

Seismic vulnerability assessment of existing buildings relies on the evaluation of capacity and demand. The demand is characterized either by ground motion accelerograms or response spectra while capacity is obtained from the analysis of mathematical model of the structure. The nonlinear response can be analyzed either by pushover or time-history analysis. When the vulnerability of a large stock of structures is the object – i.e. in the case of territorial seismic risk analyses - time-history analysis is not generally the suitable way, for being time-consuming and needing a depth knowledge of each structure to be performed. In such cases, the use of simplified pushover analyses can be instead suggested. They are generally based on predefined failure mechanisms. Some of these are particularly suitable to be applied to analyse portfolio of buildings for ease of use (Kilar and Fajfar 1996; Calvi 1999; Cosenza et al. 2005; Borzi et al. 2008), others - even interesting and innovative in their approach - are probably still complicated to be

© Springer International Publishing AG, part of Springer Nature 2018
M. di Prisco and M. Menegotto (Eds.): ICD 2016, LNCE 10, pp. 248–260, 2018.
https://doi.org/10.1007/978-3-319-78936-1_18

practically adopted for such special application (Fajfar and Gaspersic 1996; Elnashai 2001; Chopra and Goel 2002; Papanilolaou and Elnashai 2005).

An important issue especially in large-scale projects is how to derive the capacity curve by means of structural analytical tools. When dealing with large–scale project, size of analyses affects the methodology and demands careful commitment for adopting suitable strategy. This is the case of management of seismic risk of large areas. In this regards, type of analysis becomes a crucial concern. This effort is concentrated on the any of application of a system to acquire the capacity of the wide range of real sample data to be simulated. To do so, it is worth trying such a technique neither to be less precise like empirical methods nor so time-consuming such a detailed analytical methods so that the result would be a trade-off of accuracy and simplicity.

Hence, it is worth to study different approaches to make simple pushover analysis aiming to implement for several case studies in a reasonable time. The combined use of two approximate procedures from literature (Cosenza et al. 2005; Miranda and Reyes 2002), each addressed to a different aim, is herein proposed to estimate the capacity curve of a building for the life safety limit state (SLV in NTC08 2008). This and other two simplified methods addressed to the same objective (Kilar and Fajfar 1996; Borzi et al. 2008) have been compared making applications to RC 2D frame structures and looking at the "exact" results of FEM model-based analyses made by SAP2000 (2014).

1.1 Approximate Non-linear Static Analyses

There are some proposed methods in the literature which calculate the capacity of buildings in the coordinate of shear capacity at the base and displacement capacity at a given damage limit states. In this study the attractiveness and effectiveness of two proposed simplified methods (i.e. Kilar and Fajfar 1996; Borzi et al. 2008) have been aimed. Besides, another method has been suggested by the authors using the simplified idea of structural analysis (Miranda and Reyes, 2002; Cosenza et al. 2005) to derive the structural capacity of frame buildings. The results of their approximation of the lateral capacity of a reference 2D frame have been compared to the analysis done by SAP2000.

These methods observe the mechanical behavior of structural components to simply obtain the lateral capacity of building. Certainly using these methods could not consider all the detailed requirements in structural design; nevertheless they can approximate the seismic capacity with acceptable tolerance.

To calculate the capacity different failure mechanisms are accounted to derive the most predominant mode of failure. Different combinations of failure in terms of global or local mechanism are evaluated. The shear capacity defines by the flexural capacity of beams and columns at each storey and displacement capacity is derived by the chord rotation of columns multiplies to the height of building which plasticization occurs. The bi-linear capacity curve is then obtained from the minimum resulted capacity among all the considered modes.

These simple approaches have the advantage of limiting the calculations and consequently saving the time of process which is very crucial in large-size problems. Basically, the methods share similar idea when calculate the capacity but the only minor difference is in defining the mechanisms.

Kilar and Fajfar (1996) presented a simplified method for nonlinear static analysis of building structures. Basically, the pushover curve is illustrated by base shear versus yield capacity and ultimate displacement capacity. An approximate relationship between the global base shear and top displacement is computed. Top displacement is determined as:

$$D = \sum_{i=1}^{n} F_i \cdot d_{ni} \tag{1}$$

where F_i is the horizontal force in i^{th} storey and d_{ni} are the coefficients of the condensed flexibility matrix that represents top displacement due to horizontal force in i^{th} storey. Simple formulae are used to approximate the flexibility of the frame (Kilar and Fajfar 1996):

$$d_{11} = \frac{h_1^2}{12}\left(\frac{1}{c_1} + \frac{1}{4b_1 + 0.33c_1}\right) \tag{2a}$$

$$d_{12} = d_{11} + \frac{h_1 h_2}{48b_1 + 4c_1} \tag{2b}$$

$$d_{22} = d_{11} + \frac{h_2^2}{12}\left(\frac{1}{c_2} + \frac{1}{4b_1} + \frac{1}{4b_2}\right) + \frac{h_1 h_2}{24b_1 + 2c_1} \tag{2c}$$

$$d_{ii} = d_{i-1,i-1} + \frac{h_i^2}{12}\left(\frac{1}{c_i} + \frac{1}{4b_{i-1}} + \frac{1}{4b_i}\right) + \frac{h_{i-1} h_i}{24b_{i-1}} \tag{2d}$$

$$d_{ij} = d_{ii} + \frac{h_i h_{i+1}}{48b_i} \tag{2e}$$

where $c_i = \sum(EI/h_i)$ indicates the flexural stiffness of columns, $b_i = \sum(EI/L_k)$ the flexural stiffness of beams summed over all the elements at i^{th} storey, h_i is interstorey height of storey and L_k is the beam span.

The yield displacement capacity is calculated using Eq. 1 while the same equation is used to obtain ultimate displacement capacity. Moreover the moment of inertia of the those elements suffered from plasticization due to nonlinearity of material has to be reduced. It is assumed that when a failure mechanism occurs the stiffness of all beams and columns are decreased to a small percentage of their initial stiffness (Kilar and Fajfar 1996). In this study the post-yielding stiffness of beams and columns are reduced to the 10% of their initial value. To calculate the top displacement when any failure mechanism shows up first type of failure has to be figured out. Three main type have been introduced as each of those may have different possible type of occurrence depends on number of floors. They are defined based on the multipliers of the horizontal forces.

$$\alpha_k^I = \frac{\sum M_{c1} + \sum M_{ck} + \sum_{i=1}^{k-1} \sum M_{bi}}{\sum_{i=1}^{k} F_i H_i + H_k \sum_{i=k+1}^{n} F_i} \quad (2 \leq k \leq n) \tag{3}$$

$$\alpha_k^{II} = \frac{\sum M_{ck} + \sum_{i=k}^{n} \sum M_{bi}}{\sum_{i=k}^{n} F_i (H_i - H_{k-1})} \quad (1 \leq k \leq n) \tag{4}$$

$$\alpha_k^{III} = \frac{2 \sum M_{ck}}{\sum_{i=k}^{n} F_i (H_i - H_{k-1})} \quad (1 \leq k \leq n) \tag{5}$$

In the above equations F_i is the horizontal forces at the levels H_i above the ground level, $\sum M_c$ is the sum of yield moments of all columns and $\sum M_b$ is the sum of yield moments of all beams in the ith storey. The resisting bending moments of beams and columns section is obtained using the available program (VCASLU 2011).

The mechanic-based Simplified Pushover-Based Earthquake Loss Assessment so-called SP-BELA (Borzi et al. 2008) observes the structural capacity through the simplified assumption of failure of structural components based on the formation of pre-defined collapse mechanisms. The capacity curve is plotted as a function of the parameters of collapse multiplier and displacement capacity. The collapse multiplier is the ultimate base shear capacity divided by the seismic weight of building. Also, the displacement capacity is specified as the ultimate roof displacement to be a function of chord rotation of column and building height for different damage limit states at each collapse failure mechanism. The collapse multiplier is calculated by:

$$\lambda^i = \frac{V_c^i}{W_T} \cdot \frac{\sum_{j=1}^{n} w_j z_j}{\sum_{k=i}^{n} w_k z_k} \tag{6}$$

where V_c^i is the shear capacity of ith storey which is the smallest value of shear capacity of column, flexural capacity of column and flexural capacity of beams, W_T is the building weight, w_j, w_k are the weight of jth and k^{th} storey and finally z_j, z_k are building height at the level of jth and k^{th} storey to the base level. Then, the collapse multiplier for a building type is determined as the minimum value of multipliers among all the stories. The value of minimum collapse multiplier multiplies to the total weight of building gives the shear capacity.

To calculate the deformation capacity the yield and ultimate chord rotation have to be used. The yield capacity is limited to the value of chord rotation of the column as:

$$\theta_y = \phi_y \frac{L_v}{3} + 0.0013\left(1 + 1.5\frac{h_c}{L_v}\right) + 0.13\phi_y \frac{d_{cl}f_{yd}}{\sqrt{f_{cd}}} \tag{7}$$

where ϕ_y is the yield curvature of the section (Circolare n. 617, 2009) h_c is the height of column section, d_{cl} is the longitudinal bar diameter, f_{yd} is the yield resistance of steel rebar and f_{cd} is the resistance of concrete, L_v is the shear span (ratio between bending moment and shear) which for the columns is considered half of interstory height. Since it is aimed to calculate the ultimate displacement of frames at SLV damage limit state which is a state before collapse so the chord rotation capacity is limited to ¾ of the ultimate rotation capacity as:

$$\theta_u = \frac{1}{\gamma_{el}} 0.016(0.3)^v \left[\frac{\max(0.01, \omega')}{\max(0.01, \omega)} f'_c\right]^{0.225}$$

$$\times \left(\frac{L_v}{h}\right)^{0.35} 25^{\left(\alpha\rho\frac{f_{yw}}{f'_c}\right)} \left(1.25^{100\rho_d}\right) \tag{8}$$

where v is axial force normalized by compressive strength of concrete section, ω and ω' are percentage of steel rebars respect to the size of concrete section in bottom and top of cross section, h is the section height, α is factor of confinement efficiency, ρ_{sx} is percentage of stirrups and ρ_d is percentage of diagonal rebars.

Then, yield and ultimate displacement capacity (D_y and D_u) is calculated for three modes of failure (i.e. shear failure of columns, beam-sway and column-sway).

$$D_y = \kappa H_T \theta_y \tag{9}$$

$$D_{u(beam-sway)} = D_y \frac{H^*_k}{\kappa H_T} + \left(0.75\theta_u - \theta_y\right)H_k \tag{10}$$

$$D_{u(column-sway)} = D_y + \left(0.75\theta_u - \theta_y\right)h \tag{11}$$

In the above equations κ is the coefficient to approximate the height of equivalent SDOF model of frame, H_T is the total height of frame, H^*_k is the height of post-elastic equivalent SDOF model of frame and H_k is the height of part of frame above the floor where the failure mode is activated. From the above equations the minimum values according to the predominant failure mechanism would be the displacement capacity.

1.2 An Alternative Approach

The following idea comes from the observation that all the approximate methods are generally more effective in predicting ultimate base shear than yield and ultimate displacements. This novel approach is just based on the preliminary evaluation of the ultimate base shear, also to calculate displacement capacity. More in detail, yield displacement is obtained by estimation of displacement of the last storey in a linear elastic building considering the contribution of the fundamental mode of vibration (Miranda and Reyes 2002):

$$\Delta_y = \beta_1.S_d \tag{12}$$

where S_d is spectral displacement evaluated at the fundamental period of the building and β_1 is a dimensionless factor obtained as following formula assuming a uniform mass distribution as:

$$\beta_1 = \frac{\int_{z/H=0}^{z/H=1} \Psi(z/H)}{\int_{z/H=0}^{z/H=1} \Psi^2(z/H)} \tag{13}$$

where $\psi(z/H)$ is normalized lateral deflected shape (Miranda and Reyes 2002) given by $\psi(z/H) = u(z/H)/u(z/H = 1)$; and $u(z/H)$ is lateral displacement evaluated at a non-dimensionless height z/H.

The spectral displacement S_d is herein proposed to be evaluated starting from the knowledge of the ultimate base shear V_b, the latter to be estimated as by Cosenza et al. (2005). More in detail, S_d can be derived from the elastic spectral acceleration S_a:

$$S_d = S_a.\frac{T^2}{(2\pi)^2} \tag{14}$$

Once the fundamental period of the structure T has been estimated. As far as T is concerned, the expression recently proposed by Salama (2015) is suggested as one of the most effective in literature:

$$T = 0.021 \, N^{0.16} \, H^{0.75} \text{ if } H \text{ is in feet} \tag{15}$$

$$T = 0.051 \, N^{0.16} \, H^{0.75} \text{ if } H \text{ is in meters} \tag{16}$$

As the required parameters to obtain the spectral displacement are:

$$S_a = \frac{F^*}{m^*}; \, F^* = \frac{V_b}{\Gamma}; \tag{17}$$

$$\Gamma = \frac{m^*}{\sum m_i.\phi_i^2} = \frac{\sum m_i.\phi_i}{\sum m_i.\phi_i^2}; \, \phi_i = \frac{m_i.z_i}{\sum m_j.z_j} \tag{18}$$

where V_b is the shear capacity of the MDOF system, F^* is shear capacity of the equivalent SDOF system, m^* is the mass of the equivalent SDOF system, ϕ_i is the i-th component of the first mode evaluated as suggested by many seismic code (e.g. Eurocode 8, NTC08, 2008) for lateral force method of analysis, Γ is the fundamental modal mass participation factor, m_i is the storey mass at the i-th level, z_i is the height of the mass m_i above the level of application of the seismic action (foundation).

The base shear is derived from a simplified method (Cosenza et al. 2005) as lateral capacity of frame. The study adopts a lumped plasticity model for the elements, in particular only flexural behavior is considered. The flexural behaviour of beams and columns are represented by an elastic-plastic M-θ law. The M-θ relation is determined by yielding moments (M_y) and by yielding and ultimate chord rotation ($θ_y$, $θ_u$). It is worth noting that since in this study different methods are being compared so unique regulation has been followed to calculate the values of M-θ of sections for all the methods unless is required (Circolare n. 617 2009).

Building capacity is determined when plastic mechanism forms for the building. The base shear is evaluated by equilibrium relations considering some pre-defined collapse mechanisms.

$$V_{b,1} = \frac{\sum M_c^k + \sum_{i=k+1}^n \sum M_b}{\sum_{i=k+1}^n H_i (H_i - H_k)} \sum_{i=1}^n H_i \tag{19}$$

$$V_{b,2} = \frac{\sum M_c^l + \sum M_c^k + \sum_{i=2}^{k-l} \sum M_b}{\sum_{i=1}^{k-l} H_i^2 + \sum_{i=k}^n H_k H_i} \sum_{i=1}^n H_i \tag{20}$$

$$V_{b,2} = \frac{2 \sum M_c^k}{\sum_{i=k}^n H_i (H_k - H_{k-1})} \sum_{i=1}^n H_i \tag{21}$$

where M_c^k is the yielding moment of column section of kth floor, M_b is the yielding moment of beam section and H_i is the height of ith storey. A linear distribution of the horizontal seismic forces is assumed. The objective base shear of the frame is then obtained from the minimum value among the different values when different possible combinations of mechanism considers with respect to the number of floors ($V_b = \min(V_{b1,n}, V_{b2,n}, V_{b3,n})$, considering n^{th} modes of failures).

On the other hand, displacement capacity is determined by displacement of the last floor (i.e. height of building which plasticization occurs). The ultimate displacement is then calculated at the predominant failure mechanism using the idea comes from the method of Cosenza et al. (2005). It is accounted in any of failure mechanism that building has the minimum resistance. Then, the ultimate displacement is determined by considering the plastic deformation of frame added on the value of yield displacement obtained by Eq. 12.

$$Δ_u = Δ_y + (0.75θ_u - θ_y).H_k \tag{22}$$

The ultimate displacement is calculated by calculating the displacement before and after yield point separately when the behaviour of the frame would be different. In this study the deformation of the frame when it experiences plasticity accounted by the length of the plastic deformation (i.e. the ultimate displacement-yield displacement) which indicates the ductility of the frame. To obtain the plastic deformation the authors adopt the formula of ultimate displacement in Cosenza et al. (2005) to calculate the value of plastic deformation of the frame only.

2   Application to Case Studies

Four 2D structural frames have been considered to validate the capability of the four different analytical tools. The frames are equal in geometry and structural detailing except the columns as each frame is characterized by different number of rebars in column sections. Regarding to NTC (2008) for each node of beam-column connection of frame, ratio between flexural moment of column and beam is limited to a certain value. Depending on ductility class of buildings, the ratio of flexural moment of column over beam is restricted to 1.30 if class "A" and 1.10 when ductility class "B" is the case. Therefore, four different types of frame with four different ductility ratios have been chosen. The frames are distinguished with their ductility ratio as for the weakest frame (frame 1) $\gamma_{R1} = R_{col}/R_{beam} = 0.77$, $\gamma_{R2} = 1.01$ for frame 2, $\gamma_{R3} = 1.26$ for frame 3 and finally $\gamma_{R4} = 1.50$ for frame 4. Thus, frames 1 and 2 do not satisfy the requirement for ductility class "B" while types 3 and 4 do. Geometry and structural details of frames are as following: 4 storeys, interstorey height 3.5 m, 3 bays, 4.5 m of bay span, storey weight 675 kN, equal cross section for all beams and columns (40 × 60 cm^2), rebars diameter 16 mm, symmetric reinforcement levels for each cross section, 8 bars for each level of reinforcement for all the beams, 4-6-8-10 rebars for columns of frames I-II-III-IV respectively, concrete strength 25 MPa, steel yield strength 370 MPa.

Each frame has been first analyzed with SAP2000 software through a detailed FEM model (Fig. 1). Afterwards, the curves have been replaced by the equivalent bilinear capacity curve according to the regulations (NTC08, 2008) which is delineated by three parameters (base shear, yield displacement and ultimate displacement), to be comparable with the approximate ones.

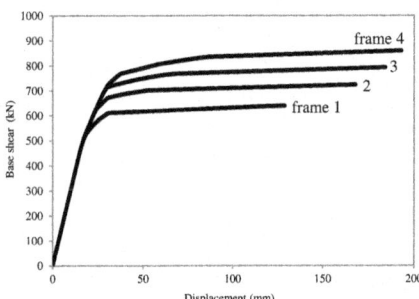

Fig. 1. Pushover curves for the 4 frames, as derived with SAP2000

The plastic hinges in SAP2000 have been modelled in the scale of moment-rotation for beam and column sections. Similar to the two previous methods the yield and ultimate moments and curvature are obtained using software VCASLU (2011) and yield and ultimate chord rotations are calculated from Eqs. 7 and 8 (Circolare n. 617 2009).

Since life safety limit state is the case of this study, the limit value for the chord rotation is $3/4\theta_u$ (NTC, 2008). Pushover analysis in SAP2000 is made for lateral distribution of load corresponding to the 1st mode of vibration. Referring to the Italian seismic design code (NTC08 2008) building has to be pushed until either a plastic hinge exceeding the ultimate objective rotation takes place or global or local failure mechanism occurs in the system.

Although the methods are following similar approach they are different to some extents. The failure mechanisms are devised so that plastic hinges are formed in beams and columns. In the simplified methods the main assumption is the hinges are formed in either beams or columns of a floor at the same time while it is not as common as in real cases. That is, when a mechanism is activated in ith floor (the weakest floor) all the beams and columns have been plasticized according to the different presumed mode of failures. Moreover, number of failure mechanism is different. In Kilar and Fajfar (1996) and proposed method the same mechanisms have been considered while in Borzi et al. (2008) fairly less number of failure mechanisms have been accounted. This could be helpful when large number of buildings have been analysed so that time of analysis could be reduced remarkably. All the three methods evaluates the shear capacity of the buildings from presumed failure mechanism at ith floor where has the least lateral capacity. Only flexural mode of failure has been taken in beam/column sections. Calculating the displacement capacity in the proposed method and Borzi et al. (2008) is done based on the ultimate chord rotation and height of part of building where failure occurs. In contrast, Kilar and Fajfar (1996) used the flexibility of the frame and the effect of moment inertia of the section to calculate the deformation capacity before and after formation of plastic hinges. Therefore, the method of calculation in Kilar and Fajfar (1996) is different respect to the two other abovementioned and seems to be more accurate.

The values of the parameters involved in the proposed procedure are $(\phi_1, \ldots, \phi_4) = (0.25, 0.50, 0.75, 1.00)$, $m^* = 229$ t, $\beta_1 = 1.286$, approximate $T_1 = 0.46$ s (the "exact" value by SAP2000 is 0.49 s).

The final result of base shear and ultimate displacement is dependent on how the plastic hinges are distributed among the deformed frame when plasticization is occurred. As been mentioned before the plasticization could occur in mode of flexural failure either in beams or in columns. Following the presumed failure modes in each adopted methods the predominant mode of failure has been found for each frame as the one yields the minimum value of shear resistance.

According to the mechanisms obtained from the analyses of the frames in SAP2000 none of the three methods could anticipate the failure mechanism, however the estimation of response for the frame 4 is closer to the analytical model in the case of methods Kilar and Fajfar (1996) and proposed method. Both methods consider the same types of modes of failure so they are similar in the prediction of the mechanisms. The slight difference comes out of the pre-defined assumption of the failure modes when only three floors plasticized then the last floor remains fully elastic while in reality the last floor deforms as well, however the relative displacement is a very small value.

The result of base shear (Figs. 1 and 2) proves this fact the difference in estimation of failure mechanism and distribution of plastic hinges do not play crucial role in estimation of the base shear. In all three approximate methods, the first floor is recognized as the weakest floor where the method of Borzi et al. (2008) the mechanism of beam-sway has been activated as plastic hinges distributed in all the beams along all the floor.

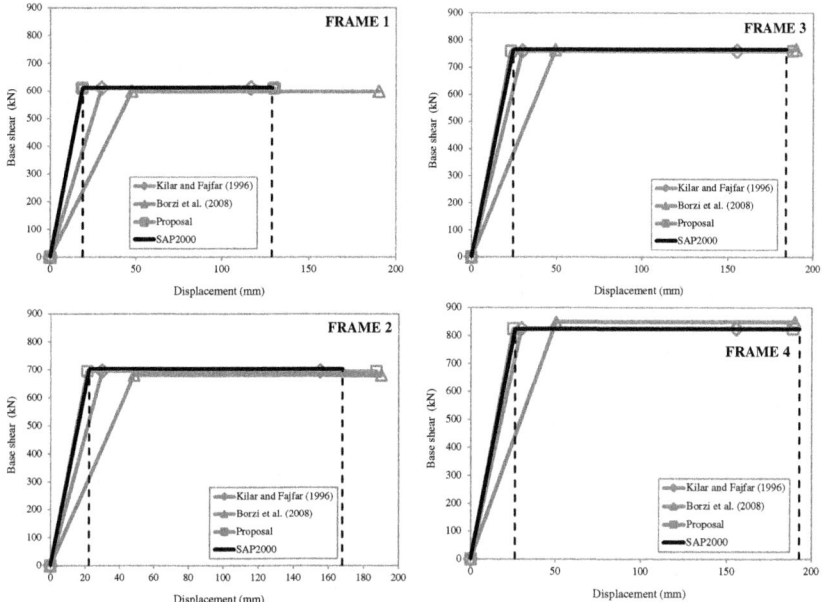

Fig. 2. Comparison of pushover curves for the 4 frames.

To verify the validity of three introduced simplified pushover methods the capacity curves of four frames have been plotted superimposed on the capacity curves derived from the modeling of frames in SAP2000 (Fig. 1) which have been bilinearized.

The results (Table 1) show that all the three methods could estimate the base shear with high accuracy while the values of displacement are not exactly as it would have been expected by the analytical model.

The method of Kilar and Fajfar (1996) and the proposed method could estimate the yield displacement with a remarkable tolerance of accuracy while the value resulted from Borzi et al. (Borzi et al. 2008) is fairly far from the reference value of yield displacement (SAP2000) which it makes underestimating the initial stiffness of the frame.

In the case of ultimate displacement Borzi et al. (2008) has overestimated the displacement capacity calculated by SAP2000 in frame 1 largely and also frame 2 with lesser rate of error since the method observed the plasticization of four stories which has been more than what has been resulted from analysis by SAP2000 (Fig. 1).

The overestimation of the ultimate displacement in frame 1 with method of Borzi et al. (2008) is due to the false recognition of failure mechanism which the estimation converges to the obtained values from SAP2000 in stronger frames (frames 2 and 3).

The error of estimation of ultimate displacement using method of Kilar and Fajfar (1996) has been observed as it has underestimated the point in all four frames. It is shown

Table 1. Comparison of the obtained results from different methods

		SAP2000	Fajfar et al.	Borzi et al.	Proposed method
Frame 1	Base shear (kN)	611.25	609.94	598.09	609.94
	Yield displacement (mm)	18.9	29.83	47.48	18.53
	Ultimate displacement (mm)	128.89	116.86	190.17	130
Frame 2	Base shear (kN)	702.16	695.63	681.4	695.63
	Yield displacement (mm)	22.3	29.83	48.38	21.14
	Ultimate displacement (mm)	167.93	155.27	190.17	187.3
Frame 3	Base shear (kN)	764.78	760.1	765	760.1
	Yield displacement (mm)	24.5	29.83	49.28	23.1
	Ultimate displacement (mm)	184.22	155.27	190.17	188.21
Frame 4	Base shear (kN)	821.92	824.77	848.98	824.77
	Yield displacement (mm)	26.2	29.83	50.17	25.06
	Ultimate displacement (mm)	192.93	155.27	190.17	189.12

that in all four frames the proposed method could estimate the ultimate displacement extremely close to the obtained values by SAP2000. It is shown that the method of Borzi et al. (2008) and proposed method could anticipate the ultimate displacement in frame 4 with mere value of 1% and 5% of error respect to the resulted value from SAP2000.

The estimation of ultimate displacement by proposed method and Borzi et al. (2008) in frame 3 and frame 4 is almost equal to the value obtained by the modeling in SAP2000 as equal as the displacement got from the proposed method. The obtained results are shown in below for four main parameters to check out the capability of the methods to estimate the base shear, ultimate displacement, initial stiffness and ductility of the nominated frames. The result endorses the ability of all three methods to evaluate the base shear with high accuracy in all four frames. In the maximum value Borzi et al. (2008) calculated the base shear in Frame 4 with the mere value 3.3% difference respect to the SAP2000. The reasonable estimation of base shear made by three simplified methods made by the keen anticipation of the failure mechanism. However, in the case of other three parameters the accuracy is not that so in all the frames.

Observing the results proves the methods of Kilar and Fajfar (1996) and Borzi et al. (2008) have not been able to anticipate the yield displacement as well as base shear and ultimate displacement. On the average Kilar and Fajfar (1996) estimated the yield displacement with 31%, Borzi et al. (2008) with large value 115% while the proposed method obtained the yield displacement with 4.3% tolerance of error on the average.

Borzi et al. (2008) estimated the ultimate displacement an equal value for all the four frame which the difference in frame 1 is 47% as the difference reduced to 1.5% in frame 4. Kilar and Fajfar (1996) estimates the ultimate displacement always less than the analytical method with the error of 13.3% on the average. The error is less relatively in proposed method when it is around 4%.

All in all, the methods are reliable in evaluation of base shear while they are not so accurate in estimation of displacement capacity when the failure mechanisms provided by the approximate methods are not so similar to the analytical method (frame 1), however the displacement values converged to the displacement obtained by SAP2000 in frame 4 where the type failure mechanisms and distribution of plastic hinges are very similar to the model of SAP2000.

3   Conclusions

Approximate methods in structural analysis are particularly suitable when preliminary, quick evaluations have to be done and/or when the analysis of a large set of structures have to be performed with reduced resources of time and money. The present study attempts to perform preliminary investigation about the effectiveness of some simplified methods for pushover analysis, that could be used to evaluate the seismic capacity of buildings instead of analytical softwares. In this study the competence of two available methods in the literature have been assessed. Besides, the authors used the idea of two related study to propose an approach to evaluate the structural capacity of frame buildings. The comparison of methods have also been done with reference to four sample 2D RC frame structures.

The advantage of the method Kilar and Fajfar (1996) is its higher computational efficiency. The method uses the matrix of flexibility of frame to calculate the top displacement using the idea of relative stiffness of beams and columns. The result of analysis shows the method is able to anticipate the base shear, yield displacement and initial stiffness of frame with a high degree of accuracy while the method underestimates the ultimate displacement. On the other hands, Borzi et al. (2008) use the chord rotation of column respect to the height of part of frame which is plasticized to calculate the displacement of last floor. In contrary with the method of Kilar and Fajfar (1996) this method could anticipate the value of ultimate displacement better than yield displacement. It makes the underestimation of elastic stiffness of building. This method have the advantage of being less complicated compare to the Kilar and Fajfar (1996) when portfolio of buildings is the case of analysis. All three methods estimated the base shear very well while the accuracy in calculation of the displacement capacities is dependent on how similar are the identified predominant mode of failure with respect to the real case. The results endorse the efficiency of the proposed method both in evaluation of base shear and yield displacement, as well as of the ultimate displacement.

It is worth noting that the above conclusions should not be considered as of general validity, since they come from the application of the approximate methods only to a small set of case study structure. The authors are working to randomly generate a large set of frame structures, to be representative of typical Italian RC structures built form 60s to date.

References

Borzi B, Pinho R, Crowley H (2008) Simplified pushover-based vulnerability analysis for large-scale assessment of RC buildings. Eng Struct 30(3):804–820

Calvi GM (1999) A displacement-based approach for vulnerability evaluation of classes of buildings. J Earthq Eng 3(3):411–438

Chopra AK, Goel RK (2002) A modal pushover analysis procedure for estimating seismic demands for buildings. Earthq Eng Struct Dyn 31(3):561–582

Circolare 2 n. 617 (2009) Istruzioni per l'applicazione delle "Nuove norme tecniche per le costruzioni" di cui al D.M

Cosenza E, Manfredi G, Polese M, Verderame GM (2005) A multilevel approach to the capacity assessment of existing RC buildings. J Earthq Eng 9(1):1–22

Elnashai AS (2001) Advanced inelastic static (pushover) analysis for earthquake applications. Struct Eng Mech 12(1):51–69

Eurocode 8, CEN (2004) design of structures for earthquake resistance. Part 1: general rules, seismic actions and rules for buildings. Standard EN-1998-1. European Committee for Standardization Technical Committee 250

Fajfar P (1975) Numerical analysis of multistory structures. In: 5th European Conference on Earthquake Engineering, Istanbul, paper 79

Fajfar P, Gaspersic P (1996) The N2 method for the seismic damage analysis of RC buildings. Earthq Eng Struct Dyn 25(1):31–46

Kilar V, Fajfar P (1996) Simplified push-over analysis of building structures. In: 11th WCEE World Conference on Earthquake Engineering, paper no. 1011

Mazzolani FM, Piluso V (1997) Plastic design of seismic resistant steel frames. Earthq Eng Struct Dyn 26(2):167–191

Miranda E, Reyes CJ (2002) Approximate lateral drift demands in multistory buildings with nonuniform stiffness. J Struct Eng-ASCE 128(7):840–849

NTC08 D.M. 14 gennaio (2008) Norme tecniche per le costruzioni. Ministero delle Infrastrutture (in Italian)

Panagiotakos T, Fardis MN (2001) Deformation of R.C. members at yielding and ultimate. ACI Struct J 98(2):135–148

Papanilolaou VK, Elnashai AS (2005) Evaluation of conventional and adaptive pushover analysis: methodology. J Earthq Eng 9(6):923–941

Salama MI (2015) Estimation of period of vibration for concrete moment-resisting frame buildings. Hous Build Nat Res Cent 11:16–21

SAP2000 (2014) Structural analysis program, Computers and Structures, Inc. University of Berkeley, CA

VCASLU (2011) Sezione generica in C.A e C.A.P: verifiche a presso-flessione stato limite ultimo, metodo n, professore Piero Gelfi, versione 7.7

A Semi-active Rocking System to Enhance the Seismic Dissipative Capability of Precast RC Columns

N. Caterino[1,3(✉)], M. Spizzuoco[2], and A. Occhiuzzi[1,3]

[1] Department of Engineering, University of Naples Parthenope, Naples, Italy
nicola.caterino@uniparthenope.it
[2] Department of Structures for Engineering and Architecture,
University of Naples Federico II, Naples, Italy
[3] Institute for Technologies of Construction (ITC), Italian National Research
Council (CNR), San Giuliano Milanese (MI), Milan, Italy

Abstract. This work is inspired by the idea of dissipating seismic energy at the base of prefabricated RC columns via semi-active (SA) variable dampers exploiting the base rocking. It was performed a wide numerical campaign to investigate the seismic behavior of a precast RC column with a variable base restraint. The latter is based on the combined use of a hinge, elastic springs, and magnetorheological (MR) dampers remotely controlled according to the instantaneous response of the structural component. The MR devices are driven by a SA control algorithm purposely written to modulate the dissipative capability so as to reduce base bending moment without causing excessive displacement at the top. The proposed strategy results to be really promising, since the base restraint relaxation, that favours the base moment demand reduction, is accompanied by an high enhancement of the dissipated energy due to rocking that can be even able to reduce top displacement in respect to the "fixed base rotation" conditions.

Keywords: Semi-active control · Rocking
Precast reinforced concrete elements

1 Introduction

The idea of a controlled rocking precast RC column is herein proposed and discussed. It is potentially suitable for seismic retrofit of existing precast RC frame structures where column-to-plinth connection, realized according to outdated technologies, can yield significant rotation in case of severe earthquakes. However the proposed technique could also be applied to optimize the lateral response of new structures, where the base joint can be specially designed so as to allow - in certain conditions and within given limits–rotations and hence energy dissipation.

The idea of exploiting unavoidable rocking mechanism between assembled precast structural elements to dissipate seismic energy has been explored during the last decades. Most research is addressed to enhance seismic capacity of precast RC structures adding energy dissipation systems at the beam-to-column connections (Spieth et al. 2004,

© Springer International Publishing AG, part of Springer Nature 2018
M. di Prisco and M. Menegotto (Eds.): ICD 2016, LNCE 10, pp. 261–271, 2018.
https://doi.org/10.1007/978-3-319-78936-1_19

Murahidy et al. 2004), less frequently the base connection of columns (Lu et al. 2016) or cantilever walls (Belleri et al. 2014) have been also involved.

Herein a semi-active (SA) control system based on the application of magnetorheological (MR) devices to realize a time-variant base restraint is investigated. The mechanical properties of such variable base restraint for precast RC columns can be driven in real time by a properly written control logic (Caterino 2015). The controller has to be programmed to instantaneously calibrate the MR devices installed at the base of the column in order to reduce the base bending moment, relaxing in selected intervals of time the base restraint. Again, the control logic has to hold the top displacement within acceptable values so as to avoid significant, detrimental second order effects. After the formulation of the above idea, a finite element model of the structure has been carried out so as to develop numerical simulations ad-dressed to optimally calibrate the control logic properly designed for such kind of applications.

2   A Variable Base Restraint for Precast RC Columns: Control Algorithm

The special base restraint is schematically shown in Fig. 1, where the uncontrolled precast RC column, fully restrained at the base, is modeled as a single degree of freedom dynamic system (Fig. 1(a)), having top mass m, stiffness k_T and inherent damping c_T. In order to control the structural demand, the authors propose to replace the perfectly rigid base restraint with a controllable one that is able to instantaneously become more or less "stiff", during the motion. Figure 1(b) just sketches the materialization of this idea by a smooth hinge, with a rotational spring (of stiffness k_ϕ) and a rotational variable damper whose damping constant c_ϕ can be driven in real time by a control algorithm. The same result can be obtained in practice by mounting two vertical

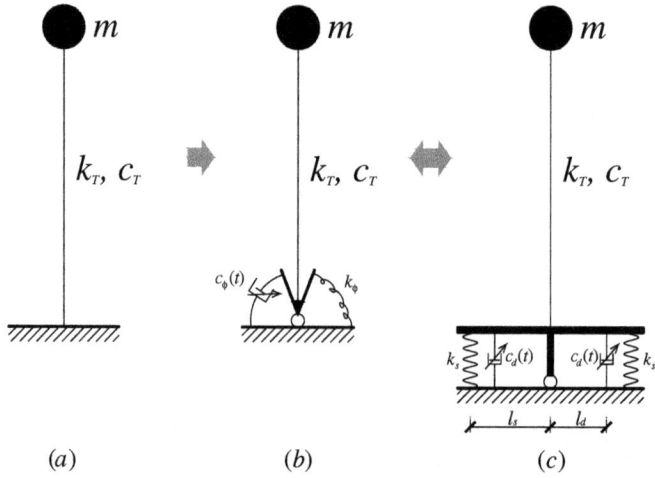

Fig. 1.   Basic idea of SA control of a precast RC column via MR dampers

linear springs (k_s) placed at a certain distance (l_s) from the hinge and two vertical SA dampers (c_d) at a distance l_d from the central hinge (Fig. 1(c)).

SA MR dampers are considered as smart devices within the proposed control system: when a low value is imposed to the base damping, the base restraint is less 'stiff', so that the structure's restraint is able to relax by converting its potential energy into kinetic energy, and the bending moment at the base is reduced. A direct consequence of controlling the demand of base bending stress could be an increase of top displacement demand; therefore, the SA base control system is thought to reduce base stress, by restraining the increase of top displacements within certain limits to control second order effects. A specific bang-bang control algorithm is formulated by the authors (Caterino 2015) to instantaneously decide the system's base configuration: it switches back and forth from an "OFF" state (intensity of current $i = i_{min}$, i.e. the minimum current set to be given to the dampers) to an "ON" state ($i = i_{max}$, i.e. the maximum assumed value for the current) according to a logic aiming to control both the base stress and the top displacement. Therefore, the control algorithm is so formulated:

$$
\begin{array}{llll}
\text{a)} & \text{if } |\sigma(t)| < \sigma_{\lim} & \rightarrow & i(t) = i_{max} \\
\text{b)} & \text{if } |\sigma(t)| \geq \sigma_{\lim} \text{ and } |x(t)| < x_{\lim} & \rightarrow & i(t) = 0 \\
\text{c)} & \text{if } |\sigma(t)| \geq \sigma_{\lim} \text{ and } |x(t)| \geq x_{\lim} \text{ and } x(t) \cdot \dot{x}(t) > 0 & \rightarrow & i(t) = i_{max} \\
\text{d)} & \text{if } |\sigma(t)| \geq \sigma_{\lim} \text{ and } |x(t)| \geq x_{\lim} \text{ and } x(t) \cdot \dot{x}(t) \leq 0 & \rightarrow & i(t) = 0
\end{array}
\tag{1}
$$

where $\sigma(t)$, $x(t)$ and $\dot{x}(t)$ are respectively the value of stress at the base, top displacement and top velocity at the instant of time t. In other words, the controller keeps 'stiffer' the base restraint until the stress exceeds the limit value σ_{lim} (expression (a) of Eq. (1)), whereas 'relaxes' it ("OFF" state of the dampers) when this limit is overpassed and the displacement falls within the limit of acceptability x_{lim} (expression (b) of Eq. (1)). When both stress and displacement are beyond the respective threshold values, the controller switches "ON" the dampers if the displacement is going towards a larger value (so trying to damp or invert the displacement's trend; expression (c) of Eq. (1)), otherwise it switches "OFF" the MR devices to make them collaborating to both stress and displacement reduction. Figure 2 schematically describes the above defined logic: the decision of the controller (switch "ON" or switch "OFF") depends on the occurrence of each of the four possible combinations regarding the value of base stress and top displacement. The application of the proposed control algorithm requires the

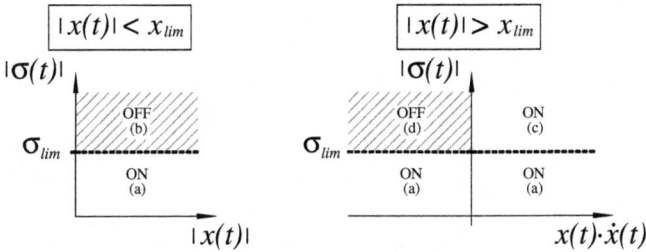

Fig. 2. The logic behind the controller (symbols refer to Eq. (1))

definition of rational criteria to optimally calibrate the parameters involved in ($i_{\min}$, $i_{\max}$, σ_{lim} and x_{lim}). An effective calibration procedure has been proposed by the authors in Caterino et al. (2016).

3   Calibration of the SA Controller: A Case Study

The calibration procedure proposed by Caterino et al. (2016) is herein applied with reference to a specific case study, to provide the optimal choice of values to be assigned to the parameters involved in the control algorithm. The first step is generating a finite

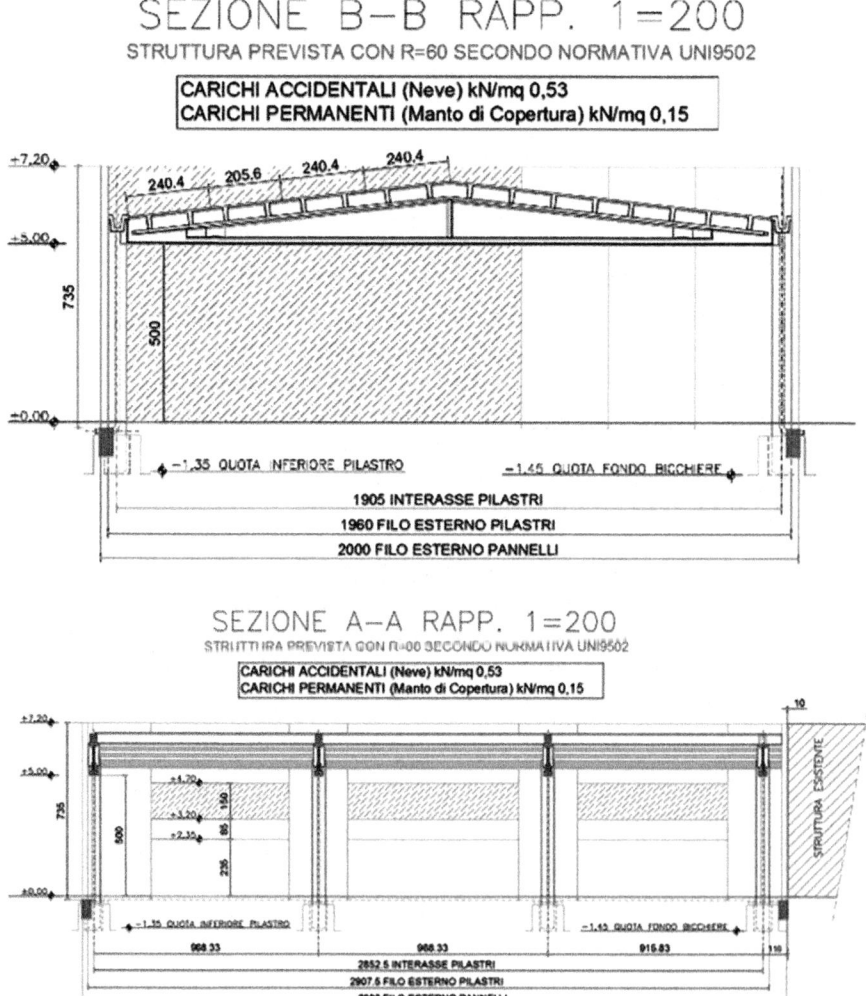

Fig. 3. Case study column's model

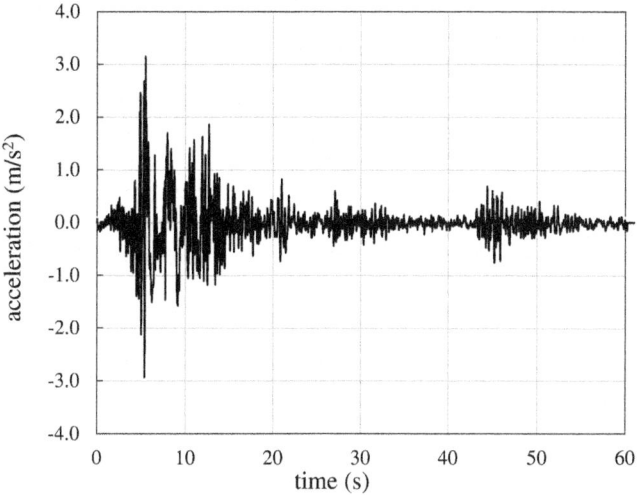

Fig. 4. Selected seismic input

element model of the structure to be examined, able to reproduce both fixed base (FB) and SA controlled configurations. With reference to a given seismic input, the structural response in the FB and passive cases has been determined. Then a small number of SA numerical simulations has been performed in order to single out the optimal configuration of the controller able to achieve the maximum reduction of base stress while not causing increasing of top displacement in respect to the FB case.

3.1   Case Study

The case study structure is a central column of a real precast RC structure (Fig. 3). The reference real structure is a precast RC structure having plan dimensions 20 m × 30 m, and a double slope covering. The columns are 5.7 m tall, with a uniform square cross section of dimensions 0.55 m × 0.55 m. The mass acting at the top of a central column is the sum of the masses of the covering elements relative to a half span at each side of the column and is equal to 25.7 tons.

The base of the model is highly stiff and is supported in the middle by a cylindrical steel hinge. On both sides of the base, one cylindrical spring and one MR damper are installed. The assembly "elastic springs + SA MR dampers", placed in parallel at the base of the tower, just represents the smart base restraint herein proposed to control the dynamic behavior of the structure.

The registration of the Campano Lucano (Italy) earthquake (Fig. 4) has been adopted for the numerical analyses (code of the seismic record 290 ya, magnitude 6.9, fault distance 32 km, date 23/11/1980, station ID ST96).

3.2   Numerical Model

A finite element model has been generated in Matlab environment to simulate the dynamic behavior of the case study structure. It consists in 37 elements: 36 elements simulate the column with uniform cross Section (55 cm × 55 cm), while the last element (37th) is more rigid and represents the connection of the top of the column to the structural covering. The part of the double slope covering acting on the considered column is simulated by a concentrated mass at the top of the column. Such mass is added in the global mass matrix at the translational degree of freedom at the top of the tower.

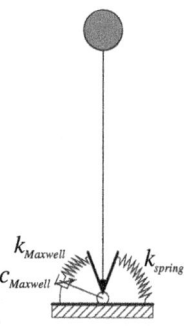

Fig. 5. Representation of the base restraint within the FE model of the SA controlled structure

The base support has been modeled as in Fig. 5, that is by a rotational spring k_{spring} and a Maxwell element (representing the MR dampers) working in parallel. The value for k_{spring} (2.1e7 Nm/rad) can be easily derived from the stiffness of the two linear springs and their distance from the center of rotation (hinge).

The Maxwell element, as known, consists of a spring $k_{Maxwell}$ and a linear viscous damper $c_{Maxwell}$ in series. The controllable part of this device is represented by the constant $c_{Maxwell}$, while $k_{Maxwell}$ has been simply assumed high enough (3e8 Nm/rad) so as to behave like a rigid link. Two different values of $c_{Maxwell}$ (c_{on}, c_{off}) have been determined so as to reproduce the dissipative capability of MR dampers respectively in the "ON" and "OFF" states. These two opposite configurations of the MR dampers are assumed to be those of the experimental campaign cited above, respectively corresponding to $i = i_{min} = 0$ A and $i = i_{max} = 1$ A. The MR dampers considered to calibrate the Maxwell device properties are those adopted in Caterino et al. (2016). The values of c_{on}, c_{off} have been calibrated as follows: $c_{on} = $ 1e10 Nms/rad and $c_{off} = $ 2e6 Nms/rad.

3.3   Numerical Simulations

A limited number of numerical analyses have been performed with reference to the above FEM model in SA configuration. Each of them corresponds to a selected combinations of stress (σ_{lim}) and displacement (x_{lim}) limits. The constrained optimization of the controller has been performed according to the condition aiming to achieve the greatest reduction of the base stress (objective function) and, at the same time, a top displacement (constraint function) no higher than that in uncontrolled FB conditions:

$$\min\left(\sigma_{\max}\big/\sigma_{\max,FB}\right) \text{ subject to } x_{\max}\big/x_{\max,FB} \leq 1 \tag{2}$$

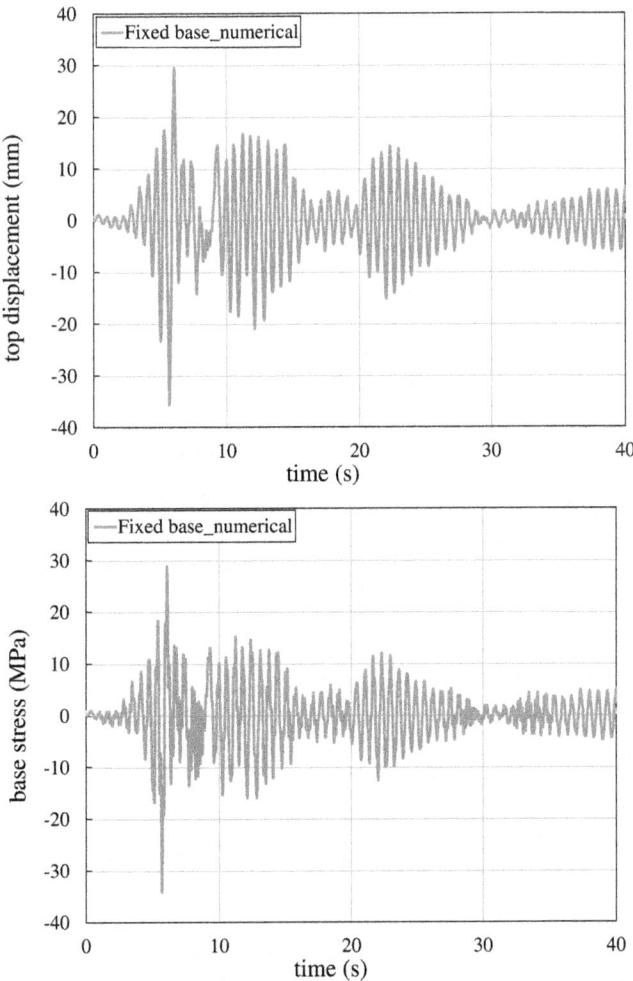

Fig. 6. Fixed base configuration (FB)

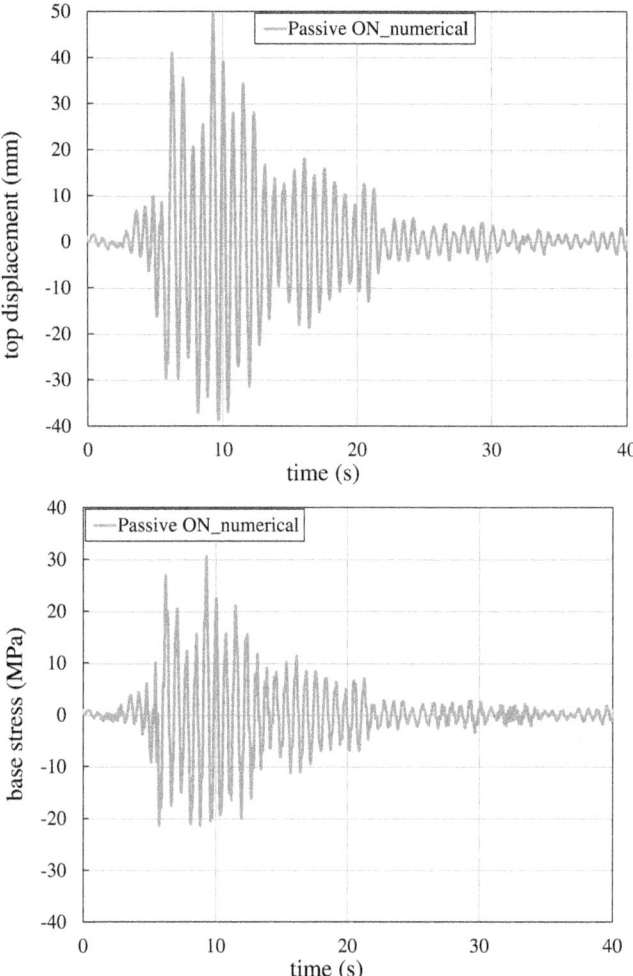

Fig. 7. Passive ON configuration

Its result is to assume, for σ_{lim} and x_{lim}, values respectively around $0.1\sigma_{max,FB}$ and $0.5x_{max,FB}$, leading to significant reduction of both base stress and top displacement, due to a sharp increase of dissipated energy due to a larger rocking of the base.

According to the criterion defined in the condition above, the optimal configuration of the control algorithm corresponds to the case $(\sigma_{lim}, x_{lim}) = (3\ MPa, 10\ mm)$: it leads to the maximum response reduction (about 48%) in base stress, without increasing the displacement in respect to the FB case.

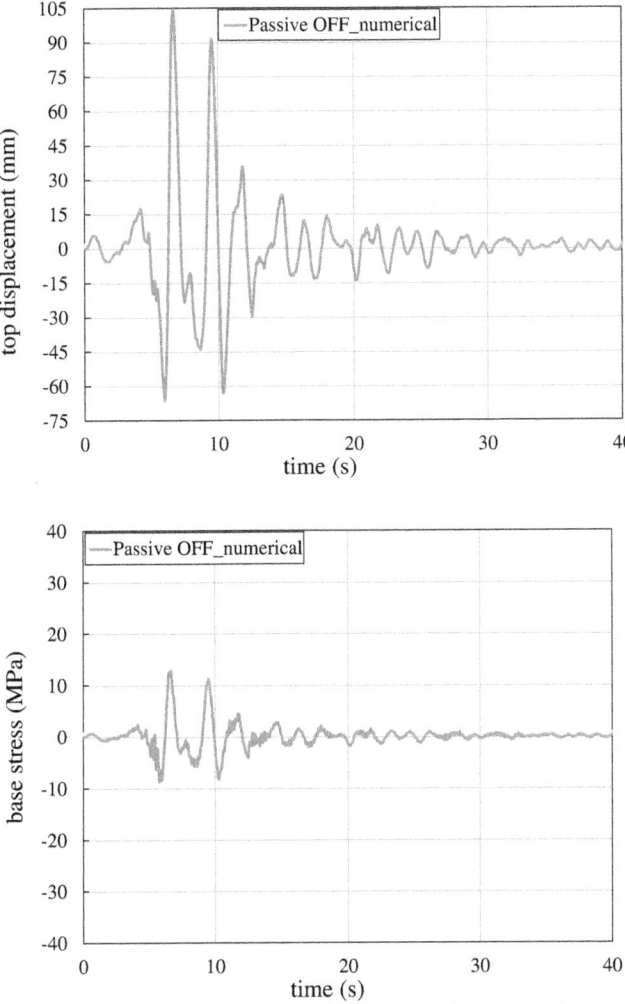

Fig. 8. Passive OFF configuration

The following Figs. 6, 7, 8 and 9 show the time history response of the structure in FB configuration, in Passive ON and Passive OFF configurations, and when semi-actively controlled with the above parameters $(\sigma_{lim}, x_{lim}) = (3$ MPa, 10 mm). The results are also summarized in Table 1. The reason behind the performance exhibited by the controller calibrated with $(\sigma_{lim}, x_{lim}) = (3$ MPa, 10 mm) is the significant number of instants where the rotations of the base are larger, so to determine a higher dissipation of energy.

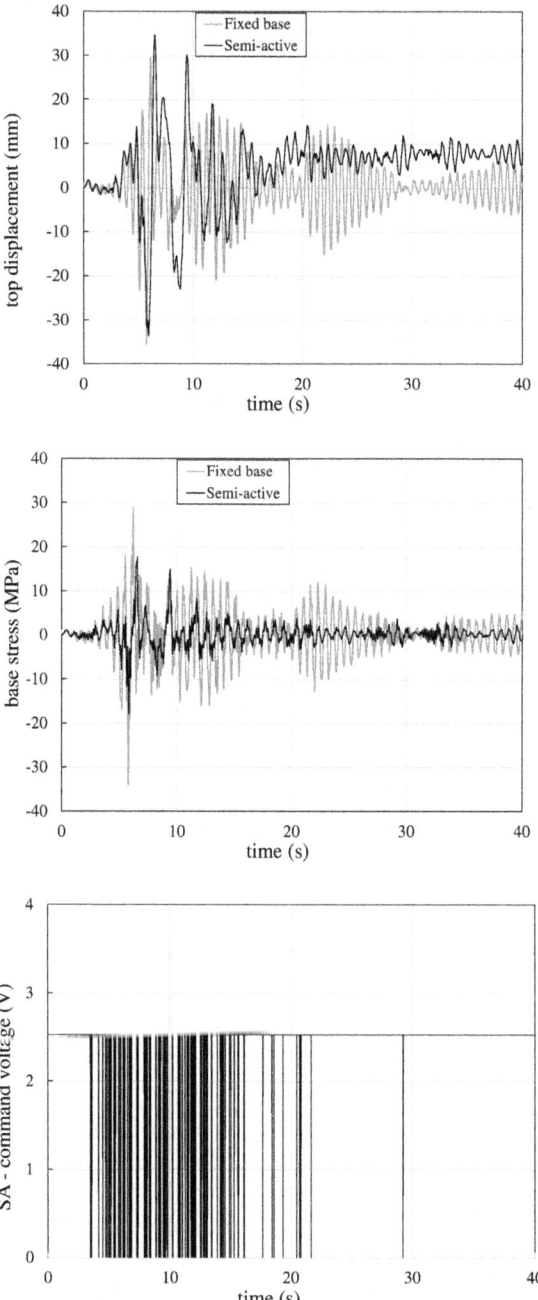

Fig. 9. Fixed base versus Optimal SA configuration, and command voltage as given by the algorithm.

Table 1. Maximum base stress and top displacement in the analysed configurations.

Setting	Base stress	Top displacement
	MPa	mm
FB	34.07	35.56
Passive ON	30.66	49.50
Passive OFF	13.00	104.31
Optimal SA	17.89	34.67

4 Conclusions

The idea to instantaneously remote control base stiffness and damping of a precast RC column to mitigate structural demand due to strong earthquakes has been discussed herein. The reduction of stiffness at the base restraint itself would imply reduction of base bending moment, but at the cost of a significant, undesired increase of displacement demand at the top of the column. This is no longer true when the change of stiffness is accompanied by a change of damping too. The greater rocking of the base can be not so harmful for displacement demand if it is coupled with a significant dissipation of energy. This is the main concept achieved by the authors and confirmed by the simulations above described. The semi-active control via magnetorheological dampers proposed for precast RC column is based on a 2-parameters control algorithm. The optimal couple of values (σ_{lim}, x_{lim}) for such parameters has been found according to a specific calibration procedure. In particular, these limit values for base stress and top displacement result to be respectively about 10% and 50% of the corresponding peak response values registered in the fixed base conditions. The so calibrated control system allowed high reduction of base stress, that results to be roughly halved in respect to the "fixed base" case, without increasing the top displacement response. The issue of recentering the system after the excitation is under study, it will be one of the focuses of the future developments of this work.

References

Spieth HA, Arnold D, Davies M, Mander JB, Carr AJ (2004) Seismic performance of post-tensioned precast concrete beam to column connections with supplementary energy dissipation. In: Proceedings of the 004 NZSEE conference

Murahidy AG, Spieth HA, Carr AJ, Mander JB (2004) Design, construction and dynamic testing of a post-tensioned precast reinforced concrete frame building with rocking beam-column connections and ADAS Elements. In: Proceedings of the 004 NZSEE conference

Lu L, Liu X, Chen J, Lu, X (2016) Seismic performance of a controlled rocking reinforced concrete frame. In: Advances in Structural Engineering, Early view (published on line)

Belleri A, Schoettler MJ, Restrepo JI, Fleischman RB (2014) Dynamic behavior of rocking and hybrid cantilever walls in a precast concrete building. ACI Struct J 111(3):661–672

Caterino N (2015) Semi-active control of a wind turbine via magnetorheological dampers. J Sound Vib 345:1–17. https://doi.org/10.1016/j.jsv.2015.01.022

Caterino N, Georgakis CT, Spizzuoco M, Occhiuzzi A (2016) Design and calibration of a semi-active control logic to mitigate structural vibrations in wind turbines. Smart Struct Syst 18(1):75–92. https://doi.org/10.12989/sss.2016.18.1.000

Improved Gradient-Based Equivalent Linear Procedure for Probabilistic Displacement-Based Design of RC Structures, Accounting for Damage-Induced Stiffness Degradation

P. Franchin$^{(\boxtimes)}$, F. Mollaioli, and F. Petrini

Department of Civil and Structural Engineering, Sapienza Università di Roma,
Rome, Italy
paolo.franchin@uniromal.it

Abstract. Starting from a novel gradient-based method for the performance-based seismic design (PBSD) of framed RC structures, proposed by one of the authors, some improvements are presented in order to take into account the damage-induced stiffness degradation at larger response values. The method allows for the computation of the values of selected independent design variables (beams and column sections and reinforcement) leading to the attainment or non-exceedance of pre-defined mean annual frequencies for multiple limit states, i.e. incorporating control of the seismic risk during the design phase. The improvement is accomplished by doubling the number of equivalent linear analyses of seismic performance at each point in the design space, implementing an intermediate element-by-element damage assessment after the first analysis. The procedure is exemplified with reference to a 15-storey RC plane frame building.

Keywords: Seismic design · Inelastic analysis · Mean annual rate
Risk · SAC

1 Introduction

Starting from the well-known SAC-FEMA closed-form approach for performance-based assessment of structures (Cornell et al. 2002), a novel gradient-based method has been proposed for the performance-based seismic design (PBSD) of framed RC structures by Franchin and Pinto (2012).The method is characterized by a direct understanding of the relationship between the performance requirements and the variation of design structural variables, and it is consistent with the original aim pursued by Cornell et al. (2002) of maintaining the performance-based earthquake engineering procedures at an affordable level of complexity for practitioners and, at the same time, of describing the aleatory nature of the problem in an explicit and sufficiently rigorous way.

© Springer International Publishing AG, part of Springer Nature 2018
M. di Prisco and M. Menegotto (Eds.): ICD 2016, LNCE 10, pp. 272–285, 2018.
https://doi.org/10.1007/978-3-319-78936-1_20

The initial method formulation was based on a number of assumptions, mainly borrowed by the original SAC-FEMA analytical architecture. The method has been shown to be efficient and reliable in characterizing the peak response parameters (EDPs, e.g. peak interstorey drift over height) for performance evaluation, but not to provide the same accuracy in evaluating response at higher scales (e.g. position of the peak interstorey drift along the building height, especially in case of collapse-related limit states (LSs)). Reasons for this partial lack of accuracy have been identified in two key components of the method: (i) the assumption of linear interpolation of the seismic hazard curve in the log-log space; (ii) the use in the equivalent linear response analysis of the original modal shape seven for high damage states, by means of a single global stiffness scaling factor.

This paper presents the state of advancement of the research that is currently under development to improve the original method. In this paper the second of the above assumptions is removed by implementing a modified version of an existing procedure for element-by-element damage assessment and associated element-based secant stiffness evaluation (Gunay and Sucuoglu 2009).

This improved PBSD method is applied to a 15-storey RC frame building. Preliminary results show that the improved method is suitable for a detailed PBSD of structures with an explicit achievement of the desired risk levels. Assessment of the designed structure by inelastic response-history analysis (IRHA) shows that the method, though some improvements are still needed, the deformation pattern of the structure is predicted much better even at higher damage levels.

2 Original Method Formulation

The method proposed in Franchin and Pinto (2012) allows the determination of the values of a set of independent design variables $\mathbf{d}$ in order to meet pre-defined performances requirements for a number n_{LS} of chosen LSs (namely λ_{LSi} for i = 1,2, ... n_{LS}) in terms of MAFs or corresponding arbitrary selected return periods Tr_{LSi} for the seismic action. The target MAF λ^*_{LSi} for the ith LS can be set for instance equal to the MAF of the associated seismic intensity, then $\lambda^*_{LSi} = 1/Tr_{LSi}$. The procedure allows for the direct control of the seismic risk (MAF of failure) associated with the specific structural configuration and site. Such a design configuration is obtained by the iterative solution of a constrained minimization problem expressed as:

$$
\begin{cases}
\min_{\mathbf{d}} \tilde{\lambda}^2 \\
\text{subjected to } \mathbf{c} \leq 0
\end{cases}
\tag{1}
$$

where $\mathbf{c}$ = vector of linear and nonlinear constrains on the design parameters, and $\tilde{\lambda}$ is defined at each iteration by

$$
\tilde{\lambda} = \max\left\{\lambda_{LSi}/\lambda^*_{LSi} - 1\right\} \qquad i = 1, 2, \ldots n_{LS}
\tag{2}
$$

The last equation allows the governing LS at each iteration to be defined as the one having the largest value of $\lambda_{LS}/\lambda^*_{LS}$. Obviously, for the considered LSs, the acceptable solution will have one LS for which $\lambda_{LS} = \lambda_{LS}^*$ and $\lambda_{LSi} < \lambda_{LSi}^*$ for the other n_{LS}–1. The square operator on $\tilde{\lambda}$ lets the objective function to be always positive and to have its minimum equal to zero, corresponding to the above-mentioned situation when one MAF of the n_{LS} considered LSs is equal to its goal value. The constrains **c** can represent requirements such as, e.g., maximum column cross-section reduction with the height or minimum column-to-beam capacity ratio): They are expressed in the general form

$$\mathbf{c}(\mathbf{d}) = \mathbf{A} \times \mathbf{f}(\mathbf{d}) + \mathbf{b} \tag{3}$$

where **f(d)** is a functional non-linear vector.

The introduced optimization problem is solved by a gradient-based approach where the gradient of (1) with respect to **d** implies the evaluation of the gradient $\nabla_{\mathbf{d}}\lambda_{LSi}$ with respect to **d** of each LS considered in (2).

This is made possible analytically by the availability of the well-known SAC-FEMA formula proposed by Cornell et al. (2002) for the MAF evaluation under the assumptions that: demand, D, and capacity, C, are lognormally distributed; median demand varies with the intensity measure (IM) according to the power law $\hat{D} = aIM^b$; seismic hazard (MAF of the IM) can be approximated (around the region $IM_{\hat{D}=\hat{C}}$) by the expression $H(IM) = k_0 IM^{-k}$, i.e. linear interpolation on a log-log space:

$$\begin{aligned}
\lambda_{LS} &= H[IM_{D=C}]exp\left[\frac{1}{2}\frac{k^2}{b^2}\left(\beta_{D,T}^2 + \beta_{C,T}^2\right)\right] \\
&= k_0\left(\frac{\hat{C}}{a}\right)^{-\frac{k}{b}}exp\left[\frac{1}{2}\frac{k^2}{b^2}\left(\beta_{D,T}^2 + \beta_{C,T}^2\right)\right]
\end{aligned} \tag{4}$$

where $H[IM_{D=C}]$ is the hazard, i.e. the MAF of the IM occurrence, evaluated at the IM value that induces a median demand $\hat{D}$ equal to the median limit-state capacity $\hat{C}$, while $\beta_{D,T}$ and $\beta_{C,T}$ are the total demand and capacity dispersions.

With these premises, the above mentioned gradient $\nabla_{\mathbf{d}} \lambda_{LSi}$ of the generic λ_{LSi} with respect to the design variables **d** is evaluated starting from Eq. (4) under the assumptions that the variation of $\beta_{D,T}$ and $\beta_{C,T}$ in the design space is negligible, something leading to the following equation for $\nabla_{\mathbf{d}} \lambda_{LSi}$

$$\begin{aligned}
\nabla_{\mathbf{d}}\lambda_{LSi} = &\frac{\partial\lambda_{LSi}}{\partial k_0}\frac{\partial k_0}{\partial \mathbf{d}} + \frac{\partial\lambda_{LSi}}{\partial k_1}\frac{\partial k_1}{\partial \mathbf{d}} + \\
&\frac{\partial\lambda_{LSi}}{\partial a}\frac{\partial a}{\partial \mathbf{d}} + \frac{\partial\lambda_{LSi}}{\partial b}\frac{\partial b}{\partial \mathbf{d}} \qquad i = 1, 2, \ldots, n_{LS}
\end{aligned} \tag{5}$$

where k_{0i}, k_i = hazard linear interpolation parameters for the ith LS, and a,b = parameters of the D-IM law for the structural response evaluation. In Franchin and Pinto (2012), terms of the Eqs. (4) and (5) are evaluated by closed form equations, thanks to the use of an equivalent linear structural model for response evaluation. In this respect the method rests on the validity of the so-called empirical equal-displacement approximation.

3 Proposed Modification

3.1 Damage-Induced Stiffness Degradation for Equivalent Linear Analysis

Parameters k_0 and k_1, as well as a and b of Eq. (4) are usually established by linear regression. The former on the hazard curve in terms of the chosen IM, and the latter on the results of inelastic time-history analyses, carried out with recorded motions appropriately selected to cover the IM range of interest and represent the expected seismic input at the site. In order to limit computational effort of the optimization procedure, in the original method b is assumed equal to one (a sufficiently reasonable assumption for displacement EDPs) and a is evaluated by a single linear response spectrum analysis at each iteration in the design space, with a unique global stiffness reduction factor irrespective of the considered LS. For the same reason, the secant structural vibration period is LS-independent and k_0 and k_1 are constant within one iteration.

A valuable addition to the initial procedure is thus the introduction of a model for the element-based damage-induced stiffness reduction. The introduction of such a secant model, leads the above-mentioned k_0, k_1, a and b damage-dependent and thus LS-dependent parameters. This changes Eq. (5) to

$$\nabla_{\mathbf{d}} \lambda_{LSi} = \frac{\partial \lambda_{LSi}}{\partial k_{0i}} \frac{\partial k_{0i}}{\partial \mathbf{d}} + \frac{\partial \lambda_{LSi}}{\partial k_i} \frac{\partial k_i}{\partial \mathbf{d}} +$$
$$\frac{\partial \lambda_{LSi}}{\partial a_i} \frac{\partial a_i}{\partial \mathbf{d}} + \frac{\partial \lambda_{LSi}}{\partial b_i} \frac{\partial b_i}{\partial \mathbf{d}} \qquad i = 1, 2, \ldots, n_{LS} \tag{6}$$

where the dependence of k_0, k_1, a and b from the LS is highlighted.

Since the design method relies on linear equivalent representation of structural response, one good option is to adopt the equivalent linear-elastic procedure proposed in Günay and Sucuoglu (2009–2010) for framed structures in order to consider the influence of the damage-induced stiffness reduction to the structural performances.

For each structural member the bending stiffness reduction factor (RF) is inversely proportional to a bending moment ratio (RM), which takes values between 1 (no stiffness reduction) and infinity (element complete loss of stiffness). The RM is evaluated for all structural members as the ratio between the sum of seismic bending moments at the member ends to the sum of the corresponding residual capacity bending moments (bending moment capacity reduced by non-seismic, i.e. gravity, effects).

Initial cracking is taken into account by an initial reduction factor (IRF) applied to the gross uncracked bending stiffness of the elements, assumed equal to 0.6 for the columns and 0.4 for the beams.

Structural response analysis is carried out in two steps: (i) response is first evaluated under non-seismic (static analysis) and seismic (modal response spectrum analysis) loads with initial stiffness (IRFs); the resulting seismic and non-seismic external moments are then used to evaluate the actual response-dependent RFs; (ii) a second analysis is carried out to evaluate the response of the structure having the stiffness additionally reduced by the RFs calculated at the previous step. In principle, the second analysis should be carried out with secant stiffness and increased equivalent viscous damping, to account for damage. Here, however, the authors take a different path, in order to avoid the difficulty of evaluating an equivalent global damping as a function of non-uniform inelastic deformation. In the second analysis seismic action at each mode is taken equal to the spectral displacement at the period of the cracked-but-undamaged structure evaluated in step i, i.e. with the same damping, while the deformed shape for each mode is the shape obtained for the damaged structure. Neglecting the increase damping leads to response overestimation, while neglecting the stiffness increase leads to response underestimation. The two effects to some extent cancel out, but in an uncontrolled, structure-dependent fashion (see later).

3.2 Capacity Design

The original method by Franchin and Pinto (2012) did not take into account capacity design. While drifts were directly controlled, the actual plastic mechanism could and actually did include hinges in columns. Capacity design is regarded as a paramount aspect of safety in structures designed to enter the inelastic range, thus it is here added to the method. Günay and Sucuoglu (2009) first introduced their procedure for capacity-designed structures, meaning that the analyst would know beforehand where plastic hinge could form (in beams). They then extended their procedure to existing structures (2010) and introduced a formal check of column-to-beam capacity ratios (CBCRs) at all joints in order to first determine potential plastic hinge zones. In the proposed improved design procedure this check on CBCRs is introduced as an additional constraint on the design variables in Eq. (3), so that all feasible points in the design space are capacity-design compliant.

The CBCR at a structural joint is defined as shown in Fig. 1 as:

$$CBCR = \frac{M_{c\_up} + M_{c\_bot}}{M_{c\_right} + M_{c\_left}} \qquad (7)$$

where M_{c_up} and M_{c_bot} = bending moment capacities of the upper and bottom column, and M_{c_right} and M_{c_left} = bending moment capacities of the right and left beams respectively.

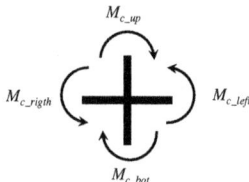

Fig. 1. Bending moments at the joint for the CBCR definition

According to Günay and Sucuoglu (2010) only the attached beams or columns ends, for joints having CBCRs greater/equal to 1.2 or lower/equal to 0.8 respectively, are assumed to be potential yielding regions, while for joints having CBCRs values between 0.8 and 1.2, both columns and beams ends are treated as potential yielding regions. Expressions used for the evaluation of the RM for the generic beam or column are selected depending on the number of ends that are considered as potential yielding regions after the evaluation of all the CBCRs:

– Both ends are not potential yielding regions

$$RM = 1 \tag{8}$$

– Only one end, e.g. i, is a potential yielding region

$$RM = \frac{M_{E,i} + M_{E,j}}{M_{RC,i} + \alpha M_{RC,j}}$$
$$\text{with } \begin{cases} \alpha = CBCR & \text{if } CBCR < 1 \\ \alpha = 1/CBCR & \text{if } CBCR > 1 \end{cases} \tag{9}$$

– Both ends are potential yielding regions

$$RM = \frac{M_{E,i} + M_{E,j}}{M_{RC,i} + M_{RC,j}} \tag{10}$$

In the above Eqs. (8, 9 and 10), $M_{E,i;j}$ is the seismic demand bending moment at the i–th or j–th end of the member, $M_{RC,i;j}$ the corresponding residual bending moment capacity.

In the modified design method, the new constraints ensure that CBCR is larger than 1.2 (or any other chosen value) at all nodes but for the base of ground floor columns. Thus Eq. (9) is used for the latter columns, Eq. (8) for all other columns and Eq. (10) for beams. Introduction of the new constraints requires of course to consider among the design variables also the reinforcement ratios, which were not included in the original method formulation.

3.3 Consistency of Seismic Action Representations in Probabilistic Design and Inelastic Response-History Analysis for Validation

Without any loss of generality, the spectral displacement at the first natural period of the structure $S_d(T_1)$ is selected as the *IM*, and the seismic input for the equivalent-linear modal response spectrum analyses is defined in terms of uniform hazard spectra(UHS) (Loh et al. 1994).

Validation of the method requires an independent assessment of the seismic risk (MAF of exceedance for each LS) by means of IRHA. It should be noted that Franchin and Pinto (2012) carried out the IRHA in a now outdated manner, though acceptable at the time. Motions were artificial spectrum-compatible with the UHS used in design.

Without entering into the details, this is not correct and methods are available today to select recorded motions that are truly compatible with the considered site seismicity, e.g. the conditional spectrum method (Lin et al. 2013). When motions are correctly selected according to such methods, the resulting estimate of the MAF of any LS is unbiased and converges with larger number of records. When motions are not correctly selected, MAF estimates are biased and changing records leads to different and more dispersed MAF estimated. For this reason, a bootstrap was carried out in Franchin and Pinto (2012) which is not necessary here. What is important instead, is to be fair to the design method in its validation and to use a UHS truly consistent with the motion selected for validation. Thus, the UHS for design is obtained for the site of interest on the basis of the set of 200 ground motions (a total of 20 motions for $N_s = 10$ levels of seismic intensity), selected by using the conditional spectrum method with a conditioning period $T = 2$ s:

- Conditional spectra of the 200 records are obtained
- For each $S_a(T^*)$ with T^* different from the conditioning $T (= 2$ s), H_{Sa*} is evaluated as

$$H_{Sa*}(s^*) = P(S_a^* > s^*) \cong \sum_{t=1}^{Ns} \hat{G}_{S_a^*|S_a}(s^*|s_t)|\Delta H_{S_a}(s_t)| \tag{11}$$

Where the complementary CDF of $S_a(T^*)$ is

$$\hat{G}_{S_a^*|S_a}(s^*|s_t) = P(S_a^* > s^*|S_a = s_t) = \frac{N_R(S_a^* > s^*)}{N_R^{tot}} \tag{12}$$

where $N_R(S_a^* > s^*)$ is the number of records with intensity level s_t having $S_a^* > s^*$, $N_R^{TOT} = = 20$ for all intensities).

- UHS are then obtained by intersection of hazard curves as qualitatively represented in Fig. 2.

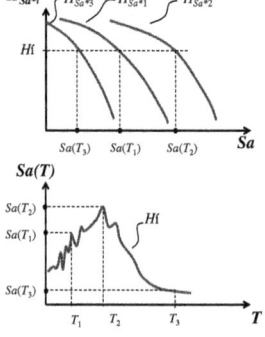

Fig. 2. Extrapolation of the UHS

4   Applicative Example

The improved method is applied to a 15-storey RC building frame shown in Fig. 3. Overall building dimensions are shown in the figure. In addition to self-weight, a vertical uniformly distributed load of 34.5 kN/m is applied to the beams. The main response parameter for the performance evaluation is the peak interstorey drift θ_{max}. The gradient-based constrained optimization is solved in Matlab$^{\copyright}$ by means of the "fmincon" function. The purpose-made code evaluates also the response of 2D frame models. Two design LSs are considered: Light Damage (LD) LS, corresponding to a median $\theta_{max,LD} = 0.004$, and Collapse Prevention (CP) LS, corresponding to a median $\theta_{max,CP} = 0.015$. Dispersions are specified later. The two corresponding target MAFs are arbitrarily set to $\lambda^*_{LD} = 1/(100\,\text{years})$ and $\lambda^*_{CP} = 1/(2470\,\text{years})$. For this application b is still set to 1 in previous equations for both LSs (see comments).

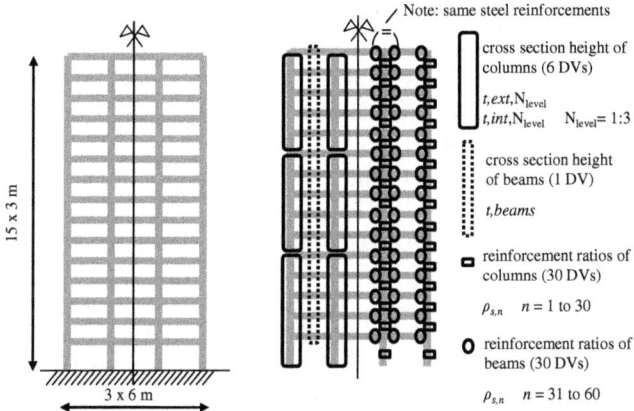

Fig. 3. Overall dimensions of the building (left) and design variables (right).

4.1   Design Variables and Constraints

A total of 67 design variables are defined, including the in-plane dimensions (cross-sections heights "t") of the columns (cross-section height variation in the same column alignment is allowed every five floors and differentiation between external and internal column alignments is allowed, for a total of 6 DVs) and beams (all the beams are assumed to have the same cross-section height, for a total of 1 DV), as well as the reinforcement ratios ρ_s of all beams (top reinforcement) and columns, for a total of 60 variables (different ρ_s at each floor, ρ_s of beams converging to the same joint forced to be the same; plus global symmetry of the building,). For all columns, top and bottom reinforcements are the same, while for all beams the bottom reinforcement is assumed to be equal to half the top reinforcement.

Table 1. Design variables and their min and max values.

DV	Description	Min [m or %]	Max [m or %]
$t,ext,1$	t external columns first 5 floors	0.7 m	1.4 m
$t,ext,2$	t external columns 6th to 10th floor	0.6	1.2
$t,ext,3$	t external columns 11th to 15th floor	0.5	1.0
$t,int,1$	t internal columns first 5 floors	0.75	1.5
$t,int,2$	t internal columns 6th to 10th floor	0.65	1.3
$t,int,3$	t internal columns 11th to 15th floor	0.55	1.1
$t,beams$	t beams, all elements	0.6	1.2
$\rho_{s,1:15}$	reinf ratios external columns 1st to 15th floor	0.8%	3%
$\rho_{s,16:30}$	reinf ratios internal columns 1st to 15th floor	0.8	3
$\rho_{s,31:60}$	top reinf ratios beams 1st to 15th floor	0.15	3

Note: "t" indicates the cross-section in-plane dimension

The DVs are shown in Fig. 3 and in Table 1, where the allowed maximum and minimum values are also reported. Initial values are set to 1.1 times the minimum ones. The out-of-plane cross section dimensions are kept constant for all members and equal to 0.5, 0.4 and 0.3 m for the columns (both internal and external) of the first ten floors, of the last five floors, and for all floor beams respectively.

4.2   Seismic Hazard and Uncertainties Treatment

The site of L'Aquila (Italy) is chosen for the example. UHS are obtained as described from a set of 200 recorded motions selected through conditional spectrum. The UHS are obtained starting from the conditional spectra of the signals with a fixed conditioning period equal to 2 s. A set of UHS obtained for the reference site at different return periods is shown in Fig. 4.

In applying the SAC-FEMA approach for performance-based assessment or design, a crucial point is the quantification of the uncertainty involved in the analysis, identified by the demand and capacity dispersions $\beta_{D,T}$ and $\beta_{C,T}$ in Eq. (4). As well known (CEB-FIP 2012, Vamvatsikos 2013) the total dispersions of both demand D and capacity C can be

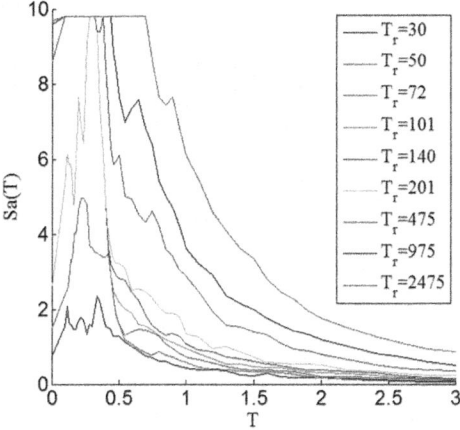

Fig. 4. Uniform hazard spectra for the selected site (L'Aquila-Italy)

modelled as composed by dispersions due to aleatory randomness ($\beta_{D,R}$ and $\beta_{C,R}$ for D and C respectively) and due to epistemic uncertainty ($\beta_{D,U}$ and $\beta_{C,U}$ for D and C respectively). Under the common assumption that there is no correlation between R and U, total dispersions are obtained as:

$$\beta_{D,T} = \sqrt{\left(\beta_{D,R}^2 + \beta_{D,U}^2\right)} \tag{13}$$

$$\beta_{C,T} = \sqrt{\left(\beta_{C,R}^2 + \beta_{C,U}^2\right)} \tag{14}$$

The following values are adopted for the dispersions:

- LD: $\beta_{D,R} = 0.3$; $\beta_{D,U} = 0.2$; $\beta_{C,R} = 0$; $\beta_{C,R} = 0$
- CP: $\beta_{D,R} = 0.3$; $\beta_{D,U} = 0.2$; $\beta_{C,R} = 0.3$; $\beta_{C,R} = 0.2$

5   Results and Validation

The final design configuration is shown in Fig. 5, while the modal periods (first 5 modal shapes, with a total modal participating mass of more than 90% of the total) of the structure "after the gravity loads" (the one obtained by the application of the stiffness IRF as described in 3.1) are: $T_{1-5} = 2.08$ s; 0.70 s; 0.40 s; 0.27 s; 0.20 s. It is important to note that the first natural period of the structure $T_1 = 2.08$ s, something very near to the conditioning period (2 s) used in obtaining the UHS by the conditional spectrum method as described in the previous section. This ensures that the chosen IM is efficient for the designed structure.

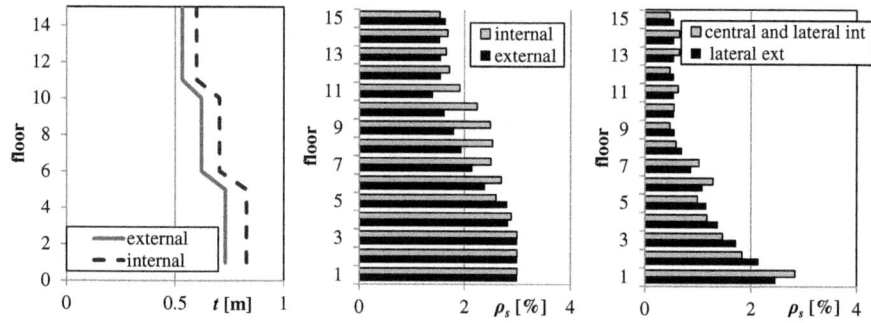

Fig. 5. Final configuration: column in-plane cross section dimensions (left), columns reinforcement ratios (central), beams top reinforcement ratios (right)

The convergence of the procedure is obtained with about 20 iterations, and the final obtained MAFs are:

- $\lambda_{LD} = 0.0025 \left(\lambda^*_{LD} = 0.01 \right)$
- $\lambda_{CP} = 3.97 \times 10^{-4} \left(\lambda^*_{CP} = 1/2470 = 4.04 \times 10^{-4} \right)$

From the result it is evident that the CP-LS is dominant for this design, and the obtained value of λ_{CP} is very near to the target value λ^*_{CP}, the difference between the two being due to the numerical convergence tolerances of the optimization procedure.

A useful representation of the damage state evaluated by the equivalent linear procedure is the one shown on the left side of Fig. 6 for the final configuration subjected to the UHS at the CP-LS (*Tr* = 2470 years): circles indicate the potential plastic hinges), their diameter being proportional to the ratio IRF/RF as evaluated for the elements during the analysis, meaning that small values of RF are interpreted as large excursions of the element in the plastic domain due to the seismic action. Black (bold)

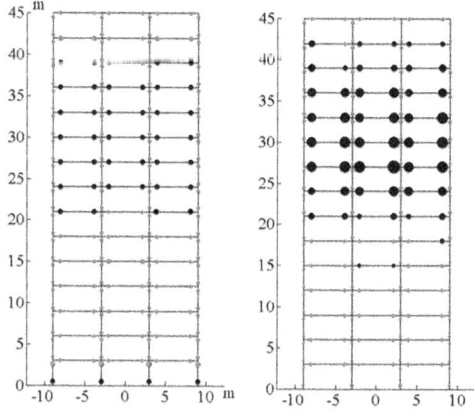

Fig. 6. Plastic hinges for the collapse prevention LS. Equivalent linear procedure (left); NLA signal 1 (right).

colour indicates that the plastic hinge is activated (RF < 1), while light grey colour indicates that the plastic hinge is not activated (RF = 1). From the figure can be also appreciated that the capacity design is correctly realized (plastic hinges are activated only in beams and at the base) by means of the constrains on the CBCR described in previous Sect. 3.2.

Validation through IRHA is carried out in OpenSees (McKenna and Fenves 2001). A model is set up with elastic frame elements and zeroLength elements representing plastic hinges having a modified Ibarra-Medina-Krawinkler constitutive law (Ibarra et al. 2005, Lignos and Krawinkler 2011). Parameters for the latter are set in order to have a bi-linear hinge behavior (with a slight hardening of 0.001 of the elastic stiffness for numerical stability purposes) without any cyclic stiffness degradation. The hinges activation values for elements are the ones obtained from the gross stiffness reduced by means of the same IRF adopted in the previous described linear equivalent analysis, taking into account for the difference between the top and bottom cross sections reinforcements in the beams.

The comparison between the maximum drift profiles obtained by the equivalent linear procedure and the UHS with $Tr = 2470$ years, with the 20 IRHAs corresponding to the seismic intensity $Tr = 2500$ years is shown in Fig. 7, where the mean and median of the IRHA maximum drift profiles are also reported, together with the median drift identifying the CP-LS ($\theta_{max,CP} = 0.015$). Both the mean and median of IRHAs and the result of the equivalent linear analysis are smaller than the CP-LS threshold. This is correct since it accounts for dispersion in both demand (visible from the individual motion results) and capacity. It is apparent that the equivalent linear model is able to correctly evaluate the shape of the max drift profile, also if the maximum is slightly underestimated and translated with respect to the mean and/or median of the IRHA results. The profile shape is much better captured than with the original method formulation (see Franchin and Pinto 2012). The discrepancy on the peak value, on the other hand, is larger than in the original formulation. This may be due to the uncontrolled counterbalancing effects described in Sect. 3.1, or to the choice of the demand dispersion in Sect. 4.2 (for instance a reduction in $\beta_{D,U}$ would increase drift profile values). The response underestimation is confirmed by the comparison (Fig. 6) of the plastic hinges configuration obtained by the equivalent linear analysis (left) subjected to the UHS at the CP-LS ($Tr = 2470$ years) with the one obtained with the IRHA (right) under one of the 20 motions corresponding to $Tr = 2500$ years. In the case of IRHA, the circles diameter (activated plastic hinges) is proportional to the ratio between the maximum curvature experimented by the hinge section to the yielding curvature.In any case, the good match of plastic hinge distribution is encouraging.

Finally, the MAF obtained with the final building configuration for the CP-LS ($\lambda_{CP} = 3.97 \ 10^{-4}$) is compared with the one estimated by a multiple stripe analysis (Jalayer and Cornell 2009) referred to as λ_{CP}^{NLA} , carried out with the $20 \times 10 = 200$

selected motions. The expected value $\bar{\lambda}_{CP}^{NLA}$ of λ_{CP}^{NLA} can be evaluated by the convolution of the structural fragility $P(\theta_{max} > \theta_{max,CP}|IM)$ with the hazard (H) by following equation

$$\bar{\lambda}_{CP}^{NLA} = \int_0^\infty P(\theta_{max} > \theta_{max,CP}|IM = x)dH(IM = x) \qquad (15)$$

where θ_{max} is the maximum drift obtained by NLA and $P(\bullet \mid \bullet)$ indicates a conditional probability, and IM is the intensity measure taken equal to the spectral acceleration at the first structural period $S_a(T_1)$ for the NLA.

In order to have an idea of the reliability of the results, the 99% confidence interval of the $\bar{\lambda}_{CP}^{NLA}$ has been also evaluated as $[4.11 \times 10^{-4} - 8.81 \times 10^{-4}]$. It is noted that both λ_{CP} and λ_{CP}^* are slightly outside this interval, as expected looking at the maximum drift profiles in Fig. 7. It is still necessary to be put in place some correction regarding the linear equivalent model and/or the uncertainty treatment.

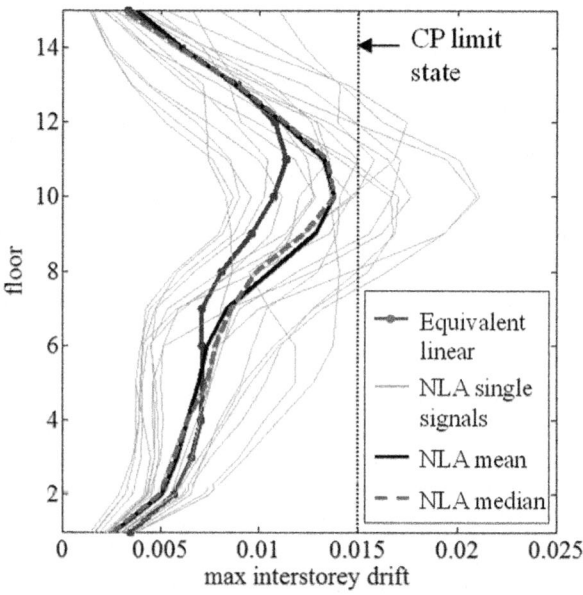

Fig. 7. Max interstorey drift profiles for collapse prevention LS. Comparison between equivalent linear procedure and nonlinear analysis (NLA).

6 Conclusions

A method for the probabilistic PBSD of RC frames is presented. The method aims at allowing for the direct control of the resulting structural seismic risk of multiple LSs, while easily imposing both construction (tapering, symmetry, minimum and maximum reinforcement, geometric conditions in general) and performance (e.g. CBCR) constrains during the design step. The implementation in the method of available equivalent linear models for seismic analysis of damaged structures allows for better consideration of stiffness degradation at higher response levels. The results of the method (MAF and peak interstory drift prediction) are compared with inelastic response history analysis with recorded ground motions. The results indicate that the method is potentially suitable for direct probabilistic design of RC frames but further improvement of the equivalent linear analysis procedure and a better quantification of the uncertainties involved in the problem are still needed.

References

CEB-FIP (2012) Probabilistic performance-based seismic design. Fib bulletin 68, Technical report of the Task Group 7.7 (Convener P. E. Pinto)

Cornell AC, Jalayer F, Hamburger RO, Foutch DA (2002) Probabilistic basis for 2000 SAC federal emergency management agency steel moment frame guidelines. J Struct Eng 128 (4):526–533

Franchin P, Pinto P (2012) Method for probabilistic displacement-based design of RC structures. J Struct Eng 138(5):585–591

Günay MS, Sucuoglu H (2009) An improvement to linear-elastic procedures for seismic performance assessment. Earthq Eng Struct Dynam 39:907–931

Günay MS, Sucuoglu H (2010) An improvement to linear-elastic procedures for seismic performance assessment. Earthq Eng Struct Dynam 39(8):907–931

Ibarra LF, Medina RA, Krawinkler H (2005) Hysteretic models that incorporate strength and stiffness deterioration. Earthq Eng Struct Dynam 34(12):1489–1511

Jalayer F, Cornell CA (2009) Alternative non-linear demand estimation methods for probability-based seismic assessments. Earthq Eng Struct Dynam 38(8):951–972

Lignos DG, Krawinkler H (2011) Deterioration modeling of steel components in support of collapse prediction of steel moment frames under earthquake loading. J Struct Eng, ASCE 137(11):1291–1302

Lin T, Harmsen SC, Baker JW, Luco N (2013) Conditional spectrum computation incorporating multiple causal earth-quakes and ground motion prediction models. Bull Seismol Soc Am 103 (2A):1103–1116

Loh CH, Jean WY, Penzien J (1994) Uniform-hazard response spectra–An alternative approach. Earthq Eng Struct Dynam 23(4):433–445

Vamvatsikos D (2013) Derivation of new SAC/FEMA performance evaluation solutions with second-order hazard approximation. Earthq Eng Struct Dynam 42:1171–1188

Comparison Between Non-linear Numerical Models for R.C. Shear Walls Under Cyclic Loading

G. Mancini, G. Bertagnoli, D. La Mazza, and D. Gino[(✉)]

Department of Buildings, Structural and Geotechnical Engineering,
Politecnico di Torino, Turin, Italy
diego.gino@polito.it

Abstract. The non-linear behaviour of concrete structures is the result of a series of phenomena, as material non-linear constitutive law and cracking process. As a consequence, in order to understand the behaviour of reinforced concrete members from elastic field to ultimate condition, is necessary to use instruments able to simulate the material damaging evolution under growing loads. Commercial non-linear finite elements codes are generally able to simulate concrete behaviour with good approximation when a progressive incremental load is applied. However, the same result could not be reached under a cyclic loading. In this work two commercial non-linear finite element codes have been considered in order to assess the skill of these codes to simulate non-linear concrete behaviour under cyclic loading. The results of six laboratory tests on shear walls have been compared with the ones obtained by means of numerical models and some conclusions on the numerical predictions are presented.

Keywords: Reinforced concrete · Non-linear · Finite elements
Cyclic loading · Shear walls

1 Introduction

The non-linear behaviour of reinforced concrete structural members, which is expressed by the non-proportionality between actions and structural response, it is the result of a series of phenomena as non-linearity of the materials constitutive law, cracking process and second order effects into slender structures.

It is then necessary to use instruments able to follow this complex behaviour and to simulate the damaging process that occur progressively into the concrete matrix.

The Model Code 2010 allows designers to assess the structural reliability by using a safety format applicable to results coming from non-linear analysis. The structure can be considered safe if the subsequent relation is satisfied:

$$F_d \leq R_d \tag{1}$$

Where F_d = design agent action; and R_d = design strength of the structural member.

© Springer International Publishing AG, part of Springer Nature 2018
M. di Prisco and M. Menegotto (Eds.): ICD 2016, LNCE 10, pp. 286–297, 2018.
https://doi.org/10.1007/978-3-319-78936-1_21

The structural response and strength can be evaluated by using GMNA (Geometrical, Material Non-linear Analysis) methods in order to take into account the real behavior under severe loading conditions.

The same analysis performed with different finite element codes can lead to discordant results, which may also show strong deviations from the experimental ones, when these are available.

Therefore, a non-linear finite element code needs to be accurately tested in order to let designers know the level of accuracy of the GMNA they will perform.

Six laboratory tests on shear walls subjected to cyclic loads up to failure are considered in this paper: Pilakoutas and Elnashai (1995), Lefas and Kotsovos (1990).

Bertagnoli et al. (2015) presented a comparison between three commercial finite element codes named A, B and C, testing plane stress structures under monotonically increasing loads.

Software B and C demonstrated a better accuracy therefore have been chosen for the present evaluation under cyclic loads.

The shear-walls have been numerically reproduced using plane stress models, choosing the same kind of elements and the same mesh dimension in both softwares.

Commercial non-linear finite elements codes generally provide better results when used to reproduce experimental tests with monotonically increasing loads, rather than with cyclic loading.

In fact, the way the code manages cracks opening and reclosing process acquires fundamental relevance under a cyclic load configuration.

Furthermore, in reinforced concrete structures subjected to cyclic loads, becomes relevant the effect of concrete confinement provided by the member geometry and stirrups or transverse reinforcement.

Shear walls can be modelled using three-dimensional or plane stress models. Both the choices need a careful interpretation of confinement phenomena.

In particular, in plane stress models, the in-plane confinement is always managed by the f.e. code, but out-of-plane confinement effect is not taken into account automatically if the user does not introduces it artificially.

The model proposed by Eurocode 2, that will be explained in Sect. 3.2, has been adopted in the present paper in order to consider out-of-plane confinement contribute.

2 Case Studies

2.1 Lefas and Kotsovos Walls

The specimens realized by Lefas and Kotsovos (1990) are 650 mm wide, 1300 mm high and 65 mm thick. The walls named SW31, SW32 and SW33 have been analysed in this paper.

All specimens, tested as isolated cantilever, are monolithically connected to an upper and a lower beam, the former is used to transfer the load coming from the jack, while the latter is used to simulate a rigid foundation; in addition both elements are used to anchor vertical bars.

Figure 1 shows the nominal dimensions of the walls and the arrangement of the reinforcements. Vertical and horizontal bars with a diameter of 8 and 6.25 mm respectively have been used, furthermore the vertical edges of the walls are provided with 4 mm stirrups in order to confine the concrete.

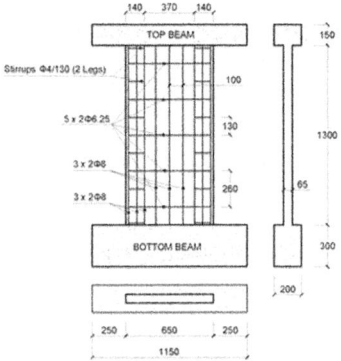

Fig. 1. Geometry and reinforcement details of walls SW31, SW32 and SW33.

In Tables 1 and 2 are summarized the main properties of the material used for each wall, where: f_{cm} is the concrete mean cylinder compressive strength, ε_{c1} is the strain at peak stress, ε_{cu1} is the ultimate strain in uniaxial loading, whereas f_y is steel yielding stress, ε_y is yielding strain, and f_u is steel failure stress and ε_u is ultimate strain.

Table 1. Concrete properties

Specimen	f_{cm} (MPa)	ε_{c1} (‰)	ε_{cu1} (‰)
SW31	29.2	2.15	3.50
SW32	44.5	2.39	3.50
SW33	40.8	2.34	3.50

Table 2. Reinforcement properties

Diameter (mm)	f_y (MPa)	f_u (MPa)	ε_y (%)	ε_u (%)
Φ 4	420	490	0.21	7.50
Φ 6.25	520	610	0.26	7.50
Φ 8	470	565	0.24	7.50

Figures 2, 3 and 4 show the three simplified types of horizontal cyclic loading adopted respectively for the specimens SW31, SW32 and SW33. After the cyclic phase all specimens have been incrementally loaded up to failure.

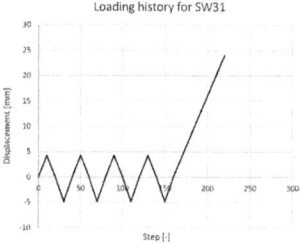

Fig. 2. Loading history for wall SW31.

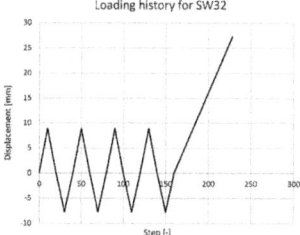

Fig. 3. Loading history for wall SW32.

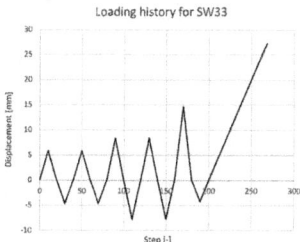

Fig. 4. Loading history for wall SW33.

2.2 Pilakoutas and Elnashai Walls

The specimens realized by Pilakoutas and Elnashai (1995) are 600 mm wide, 1200 mm high and 60 mm thick; the walls named SW4, SW6 and SW8 have been analysed in the present work.

The horizontal force coming from the jack has been applied to a top beam designed to diffuse it uniformly into the wall.

The load scheme presented in Pilakoutas paper shows also the presence of two rollers placed at the sides of the upper beam and connected to a vertical steel frame in order to control the horizontal movement of the top beam.

Detailed information about these devices are not given in the paper, therefore the authors performed several simulations with different constraint conditions at the top of the walls.

The best fitting with experimental results has been obtained with nil vertical constraints applied to the top beam.

The walls are fully restrained and anchored to a lower beam used to simulate a rigid foundation; the upper and lower beams are also used to anchor the vertical reinforcement bars.

Figures 5, 6 and 7 show the nominal dimensions of the walls and the arrangement of the reinforcements. Vertical and horizontal bars with a diameter between 4 and 12 mm have been used; two "pillars" at the sides of the walls are provided with stirrups with a diameter of 4 and 6 mm, each walls has a different disposition of the stirrups.

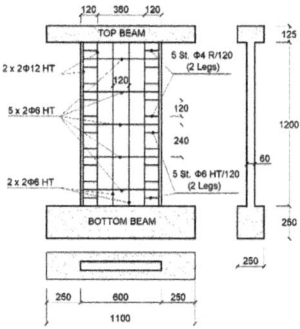

Fig. 5. Geometry and reinforcement details of wall SW4 /.

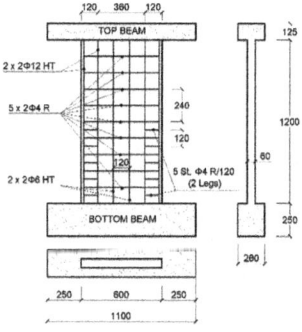

Fig. 6. Geometry and reinforcement details of wall SW6 /.

In Tables 3 and 4 are summarized the main properties of the material used for each wall. The meaning of the symbols has been already described in the previous paragraph.

Figure 8 shows the loading history adopted for SW4, SW6 and SW8. A single load level is composed by two full cycles with the same maximum top displacement. When one load level is completed the top displacement has been incremented of 2 mm in both directions. All specimens have been tested up to failure.

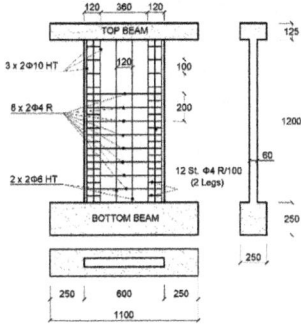

Fig. 7. Geometry and reinforcement details of wall SW8.

Table 3. Concrete properties.

Specimen	f_{cm} (MPa)	ε_{c1} (‰)	ε_{cu1} (‰)
SW4	36.9	2.28	3.50
SW6	38.6	2.30	3.50
SW8	45.8	2.41	3.50

Table 4. Reinforcement properties.

Diameter (mm)	f_y (MPa)	f_u (MPa)	ε_y (%)	ε_u (%)
Φ 4	400	460	0.20	6.00
Φ 6	545	590	0.27	2.00
Φ 10	530	660	0.27	4.20
Φ 12	500	660	0.25	8.50

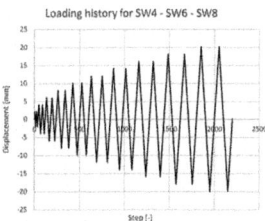

Fig. 8. Geometry and reinforcement details of wall SW4.

3 Finite Elements Models

3.1 Mesh and Material Models

Numerical models of the shear walls have been realized using four-node quadrilateral isoparametric plane stress finite elements, based on linear polynomial interpolation and

2×2 Gauss point's integration scheme. The dimension of each element is about of 0.05×0.05 m, this size has been chosen after a calibration procedure.

The main mechanical characteristics of the material models used in the codes are the same defined in Bertagnoli et al. (2015), the only differences are:

- In Software C, the constitutive law adopted for concrete follows EN1992-1-1 instead of the Thorenfeld law;
- In Software B and C, the shear stiffness reduction after cracking has been considered by means of a shear retention factor $\beta = 0.15$.
- In Software B and C, the concrete tensile behaviour has been modelled with a linear tension softening law only. The latter presents the ultimate strain of softening branch as $10 \cdot \varepsilon_1$. Where ε_1 is the strain corresponding to the peak concrete tensile strength.

3.2   Out-of-Plane Confinement

The out-of-plane confinement action given by the closed stirrups located along the external chords of the shear walls, can provide an important stiffening contribute, particularly in presence of cyclic loading.

In the present work has been adopted the model proposed by Eurocode 2 in order to consider this favourable effect as already proposed by the authors (Bertagnoli et al. (2011)).

Out-of-plane confinement has been taken into account by increasing the peak strength, the corresponding strain and the ultimate compression strain in the constitutive law for concrete in compression. The increase in strength and both in ultimate and peak strains of concrete depends on the effective transverse compressive stress that has been estimated as a function of the amount of the out-of-plane reinforcement ρ_z and its deformation level ε_z.

Out-of-plane reinforcement can be evaluated as the shear link leg area A_{st} smeared on the confined chords width cs ($\rho_z = A_{st}/cs$).

The stress $\sigma_{s,z}$ in the transverse reinforcement can be evaluated by an approximated approach using an elastic-plastic law:

$$\sigma_{s,z} = \begin{cases} E_s' \cdot \varepsilon_z & \text{if } \varepsilon_z < \varepsilon_{s,y} \\ f_y & \text{if } \varepsilon_z \geq \varepsilon_{s,y} \end{cases} \tag{2}$$

Where ε_z = out-of-plane strain from plane stress theory. Thus, the confining stress σ_z in concrete has been calculated as follows:

$$\sigma_z = -\rho_z \cdot \sigma_{s,z} \tag{3}$$

The effect of confining stresses is a strength enhancement:

$$f_{cm,c} = f_{cm}(1 + 5\sigma_z/f_{cm}) \quad \text{if} \quad \sigma_z < 0.05 f_{cm} \tag{4}$$

$$f_{cm,c} = f_{cm}(1.125 + 2.5\sigma_z/f_{cm}) \quad \text{if} \quad \sigma_z > 0.05 f_{cm} \tag{5}$$

Where $f_{cm,c}$ = average cylindrical strength of confined concrete; f_{cm} = average cylindrical strength from a uniaxial test.

The strain at peak strength $\varepsilon_{c0,c}$ and the ultimate strain $\varepsilon_{cu,c}$ in the confined state are related to the respective unconfined values as follows:

$$\varepsilon_{c0,c} = \varepsilon_{c0}\left(f_{cm,c}/f_{cm}\right)^2 \tag{6}$$

$$\varepsilon_{cu,c} = \varepsilon_{cu} + 0.2\sigma_z/f_{cm} \tag{7}$$

4 Results Discussion

The results obtained respectively with software B, software C and the experimental ones are shown side by side in Figs. 9, 10, 11, 12, 13, 14, 15, 16, 17, 18, 19, 20, 21, 22, 23, 24, 25 and 26 in order to allow an easy interpretation.

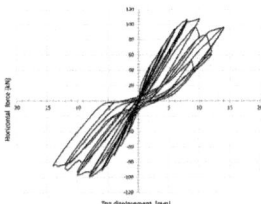

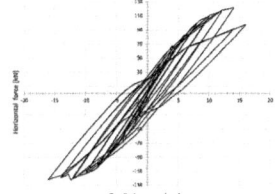

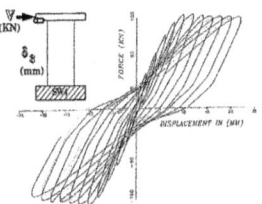

Fig. 9. Software B: force vs displacement SW4.

Fig. 10. Software C: force vs displacement SW4.

Fig. 11. Experimental: force vs displacement SW4.

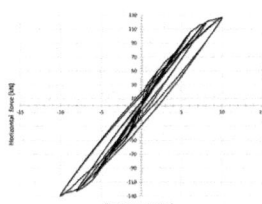

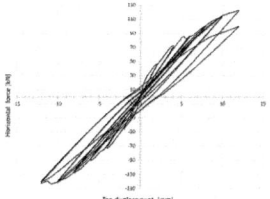

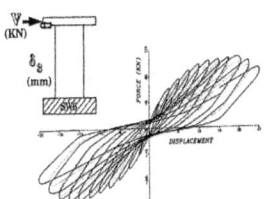

Fig. 12. Software B: force vs displacement SW6.

Fig. 13. Software C: force vs displacement SW6.

Fig. 14. Experimental: force vs displacement SW6.

All numerical models have been loaded applying an imposed horizontal displacement to the top beam and the comparison between experimental and numerical results is done in terms of horizontal force corresponding to the imposed displacement (jack force).

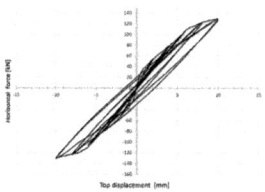

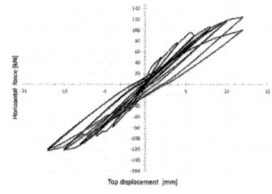

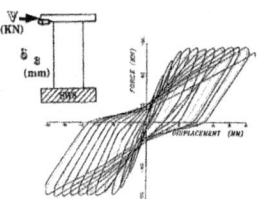

Fig. 15. Software B: force vs displacement SW8.

Fig. 16. Software C: force vs displacement SW8.

Fig. 17. Experimental: force vs displacement SW8.

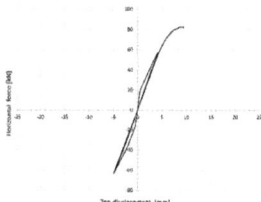

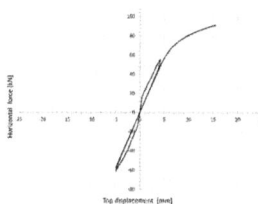

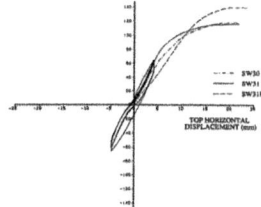

Fig. 18. Software B: force vs displacement SW31.

Fig. 19. Software C: force vs displacement SW31.

Fig. 20. Experimental: force vs displacement SW31.

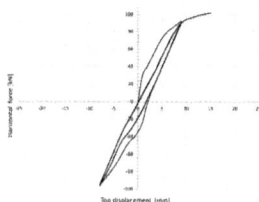

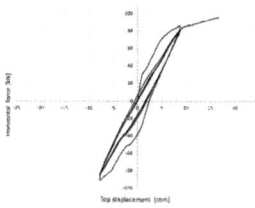

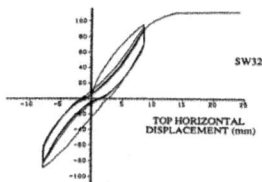

Fig. 21. Software B: force vs displacement SW32.

Fig. 22. Software C: force vs displacement SW32.

Fig. 23. Experimental: force vs displacement SW32.

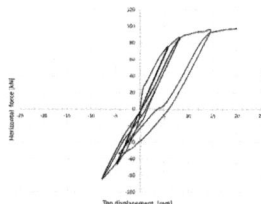

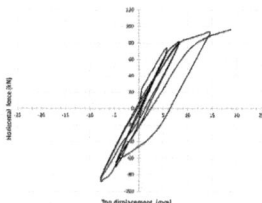

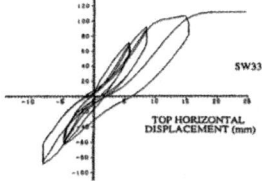

Fig. 24. Software B: force vs displacement SW33.

Fig. 25. Software C: force vs displacement SW33.

Fig. 26. Experimental: force vs displacement SW33.

The results of walls SW31, SW32, SW33 turn out to be more accurate and faithful to experimental ones than walls SW4, SW8, SW10.

The main difference between the two families of walls is the loading processes as described in Sect. 2.

4.1 Software B Results

– SW4: the model is able to reproduce adequately the experimental behaviour only up to an imposed displacement of the top beam of ±14 mm (maximum displacement applied during the experimental test = ±20 mm). For higher displacements software C is not able to provide reliable results.

   However, especially for high displacement levels, the numerical prediction of the applied force differs from the experimental one only of ±5%.

– SW6: the model is able to reproduce adequately the experimental behaviour only up to an imposed displacement of the top beam of ±14 mm (maximum displacement applied during the test = ±20 mm). The applied load is overestimated of about 20% for higher imposed displacements.

– SW8: the model is able to reproduce adequately the experimental behaviour only up to an imposed displacement of the top beam of ±10 mm (maximum displacement applied during the test = ±20 mm). The load is overestimated again of about +25%. The model is stiffer in the initial stages but becomes too deformable for higher displacements.

– SW31: the model is able to reproduce adequately the experimental behaviour for the whole test. As regards the cyclic stage, during the loading phases it can be noted a good correspondence with the actual values, while in the unloading phases the response of the software B can be considered elastic. Nevertheless an underestimation of the ultimate load of about 20% has to be declared.

– SW32: the model is able to reproduce adequately the experimental behaviour for the whole test. In last stage up to the failure, software B is able to reproduce the actual behaviour, but also in this test there is an underestimation of the ultimate load of about 10%;

– SW33: Similar considerations to the previous case can also be done for this wall. It can be noticed a small underestimation of the experimental load of about 10% during the cycles, whereas in the final loading up to failure there is an underestimation of the ultimate load of about 15%. The global behaviour is although well reproduced.

4.2 Software C Results

– SW4: the numerical model is able to reproduce adequately the experimental behaviour only up to an imposed displacement of the top beam of ±16 mm (maximum displacement applied during the experimental test = ±20 mm). For higher displacements software C is not able to provide reliable results.

   The amplitude of the cycles, as well as the slope of the loading and unloading curves, are approximated in a satisfactory way. However, especially for high displacements, the corresponding loads are greater than the experimental ones (deviation between

+5% and +20%). The numerical model seems to be stiffer than the experimental one especially in the first part of the analysis.

- SW6: as for wall SW4, the numerical model is able to reproduce adequately the experimental behaviour only up to an imposed displacement of the top beam of ±16 mm (maximum displacement applied during the test = ±20 mm). For higher displacements with software C it can be noted the presence of high shear deformation localized at the base which is not in accordance with the experimental results.
- SW8: the numerical model is able to reproduce adequately the experimental behaviour only up to an imposed displacement of the top beam of ±12 mm (maximum displacement applied during the test = ±20 mm). A progressive softening can be noted for higher imposed displacements, resulting in flattening of loading and unloading cycles; this phenomenon is not in accordance with the experimental behaviour. The prediction of the applied horizontal force is correct up to ±12 mm of displacement, whereas the numerical model becomes too deformable for higher imposed displacements.
- SW31: the numerical model is able to reproduce adequately the experimental behaviour in the whole test. A good correspondence with the experimental values of force and displacement can be observed in the cyclic stage in the loading curves, while some disagreement between numerical and experimental can be seen in the unloading curves as the response of the software C can be considered elastic during unloading. In the last part of the test, software C is able to reproduce the experimental behaviour in a satisfactory way up to failure, however it underestimates the ultimate load of about 20%.
- SW32: the numerical model is able to reproduce adequately the experimental behaviour in the whole test. In the cyclic stage, the model response is very good both during the loading phases and the unloading ones, unlike SW31. Finally, in last stage up to the failure, software C is able to reproduce the actual behaviour, also in this test there is an underestimation of the ultimate load, but the difference is less than 10%.
- SW33: the same considerations seen in the previous case can be drawn also for this wall. The amplitude of the cycles, as well as the slope of the loading and unloading curves, are approximated in a satisfactory way for the whole duration of test. Again, a small underestimation (about 10%) of the ultimate load can be noted. During the last loading curve up to the failure, software C is able to reproduce the actual behaviour up to 20 mm of imposed displacement, whereas the experimental test reached 25 mm. A slight underestimation of the ultimate load (10%) is found again.

Software B and Software C show a substantial equivalence in the simulation of the structural behaviour.

A relevant parameter to evaluate performance of the numerical analyses seems to be the ratio between the top horizontal displacement and the height of the wall, which can be called "shear deformation".

In fact, it represents the global average level of angular distortion γ that occur into the wall during the test.

Solutions proposed by Software B and C are satisfactory until the parameter γ reaches a value around $8 \cdot 10^{-3}$, which, being all the walls height almost equal, corresponds to a top horizontal displacement of about 10 mm.

Such value of γ implies a principal deformation ε_2 magnitude which is close to the deformation at failure of concrete ($\approx 3.5 \cdot 10^{-3}$).

When γ level of angular distortion is passed, both software are unable to reproduce accurately the actual structural behaviour.

For high displacement levels, the phenomena connected to confinement effect due to both reinforcements and two-dimensional behaviour become relevant and material non-linearities are particularly uncertain.

A second parameter which plays an important role in governing the numerical results is the shear retention factor β.

The factor β is the ratio between the shear modulus G in cracked and uncracked state. High values of β ($0.2 < \beta < 0.5$) imply big hysteresis cycles, whereas low values ($\beta < 0.15$) give rise to "slimmer" curves and low levels of dissipated energy. Parameter β can be variable according to the strain level, but it has been kept constant in this work in order to compare software B and C results more accurately.

5   Conclusions

The present work has analysed the uncertainties of the outcome of different non-linear analysis of reinforced concrete shear walls subjected to horizontal cyclic loading. Two different finite element software have been used to simulate the behaviour of two distinct triplets of walls.

Quite good accuracy and small scattering of the results have been achieved for cycles having a shear deformation up to $8 \cdot 10^{-3}$, whereas for higher loading levels it has been found a loss of accuracy of the results. Shear retention factor deeply governs the amount of dissipated energy controlling the hysteretic behaviour.

Further development of the research will consider the influence of more FE model parameters on the prediction of the structural response.

References

Pilakoutas K, Elnashai A (1995) Cyclic behaviour of reinforced concrete cantilever walls, part I: experimental results. ACI Struct J 92(3):271–281. Technical paper, Title No. 92-S25

Lefas ID, Kotsovos MD (1990) Strength and deformations characteristics of reinforced concrete walls under load reversals. ACI Struct J 87(6):716–726. Technical paper, Title No. 87-S74

Bertagnoli G, La Mazza D, Mancini G (2015) Effect of concrete tensile strength in non linear analyses of 2D structures - a comparison between three commercial finite element softwares

Bertagnoli G, Mancini G, Recupero A, Spinella N (2011) Rotating compression fields for reinforced concrete beams under prevalent shear actions. Struct Concr 12(3):178–186

CEB-FIP Model Code (1990) First Draft. Committee Euro-International du Beton, Bulletin d'information No. 195, 196

Eurocode 2, EN 1992-1-1 (2004) Design of concrete structures – part 1: general rules and rules for buildings

Experimental Investigation on the Seismic Behaviour of a New Pier-to-Deck Connection for Steel Concrete Composite Bridges

S. Alessandri[1], R. Giannini[2], F. Paolacci[1(✉)], and Nam H. Phan[1]

[1] Department of Engineering, University Roma Tre, Rome, Italy
fabrizio.paolacci@uniroma3.it
[2] Department of Architecture, University Roma Tre, Rome, Italy

Abstract. Nowadays, short-medium span steel-concrete composite I-girder bridges (SCC) are very popular, owing to their short construction time and reduced costs. Their limited weight makes their use adequate also for seismic areas, even though their seismic behaviour has not been yet adequately investigated. With this aim, within the European project (SEQBRI), the seismic behaviour of new pier-to-deck connections entailing the use of concrete cross-beams (CCB) has been recently studied. This paper shows the results of a comprehensive experimental investigation on this kind of connections conducted with the aim of characterizing the hysteretic behaviour and calibrating a component-based model for seismic analysis. Three different type of connection have been tested: one designed according to the standard DIN-FB104, generally utilized for gravity loads only, and other two, proposed for bridges located in low and medium intensity seismic prone areas. Based on different resistant mechanism, these latter have demonstrated a good behaviour in terms of strength and ductility.

Keywords: Steel-concrete bridges · Concrete cross beam · Experimental tests

1 Introduction

Steel-concrete composite solution is attractive for small and medium span bridges and competitive with respect to pre-stressed bridges. This is mainly due to a low dead weight and fast construction on site.

For continuous bridges, the composite solution presents some problems due to the inversion of the bending moment; as well known, hogging bending moment at the supports causes excessive tensile stress in the concrete slab and excessive compression at the bottom flange of the steel girder resulting in a concrete cracking and instability of the steel girder.

An alternative solution for composite bridges entails the adoption of a concrete crossbeam (CCB) connected to the steel girders by shear studs.

This paper deals with a comprehensive experimental investigation on different typologies of pier-to-deck connection. In particular, three different type of connection are tested: the first one designed according to the standard DIN-FB104 (2009), generally utilized for gravity loads only, and the other two proposed for bridges located in low and medium intensity seismic prone areas.

© Springer International Publishing AG, part of Springer Nature 2018
M. di Prisco and M. Menegotto (Eds.): ICD 2016, LNCE 10, pp. 298–312, 2018.
https://doi.org/10.1007/978-3-319-78936-1_22

Specimens in scale 1:2 of a selected subassembly have been subjected to longitudinal imposed cyclic displacement histories with the aim at characterizing the hysteretic behaviour and to further calibrate a novel component-based model for the seismic analysis of steel composite bridges endowed with CCBs (Paolacci et al. 2014).

2 Steel Concrete Composite Bridge with High Strength Steel and Concrete Cross Beam

2.1 Conceptual Design and Static Behaviour

The main girders are welded or rolled steel cross-sections completely prefabricated in the shop. The typical reinforced concrete deck consists of partially prefabricated elements and additional in-situ concrete. Only crossbeams at supports are used and these beams are normal reinforced concrete members. This solution allows a simple erection to minimize or eliminate welding or bolting on side. The reinforced concrete crossbeams over intermediate supports of multiple span bridges may be designed as splices of longitudinal girders. Continuity is achieved by the use of vertical end plates and additional reinforcing bars in the deck slab.

During concreting, loads due to the dead weight of steel girders, formwork and wet concrete are carried by simply supported beams. After the concrete has hardened, moment resistance is provided at splices and subsequent loads are supported by continuous girders. Thus hogging bending is produced at supports only by super-imposed dead loads and variable actions.

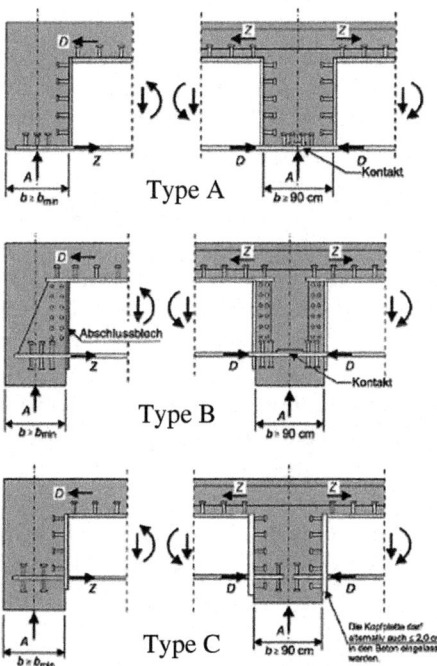

Fig. 1. DIN Fachberichte 104 Concrete Cross Beam variants

Different types of CCB configurations have been applied in several occasions (Hanswille, 2007). Many of them are variants of the three typical configurations, which are presented at the DIN Fachberichte 104 (2009) (Fig. 1).

For each type of the DIN Fachberichte 104 (2009), the steel girder ends to a head plate along the whole height of the cross section and the bottom flange of the girder continues inside the CCB.

The predominant (usually tensile) forces at the top flange are transferred through the shear studs to the slab, whereas the corresponding compression through the contact between the two opposite bottom steel flanges (types A and B) or through concrete compression (type C). On the other hand, if tensile forces is developed at the bottom flange, these should be transferred through the welded connection between the flanges' extensions (type A and B) or vertical shear studs (type C for the intermediate CCBs and all types for the edge CCBs). The shear forces are typically transferred through shear studs placed on the head plates, aligned to the bridge axis (types A and C). Alternatively, in type B, the web is also inserted into the CCB and transversally aligned shear studs transfer the shear forces. These configurations are proposed by the DIN FB in order to manage mainly vertical actions (i.e. dead and live loads), which produces negative moments and consequently tensile forces at the concrete slab and the top steel flange as well as compressive forces at the bottom flange.

For bridges subjected to seismic actions, significant tensile forces might be exhibited at the bottom flange of the steel girder, especially when monolithic connection between CCB and pier is formed. Considering that the CCB configuration is the most crucial detail, especially for composite bridges in seismic areas, and taking into account the aforementioned discussion, the widely used DIN FB type C is chosen to be analytically and experimentally investigated.

Though, trying to exploit the advantages and avoid the disadvantages of the previous three configurations, new variants are designed and investigated, especially for the most critical intermediate CCB. The new configurations of the intermediate CCB are closer to the DIN-FB-104 variant B, however the bottom flange of the steel girder is not inserted into the CCB and the height of the head plate is limited (Fig. 2). The idea is to transfer the forces from the composite girder to the CCB through contact and dedicated groups of shear studs. The tensile force at the top flange is transferred gradually to the longitudinal reinforcement (or reversely a potential compression force to the concrete slab) through a group of vertical studs, which are placed on the top flange of the girder before the CCB. The top flange does not enter into the CCB. The shear force is transferred to the CCB through a group of horizontal studs, transversally placed at the sides of the girder's web, which is inserted into the CCB. This group of studs is subjected only to pure shear, action that is compatible to their actual function. In the VAR-2 the tensile forces in the bottom flange are transferred through prestressed bars inserted throughout holes foreseen at the web. Finally, regarding the magnitude of the tensile stress, which might be exhibited at the bottom flange of the steel girder, the following three cases are distinguished: (a) Bottom steel flange always in compression, (b) Bottom flange in compression or in light tension, (c) Bottom flange in compression or in significant tension.

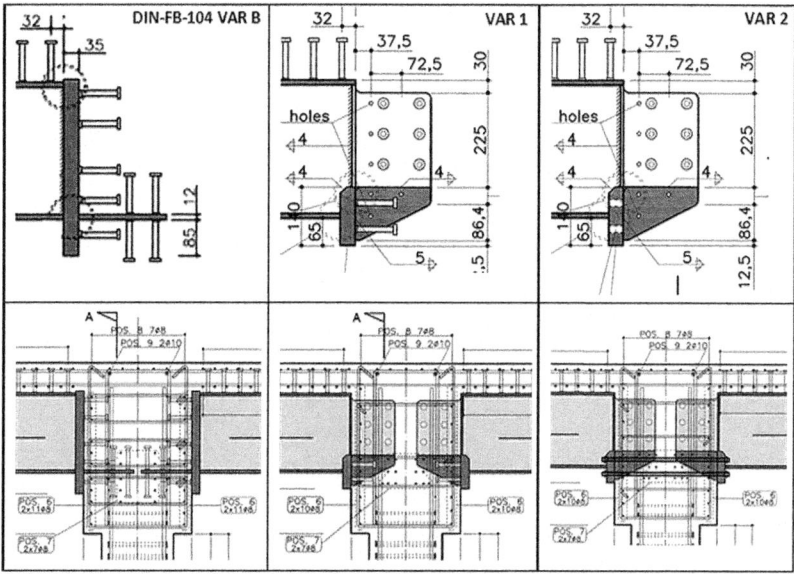

Fig. 2. CCB typologies: (a) DIN-FB-104 VAR B -, (b) Var-1, (c) Var-2

In the next section, the results a wide test campaign of these three typologies of CCB connection will be presented and commented.

3   Design of the Experimental Campaign

3.1   Subassemblies Identification

The analysed case study is a straight 2-span deck consisting of 4 main girders HE600B of S460 steel grade with 2.65 m in-between distance (Fig. 3).

The bridge is 40.00 m long and consists of 2 spans of 20.00 m, while the total width of the road cross-section is 10.60 m, with carriageway 6.50 m wide and 2 sidewalks 2.05 m wide. The thickness of the concrete slab is 25 cm.

At the abutments, the steel girders are fixed to an end reinforced concrete cross-beam 0.60 m wide. By this diaphragm, the deck is simply supported on normal damping rubber bearings. At the intermediate pier, steel girders are fixed to an intermediate reinforced concrete cross-beam 0.90 m wide. The pier (clear) height is 7.00 m. A wall type pier 0.60 m thick and 7.00 m wide is used. It is assumed that the foundation soil is categorized as type B according to EN1998 (2004). In this example the soil structure interaction effect is neglected.

From the previously defined case studies, it has been possible to identify, by using preliminary analysis with SAP2000, the I-girder subassemblies associated to the Concrete Cross Beam to be tested. Particular attention was dedicated to the interconnection arrangement of the steel girders with the CCBs on piers. For these reasons, the test subassemblies for the longitudinal direction were identified on an isostatic portion

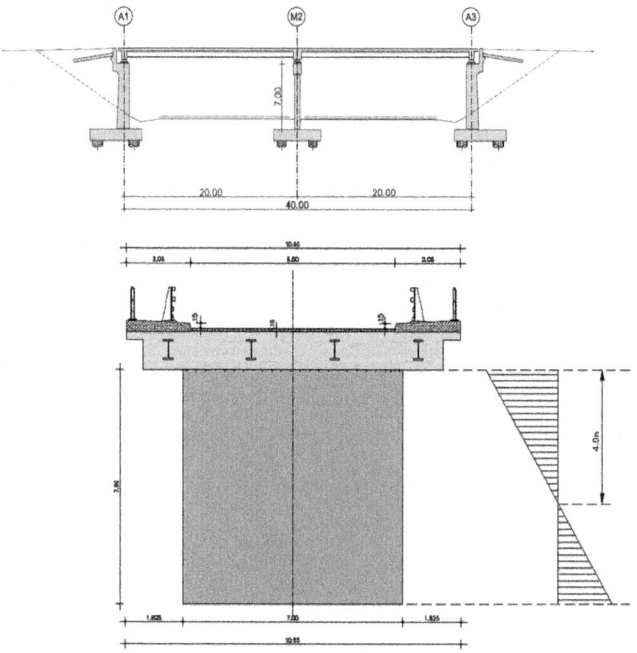

Fig. 3. Longitudinal and cross section for selected case study

of the bridge that includes the cross beam–deck joint. The specimen represents a stripe of the bridge including one steel girder. The considered static scheme is that with integral connection between CCB and pier, which represents the most critical seismic condition. The static scheme for vertical and horizontal loads is presented in Fig. 4.

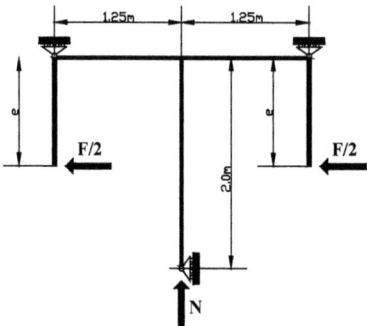

Fig. 4. Static scheme of the testing configuration

The specimen was scaled of a factor S = 2 according to the procedure proposed by Kumar et al. (1997). In particular, it was assumed that model was made of the same material as the prototype, and that stress identities were preserved.

3.2   Description of the Specimens and Testing Program

Two specimens of type DIN-FB-104 VAR B, two specimens of type VAR-1 and four specimens of type VAR-2 were built for the experimental campaign. Type and denomination of the specimens are reported in Table 1. Each connections typology was subjected to a monotonic and cyclic tests in horizontal direction.

Table 1. Type and denomination of tests

Typology of specimen	Name of test	Type of test	Loading direction
DIN-FB-104 VAR B	FBA2 M	Monotonic	Horizontal
DIN-FB-104 VAR B	FBA2C	Cyclic	Horizontal
VAR-1	D1B2 M	Monotonic	Horizontal
VAR-1	D1B1C	Cyclic	Horizontal
VAR-2	D2C1C	Cyclic	Horizontal
VAR-2	D2C2C	Cyclic	Horizontal
VAR-2	D2C4C	Cyclic	Horizontal
VAR-2	D2C3 M	Monotonic	Horizontal
VAR-2	D2C1MV	Monotonic	Vertical

The specimens are made of a single IPE330/S460 steel girder with its tributary concrete slab 1.325 m wide, 12.5 cm thick and 3.5 m long and a portion of pier of 2.0 m long and 30 cm thick. All the girders are 50 cm longer than the necessary, in order to simplify the realization of the appropriate boundary conditions in the laboratory. Consequently the specimens have an overall length of 3.50 m. Finally the CCB has a Section 45 × 70 cm.

The main results relevant to the mechanical characterization of materials are presented in Table 2.

Table 2. Mechanical properties of materials

Concrete C35/45		Steel		f_y [MPa]	E_s [MPa]	b[-]
R_{cm} [MPa]	60					
f_{cm} [MPa]	52	B450C-φ8		527	196882	0.008
f_{cu} [MPa]	42	B450C-φ10		537	198264	0.008
ε_{c0} [-]	0.0028	S460 M-flange		522	191650	0.006
ε_{cu} [-]	0.0067	S460 M-web		538	203735	0.004
E_{cm} [MPa]	36050	10.9-φ16		776	203750	0.031

The loading protocol for cyclic tests was chosen on the basis of the yielding displacement e_y^+ of a monotonic response, calculated as indicated in Bursi et al. (2002) and schematically depicted in Fig. 5. In order to conveniently define the parameter e_y relevant to the yield displacement, a yield limit state characterized by the displacement e_y^+ as well as by the corresponding reaction force P_y^+ must be defined. Such quantities have been traced on the first part of each non-linear response envelope obtained from monotonic tests. The trilinear approximation of each curve is determined on the basis of best-fitting and of the equivalence of the dissipated energy between the actual non-linear response and the idealized trilinear approximation up to (e_{max}^+, P_{max}^+). Hence, the ECCS (1996) loading protocol procedure conceived for steel structure components was applied. An example of cyclic imposed displacement is illustrated in Fig. 6 for the specimen D1B1C.

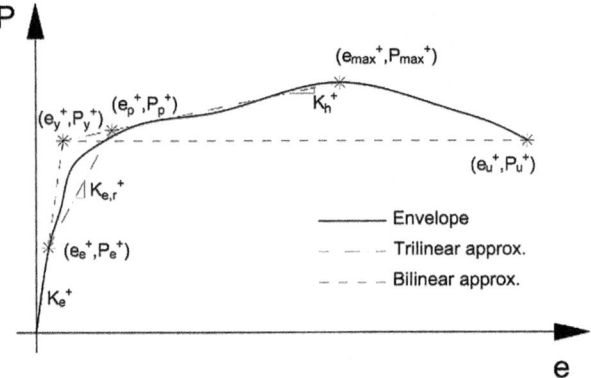

Fig. 5. Bi and trilinear fits of a force-displacement envelope

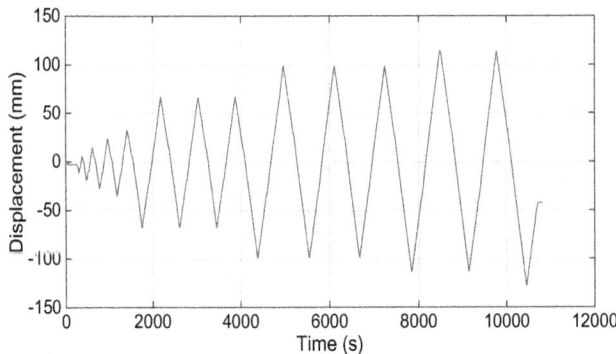

Fig. 6. Imposed displacement history according to the ECCS (1986) procedure ($e_y \approx 30$ mm) for the specimen D1B1C

3.3 Design of the Test Rig and Instrumentation

The test setup depicted in Fig. 7 was designed based on the results of a preliminary finite element analysis of test specimens. In the following subsections, the test rig for each case study is described along with the instrumentation used to acquire the response signals.

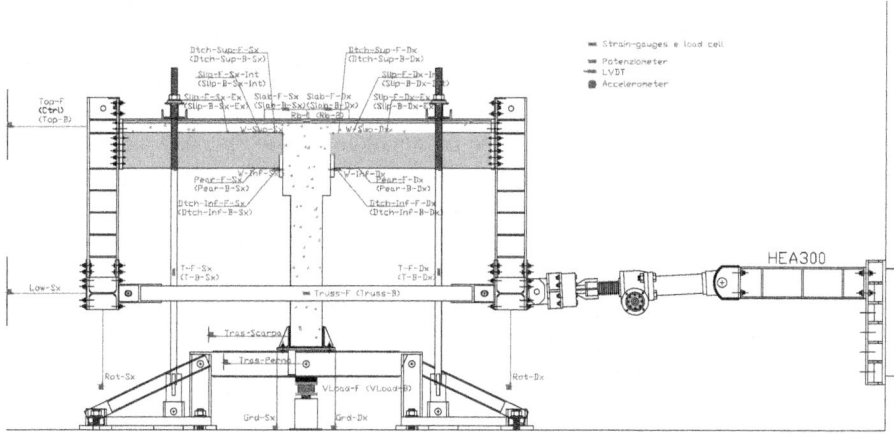

Fig. 7. Lateral view of the test setup

In this configuration the horizontal actuator is directly connected to the level arm on the right. The setup is made of a rectangular steel frame lying in the horizontal plane, whose elements are mutually hinged (horizontal trusses), as shown in Fig. 7.

This horizontal frame is bolted to two vertical steel beams (lever arms) linked to the specimen, by a connection steel plate, allowing transmitting to both the sides of axial force, shear force and bending moment. The bonding effect at the end of the reinforcing bars is reproduced using a special clamping system.

The vertical force is transmitted through a couple of hydraulic jacks (Fig. 7), aligned under the wall pier, placed at a mutual distance of 500 mm, in order to obtain a uniform diffusion of the compressive force in the pier. Before positioning the specimen, in order to allow only vertical displacements at the bottom of the pier, two couples of horizontal trusses are assembled under the steel basement, on the top of the two load cells placed to record the vertical load.

The instrumentation used in the longitudinal direction tests is composed by: (a) load cells for the measurement of the vertical and horizontal loads, (b) LVDTs, potentiometers and strain gauges for the measurements of beam-to-CCB detachment, concrete and steel deformations, respectively, (c) linear potentiometers for the measurement of the cracks opening at the top fiber of the slab, (d) internal strain gauges for the acquisition of strain in the rebars, (e) strain gauges on the vertical steel bars for the indirect measurement of the vertical loads, (f) wire sensors for the measurement of horizontal displacements at the top of the specimen and at the bottom of the two lever arms.

4   Description of the Results and Discussion

In this section the main outcomes of the experimental campaign will be presented and commented.

The testing program shown in Table 1 involves 8 specimens as described in the previous section. According to the procedure described in Sect. 3.2, one monotonic test is performed for each typology of CCB in order to calibrate the subsequently cyclic tests. In particular, for the CCB typology DIN-FB-104 VAR B and VAR-1 two specimens have been tested, the first one in monotonic conditions and the second one under cyclic loading. VAR-2 specimens where firstly tested monotonically, then three full cyclic tests were performed on the remaining specimens. The VAR-2 specimen tested under monotonic condition were also tested under vertical loads to evaluate the residual vertical load-carrying capacity. In what follows the results of both monotonic and cyclic tests will be presented. The monotonic tests consist in a first phase in which the vertical load is imposed. According to the preliminary analyses presented in Sect. 3, the maximum scaled vertical load was equal to 220 kN. Subsequently, a cyclic history was applied at the top of the specimen. In this way the control system imposed the necessary force through the electro-mechanical actuator in order to reach time by time the desired displacement.

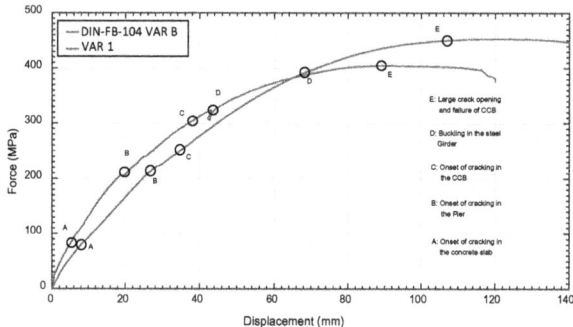

Fig. 8. Damage states in the DIN-FB-104 VAR B and VAR-1 under monotonic loading condition

Figure 8 shows the monotonic response and damage limit states for the DIN-FB-104 VAR B and VAR-1 specimens, respectively These tests provided indications on the damage condition with the growing of both lateral displacement and force and allow to calibrate the cyclic test displacement history protocol. From these tests the mean value of the yielding displacement was equal to $d_y = 30$ mm.

Concerning the cyclic tests, even though the specimen DIN-FB-104 VAR B and VAR-2 displayed a similar global behaviour in terms of maximum force and displacement (Figs. 9 and 11) a different local behaviour was identified. In particular, the CCB cracking was anticipated for the specimen DIN-FB-104 VAR B with respect to VAR-2 (Fig. 12). In addition, in the case of DIN-FB-104 VAR B specimen the failure of the CCB was anticipated with respect to VAR-2.

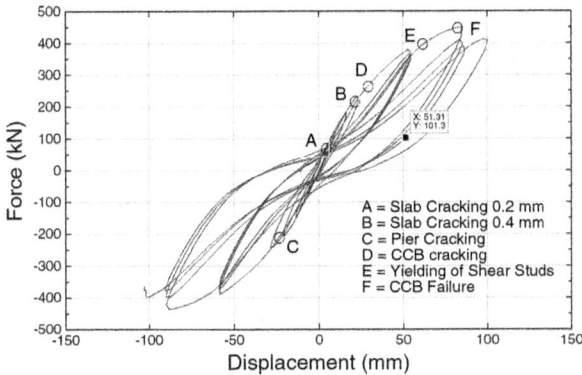

Fig. 9. Damage states in the DIN-FB-104 VAR B specimen under cyclic loading condition

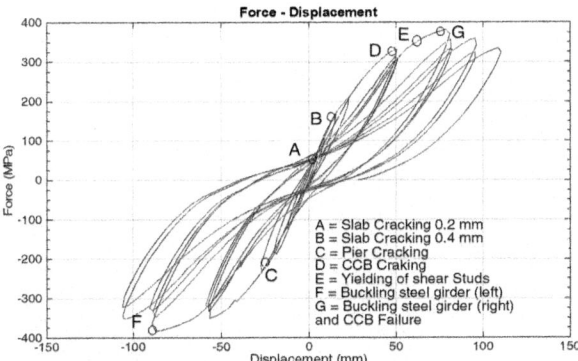

Fig. 10. Damage states in the VAR-1 specimen under cyclic loading condition

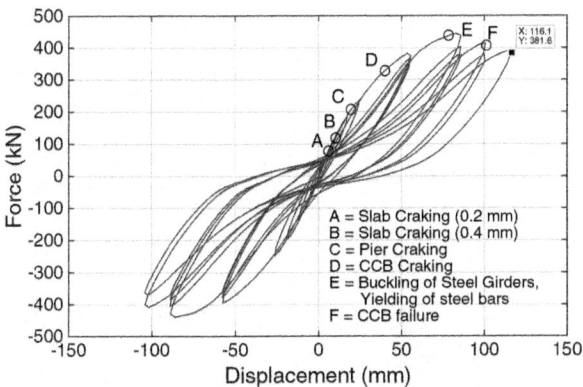

Fig. 11. Damage states in the VAR-2 specimen under cyclic loading condition (D2C1C)

The presence of pre-stressing bars in the VAR-2 specimen avoided any pull-out phenomenon in the CCB in correspondence of the bottom flange of the steel girders. This latter was clearly present in the DIN-FB-104 VAR B and VAR-1 specimens with an irreversible damage condition, with detachment of the steel girder from the CCB (Fig. 13). The possibility to recover the pre-stressing level or substitute the pre-stressing bars after a seismic event makes the VAR-2 joint superior in case of SCC bridges located in high seismic zones.

VAR-1 solution has shown a less effective behaviour both with respect to specimen VAR-2 and DIN-FB-104 VAR B, as shown in Fig. 10. In fact, an anticipated failure condition of the CCB was noticed. In addition, the longitudinal shear studs yielded earlier than the shear studs in DIN-FB-104 VAR B specimen and it is not negligible the obtained lower strength with VAR-1 specimen both in monotonic and cyclic conditions.

A phenomenon common to all specimens is the buckling in the bottom flange of the steel girders (Fig. 14). However, no clear indications were identified about the conditions in which it develops. In fact, it depends on the level of imperfection of the steel plates.

Fig. 12. Crack pattern in the CCB

Fig. 13. Detachment of the steel girder end plate

In any case, a more pronounced level of buckling for both VAR-1 and DIN-FB-104 VAR B specimens was clearly noticed. This is certainly another positive aspect in favour of VAR-2 joints. The development of cracks in the slab was instead similar for all three CCB typologies. Values of force and displacements for each identified damage state are reported in Table 3.

Table 3. Damage states for the different joint typologies

Damage state	DIN-FB-104 VAR B		VAR-2		VAR-2 – C1		VAR-2 C4	
	δ(mm)	F(kN)	δ(mm)	F(kN)	δ(mm)	F(kN)	δ(mm)	F(kN)
Crack Slab (0.2 mm)	7	50	5	45	7.5	23	5	15
Crack Slab (0.4 mm)	22	210	15	170	22	120	10	102
Onset of Cracking in the Pier	25	220	23	212	20	200	15	150
Onset of Cracking in the CCB	30	260	49	331	40	320	50	350
Yielding of Shear Studs	64	400	60	350	-	-	-	-
Buckling of Steel Girders	84	440	75	382	75	450	70	390
Failure of CCB	84	440	75	382	100	450	100	380

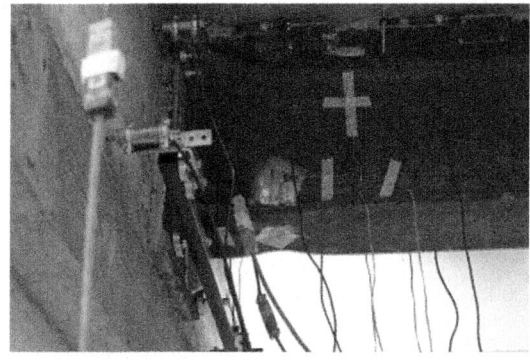

Fig. 14. Buckling phenomenon in the steel girder (VAR-1)

5   Identification of Damage Measures and Limit States

In this specific case the parameters selected to perform the damage analysis are: (a) the relative joint rotation θ between the steel girder end plate and the CCB, (b) the maximum pier drift D. The first one, whose definition is provided in Fig. 15, concerns the girder-CCB connection, whereas the second one is relative to the slab condition. The corresponding Damage Measures are: (a) the yielding and ultimate rotations, (b) the crack opening and the yielding conditions of rebars of the slab. This choice derives from the idea of having both response parameters and Damage Measures easily measurable. In Fig. 16 the experimental moment-rotation cyclic response of the pier-to-deck joint of D1B1C specimen is showed. In the same figure, the limit states analytically defined in (Paolacci et al 2014) are also shown. In particular, the yielding and ultimate joint rotation θ are defined as the ratio between the shear stud deformation Δx (Fig. 15), corresponding to the same limit conditions, and the length of the steel girder end plate Δz.

Tables 4 and 5 report the mean value of the selected damage measure (DM) and parameter (EDP) for each identified damage state evaluated from the cyclic tests both for the CCB and Slab.

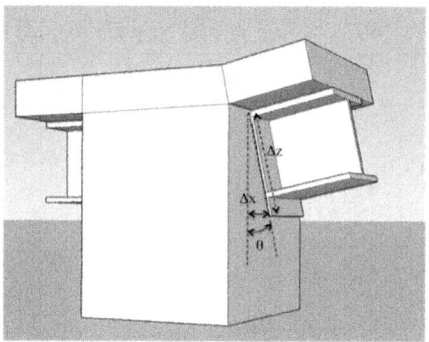

Fig. 15. Definition of the Pier-to-Deck Joint rotation (θ)

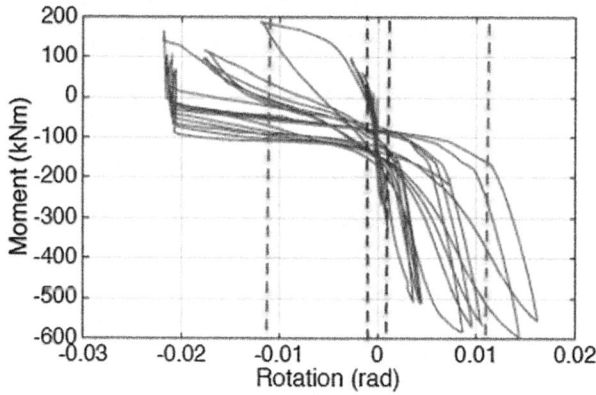

Fig. 16. Moment-Rotation of the pier-to-deck joint and comparison with damage states (green line: yielding, red line: ultimate condition) (Color figure online)

These limit conditions have been evaluated based on the results of experimental tests executed by several authors (Viest (1956), Gattesco and Giuriano (1996), Shim et al. (2004), Lee et al (2005)); the mean value and the dispersion of the yielding and ultimate slip deformation of the shear studs have been evaluated, equal respectively to 5.31·10-4 and 1.1·10-2 rad; a very limited correlation between the two variables has been found (Fig. 16).

Table 4. EDP and DM for the CCB

	DM	Joint rotation (θ) (mrad)
DIN-FB-104	Yielding	0.10
VAR B	Ultimate	23
VAR-2	Yielding	0.25
	Ultimate	20
VAR-2	Yielding	0.3
	Ultimate	19

Table 5. EDP and DM for the slab

DM	Drift (%)		
	DIN-FB-104 VAR B	VAR-2	VAR-2
Crack opening 0.2 mm	0.35	0.25	0.38
Crack opening 0.4 mm	1.10	0.75	1.10
Yielding of rebars	4.50	4.50	5.0

It can be noticed form Table 4 that the experimental results are substantially confirmed being the mean values of the yielding and ultimate rotation equal to $2.1 \cdot 10^{-4}$ and $2.0 \cdot 10^{-2}$, respectively

6 Conclusions

This paper shows the results of a comprehensive experimental investigation conducted on a new pier-to-deck connection, entailing the use of concrete cross-beams, with the aim of characterizing its hysteretic behaviour towards longitudinal seismic actions. The tests have been executed on a connection designed according to the DIN code and on two variations. The cyclic tests have highlighted a good seismic behaviour of all the connections. The DIN-FB-104 VAR B and VAR-2 displayed a similar global beha-viour in terms of maximum force and displacement. Moreover CCB cracking and failure was anticipated for the specimen DIN-FB-104 VAR B with respect to VAR-2. The VAR-2 also displayed a less pronounced level of buckling respect to the DIN-FB-104 VAR B while the pull-out phenomenon in the CCB in correspondence of the bottom flange of the steel girders was avoided by the presence of pre-stressing bars. The VAR-1 solution has shown a less effective behaviour in terms of both maximum force and displacement.

Acknowledgements. This work was carried out with a financial grant from the Research Fund for Coal and Steel of the European Community, within the SEQBRI project: "Performance-Based Earthquake Engineering Analysis of Short- Medium Span Steel-Concrete Composite Bridges", Grant RFSR-CT-2012-00032.

References

Bursi OS, Ferrario F, Fontanari V (2002) Non-linear Analysis of the Low-cycle Fracture Behaviour of Isolated Tee Stub Connections. Comput Struct 80:2333–2360

DIN-Fachbericht 104 (2009) Verbundbrücken. Berlin: Beuth Verlag

ECCS (1986) Recommended Testing Procedure for Assessing the Behaviour of Structural Elements under Cyclic Loads. Technical Committee 1 – Structural Safety and Loading

EN 1998 (2004) Eurocode 8: Design of structures for earthquake resistance

Gattesco N, Giuriani E (1996) Experimental study on stud shear connectors subjected to cyclic loading. J Constr Steel Res 38(1):1–21

Hanswille G, Porsch M, Ustundag C (2007) Resistance of headed studs subjected to fatigue loading, Part I: Experimental study. J Constr Steel Res 63(4):475–484

Kumar S, Itoh Y, Saizuka K, Usami T (1997) Pseudodynamic testing of scaled models. J Struct Eng 123(4):524–526

Lee PG, Shim CS, Chang SP (2005) Static and fatigue behavior of large stud shear connectors for steel-concrete composite bridges. J Constr Steel Res 61(9):1270–1285

Paolacci F, Giannini R, Malena M, Alessandri S, Paglini A (2014). Toward a performance-based earthquake engineering analysis of short-medium span steel-concrete composite bridges. 2ECEE. 2nd European conference on earthquake engineering and seismology. Istanbul, 25–29 August 2014

SAP2000. Structural Analysis Program ©. 1976–2014 Computer & Structures, Inc

Shim CS, Lee PG, Yoon TY (2004) Static behavior of large stud shear connectors. Eng Struct 26 (18):53–60

Viest IM (1956) Investigation of Stud Shear Connectors for Composite Concrete and Steel T-Beams. J Am Concr Inst 27(8):875–891

Seismic Prevention and Mitigation in Historical Centres

S. Biondi[(✉)], N. Cataldo, C. Sulpizio, and I. Vanzi

Dipartimento di Ingegneria e Geologia,
Università G. d'Annunzio di Chieti-Pescara, Pescara, Italy
pricos@unich.it

Abstract. After recent seismic events, the topic of seismic prevention and mitigation in historical centres is become very important, in particular for a seismic prone areas, like Italy, Greece, Portugal, in which a lot of historical towns, for high quantity of old buildings and for their urban structure, suffered a lot of damages. In the past, for this reason, the Authors described and discussed a strategy for seismic prevention and mitigation of historical centres, analyzing them in terms of structural safety. This approach was based on two relevant steps: the first is to study the Urban Risk, the second is to program the Post-Earthquake Activity. The first activity is the analysis of a complex system aiming to individuate the nodal fragility, the second tends to evaluate the buildings safety and the occupancy conditions for these buildings.

In this paper a particular aspect of this topic will be discussed, i.e. how different typologies of buildings, that coexist in historical towns, could influence the reconstruction strategy. In fact, when one thinks to an old building isn't only a masonry historical building, an old building could be a more recent r.c. building designed without any seismic provisions or, as unfortunately usual, realized with poor quality material. Considering that in recent seismic events often this class of buildings caused a lot of damages and deaths, an efficient procedure to approach the impact of r.c. structures on seismic prevention and mitigation in historical centres has to be considered a fundamental goal to reach.

Keywords: AeDES classification method · Damage survey · In-situ survey
Reconstruction plans · Urban Minimum System · Urban Risk Assessment

1 Introduction

Urban seismic risk prevention deals with the effects of territorial transformation, in order to evaluate the impact these ones may have in modifying the functions of different parts of a settlement. Unlike ordinary buildings, urban vulnerability depends not only on the structure characteristics but also on the functional systems that compose a city. Urban prevention, therefore, has a wider vision as compared to a single building and is designed to maintain the vita settlement functions.

The key issue is to identify the essential parts of the urban structure, which must remain operational even after the earthquake. This Urban Minimum System (UMS) is conditioned by settlement strategic role as compared to the surrounding area and with due consideration of the different elements that compose it. This approach, i.e. selection

© Springer International Publishing AG, part of Springer Nature 2018
M. di Prisco and M. Menegotto (Eds.): ICD 2016, LNCE 10, pp. 313–328, 2018.
https://doi.org/10.1007/978-3-319-78936-1_23

of some elements only, is justified by the fact that it is impossible to protect the entire settlement, for reasons of costs and time. It is therefore natural to make a choice: which structures, and to which level, to protect first. Prevention planning is based upon the need to maintain the vital functions that make up a city. The idea of minimum urban structure is linked to the strategic role, of the different elements, on the ordinary life of a city.

One needs to understand which are, at any given time, components of the Urban Minimum System (UMS), with the final goal of identifying the set capable of having a city work after an earthquake.

The Authors discussed in the past of this topic. In particular after a lot of papers dealing with vulnerability analysis of structures and infrastructures (Nuti et al. 2001; Nuti and Vanzi 2005; Vanzi et al. 2005; Nuti et al. 2007; Nuti et al. 2010), they focused their attention on Urban vulnerability analysis both in structural and in urban-planning point of view (Biondi et al. 2011; Biondi and Vanzi 2012; Sepe et al. 2016; Vanzi et al. 2016).

After the 2009 Abruzzo earthquake, they were involved, under supervision of an Italian Government Department, in the reconstruction phases. The Abruzzo municipalities characteristics, with low population density and low property values, but important historical and aesthetic values, were the focus of the public intervention policy. In this activity the idea of the Urban Minimum System (UMS) was improved in order to define a strategy for the reconstruction (Biondi and Vanzi 2011; Biondi et al. 2012).

The main initial choice was to co-ordinate reconstruction activities among different towns and necessities (economic, urban, artistic, historical, structural) in a geographical and functional coordination. The work consists mainly in preparation of reconstruction plans and in definition of the main guidelines, together with the preparation of pilot retrofitting projects on important structures (mainly public) having an exemplary character. With a special view on the current regulations, the attention is focused on the main structural design approaches, in synergy with all the other design specialties (i.e. architectural and urban planning), in order to implement design criteria that are both safe and respectful of history and aesthetics.

The principal goals of this past activity were to carry out a multi-disciplinary approach considering both structural and urban-planning point of view and to test the reliability of the post-seismic rehabilitation and seismic improvement procedure as defined in the Italian Code.

The first procedure, capable to define the most efficient structural improvement strategy within a urban centre, has been set up. The system (a portion of a municipality) is modelled via its cut sets and at each element is assigned a fragility curve specifically computed. An optimization procedure, aiming at maximizing the global system safety and minimizing retrofitting costs, is then set up. Results clearly indicate the best seismic retrofitting strategy.

The second aspect tends to evaluate the coherence of the damage survey obtained by means of a simple abacus (AeDES chart) with the local earthquake effects (in terms of maximum peak ground acceleration for example); to define a code of practice for the assessment of historical patrimony. For this aim regional attenuation relationships (Sabetta and Pugliese 1987; Zonno and Montaldo 2002) have to be taken into account and to be compared with damage survey results.

These activities pointed out a different approach regarding the old existing buildings. When these buildings were built with masonry, the original architectonic characteristics have to be preserved. This goal is prevalent and taking into account this idea the structural material characterization is carried out in a parametric approach according to Seismic Code, Biondi and Vanzi (2012).

When these buildings were built with reinforced concrete, the original architectonic characteristics have a little relevance and the preservation goal has an attenuation. So in this case a complete and severe structural restoration has to be considered and a more detailed characterization of structural materials has to be carried out. The in-situ concrete strength assumes a fundamental role (Breysse et al. 2017) and these uncertainties have to be taken into account in the vulnerability approach.

2 A Strategic Approach to Historical Centres

An historical centre is like a general infrastructure a complex mix of different functions; these functions are in part in series and in part in parallel. This distinction is very relevant.

A series system is a configuration such that, if any one of the system components fails, the entire system fails. Conceptually, a series system is one that is as weak as its weakest link. A parallel system is a configuration such that, as long as not all of the system components fail, the entire system works. Conceptually, in a parallel configuration the total system reliability is higher than the reliability of any single system component.

These conceptual approaches have to be redefined for the historical centres, above all if little towns in marginal territories are considered, like those of previous Abruzzo Region experiences. In this case historical centres show low inhabitant density, a great part of uninhabited or partially inhabited buildings, a poor maintenance of those buildings.

In this case is not possible to define if a building is a part of a series system or is a part of a parallel system: probably those buildings are out of any system from the functional point of view and it isn't so clear how manage their failure.

On the contrary these buildings have a great value from urban point of view. They may be particularly relevant in terms of architectural content, may be particularly interesting in terms of touristic use, may be particularly usefulness in terms of avoiding the soil use. Finally it could be extremely complex to individuate the owners of this existing estate patrimony, so it could be extremely complex to characterize these buildings in terms of fragility.

Generally these old, masonry, buildings are now uninhabited buildings without any maintenance effort. Their inhabitants live or in a greater town or in a neighbour place of the same town. They live in more recent, r.c.?, building that it isn't sure that guarantee higher structural security level.

For this complex of reasons a strategic approach to urban historical centers needs an additional level of analysis.

In a historical center it isn't mandatory to investigate the actual security level according to the actual functional distribution, in a historical center could be mandatory to investigate how to reactivate the original use of buildings, despite of the actual security level. Practically in a historical center the choice if restore or not a building (or if seismically improve or not) could be devoted to urban or architectonical considerations and not to economical or purely structural evaluations.

Often in the past structural difficulties caused uncorrected urban choices. For example, if a primary school was located in an old friar building, it was difficult to create a gymnasium in this complex. So the gym dome was built in the suburbs. Now it is mandatory to reunify those two functions, school and sport, and surely the prevalent issue is to rebuilt the gymnasium dome near the school and not the contrary.

Again a urban planning that provides to rearrange the residential building position could prefer to relocate those buildings in the historical center, could prefer to restore old existing buildings and to abandon more recent (r.c.?) buildings designed without any seismic provisions.

For these reasons when an engineering approach to urban safety has to be carried out for a historical (minor) center an accurate evaluation of the historical evolution of urban pattern can be avoided.

In this case the population size trend could be a fundamental parameter. As shown by the recent experiences in post L'Aquila Earthquake Reconstruction Planning, population size trend is linked to damage response for historical center.

In fact population size determines both building construction and building maintenance. If in a certain period a town has a great population, it needs a high number of buildings for home and service. If the same town later losses population, those buildings will be not maintained or will be abandoned. In case of an earthquake this town will be more fragile than another town with constant population size trend.

A comparison between population trend and post-earthquake damage survey will be shown in following chapters.

3 Urban Risk Assessment and Reduction

A particular relevance assumes the analysis of urban seismic vulnerability; such kind of analysis has been developed in the past by the Authors. In particular a procedure for safety evaluation was improved for network systems like electric power, road, water, hospital regional systems or for hospitals, bridges or strategic buildings as a single structure, Nuti et al. (2010).

In the specific case of Urban Risk Assessment and Reduction a new system is considered: the so-called Urban Minimum System (SUM) i.e. an urban system composed of buildings, open spaces and public ways (Biondi and Vanzi 2011; Biondi et al. 2011). If this system is composed with infrastructural networks and external risks (environmental and geological risks) it is possible to analyze a complex system. From a mathematical point of view, considering that aleatory quantities are involved, as structural strength, the approach has to be probabilistic; on the other hand if a Urban Plan has to be approved, practical and operational decisions have to be assumed.

Generally when a seismic safety evaluation is carried out a procedure to maximize safety of selected nodes and minimize economic expenses has to be constructed, allowing identification of which components, within each part of the system, have to be upgraded to obtain the maximum economic convenience. In the case of a Urban System the approach has to be revisited in order to take into account functional, and social, role of the different part of a city.

So, as above discussed, the evaluation of urban vulnerability doesn't only depend on the constructive characters of each structure but it is strictly connected with city identity. So, for example, in a historical towns isn't only important that inhabitants will be safe during an earthquake but it's important that they will remain in the historical center, that shops and public offices will be re-open, that schools will guarantee their lessons, that monumental buildings will not damaged and touristic activity will continue.

Or it could be preferable that at the end of Reconstruction Period many people abandon the suburbs in order to repopulate the historical center.

Again when a Urban Minimum System (UMS) is analyzed, it has to be clear that it generally plays a fundamental role not only in municipal range but also in territorial range. For example if some public or private services are located in every municipality (as town office, postal service, primary school, Pharmacia or food store), other services are territorial (as hospitals, police stations, fire departments, superior schools). This territorial approach was deeply discussed in previous papers. In this approach attention is paid on a smaller portion of territory: a historic centre of a little town or of a small village with its social life and its necessity of safety. In this centre often buildings have low maintenance, inhabitants are generally older and poor and, in some cases, the building owners are unknown and a large part of estate patrimony is abandoned. For these reasons fragility assumptions have to be more conservative than for similar building that have a regular and continuative maintenance; i.e. when a fragility curve is selected for these buildings, a more probable lack of capacity has to be assumed and population size trend has to be kept in mind.

The logical scheme for an Urban Minimum System is shown in Fig. 1(a). This scheme, for Montebello di Bertona (Vanzi et al. 2016), is composed of four sub-systems (strategic buildings, open spaces, external risks, public ways) arranged in series; each of these sub-system is arranged in series too.

When a system is arranged in series it means that each element has to be safe if global safety has to be preserved, Fig. 1(b). So if a strategic building is considered, for example a primary school, it is safe if open spaces near the school are accessible, if electric power is at disposal, if water network is operative, if eventual ground sliding remains in a quiescent stage, if public ways preserve their accessibility to the entire community and, above all, by ambulances or civil protection and fire trucks.

On the other hand when an element class shows some redundancy, the component can be assumed as arranged in parallel. So if the same primary school can be reached by means of two different road ways, these two ways are in parallel and one of these can collapse if the other remains full efficient. In order to guarantee this equilibrium a probabilistic approach has to be carried out. Fragility curves of each component have to be selected, fragility behavior of the system has to be defined via Montecarlo and target safety level has to be selected. That it is with drastic decisions too: if a building can collapse on an important way, it would be better if the building could be demolished. Any macro sub-system has to be in series with the others while a punctual analysis permits to decide what element of the sub-system is in series and what in parallel, Fig. 1(c).

In previous papers the case of Montebello di Bertona was discussed basing on both sub-systems and components arrangement (in series or parallel).

The fragility curves for the study Urban Minimum System are shown in Fig. 2. Almost 40 elements are considered and red thick line is the actual fragility curve of the system. It is possible to note that actual failure probability is PF = 50% for MMI ≈ 5,70, i.e. this little town is too much fragile on respect to its local seismicity. A retrofitting procedure has to be

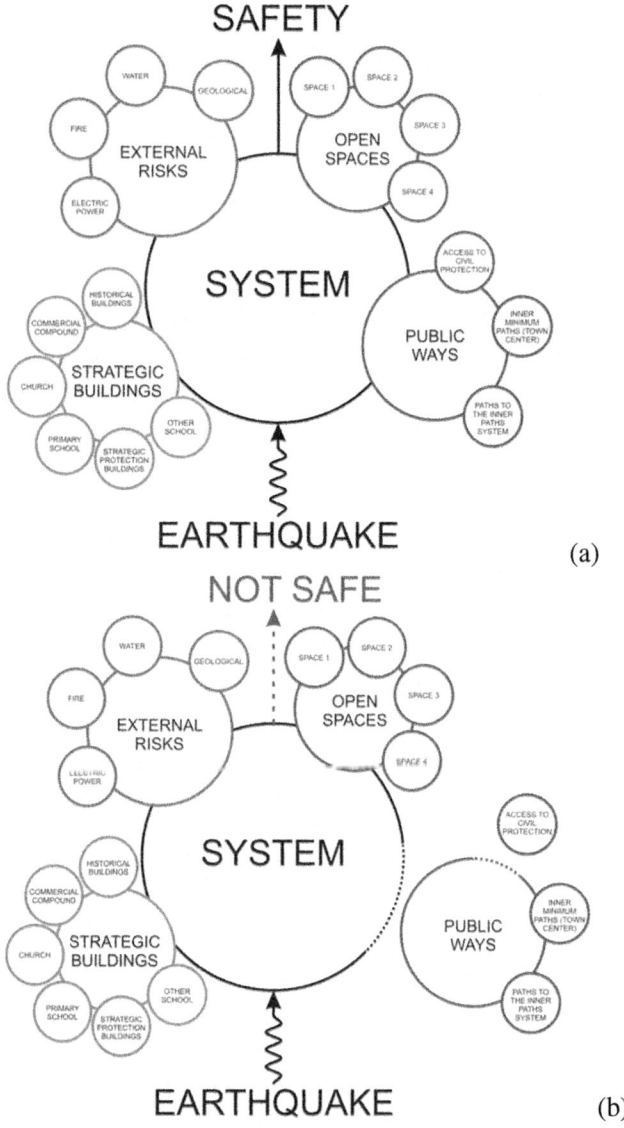

Fig. 1. Logical schemes for Montebello di Bertona Urban Minimum System with different hypothesys for sub-systems

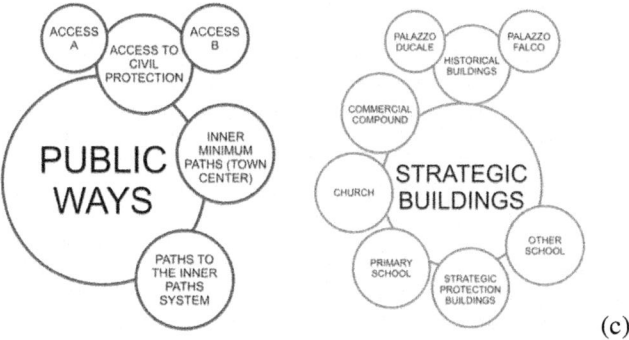

(c)

Fig. 1. (*continued*)

carried out in order to obtain an acceptable Security Level. Considering the nature of a historical town with masonry building as a prevalent building typology, a MMI = 10 level can be assumed as acceptable Risk Level target.

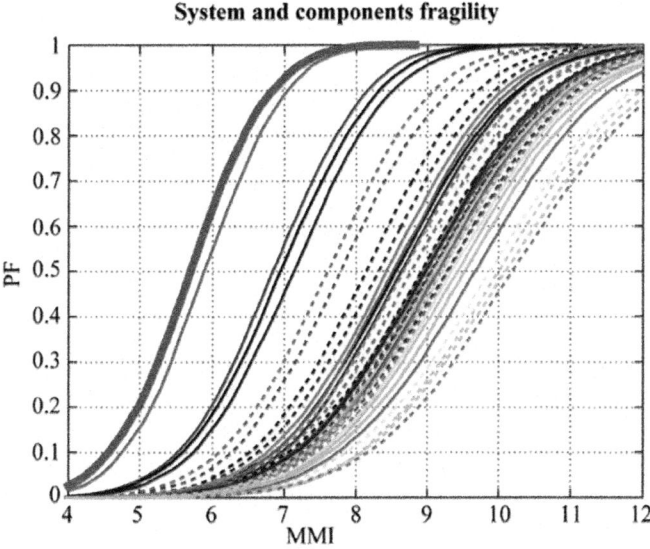

Fig. 2. Fragility curves for Montebello di Bertona Urban Minimum System (Color figure online)

It is possible to study the MMI failure level for each component at every retrofitting step procedure, Fig. 3, and it is possible to note that system fragility depends mostly on a few number of components that show high fragility levels. It is a priority to retrofit these elements in order to obtain an improvement of system behavior i.e. in order to obtain an acceptable Security Level.

Component state for each step

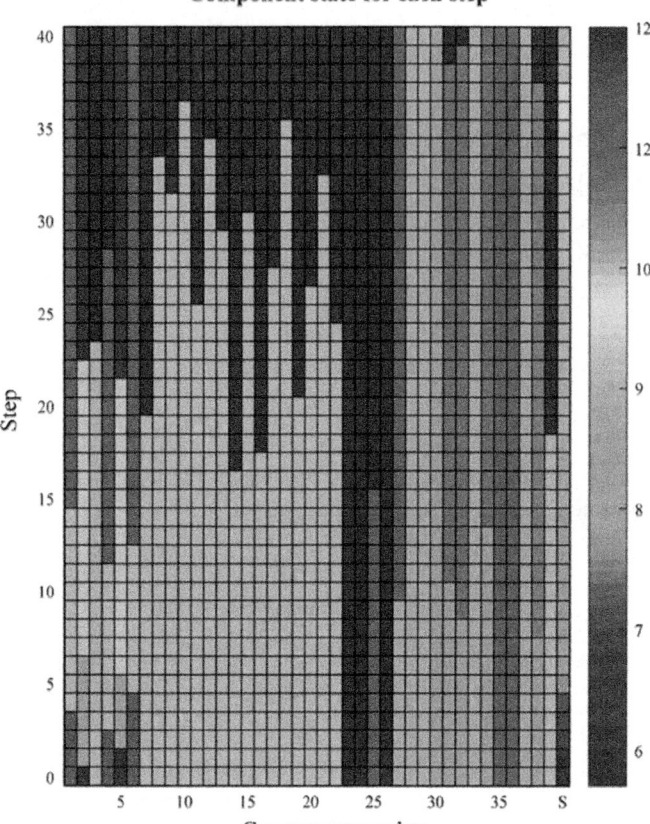

Component number

Fig. 3. MMI failure intensity level for each component at every retrofitting step procedure

If a retrofitting procedure is carried out, it is possible obtain that, in about 40 steps of a single element independent improvements. In Fig. 4 cumulative PF average value for each component at every retrofitting step procedure is shown.

It is possible to note that at the end of those 40 steps the average failure probability (PF = 50%) reaches a more acceptable value: MMI ≈ 10,20.

An important safety gain for the Urban System due to the retrofitting of some particular elements.

In Fig. 5 the MMI failure intensity level for each component for the most relevant retrofitting steps is shown while in Fig. 6 the cumulative PF average value for each component for the most relevant retrofitting steps is depicted. It is possible to note that the retrofitting of some particular elements (i.e. some particular retrofitting steps) have a greater influence on system security level.

E (PF)

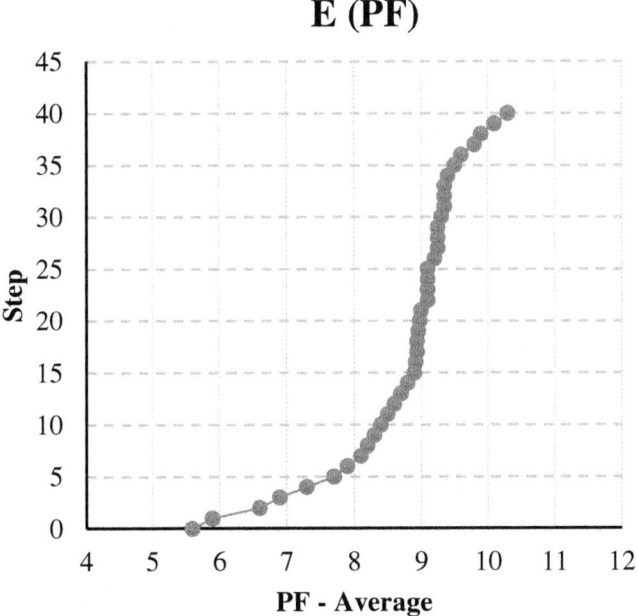

Fig. 4. Cumulative PF average value for each component at every retrofitting step procedure

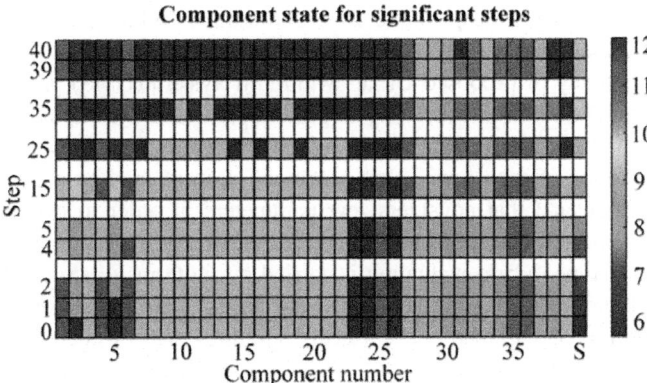

Fig. 5. MMI failure intensity level for each component for the most relevant retrofitting steps

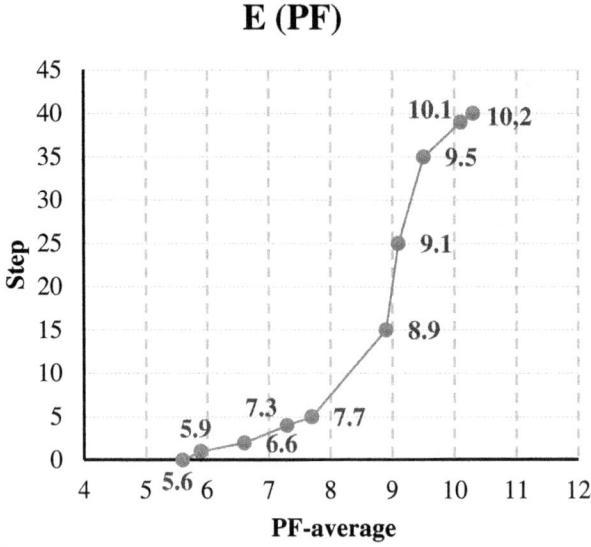

Fig. 6. Cumulative PF average value for each component for the most relevant retrofitting steps

3.1   Post-earthquake Safety Survey: The Case of 2009 L'Aquila Earthquake

As discussed in previous papers the Post-Earthquake Building Safety Survey and Occupancy Evaluation after the 2009 L'Aquila Earthquake represented a good occasion to test the coherency of proposed Urban Minimum System (UMS) procedure.

In fact after the Main-Shock, a lot of professional teams, composed by structural engineers and architects, are created in order to evaluate structural damages and to report occupancy situation in each town. This judgment was obtained by means of a simple abacus (AeDES chart) that considers few parameters in order to evaluate structural damage. After a review of general data (location, construction type, age, height and plan area, occupancy type) a risk evaluation is carried out in terms of structural, non structural, external and geotechnical risks, Fig. 7. In terms of structural configuration both masonry buildings and framed (r.c. or steel) buildings are considered in the AeDES

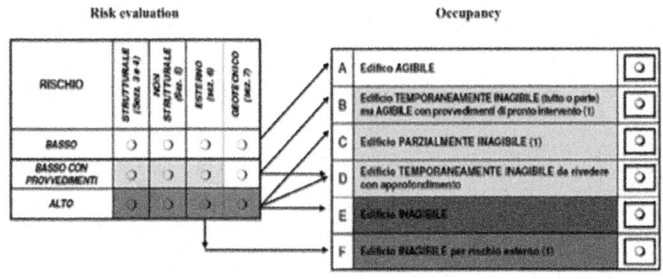

Fig. 7. Risk evaluation (left) and occupancy judgment (right) in the AeDES chart

Fig. 8. Damage survey plans for Civitella Casanova, Montebello di Bertona e Ofena (Color figure online)

chart. Structural (on vertical and horizontal elements) and non-structural damages have to be combined in order to obtain the occupancy judgment finally.

Six categories of occupancy judgment can be selected: -A- immediate occupancy without temporary measures, -B- immediate occupancy with temporary measures, -C- partial unoccupancy due to damage, -D- partial unoccupancy due to insufficient structural information, -E- full unoccupancy for building strong damage or collapse, -F- full unoccupancy for external risk.

Basing on this data base, damage survey plans have been drawn for different little historical centers, Fig. 8, (red buildings are those whit E occupancy judgments).

In Fig. 9 cumulative frequencies of these different occupancy judgments (A-B-C-D-E-F) are shown. These frequencies are collected considering, for every town, the ratio between the number of building having an occupancy judgment to the total of buildings (up) or the same ratio if the gross area of each building is considered (down).

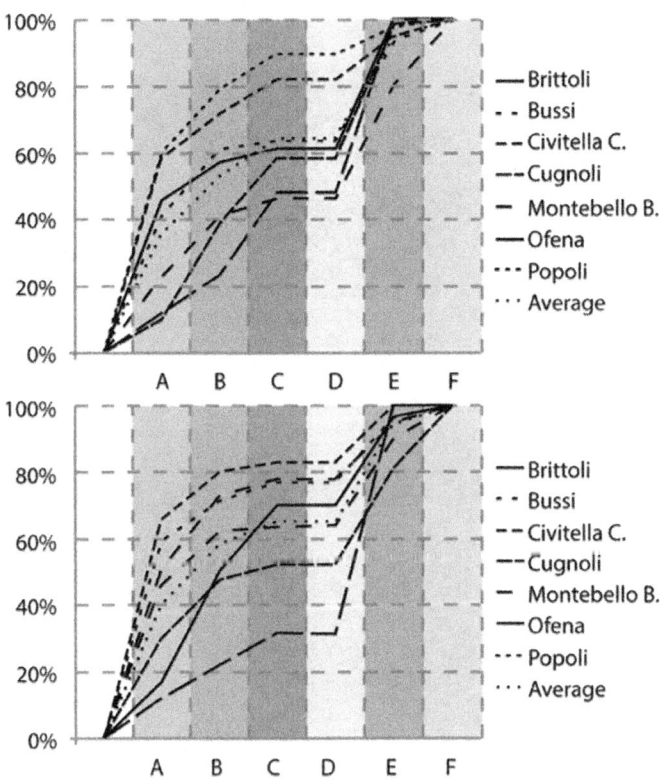

Fig. 9. Cumulative frequencies of building number (up) or building gross area (down) for different occupancy judgments (A-B-C-D-E-F)

As previously discussed (Biondi et al. 2012; Biondi and Vanzi 2012), in order to evaluate the AeDES Chart availability the ratio $S_{aBV/SLD}$ (1) between the maximum local peak ground acceleration estimated via the original Biondi-Vanzi attenuation relationship, $\alpha|_{BV} = a_g/g|_{BV}$, and the design peak ground acceleration at SLD limit state as determined for each site, $\alpha|_{dSLD} = a_g/g|_{dSLD}$, is considered.

$$S_{aBV/SLD} = \frac{a_g/g|_{BV}}{a_g/g|_{d(SLD)}} \tag{1}$$

It was possible to note a good fitting, in terms of increasing E frequency, between AeDES Chart responses and spectral ratio defined in (1). In fact Ofena showed the highest ratio $S_{aBV/SLD}$ and it suffered an high level of damage (the second place in terms of E judgment).

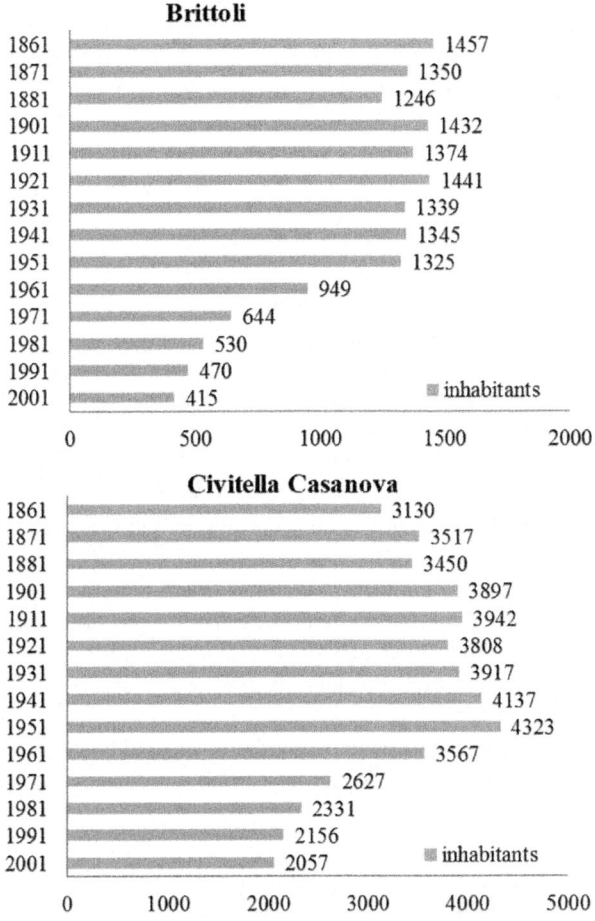

Fig. 10. Population size trend according to official census

So the AeDES Chart approach can be assumed as both a good operative procedure in post-earthquake activity and a good basis for Urban Minimum System (UMS) procedure in order to calibrate actual fragility. On the other hand, historical center that suffered highest level of damage (the first place in terms of E judgment) was Biondi et al. (2012), despite a medium spectral ratio $S_{aBV/SLD}$. Brittoli spectral ratio in fact was similar to Civitella Casanova ratio while Brittoli population size trend showed a greater population loss than Civitella Casanova, Fig. 10. For these reasons the highest level of damage in Brittoli could be expected: a Urban Minimum System (UMS) procedure approach has to take into account population size trend (or another building maintenance parameter) in order to define building fragility.

4 Impact of R.C. Structures

In order to define hypotheses for fragility curves of structures selected in an Urban Minimum System (UMS) procedure, both structural behavior of buildings and retrofitting activities according to Seismic Code have to be kept in mind.

In Biondi and Vanzi (2012) a so-called Code Flow-Chart was presented and discussed for old masonry building according to 2008 Italian Code Chap. 8.

This Flow-Chart focused its attention on stiffness (elastic modulus) and strength (compressive) values derived from Code provisions, structural Model, Sectional analysis and material Tests, respectively C, M, S and T in (2) and (3).

$$E_d(X_C, X_T, \sigma) = E_m(X_T) \cdot g_0(\sigma) \cdot g_2(X_C) \tag{2}$$

$$\textbf{Error!} \tag{3}$$

For masonry design values, d, can be defined basing on average test values, m, and using different Code parameters, 1, 2, 3. For r.c. structures the procedure is quite different and a little bit complicated.

First of all (2) and (3) relationships have to be defined for two different components: concrete and steel. But steel characteristics are stable and depending only on steel class (FeB24k, FeB38k … and so on according Italian Code). So this dependence isn't very relevant.

A second aspect is more significant: for an r.c. structure both regularity in structural arrangement and non structural element behavior can influence global response and define structural fragility. On this regard experiences of recent earthquakes show how r.c. structures designed without or with older seismic provisions behave worse than old masonry buildings.

For these reasons, more conservative hypotheses have to be taken into account if an Urban Minimum System (UMS) procedure is carried out considering the r.c. structures impact too.

5 Conclusions

The paper shows a system reliability model to assess the seismic safety of a whole historical centre. Applications are given for typical Italian cases, making reference to Abruzzo, which has been recently hit by the major L'Aquila earthquake. The whole system is simplified using the Urban Minimum Structure concept, which is excerpted from town planning sciences, and adapted to structural engineering.

It is shown that a sensible prioritization and model optimization, even for a complex system like an historical center, is feasible; the results allow to give a clear indication of the system component to retrofit first. So retrofitting process has to be well calibrated and fragility hypotheses have to be selected according to actual situation that depends on both local seismicity and building maintenance. The first could be evaluated basing a post-earthquake survey, the second keeping in mind population size trend in the past.

References

Biondi S, Vanzi I (2011) L'esperienza dei Piani di Ricostruzione della Facoltà di Architettura di Pescara. In: Atti XIV Convegno L'Ingegneria Sismica in Italia. Dvd paper, Ed. Digilabs, Bari (in Italian)

Biondi S, Fabietti V, Vanzi I (2011) Modelli di valutazione per la vulnerabilità sismica urbana. Urbanistica 147:89–99 (in Italian)

Biondi S, Fabietti V, Sigismondo S, Vanzi I (2012) 2009 Abruzzo earthquake reconstruction plans: a multidisciplinary approach. In: Proceedings 15th world conference on earthquake engineering, Paper No. 3402, September 2012

Biondi S, Vanzi I (2012) Ingegneria delle Strutture. In: Clementi A, di Venosa M (ed) Pianificare la ricostruzione, Sette Esperienze dall'Abruzzo, Marsilio Editori S.p.a., Venezia, pp 70–93 (in Italian)

Breysse D, Balayssac JP, Biondi S, Borosnyoi A, Candigliota E, Chiauzzi L, Garnier V, Grantham M, Gunes O, Luprano VAM, Masi A, Pfister V, Sbartai ZM, Szilagyi K, Fontam M (2017) Non Destructive Assessment of In-situ Concrete Strength: comparison of approaches through an international benchmark. Mater Struct (Materiaux et Constructions) 50:133

Nuti C, Santini S, Vanzi I (2001) Seismic risk of the Italian hospitals. Eur Earthq Eng 15(1):1–19

Nuti C, Vanzi I (2005) A method for the (fast) evaluation of the seismic vulnerability of hospitals. In: Proceedings of the 9th international conference on structural safety and reliability. CD-ROM proceedings

Nuti C, Rasulo A, Vanzi I (2007) Structural safety evaluation of electric power supply at urban level. Earthq Eng. Struct Dyn 36(2):245–263 Special issue on earthquake engineering for electric power equipment and lifeline systems

Nuti C, Rasulo A, Vanzi I (2010) Seismic safety of network structures and infrastructures. J Struct Infrastruct Eng 6(1–2):95–110

Sabetta F, Pugliese A (1987) Attenuation of peak horizontal acceleration and velocity from Italian strong-motion records. Bull Seismol Soc Am 77(5):1491–1513

Sepe V, Biondi S, De Matteis G, Spacone E, Vanzi I (2016) La sicurezza del patrimonio edilizio pubblico. In: VersoPescara2027 1.2 dossier di ricerca, Gangemi Editore, pp 33–49. ISBN 88-492-1020-5

Vanzi I, Bettinali F, Sigismondo S (2005) Fragility curves of electric substation equipment via the Cornell method. In: Proceedings of the 9th international conference on structural safety and reliability. CD-ROM proceedings

Vanzi I, Biondi S, Cataldo N, Sulpizio C (2016) An innovative approach to assess seismic safety of historical centres. Appl Mech Mater 847:290–298 ISSN: 1662-7482

Zonno G, Montaldo V (2002) Analysis of strong ground motions to evaluate regional attenuation relationships. Ann Geophys 45(3/4):439–453

Non Linear Static Analysis: Application of Existing Concrete Building

A. V. Bergami[1(✉)], A. Forte[1], D. Lavorato[1], and C. Nuti[1,2]

[1] Department of Architecture, University of Roma Tre,
Largo G. Marzi 10, Rome, Italy
alessandro.bergami@uniroma3.it
[2] College of Civil Engineering Fuzhou, University of Fuzhou, Fuzhou, China

Abstract. Recently, as a result of seismic events occurrences in Italy, building safety has become a topic of considerable interest. Most of existing reinforced concrete frame buildings designed for vertical loads only and without construction details, which guarantee ductility and dissipative capacity. Then it is necessary to develop a reliable and practical analysis procedure for professional use.

This is confirmed by intense research activities to identify the capacity and the safety level of existing structures. Between existing approaches, the Incremental Dynamic Analysis (IDA) is considered to be one of the most accurate methods to estimate the seismic demand and capacity of structures. However the executions of many nonlinear response history analyses (NL_RHA) are required, therefore approaches non-linear static analyses based are studying to aim less onerous methods. The research discussed in this paper deals with the proposal of an efficient Incremental Modal Pushover Analysis (IMPA) to obtain capacity curves by replacing the nonlinear response history analysis of the IDA procedure with Modal Pushover Analysis (MPA). Finally, these approaches were applied to an existing case study that is a concrete framed building of strategic relevance.

Keywords: Modal Pushover Analysis
Existing irregular framed building · Capacity curve

1 Introduction

Recently, as a result of seismic events occurrences in Italy, building safety has become a topic of considerable interest. Therefore, the development of fast and reliable analysis procedures determining the safety level of existing structures has become essential. Incremental dynamic analysis (IDA) is a method for estimating the seismic response and capacity of structures over the entire range of structural responses, from elastic behavior to global dynamic instability. The most accurate way to compute the seismic demands of a structure under a given seismic action is to carry out a nonlinear response history analysis (NL_RHA) on a detailed three-dimensional (3D) mathematical model

C. Nuti—Visiting Professor, College of Civil Engineering, University of Fuzhou, Fuzhou, China.

M. di Prisco and M. Menegotto (Eds.): ICD 2016, LNCE 10, pp. 329–340, 2018.
https://doi.org/10.1007/978-3-319-78936-1_24

of the structure. IDA requires the execution of NL_RHA for an ensemble of ground motions, each scaled for various intensity levels, selected to cover a wide range of structural responses, from elastic behavior to global instability. From the results of this computation, it is possible to determine the structural capacities corresponding to the various limit states (Vamvatsikos 2002; Vamvatsikos and Cornell 2002).

However, IDA is onerous for practicing engineers since it requires an intensive computation of many NL_RHAs (Vamvatisikos and Cornell 2005); recognizing that IDA of practical structures is computationally extremely demanding, even the developers of IDA have devised a simplified, approximate method (Fajfar 2000).

Hence, Nonlinear Static Procedures (NSPs) attract the attention of both practicing engineers and the research community since it is more practical and of faster implementation. Different NSPs have been developed and used for their conceptual simplicity, computational attractiveness and capability of providing satisfactory predictions of the building's seismic response: the N2 method (Freeman 1998), the Capacity Spectrum Method (CSM) (Casarotti et al. 2007), the Adaptive Capacity Spectrum Method (ACSM) (Chopra and Goel 2002), and the Coefficient Method of Displacement Modification (Goel and Chopra 2004), etc. Among the current nonlinear static analysis methods, Modal Pushover Analysis (MPA) was developed by Chopra and Goel (2002) to take into account the higher mode contributions to the total response; later, Goel and Chopra (2004), Chopra and Chintanapakdee (2004) reported that MPA yields better results compared to a traditional pushover analysis. However, most of the researches dealing with nonlinear static analysis procedures are limited to planar structures.

Presently a focus on existing buildings, having a remarkable architectural and structural complexity, showed the importance of extending NSP to the entirety of the 3D structure to examine the 3D effects of irregularities, and the relevance of the control point position (both in plan and elevation).

The simplified IDA previously cited (Vamvatsikos and Cornell 2002), requires to replace NL_RHA with NSP to evaluate seismic demand, for each given seismic intensity level, reducing the computational effort required for the standard IDA: it requires performing several pushover analysis with different load distribution in order to select the most conservative one and, moreover, estimating the elastic stiffness of the SDF system from the IDA curve.

The aim of this paper is to develop the opportunity of an approximate IDA procedure based on MPA realized by Han and Chopra (2006), proposing a procedure named Incremental Modal Pushover Analysis (IMPA), for the analysis of complex 3D structures and in particular concrete frame structures. The novelty of this approach is in the evaluation of a multimodal performance point in terms of displacement and base shear and therefore in the evaluation of a capacity curve than can replace the standard capacity curve from pushover, already introduced in previous papers of Bergami et al. (2015a, b).

The capacity curve obtained defines the relationship between the base shear and top displacement of the building and can be used to evaluate seismic performance: this is a novelty in fact MPA has been developed and applied to determine structural response in terms of displacements and drift distribution. Is opinion of the authors that the evaluation of global forces and displacement is an important information dealing with retrofitting of existing structures, for example with additional energy dissipation devices (Bergami and Nuti 2013, 2014), as long as foundations are usually barely to be characterized and therefore the designer needs to evaluate if the global actions can be supported and if the retrofitting process reduces or not e.g. the base shear.

Therefore IMPA is a proposal of a new approach to be used as standard pushover procedures as commonly used. In the following report, first the MPA is discussed and than, an existing building, presenting both vertical and plan irregularities, was selected as a case study to verify whether the MPA procedure for asymmetric structures could be successful, even in the case of very complex irregular conditions and develop the building's capacity curve through IMPA.

2 Incremental Modal Pushover Analysis (IMPA)

The IMPA procedure to determine the capacity curves uses MPA procedures rather than NL_RHA to estimate seismic demands for each intensity level of earthquake motions. The MPA procedure is described in a convenient step-by-step form (Chopra and Goel 2002).

The incremental modal pushover analysis (IMPA) proposed is a pushover based procedure that requires the execution of MPA and an evaluation of structural performance within a range of different seismic actions and intensity. Data resulting from MPA application within an identified range of seismic intensity provides all information necessary to estimate the seismic response from different intensity levels. Differently from MPA, this approach is finalized to develop a multimodal capacity curve in terms of base shear and top displacement: the MPA has been developed and used to analyze displacements and drifts distribution. Therefore, dealing with MPA, the evaluation of drifts has to be related with other damage index in order to evaluate the structural performance. With IMPA the author's want to develop a new pushover procedure useful for the same targets of other pushover methods but more suitable for buildings sensitive to higher order modes. For each seismic intensity level, the corresponding Performance Point (P.P.) for the multi-degree-of-freedom (MDOF) system, in terms of roof displacement and corresponding base shear, can be obtained by combining the P. P. determined through the application of many different procedures: in this paper the Capacity Spectrum Method (CSM) has been used but other approaches could be evaluated if considered more suitable. Therefore, according to the procedure selected to be applied for each significant mode: the P.Ps will be combined through the Square Root of the Sum of Squares (SRSS) rule. It is thus possible to obtain a range of multimodal performance points (P.P.mm), each one corresponding to a specific seismic intensity level: CSM (or other approaches as well) are applied using Response Spectrums (RS) for all the intensity level considered (the RS will be scaled up to obtain a range of intensities such as in IDA with the time histories) By connecting all the P.P.mm, a curve can be obtained: this curve is called the "Multimodal Capacity Curve" (MCC). The detailed step-by-step implementation of the IMPA procedure is presented below:

1. Compute the natural frequencies, w_n and modes, φ_n for the linear elastic vibration of the building;
2. Select the ground motions and the RS for a range of intensity levels;

3. For the intensity level i, represented by Peak Ground motion Acceleration (PGA), CSM is adopted to find the P.P. for the predominate modes: for the nth mode, transform the capacity curve, which is defined in terms of base shear and roof displacement, into a capacity spectrum and transform the RS into an Acceleration Displacement Response spectrum (ADRS) format, and plot them on the same chart. Their intersection is taken as the P.P., as shown in Fig. 1(a). Obtain the corresponding P.P. from the capacity curve, as shown in Fig. 1(b). It is important to note that, for the nth mode, if the structure enters a nonlinear plastic stage, then the demand spectrum should be reduced by the spectral reduction factor which depends on the effective viscous damping of structure ξ_{ni}:

$$\xi_{ni} = \xi_0 + k\frac{1}{4\pi}\frac{E_{dni}}{E_{S0ni}} = \xi_0 + k\xi_{eqni} \tag{1}$$

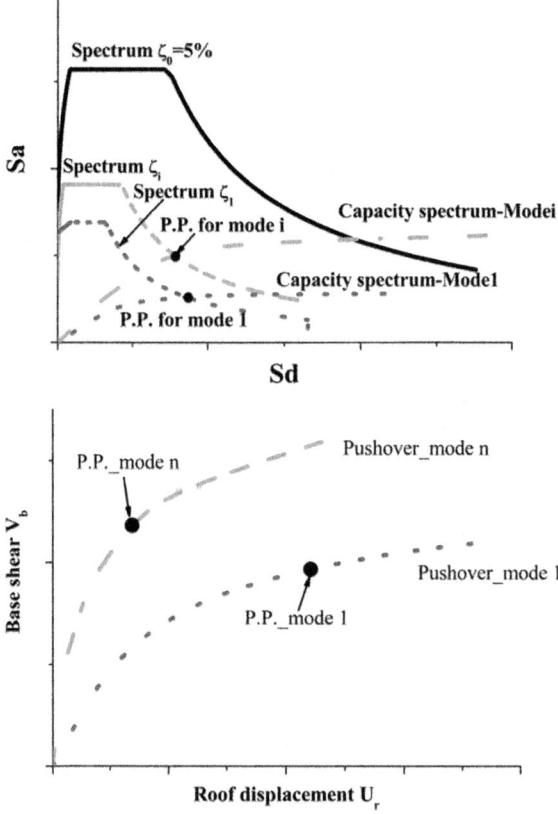

Fig. 1. Evaluation of the performance points (P.P.) for each capacity curve that belongs to the pushover analysis with the selected load distributions: proportional to Mode1...Mode n. (a) for each capacity curve the P.P. is determined via C.S.M. (b) P.P. can be plotted in the V-U plane

where ξ_{ni} is the effective damping for nth mode, ξ_0 is the inherent damping of the elastic structure, about 5% for reinforced concrete structures; E_{dni} is the energy dissipated in an ideal hysteretic cycle, which corresponds to the area enclosed by the hysteresis loop; E_{sOni} is the maximum strain energy dissipated by the structure corresponding to the area of the hatched triangle. The term k is the damping modification factor that is an adjustment factor to approximately account for changes in hysteretic behavior in reinforced concrete structures.

4. Determine multimodal performance point (P.P.mm) in terms of multimodal base shear V_{bmmi} and multimodal roof displacement u_{rmmi}.

   The value of the roof displacement of the selected control point is determined, for each level of earthquake intensity considered, by combining the modal displacements of the control point u_{rni} using the SRSS rule.

$$u_{rmmi} = \left(\left(\sum_n u_{rni}^2 \right)^{1/2} \right) \tag{2}$$

   Instead for the considered earthquake intensity level, to derive the base shear, the procedure adopted follows:

   (a) if the structure remains elastic the value of the base shear of the structure is determined using the same procedure;

$$V_{bmmi,el} = \left(\left(\sum_n V_{bni}^2 \right)^{1/2} \right) \tag{3}$$

   (b) if the structure enters the inelastic range a different procedure is used:

   step (1) the total value of the plastic hinge rotation, θ_{cb} at column end of the first level is estimated as the SRSS combination of the values θ_{cbi} obtained with the pushovers with each modal load distribution.

   step (2) the corresponding bending moments in the columns are estimated through the relevant moment-rotation diagram at the value of the plastic hinge rotation calculated from the SRSS combination.

   step (3) shears in the columns are calculated using the corrected bending moments, and the base shear is calculated as the sum of the column shears.

$$V_{bmmi,y} = V(\theta_{cb}); \quad \theta_{cb} = \left(\sum_n \theta_{cbi}^2 \right)^{1/2} \tag{4}$$

5. Repeat steps 2–4 for as many intensity levels needed to form the IMPA curve, as shown in Fig. 2 where P.P. are the performance points obtained with a lateral load distribution proportional to modal shape 1, ..., n.; according to MPA approach the multimodal performance point (P.P.mm) can be obtained to trace the multimodal capacity curve (MCC).

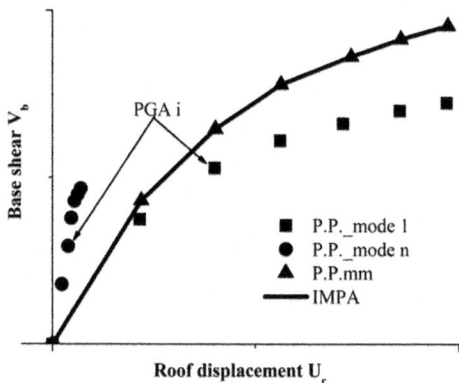

Fig. 2. Construction multimodal capacity curve (MCC) from the IMPA procedure. By applying the SRSS rule with the P.P. obtained with each load distribution (Mode1…Mode n) and for each intensity level (the response spectrum is scaled from lower to higher intensity levels) the MCC) can be obtained.

3   Case Study

3.1   Building Description

The building used as a case study is an existing nine story RC framed building located in Italy, designed for gravity loads only and built in the 1970s. Details of this building are available elsewhere (Bergami et al.), however the building consists of a ground floor, an eight-story elevation and a roof terrace: finite element model of existing framed building is shown in Fig. 3.

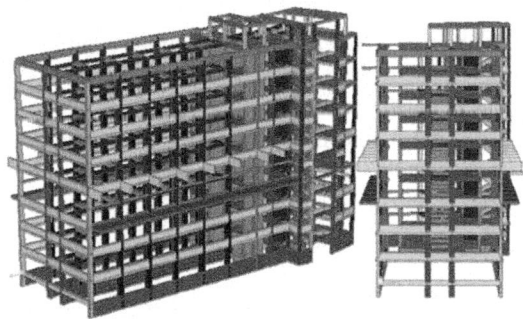

Fig. 3. Finite element model of existing framed building

From a structural point of view, the plan is an irregular polygon where the resistant elements are distributed unevenly: the concentration on one side of shear walls and the one way beam orientation cause a strong irregularity.

3.2   Seismic Input

In this study, the seismic action is defined using both the elastic response spectrum according to NTC'08 and a set of 7 natural time histories. In both cases, seismic action is described by two orthogonal components assumed as being independent and represented by the same response spectrum or by time history; the vertical component of the seismic action has been ignored.

Seismic action details are available elsewhere (Bergami et al.), nevertheless in agreement with NTC'08, seismic action has been defined according to the site and return period detected: the return period depends on the limit state and the category of the existing building. The life safety (SLV) limit state has been adopted.

According to the elastic response spectrum, a set of 7 natural time histories are defined using Rexel software (Iervolino et al. 2010). In Fig. 4, the elastic response spectrum defined by NTC'08 is shown with the response spectrum of each time history record (RS record a-b-c-4-d-e-l).

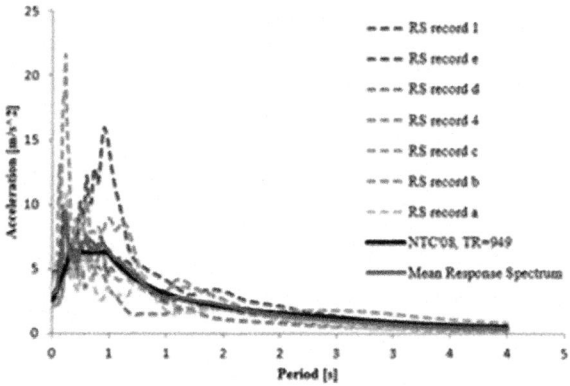

Fig. 4. Response spectra of the code-compliant set of Time Histories: RS record a...RS record l are the selected ground motion records, NTC'08 is the response spectra according to the Italian technical code for a returning period of $T_R = 949$ years.

4   Results

Modal analysis is employed to identify the dynamic behavior of the existing structure and investigate the relevance of higher modes (Fig. 5). In this paper only the Y direction will be discussed; along Y the first, fourth and seventh modes exhibit more than 78% of the participation mass and therefore these modes will be considered in the IMPA.

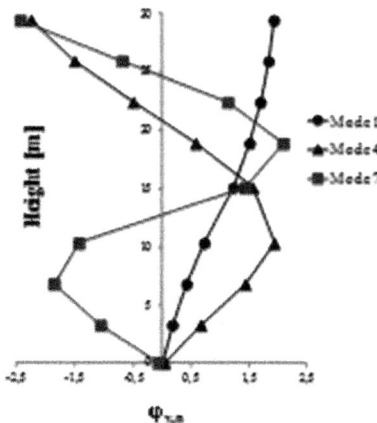

Fig. 5. Modal shapes: main three modes in terms of participation mass along Y.

Figure 6a shows the performance point (P.P.) obtained with the CSM for each one of the capacity curves obtained applying a pushover analysis with a load distribution proportional to the three modal shapes considered. The demand spectrum used has obtained according to NTC'08 (PGA = 0.25g). The structure enters in the nonlinear state for the first mode, and linear elastic state for fourth and seventh mode. According to CSM, when structure enters nonlinear plastic stage, and the spectral reduction factor depends on the effective viscous damping of equivalent Single Degree of Freedom (SDOF) system ξ_i.

By repeating this procedure for other intensity levels the Multimodal Capacity Curve (MCC) has been obtained. By combining the responses considering a range of intensity levels through SRSS rule, both for roof displacements and base shear are

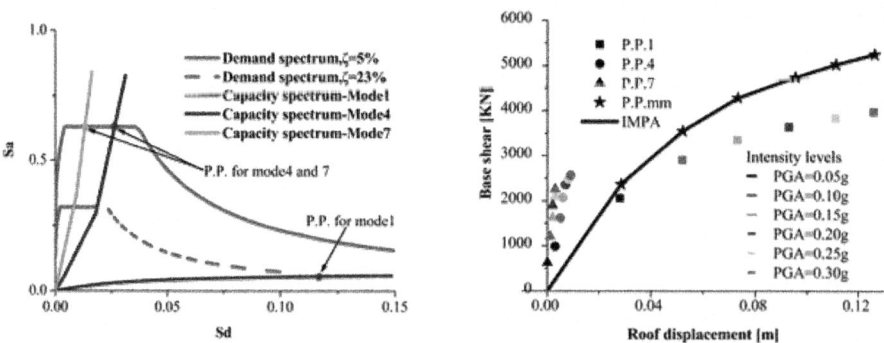

Fig. 6. (a) Evaluation of the P.P. with CSM: in the plot there is the elastic Response spectrum ζ = 5% (P.P. for Mode 4 and Mode 7 are in the elastic range) and the Response spectrum reduced according to a damping of ζ = 23% (23% is the equivalent viscous damping at the P.P. for Mode 1), (b) Construction of MCC from the IMPA procedure. The P.P.mm is obtained by applying SRSS rule with the P.P. obtained from single mode pushover (Mode 1, 4, 7).

determined. Responses are obtained applying CSM on capacity curves obtained considering each modal shape and the response spectrum is scaled from PGA = 0.05g to PGA = 0.30g so for each intensity level an SRSS combined P.P. (P.P.mm) is obtained. Connecting this sequence of P.P. the multimodal capacity curve (MCC) can be defined (Fig. 6b).

Figure 7 shows the capacity curves obtained with a standard pushover (load profile proportional to Mode 1: Push-Mode 1) or performing IMPA (connecting the P.P.mm) and NL_RHA (in the plot named IDA-Umax-Vmax); we underline that maximum displacements and maximum base shear from a NL_RHA are not contemporary, therefore this curve can be considered the upper bond of the capacity curve (any other pushover based curve will be lower).

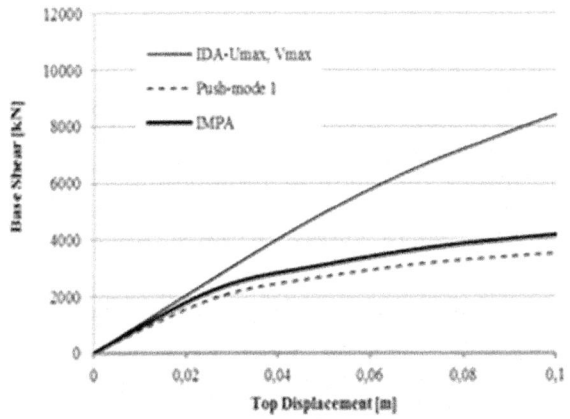

Fig. 7. Capacity curves obtained from different methods: the standard pushover analysis along Y (for the predominate mode), IMPA method and IDA method.

Comparing IMPA to standard single mode pushover curve we can observe an increase of base shear that makes the capacity curve from IMPA stiffer than Push-Mode 1, and closer to the IDA curve.

The capacity curves are almost the same, it is also another indication that the pure translation along this direction. When the structure enters into inelastic state, IMPA would underestimate the base shear compared to IDA.

If we compare IMPA and IDA procedures in terms of Intensity Measure (IM), as a Peak Ground Acceleration (PGA) or Spectral Acceleration at the structure's first mode period ($S_a(T_1, 5\%)$), and Damage Measure (DM), as maximum interstory drift or maximum roof displacement, the incremental curves are evidently closer.

Figure 8a shows the incremental curves obtained with performing IMPA (connecting the P.P.mm in terms of top displacement and PGA) and NL_RHA (in the plot named IDA-Umax, PGA). IDA curve and IMPA curve have almost the same values.

Figure 8b shows, in the same way, the incremental curves obtained with performing IMPA (connecting the P.P.mm in terms of maximum interstory drift and S_a) and NL_RHA (in the plot named IDA-Drift$_{max}$, S_a). The incremental curves, IDA and IMPA, are very close: IMPA method overestimates slightly maximum interstory drift values.

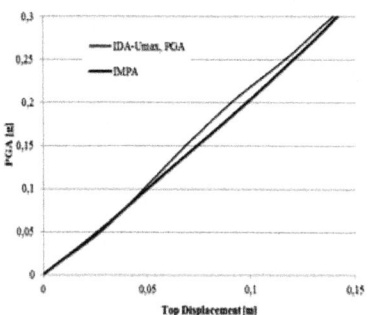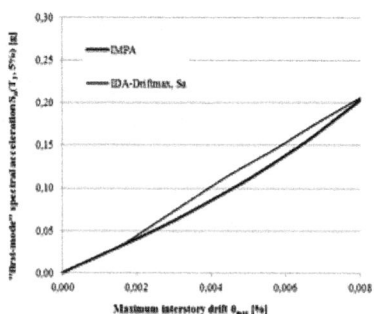

Fig. 8. Incremental curves obtained from two methods: IMPA method and IDA method, (a) in terms of Peak Ground Acceleration (PGA) and maximum top displacement (b) in terms of "first mode" spectral acceleration ($S_a(T_1, 5\%)$) and maximum interstory drift.

The structural behavior obtained from two methods is showed on Fig. 9, the story mechanism of deformation is the same of all PGA's values. However, IMPA method overestimates maximum interstory drift for the story levels lower and underestimates them for the story levels higher.

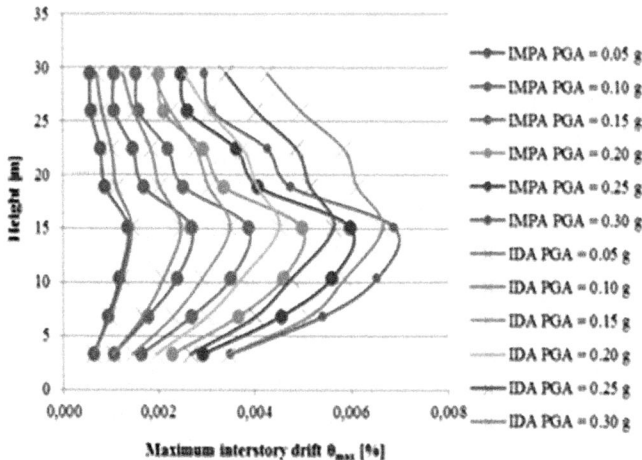

Fig. 9. Structural behavior obtained from IMPA method and IDA method.

5   Conclusion

This work proposes a non linear static procedure to evaluate the seismic capacity of buildings that are sensitive to higher modes. The procedure, named Incremental Modal Pushover Analysis (IMPA), is proposed as an alternative to the non linear Incremental Dynamic Analysis (IDA).

It is well known that IDA implies the execution of many non linear response histories that entail complex and intense computational activity (in this case we used a PC-Intel® Core™ i7-4770 CPU @ 3.40 GHz for 52 h of analysis): non linear dynamic analysis (NL_RHA) requires the preliminary definition of a set of time histories and the execution of the analysis, does not give an immediate and univocal interpretation of results. In fact the IDA curve, which represents the structure's non linear response to each of the selected and scaled time histories, can be realized with different approaches: maximum displacement and maximum base shear, maximum displacement and corresponding base shear, or maximum base shear and corresponding displacement, etc. Therefore, considering that the non linear dynamic analyses are of difficult execution, a simpler approach such as IMPA may constitute a valid alternative, especially for professional use.

IMPA is performed based on the well known modal pushover analysis (MPA) (Chopra and Goel 2002) but its objective is to obtain a capacity curve (the capacity curve is commonly obtained with a standard pushover analysis) also considering the effect of higher modes: this approach widens applicability range of the non linear static analysis to include irregular and high rise buildings.

In IMPA the MPA is used to estimate the seismic demand and capacity of structures over the entire range of structural responses: the demand curve (the response spectrum) is scaled from lower to higher intensity values starting from the definition of a design response spectrum. Using the capacity spectrum method (CSM), for each single mode, and than the combination rule SRSS, a multimodal performance point (P. P.$_{mm}$) can be defined for each intensity level: the multimodal capacity curve (MCC) is the conjunction of all the multimodal performance points obtained.

The approach was tested by applying it to an existing irregular mid-rise building, therefore IMPA is suggested to predict the capacity of structures and, in particular, in case of structures sensitive to higher modes. Finally, comparing results in terms of base shear-top displacement between IDA, standard pushover and IMPA, the effectiveness of IMPA has been demonstrated: the multimodal capacity curve obtained with IMPA results closer to the IDA curve. IMPA's effectiveness is sensitive to the PGA (the distance between IMPA and IDA increases with PGA). Both IMPA and IDA curves show a hardening behaviour, with IDA resulting stiffer in the plastic range, while the pushover curve is mostly elasto-plastic. However, for the pure translation direction, standard pushover underestimates base shear with an error of 28%, the IMPA underestimates base shear with an error of 13%.

Therefore this procedure can be considered a valid tool for professional use to estimate the capacity of structures even for non regular and high rise structures.

Acknowledgements. This work was partially supported by the Italian Consortium of Laboratories RELUIS, funded by the Italian Federal Emergency Agency, with partial funding from PE 2015–2018, joint program DPC-ReLUIS.

References

Vamvatsikos D (2002) Seismic performance, capacity and reliability of structures as seen through incremental dynamic analysis. Doctoral Dissertation, Stanford University

Vamvatsikos D, Cornell CA (2002) Incremental dynamic analysis. Earthq Eng Struct Dyn 31 (3):491–514

Vamvatisikos D, Cornell CA (2005) Direct estimation of seismic demand and capacity of multi-degree of freedom systems through incremental dynamic analysis of single degree of freedom approximation. J Struct Eng (ASCE) 131(4):589–599

Fajfar P (2000) A nonlinear analysis method for performance-based seismic design. Earthq Spectra 16(3):573–592

Freeman SA (1998) The capacity spectrum method as a tool for seismic design. In: Proceedings of the 11th European conference on earthquake engineering, Paris, January 1998

Casarotti C, Pinho R (2007) An adaptive capacity spectrum method for assessment of bridges subjected to earthquake action. Bull Earthq Eng 5(3):377–390

Chopra AK, Goel RK (2002) A modal pushover analysis procedure for estimating seismic demands for buildings. Earthq Eng Struct Dyn 31(3):561–582

Goel RK, Chopra AK (2004) Evaluation of modal and FEMA pushover analyses SAC buildings. Earthq Spectra 20(1):225–254

Chopra AK, Chintanapakdee C (2004) Evaluation of modal and FEMA pushover analyses: vertically "Regular" and irregular generic frames. Earthq Spectra 20(1):255–271

Han SW, Chopra AK (2006) Approximate incremental dynamic analysis using the modal pushover analysis procedure. Earthq Eng Struct Dyn 35(15):1853–1873

Bergami AV, Liu X, Nuti C (2015a) Proposal and application of the incremental modal pushover analysis (IMPA). In: Proceedings of IABSE conference – structural engineering: providing solutions to global challenges, 23–25 September 2015, Geneva

Bergami AV, Liu X, Nuti C (2015b) Evaluation of a modal pushover based incremental analysis. In: Proceedings of ACE, Vietri sul mare, 12–13 June 2015

Bergami AV, Nuti C (2013) A design procedure of dissipative braces for seismic upgrading structures. Earthq Struct 4(1):85–108

Bergami AV, Nuti C (2014) Design of dissipative braces for existing strategic building with a pushover based procedure. J Civ Eng Architect 8(7):815–823. Serial No. 80, ISSN 1934-7359

Iervolino I, Galasso C, Cosenza E (2010) REXEL: aided record selection for code-based seismic structural analysis. Bull Earthq Eng 8:339–362

Shear Connection Local Problems in the Seismic Design of Steel-Concrete Composite Decks

S. Carbonari[1(✉)], F. Gara[1], A. Dall'Asta[2], and L. Dezi[1]

[1] Department of Civil and Building Engineering, and Architecture,
Università Politecnica delle Marche, Ancona, Italy
s.carbonari@univpm.it
[2] School of Architecture and Design, University of Camerino, Ascoli Piceno, Italy

Abstract. This paper investigates local problems involved in the transfer mechanisms of inertia forces acting at the level of the concrete slab of typical steel-concrete composite bridge decks. In detail, the behaviour of the shear connection, usually designed with only reference to non-seismic loads at the ultimate limit state, is studied considering shear forces descending from both horizontal and transverse seismic actions. Some results concerning twin-girder steel-concrete composite bridge decks characterised by different static schemes are presented and the distribution of longitudinal and transverse forces acting on the shear connection is discussed. First results demonstrate the significance of properly considering seismic induced forces in the design of the steel-concrete shear connection and support the need of further investigations.

Keywords: Bridges · Local problems · Seismic design · Shear connection
Steel-concrete composite decks

1 Introduction

Steel-concrete composite bridges represent a very common, economical, and efficient structural solution, especially for short and medium span lengths (Collings 2005). These bridges are generally constituted by a continuous steel-concrete composite deck supported by reinforced concrete piers (Itani et al. 2004), the latter usually providing the main seismic energy dissipation source, unless the seismic isolation technique is adopted.

In bridges with dissipative piers, a restraint is often introduced at the abutments in the transverse direction in order to reduce deck bending and avoid expensive bidirectional joints. The rigid connection between the deck and the abutments can be established by means of fixed bearings, steel-plate stoppers, or special links restraining the transverse displacements (EC8 2006). In this case, a "dual load path" transverse behaviour has been observed by Calvi (2004) and by Tubaldi et al. (2010), characterised by the following two different mechanisms resisting the earthquake-induced inertia forces: (*i*) the inelastic load path constituted by the piers, designed to yield and dissipate energy;

© Springer International Publishing AG, part of Springer Nature 2018
M. di Prisco and M. Menegotto (Eds.): ICD 2016, LNCE 10, pp. 341–354, 2018.
https://doi.org/10.1007/978-3-319-78936-1_25

and (*ii*) the elastic load path formed by the deck and the abutments, designed to remain elastic according to capacity design principles (EC8 2006). The seismic response of steel-concrete composite bridges with dual load paths presents several elements of interest since the elastic load path (involving the deck) may become important, depending on the ratio of the deck-to-piers stiffness. Consequently, this bridge typology is very sensitive to earthquake loading with likelihood of damage to various components laying in the seismic load paths. Extensive damage con occur not only in the substructures, which can be designed to yield and dissipate the seismic energy, but also in the components of the superstructure (steel-concrete composite deck and bearings), involved in carrying the seismic loads. Above issues are the basis for the interpretation of the unsatisfactory seismic behaviour of several bridges in recent earthquakes (Astaneh-Asl et al. 1994; Itani et al. 2004; Kawashima 2010).

Multiple failure states must be generally considered in analysing steel-concrete composite bridges, which involve different resisting components (the deck and the piers). Parametric studies have been performed to investigate the seismic response of multispan continuous steel-concrete composite bridges with dual load path as a function of the relative deck-to-pier stiffness ratio, also considering uncertainties in seismic input and model parameters (Tubaldi et al. 2010; Tubaldi et al. 2012). It was found that the elastic load path (involving the deck) may assume an increasing importance, relative to the inelastic one (involving substructures), for increasing values of the ratio of the deck to piers stiffness. However, results have been obtained by only considering global failure mechanisms, resulting from a cross section analysis of the deck based on the assumption of conservation of the plain section and on the hypothesis of rigid steel-concrete connection. The deck failure is then defined consistently with the EC8-part 2 (EC8 2006) requirements, for which the bridge deck must remain elastic under the design earthquake, hence it must not show "significant yielding". Yielding of the deck for flexure within a horizontal plane is considered to be significant if the reinforcement of the top slab of the deck yields up to a distance from its edge equal to 10% of the top slab width, or up to the junction of the top slab with a web, whichever is closer to the edge of the top slab.

The beam approach for the deck modelling, currently adopted in the literature on the seismic evaluation or design of bridges, does not allow the investigation of the failures associated to local problems involved in the transfer mechanisms of inertia forces from the slab (where most of the bridge mass is located) to the steel components and substructures. Actually, both diaphragms (especially the end ones) and the steel-concrete connection deserve attention since they may undergo damage, as a consequence of inertia forces paths. It should be remarked that the impact of diaphragms on load paths for seismically induced loads acting on steel-concrete composite bridges has been little investigated in the literature (Zahrai et al. 1998; Zahrai and Bruneau 1999a, 1999b) despite many steel bridges have suffered diaphragm (cross frame) damage during recent earthquakes (e.g. Bruneau et al. 1996; EERI 1990; EERI 1994). Analogously, local problems involving the steel-concrete connection (currently represented by headed shear studs) are usually disregarded in performing deck verifications at dynamic conditions.

In this paper, local problems relevant to the actual transfer mechanisms of inertia forces from the concrete slab to the steel-components and bearings are identified with reference to some case studies constituted by twin-girder steel-concrete composite bridges. Particular attention is focused on the behaviour of the shear connection, usually designed with only reference to non-seismic loads at the Ultimate Limit State (ULS), subjected shear forces descending from both horizontal and transverse seismic actions. The distribution of longitudinal and transverse forces acting on the shear connection is addressed and discussed interpreting the deck behaviour.

2   Case Studies and Modelling

2.1   Investigated Steel-Concrete Composite Bridges

The analysis of the shear connection local problems is performed with reference to some case studies constituted by single span and two-span bridges, characterised by a twin-girder steel-concrete composite deck. In particular, span lengths of 48 m are assumed and the static schemes reported in Fig. 1 are considered. The deck restraints are defined to allow free elongations at service conditions and are not modified for the dynamic situation through the use of lock-up devices. Thus, seismic actions in the transverse direction of the bridges are resisted by only one bearing at each support while a couple of bearings at only one support entrusts inertia forces in the longitudinal direction. In addition, for the sake of simplicity, the compliance of the middle supports, due to the flexural behaviour of the pier, is neglected for the two-span bridges.

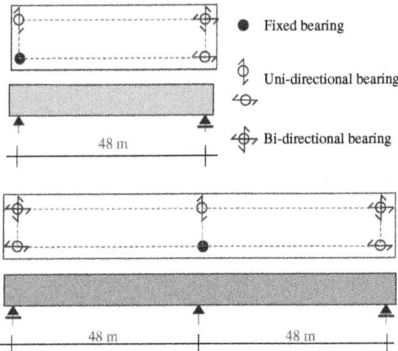

Fig. 1. Deck static schemes for single span and two-span bridges

The bridge has a composite continuous twin-girder deck constituted by a 12 m wide slab, with mean thickness of 0.30 m, sustained by two 6 m spaced 2.4 m high steel I-shaped girders (Fig. 2). Vertical loads for the seismic combination are about 140 kN/m, including self-weights.

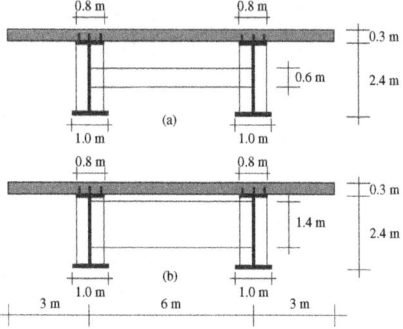

Fig. 2. Deck cross-sections at (a) middle span and (b) at supports.

Cross beams do not directly support the concrete slab and are 6 m spaced in the longitudinal direction, with height equal to 0.6 m and 1.4 m at mid span and at supports, respectively. Slab reinforcement ratio is 1% for mid span cross-sections and 2% for cross-sections at supports.

The shear connection is constituted by 180 mm high headed shear studs of diameter ϕ22 mm, organised in 3 rows and spaced 200 mm along the beam axis. According to the (EC4 2004), the shear capacity of the single stud is 109.5 kN.

The bridge deck is designed with respect to non-seismic actions at ULS and comply with requirements of EC8-2 for what concerns verifications at seismic conditions.

2.2 Seismic Action

In order to emphasize effects of seismic loads, bridges are assumed to be located in a high seismicity area characterised by a Peak Ground Acceleration (PGA) of 0.373 g, corresponding to a probability of exceedance of 10% in 100 years, which is assumed to be the bridges service life, for a type C soil profile. The pseudo-acceleration response spectrum, associated to the selected area, is reported in Fig. 3.

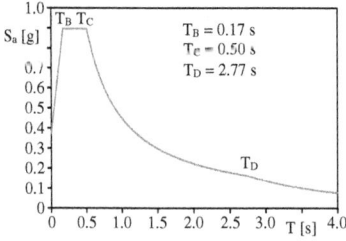

Fig. 3. Pseudo-acceleration elastic response spectrum.

2.3   Structural Modelling and Seismic Analyses

As stated above, the equivalent beam approach, usually adopted for the modelling and analysis of bridge decks, is not able to capture local mechanisms assuring the transfer of inertia forces from the slab to substructures. Thus, for the analysis of local problems, complete 3 dimensional finite element models of the decks are developed exploiting potentials of the computer structural analysis program (Straus7). All structural elements (plates of longitudinal steel girders, cross beams and stiffeners, as well as the concrete slab) are modelled through shell elements (Fig. 4a, b, c) taking into account the actual position of their mid-planes; the mean dimension of the mesh is 0.2 × 0.2 m. Materials are assumed to behave linearly with Young's modulus E_s = 200000 MPa and E_c = 34000 MPa for the steel and the concrete, respectively. Rigid constraints are used to account for the bearings dimensions, and restraints are applied to the master node of the constraints.

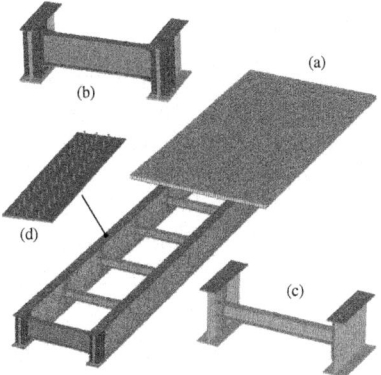

Fig. 4. 3D finite element model of the bridge deck: (a) overall view, (b) detail of the end cross beam; (c) detail of the mid-span cross beams and (d) detail of the studs modelling

The shear connection is modelled with beam elements. The actual position of studs is considered (3 rows of connectors at the top plate of each longitudinal girder) and connectors are constituted by beam elements connecting the mid-plane of the top plate of the steel girder with the mid-plane of the concrete slab. In particular, each connector is modelled by two beam elements, one of short length in which the shear deformability is lumped and one with increased stiffness properties to simulate a rigid shear behaviour (Fig. 5a). In addition, both beam elements have a rigid flexural behaviour. Studs have a nonlinear shear behaviour defined according to the model of Ollgaard et al. (1971); an equivalent secant shear stiffness (at 40% of the maximum shear capacity) is also provided for the need of performing linear analyses (Fig. 5b) in conjunction with nonlinear ones.

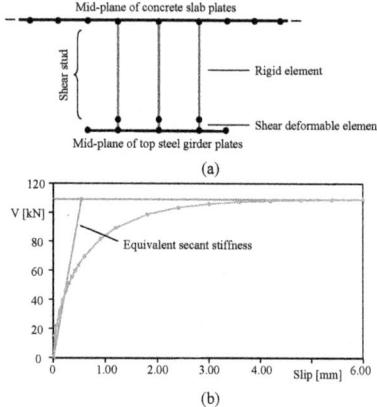

Fig. 5. (a) Modelling of the steel-concrete connection; (b) shear force-slip relationship of the single stud.

Linear and nonlinear static analyses are performed simulating the seismic actions through a mass proportional distribution of horizontal forces. The intensity of forces are determined with reference to the highest pseudo-spectral acceleration, corresponding to 0.9 g (Fig. 3). Both linear and nonlinear seismic analyses include permanent loads at dynamic condition.

3   Main Results

In this section main results of seismic analyses performed on the selected case studies are reported. In particular, distributions of shear forces on the steel-concrete connection on the top plate of the two steel girders due to seismic loads acting in the longitudinal and transverse directions are presented and discussed. The significance of seismic induced shear forces on the steel-concrete connection is evaluated comparing results with those obtained for the ULS.

3.1   Single Span Bridge

Results obtained for the simply supported bridge deck are firstly presented. Figure 6a shows the deformed shape of the deck subjected to the transverse seismic action; mid span cross sections undergo an overall in-plane rotation due to the eccentricity of inertia forces (mainly acting at the level of the concrete slab) with respect to the shear center of the composite open cross-section. On the contrary, deck cross-sections nearby the seismic resistant supports locally deform and distort as a consequence of the transfer mechanism of inertia forces from the slab, where most of the mass is located, to the fixed support (Fig. 6b). Above effects induce a flexural behaviour of longitudinal beams in the web plane to which shear forces is associated.

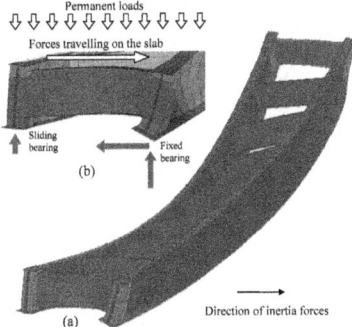

Fig. 6. (a) Deformed shape of the simply supported deck subjected to transverse seismic action; (b) local transfer mechanism at supports inducing the cross-sections distortion

Figure 7 plots the transverse shear forces on the steel-concrete connection (obtained by summating forces acting in the three aligned connectors) along the top plates of the two steel girders due to the transverse seismic action. Results from linear analyses are depicted in Fig. 7a while those of nonlinear applications are reported in Fig. 7b.

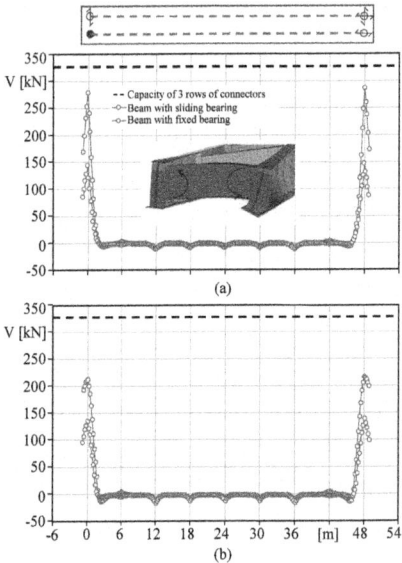

Fig. 7. Distribution of transverse shear forces on the steel-concrete connection due to the transverse seismic action: (a) results of linear analysis and (b) results of nonlinear analysis.

It can be observed that the shear connection is almost unstressed along the whole beams, excepting for the end beam sections, about 2 m long, where transverse shear forces increase rapidly up to a maximum value in correspondence of the support that depends on the connection behaviour (linear or nonlinear). This result confirms that inertia forces, mainly acting on the concrete slab, travel on the slab itself, which is much more stiff than the steel counterpart. However, at supports, due to equilibrium considerations (Fig. 6b), high shear forces are concentrated in a limited number of studs. For the selected case studies, forces do not exceed the connection capacity.

The end cross beam plays an important role in distributing the seismic inertia forces on the steel-concrete connections of the two steel girders. The local transfer mechanism, developing as a consequence of the un-symmetric boundary conditions, induces a significant distortion of the end cross sections with the cross beam deforming flexurally. The cross beam exerts recall forces on the steel girder with sliding bearing that limit the beam rotation and produce an increment of shear forces acting at the level of the steel-concrete connection.

Finally, as expected, shear forces from nonlinear analyses are lower with respect to those obtained from linear applications, as a consequence of the connection plasticization. However, the redistribution of forces appears to be confined in a limited beam end section.

Figure 8 shows the longitudinal shear forces on the steel-concrete connection along the top plates of the two steel girders due to the transverse seismic action. These are consequences of the overall flexural behaviour of the steel girders, produced by torsion resulting from the eccentricity of inertia forces with respect to the shear centre of the deck cross-section. For the sake of brevity, only results of linear applications are reported (in view of the relatively moderate induced forces). Differently form transverse shear forces, maximum longitudinal shear forces on the steel-concrete connection are attained at a certain distance from the supports (about 2 m).

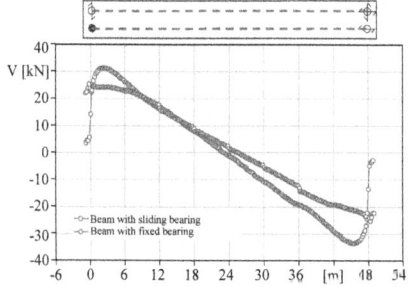

Fig. 8. Distribution of longitudinal shear forces on the steel-concrete connection due to the transverse seismic action

Finally, Fig. 9 shows the absolute values of the resulting shear forces on the steel-concrete connection of the two steel girders, obtained from the vector sum of longitudinal and transverse shear forces induced by the transverse seismic action. Results from linear analyses are depicted in Fig. 9a while those of nonlinear applications are reported in Fig. 9b.

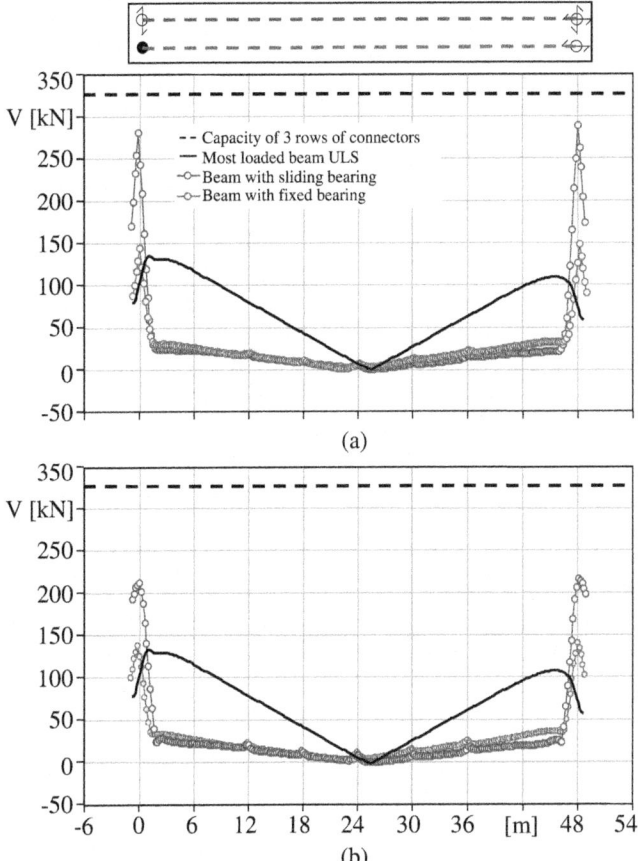

Fig. 9. Distribution of total shear forces on the steel-concrete connection due to the transverse seismic action: (a) results of linear analysis and (b) results of nonlinear analysis

The combined shear forces are compared with those acting at the ULS, resulting from the application of non-seismic loads (including traffic loads). It can be observed that maximum shear forces due seismic and non-seismic actions are attained at different locations. Despite the shear connection capacity is never exceeded, shear forces due to transverse inertia forces are sensibly higher than those relevant to non-seismic actions (almost twice for the linear applications and 40% higher for the nonlinear applications), due to the high intensity of the seismic loads.

Concerning the longitudinal direction, Fig. 10a shows the deformed shape of the deck subjected to both seismic and permanent loads. Seismic actions increase or decrease (depending on the direction of loading) the vertical deflection of the bridge deck due to vertical loads, as a consequence of the overall bending moment produced by the eccentricity of the horizontal inertia forces (mainly located at the concrete slab level) with respect to the horizontal reaction at the fixed support (Fig. 10b).

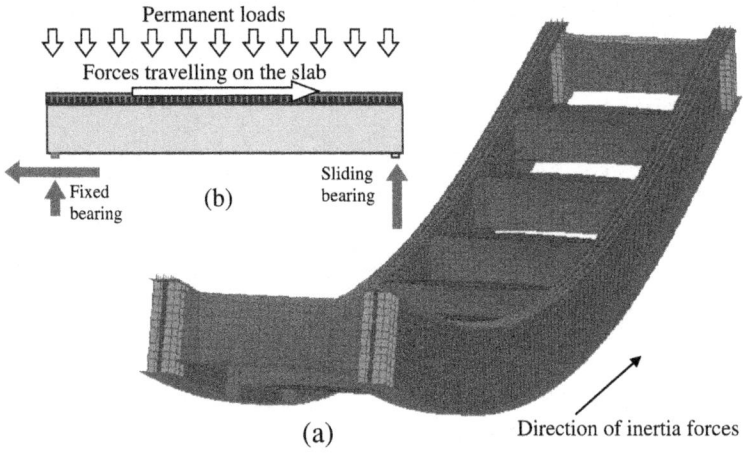

Permanent loads

Forces travelling on the slab

Fixed bearing   (b)   Sliding bearing

(a)   Direction of inertia forces

Fig. 10. (a) Deformed shape of the simply supported deck subjected to longitudinal seismic action; (b) equilibrium considerations

Figure 11 plots the longitudinal shear forces in the steel-concrete connection obtained from the linear applications; results of nonlinear analyses are not reported for the sake of brevity, also considering the low intensity of forces. Contributions of vertical loads at dynamic conditions and of seismic actions are plotted separately as well as combined, for a better understanding of phenomena. Since in this case the behaviour of the steel-concrete connection over the two girders is the same (due to symmetry), results of Fig. 11 only refer to one of the two girders.

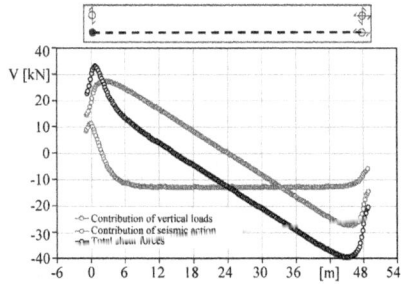

Fig. 11. Distribution of longitudinal shear forces on the steel-concrete connection due to the longitudinal seismic action (single-span bridge) (Color figure online)

It can be observed that, while the contribution of vertical loads assumes the typical "butterfly" shape (red dots), the contribution of seismic loads (blue dots) results from the superposition of two effects: (*i*) one associated to the overall shear force in the steel beam resulting for the equilibrium of bending moment arising for the longitudinal equilibrium of inertia forces (Fig. 10b); and (*ii*) another one produced by the local transfer mechanism of horizontal forces through the steel-concrete connection in correspondence of the seismic resistant bearings. Despite for the presented applications the whole shear forces (black dots) are much lower than the shear connection capacity, it should be remarked that, depending on the span length and the steel girder height, effects of longitudinal inertia forces may become important.

3.2   Two-Span Bridge

This section reports the results obtained for the two-span bridge. Since phenomena have been detailed in the previous section, only minor comments are herein addressed.

Figure 12 plots the transverse shear forces on the steel-concrete connection along the top plates of the two steel girders due to the transverse seismic action.

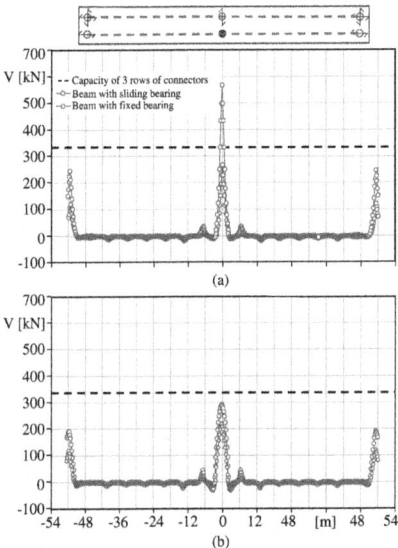

Fig. 12. Distribution of transverse shear forces on the steel-concrete connection due to the transverse seismic action: (a) results of linear analysis and (b) results of nonlinear analysis

Results from linear analyses are depicted in Fig. 12a while those of nonlinear applications are reported in Fig. 12b.

Previous considerations relevant to the distribution of shear forces along the longitudinal axes of beams with fixed and sliding bearings hold. However, higher shear forces arise in the steel-concrete connection in proximity of the fixed central bearing, as a consequence of inertia forces deriving from masses of two spans. For the linear applications, transverse shear forces largely exceed the shear connection capacity; allow the shear forces redistribution among connectors at supports to be captured.

Like for the simply supported deck, torsion resulting from the eccentricity of transverse inertia forces with respect to the shear centre of the deck cross-section promotes the development of global bending moments and shear forces on the longitudinal beams and consequently on the shear connection. Figure 13 shows the absolute values of the total shear forces on the steel-concrete connection of the two steel girders, obtained from

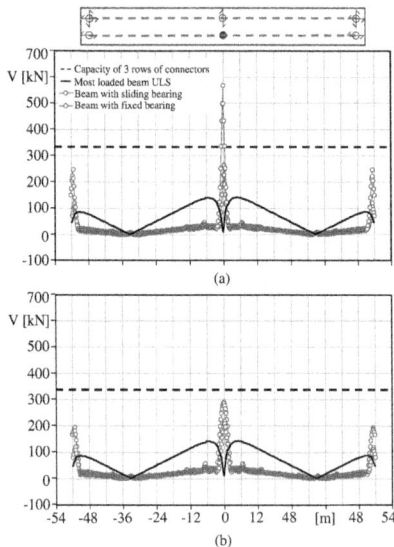

Fig. 13. Distribution of total shear forces on the steel-concrete connection due to the transverse seismic action: (a) results of linear analysis and (b) results of nonlinear analysis

the vector sum of longitudinal and transverse shear forces induced by the transverse seismic action. Results from linear analyses are depicted in Fig. 13a while those of nonlinear applications are reported in Fig. 13b.

The combined shear forces are compared with those acting at the ULS, resulting from the application of non-seismic loading (including traffic loads). Even in this case, maximum shear forces due seismic and non-seismic actions are attained at different locations. It should be remarked that shear forces induced by non-seismic loads are almost the same for the single span and two-span bridges while effects of seismic actions are sensibly higher in the latter case, since the transfer mechanism of forces at the middle support requires the equilibrium of horizontal inertia forces relevant to two spans.

Finally, Fig. 14 plots the longitudinal shear forces in the steel-concrete connection at the top plates of the steel girder equipped with the fixed bearing due to the longitudinal seismic action. Considering the low force intensity, only results of linear applications are reported. Like for the simply supported beam, contributions of vertical loads at dynamic conditions and of seismic actions are reported separately, as well as combined, for a better understanding of their relevant effects.

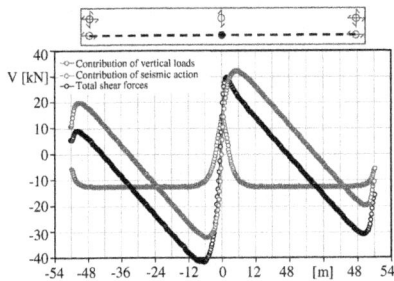

Fig. 14. Distribution of longitudinal shear forces on the steel-concrete connection due to the longitudinal seismic action (two-span bridge)

4   Conclusions

Local problems involved in the transfer mechanisms of inertia forces acting at the level of the concrete slab of typical steel-concrete composite bridge decks have been investigated in this paper. In particular, the shear connection behaviour, usually designed with only reference to non-seismic loads at the ultimate limit state, is studied through applications to some case studies constituted by twin-girder steel-concrete composite bridge decks characterised by different static schemes. The analysis of the distribution of longitudinal and transverse shear forces acting on the connectors demonstrates that the behaviour of the steel-concrete connection in proximity of cross frames at the seismic resistant restraints is not as easy to understand as expected: transverse actions are concentrated at a limited number of studs and, for high intensity earthquakes, may exceed the connectors capacity. Despite results descend from few applications, analyses demonstrate that the seismic induced forces should be properly considered in the design of the shear connection, and support the need of further investigations.

Acknowledgements. This research is supported by the DPC/ReLUIS 2015 grant (Line 1: Steel structures and composite steel-concrete structures).

References

Collings D (2005) Steel-concrete composite bridges. Thomas Telford, London

Itani AM, Bruneau M, Carden LP, Buckle IG (2004) Seismic behavior of steel girder bridge superstructures. J Bridge Eng 9(3):243–249

European Committee for Standardization (ECS) (2006) Eurocode 8 - Design of structures for earthquake resistance Part 2: Bridges (EN1998-2), Brussels, Belgium

Calvi GM (2004) Recent experience and innovative approaches in design and assessment of bridges. In: Proceedings of the 13th world conference on earthquake engineering, Vancouver, Canada, 1–6 August 2004, Paper No. 5009

Tubaldi E, Barbato M, Dall'Asta A (2010) Transverse seismic response of continuous steel-concrete composite bridges exhibiting dual load path. Earthquakes Struct 1(1):21–41

Astaneh-Asl A, Bolt B, Mcmullin KM, Donikian RR, Modjtahedi D, Cho SW (1994) Seismic performance of steel bridges during the 1994 Northridge earthquake. Report No. UCB/CE-Steel-94/01, Dept. of Civil Engineering, Univ. of California, Berkeley

Kawashima K (2010) Seismic damage in the past earthquakes, seismic design of urban infrastructure. Lecture note

Tubaldi E, Barbato M, Dall'Asta A (2012) Influence of model parameter uncertainty on seismic transverse response and vulnerability of steel–concrete composite bridges with dual load path. J Struct Eng 138(3):363–374

Zahrai SM, Bruneau M (1998) Impact of diaphragms on seismic response of straight slab-on-girder steel bridges. J Struct Eng 124(8):938–947

Zahrai SM, Bruneau M (1999a) Ductile end-diaphragms for seismic retrofit of slab-on-girder steel bridges. J Struct Eng 125(1):71–80

Zahrai SM, Bruneau M (1999b) Cyclic testing of ductile end diaphragms for slab-on-girder steel bridges. J Struct Eng 125(9):987–996

Bruneau M, Wilson JW, Tremblay R (1996) Performance of steel bridges during the 1995 Hyogo-ken Nanbu (Kobe, Japan) earthquake. Can J Civ Engrg 23(3):678–713 Ottawa, Canada

Earthquake Engineering Research Institute (1990) Lorna Prieta earthquake reconnaissance report. Spectra. supplement to vol. 6, Oakland, Calif

Earthquake Engineering Research Institute (1994) Northridge Earthquake Jan. 17, 1994, preliminary reconnaissance report. Oakland, Calif

Eurocode 2004. CEN4: Design of composite steel and concrete structures. Part 1.1: General rules and rules for buildings. Comité Européen de Normalisation: Brussels, Belgium

Straus7 Software. Finite element analysis system. G + D Computing, Padova

Ollgaard J, Slutter R, Fisher J (1971) Shear strength of stud connectors in lightweight and normal-weight concrete. AISC Eng J 8:55–64

Cloud to IDA: A Very Efficient Solution for Performing Incremental Dynamic Analysis

A. Miano[✉], F. Jalayer, and A. Prota

Department of Structures for Engineering and Architecture,
University "Federico II", Naples, Italy
andrea.miano@unina.it

Abstract. Incremental dynamic analysis (IDA) is a procedure in which a structure is subjected to a suite of ground motion records, scaled to multiple levels of intensity and leading to corresponding curves of response versus intensity. However, implementation of IDA usually involves a significant computational effort. In this work, a simple and efficient solution for IDA analysis with only few points, based on the structural response to un-scaled records (a.k.a the "cloud"), has been implemented. The transverse frame of a shear-critical seven-storey older RC building in Van Nuys, CA, which is modeled in Opensees with fiber-section considering the flexural-shear-axial interactions and the bar slip, is employed. It is demonstrated that the simplified IDA, obtained based on a significantly lower computational effort with respect to the full IDA, provides reliable results in terms of the statistics of structural response (e.g., mean and mean plus/minus one standard deviation) versus intensity and structural fragility.

Keywords: Seismic fragility · Existing RC frames
Non-linear dynamic analysis procedures

1 Introduction

Many existing reinforced concrete (RC) moment-resisting frame buildings in regions with high seismicity were built without adequate seismic-detailing requirements and are particularly collapse-prone buildings. Identifying accurately the level of performance can facilitate an efficient seismic assessment and classification of these buildings. In this context, analytic structural fragility assessment is one of the fundamental steps in the modern performance-based engineering (Cornell and Krawinkler 2000). The structural fragility can be defined as the conditional probability of exceeding a prescribed limit state given the intensity measure (IM). There are alternative non-linear dynamic analysis procedures available in the literature for characterizing the relationship between engineering demand parameters (EDPs) and IM based on recorded ground motions, such as, the Incremental Dynamic Analysis (IDA, (Vamvatsikos and Cornell 2004) Multiple-Stripe Analysis (MSA, see (Jalayer and Cornell 2009)) and the Cloud Method (Bazzurro et al. 1998; Cornell et al. 2002; Jalayer 2003; Jalayer and Cornell 2003; Jalayer et al. 2015). The nonlinear dynamic methods such as IDA and MSA are suitable for evaluating the relationship between EDP and IM for a wide range of IM values; however, their

© Springer International Publishing AG, part of Springer Nature 2018
M. di Prisco and M. Menegotto (Eds.): ICD 2016, LNCE 10, pp. 355–368, 2018.
https://doi.org/10.1007/978-3-319-78936-1_26

application can be quite time-consuming as the non-linear dynamic analyses are going to be repeated (usually for scaled ground motions) for increasing levels of IM.

In this context, it can be very useful to find a way to reduce the computational effort of IDA analysis, keeping the same accuracy of the results. Herein, an alternative and more quick way to implement IDA analysis is presented, starting from the results of *Cloud Analysis*. Cloud analysis is based on a simple regression in the logarithmic space of the structural response versus the seismic intensity for a set of registered records. Cloud is particularly useful and efficient since it involves the non-linear analysis of the structure subjected to a set of un-scaled ground motions. It is shown herein that, exploiting the cloud results in term of predicted median and standard deviation, the IDA analysis can be performed in an efficient manner without significant loss of accuracy (with respect to a complete IDA). This method, which is called *Cloud to IDA*, considers only few spectral acceleration levels (i.e., data points) for each record.

As a numerical example, the transverse frame of a seven-story existing RC building in Van Nuys, CA, modeled in Opensees modeled by considering the flexural-shear-axial interactions in the columns, is employed. Because of the old construction philosophy, column members are sensible to possible shear failure during earthquakes; hence, a non-linear model is used to predict an envelope of the cyclic shear response (Setzler and Sezen 2008; Sezen 2008). This envelope includes the shear displacements and corresponding strength predictions at the peak strength, onset of lateral strength degradation, and loss of axial-load-carrying capacity. In addition, the total lateral displacement of the members includes also the consideration of the deformability due to bar slip contribution. The adopted engineering demand parameter (EDP) is the critical demand to capacity ratio (Jalayer et al. 2007) corresponding to the component or mechanism that leads the structure closest to the onset of near collapse limit state. This structural response parameter, that is equal to unity at the onset of the desired limit state, can encompass both ductile and fragile failure mechanisms.

It is demonstrated that, for the case-study structure considered, the *Cloud to IDA* procedure provides reliable results in terms of the capacity curves that are very close to those based on a complete IDA and with smaller computational effort.

2 Methodology

2.1 Structural Performance Variable

The EDP herein is taken to be the critical demand to capacity ratio (Jalayer et al. 2007; Jalayer et al. 2015) denoted as Y_{LS} and defined as the demand to capacity ratio for the component or mechanism that brings the system closer to the onset of limit state LS (herein, the near collapse limit state). The formulation is based on the cut-set concept (Ditlevsen and Madsen 1996), which is suitable for cases where various potential failure mechanisms (both ductile and fragile) can be defined a priori. Y_{LS}, which is always equal to unity at the onset of limit state, is defined as:

$$Y_{LS} = \max_{l}^{N_{mech}} \min_{j}^{N_l} \frac{D_{jl}}{C_{jl}(LS)} \tag{1}$$

where N_{mech} is the number of considered potential failure mechanisms; N_l the number of components taking part in the lth mechanism; D_{jl} is the demand evaluated for the jth component of the lth mechanism; $C_{jl}(LS)$ is the limit state capacity for the jth component of the lth mechansim. The capacity values refer to the near collapse limit state in this work, but the procedure can be repeated for any other prescribed limit state. In the context of this work, the demand is expressed in terms of maximum chord rotation for the component, denoted as θ_{max}, and computed based on the nonlinear dynamic analysis. The component chord rotation capacity is denoted as $\theta_{ultimate}$ corresponding to the ultimate capacity of the member. In particular, $\theta_{ultimate}$ corresponds to the point on the softening branch of the force-deformation curve of the member, where a 20% reduction in the maximum strength takes place.

The possible failure mechanisms are associated to the limit state of near collapse $\left(Y_{near\ collapse} = Y\right)$. They correspond to ductile or brittle failures of the columns, depending on whether the column is flexural or shear critical. $Y > 1$ for a column is achieved when $\theta_{demand} > \theta_{ultimate}$ where $\theta_{ultimate}$ for each element takes into account the flexural/axial behavior, the shear behavior and the deformation due to bar slip.

When predicting non-linear response of structures, it is necessary to account for the possibility that some records may cause global "Collapse"; i.e., very high global displacement-based demands or non-convergence problems in the analysis software. It is obvious that, $Y > 1$ for the limit state of near-collapse does not guarantee the exceedance of collapse limit state. Herein, the cases of collapse are identified explicitly by verifying the following criteria for structural collapse: (1) 50% +1 of the columns of only one floor have achieved θ_{axial} (Galanis and Moehle 2015), where θ_{axial} corresponds to the point associated with the complete loss of vertical-load carrying capacity of the component on the softening branch; (2) 10% of the maximum interstory drift between all the floors has achieved.

2.2 Nonlinear Dynamic Analyses Procedure

In order to estimate the structural fragility, *Cloud*, *IDA* and *Cloud to IDA* analyses are adopted herein as alternative nonlinear dynamic analysis procedures. The cloud analysis is a procedure in which a structure is subjected to a set of ground motion records of different first-mode Sa(T) values. The cloud data encompasses pairs of ground motion IM (herein first-mode Sa(T)) and its corresponding structural performance variable Y (see Eq. 1) for each record. Cloud method provides estimates of the first two statistical moments (e.g., logarithmic mean and standard deviation) of the performance parameter Y given the first-mode spectral acceleration. Once the ground motion records are selected, they are applied to the structure and the resulting $Y = D/C$ (demand over capacity ratio, as described above) is calculated. This provides a set of values that form the basis for the cloud-method calculations. The cloud data can be separated to two parts: (a) *NoC* data which correspond to that portion of the suite of records for which the structure does not experience "*Collapse*", (b) *C* data for which the structure leads to "Collapse". In order to estimate the statistical properties of the cloud response, with respect to *NoC* data, conventional linear regression (using least squares) is applied to the response on the natural logarithmic scale, which is the standard basis for the underlying log-normal distribution model. This is equivalent to fitting a power-law curve to the cloud response in the original (arithmetic) scale. This results in a curve that predicts the median drift demand for a given level of structural acceleration:

$$\eta_{Y|Sa,NoC}(Sa) = a \cdot Sa^b$$
$$\ln(\eta_{Y|Sa,NoC}(Sa)) = \ln(a) + b \cdot \ln(Sa)$$

$$(2)$$

where *ln(a)* and *b* are linear regression constants. The logarithmic standard deviation $\beta_{Y|Sa,NoC}$ can be estimated as the root mean sum of the square of the residuals with respect to the regression prediction:

$$\beta_{Y|Sa,NoC} = \sqrt{\frac{\sum (\ln(Y_i) - \ln(a \cdot S_{a,i}^b))^2}{N_{NoC} - 2}}$$

$$(3)$$

where Y_i and $S_{a,i}$ are the demand over capacity ratio values and the corresponding spectral acceleration for record number *i* within the cloud response set and N_{NoC} is the number of *NoC* records.

The standard deviation of regression, as introduced in the preceding equation, is presumed to be constant with respect to spectral acceleration over the range of spectral accelerations in the cloud.

The fragility, expressed generally as the conditional distribution of *Y* given *Sa*, can be expanded with respect to *NoC* and *C* data as follows using Total Probability Theorem (see (Shome and Cornell 1999; Jalayer and Cornell 2009; Jalayer et al. 2017)):

$$P(Y > 1|Sa) = P(Y > 1|Sa, NoC) \cdot P(NoC|Sa)$$
$$+ P(C|Sa)$$

$$(4)$$

The probability terms in Eq. (4) are described clearly as follows:

- The *NoC* term $P(Y > 1|Sa, NoC)$ is the conditional distribution of *Y* given *Sa* and *NoC*, and can be described by a lognormal distribution (a widely used assumption that has been usually verified for cases where the regression residuals represent unimodal behavior, see e.g. (Jalayer and Cornell 2009; Jalayer and Ebrahimian 2017; Miano et al. 2017)):

$$P(Y > 1|Sa, NoC) = \Phi\left(\frac{\ln \eta_{Y|Sa,NoC}}{\beta_{Y|Sa,NoC}}\right)$$

$$(5)$$

where Φ is the standardized Gaussian cumulative distribution function (CDF) and $\eta_{Y|Sa,NoC}$ and $\beta_{Y|Sa,NoC}$ are presented in Eqs. (2) and (3). It should be noted that Eq. (4) is based on the implicit assumption that in the cases of global dynamic instability (global Collapse), the limit state LS (hereafter LS = Near Collapse) is certainly exceeded.

- The term $P(C|Sa) = 1 - P(NoC|Sa)$ is probability of global dynamic instability (Collapse), which can be expressed by a logistic regression model (a.k.a., logit) on the Sa values of the entire cloud data:

$$P(C|Sa) = \frac{1}{1 + e^{-(\beta_0 + \beta_1 . Sa)}} \tag{6}$$

where β_0 and β_1 are the parameters of the logistic regression. It is to note that the logistic regression model belongs to the family of generalized regression models and is particularly useful for cases in which the regression dependent variable is binary (i.e., can have only two values 1 and 0, *yes* or *no*, which is the case of C and NoC herein). Note that the logistic regression model described above is applied to all records; they are going to be distinguished by 1 or 0 depending on whether they lead to C or NoC.

The structural fragility from IDA analysis can be calculated using the (Log-Normal) probability density function fitted to the spectral acceleration values at $Y = 1$, $Sa_{Y=1}$:

$$P(Y > 1|Sa) = P\left(Sa_{Y=1} < Sa\right) = \Phi\left(\frac{\ln Sa - \ln \eta_{Sa_{Y=1}}}{\beta_{Sa_{Y=1}}}\right) \tag{7}$$

where $\eta_{SaY=1}$ and $\beta_{SaY=1}$ are the parameters of the Log-Normal probability density function.

The proposed *Cloud to IDA* procedure can be carried out by considering few levels of spectral acceleration for each record (limited to 4 levels in the majority of cases), in order to obtain the distribution of $Sa_{Y=1}$. In particular, four spectral acceleration levels are chosen per record based on the results of cloud analysis; namely, the original $Sa(T_1)$; median $Sa(T_1)$ at $Y = 1$ estimated based on Cloud Analysis (equal to $(1/a)^{1/b}$ per Eq. (2)); (logarithmic) mean plus one standard deviation $Sa(T_1)$ at $Y = 1$ estimated based on Cloud Analysis (equal to $(1/a)^{1/b}e^{+\beta_{Y|Sa,NoC/b}}$ per Eqs. (2) and (3)); and (logarithmic) mean minus one standard deviation $Sa(T_1)$ at $Y = 1$ (equal to $(1/a)^{1/b}e^{-\beta_{Y|Sa,NoC/b}}$ per Eqs. (2) and (3)). Obviously, other scaling points can be added as needed, by assigning a certain value of α in order to calculate (logarithmic) mean plus or minus α standard deviation $Sa(T_1)$ at $Y = 1$ estimated based on Cloud AnalysiS equal to $(1/a)^{1/b}e^{\pm\alpha\beta_{Y|Sa,NoC/b}}$. As a rule of thumb, it is important to have enough spectral acceleration levels so that the resulting IDA curve (obtained by connecting the points) covers $Y = 1$ (that is, $Sa_{Y=1}$ for each record can be obtained by interpolation). Finally, the structural fragility from *Cloud to IDA* analysis is calculated the same as that of IDA analysis (see Eq. 7).

3 Numerical Application

3.1 Building Description and Modeling

One of the transverse frames of the seven-story hotel building in Van Nuys, California, is modeled and analyzed in this study. The building is located in the San Fernando Valley of Los Angeles County (34.221° north latitude, 118.471° west longitude). The frame building was designed in 1965 according to the 1964 Los Angeles City Building Code, and constructed in 1966. The building was severely damaged in the M6.7 1994 Northridge earthquake (Krawinkler 2005).

Columns in the transverse frame are 356 mm wide by 508 mm deep, i.e., oriented to bend in their strong direction when resisting lateral forces in the plane of the transversal frame. Spandrel beams in the frame are typically 406 mm wide and 762 mm deep in the second floor, 406 mm wide and 572 mm deep in the third through seventh floors, and 406 mm by 559 mm at the roof level. Column concrete has compressive nominal strength $f'c$ of 34.5 MPa in the first story, 27.6 MPa in the second story, and 20.7 MPa in other floors. Beam and slab concrete strength $f'c$ is 27.6 MPa in the second floor and 20.7 MPa in other floors. Grade 60 ($fy = 414$ MPa) reinforcing steel is used in columns. The specified yield strength, fy, is 276 MPa (Grade 40) for the steel used in beams and slabs. The column and beam reinforcement details are provided in Krawinkler (2005). Figure 1 shows the transverse frame modeled in this research.

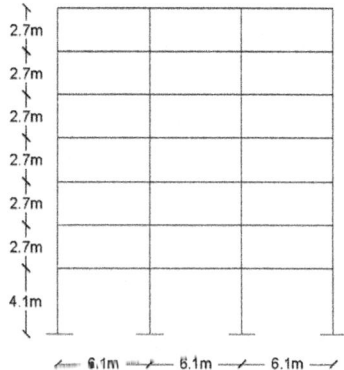

Fig. 1. Geometric configuration of the transverse frame.

3.1.1 Flexural, Shear and Bar Slip Models

The Holiday Inn hotel building experienced multiple shear failures in the columns in the fourth story during the 1994 Northridge earthquake (Krawinkler 2005) in the longitudinal perimeter frames. The amount and the spacing of the transversal reinforcement in most columns were insufficient. Therefore, it is necessary to model materials and column members to capture the shear and the flexure-shear failure modes in columns and the potential collapse of the transverse frame. About flexural model, unidirectional axial behaviour of concrete and steel are modeled to simulate the nonlinear response of beams and columns. Concrete material response is simulated using the Concrete01 material in OpenSees (http://opensees.berkeley.edu), which includes zero tensile strength and a parabolic compressive stress-strain behaviour up to the point of maximum strength with a linear deterioration beyond peak strength. Because the transverse reinforcement ratio for beams and columns in the Van Nuys building is relatively low and detailing does not meet the modern seismic code requirements, concrete is modeled more close to the unconfined model, with peak strength achieved at a strain of 0.002 and minimum post-peak strength achieved at a compressive strain of 0.006. The corresponding stress capacity at ultimate strain is $0.05*f'c$ for $f'c = 34.5$ MPa and for $f'c = 27.6$ MPa and $0.2*f'c$ for $f'c = 20.7$ MPa. Longitudinal reinforcing steel behavior is simulated using the Steel02 material in OpenSees. This model includes a bilinear stress-strain envelope with a curvilinear unload-reload response under cyclic loading. The previous research indicate that the observed yield strength of reinforcing steel exceeds the nominal strength (Krawinkler 2005, Islam 1996). As suggested by Islam (1996), yield strength of 345 MPa (50 ksi) and 496 MPa (72 ksi) are used in this research for Grade 40 and Grade 60, respectively. Both Grade 40 and Grade 60 reinforcement are assumed to have a post-yield modulus equal to 1% of the elastic modulus, which is assumed to be 200 GPa. Additional parameters required to define the Steel02 material model are taken equal to those recommended in the OpenSees User's Manual.

Flexural response of beams and columns response is simulated using fiber cross sections, representing the beam-column line elements. Uniaxial fibers within the gross cross section were assigned either concrete or steel. A typical column cross section included 30 layers of axial fibers, parallel to the depth of the section. In OpenSees, flexural beam-column members are modeled as force-based in which a specific moment distribution is assumed along the length of the member. An internal element solution is required to determine member deformations that satisfy the system compatibility. In force-based column elements, distributed plasticity model is used in OpenSees in order to allows for yielding and plastic deformations at any integration point along the element length under increasing loads. In order to characterize the numerical integration options for the force-based column element and to accurately capture plastic deformations along the members, Newton-Cotes integration (Scott and Fenves 2006) is selected. Newton-Cotes method distributes integration points uniformly along the length of the element, including one point at each end of the element (Fig. 2a). Beams member force-deformation response is computed assuming that inelastic action occurs mainly at the member ends and that the middle of the member remains typically elastic, but this is not necessary. Plastic hinge integration methods are used to confine non linear deformations in end regions of the element of specified length. The remainder of the element is assumed to stay linear elastic and it is assumed that the length of plastic region is equal to the depth of the cross-section. The modified Gauss Radau hinge integration method is used for numerical integration to capture non linear deformations near the ends of the force-based beam elements. The modified two-point Gauss-Radau integration within each hinge region is implemented at two integration points at the element ends and at 8/3 of the hinge length, $Lo = h$, from the end of the element (Fig. 2b).

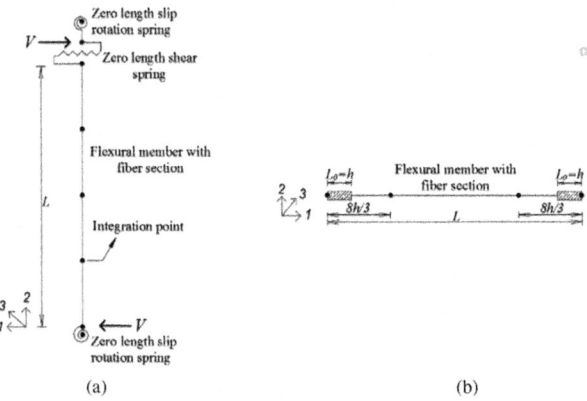

Fig. 2. Elements used for modeling (a) columns and (b) beams.

As far as it regard shear modeling, the shear model by Setzler and Sezen (2008) can capture the shear response with a lateral force-shear displacement envelope, that includes three distinct points corresponding to: (1) Maximum shear strength and corresponding shear displacement; (2) Onset of shear strength degradation and corresponding displacement; (3) Shear displacement at axial load failure. The shear strength is calculated according to the model by Sezen and Moehle (2004):

$$
V_n = V_s + V_c = k\,\frac{A_v f_y d}{s}
$$
$$
+ k\left(\frac{0.5\,\sqrt{f_c}}{a/d}\sqrt{1 + \frac{P}{0.5\,\sqrt{f_c}\,A_g}}\right)0.8\,A_g
\tag{8}
$$

where A_v is the transverse reinforcement area in the loading direction; s is the transverse reinforcement spacing; f_y is the transverse reinforcement yield strength; d is the section depth; $f'c$ is the compressive strength of concrete; a is the shear span of the element; P is the axial load; A_g is the gross area of the section; k is a factor to account for ductility-related strength degradation and it is defined to be equal to 1.0 for displacement ductility less than 2, equal to 0.7 for displacement ductility exceeding 6, and varies linearly for intermediate displacement ductility values.

Shear displacements are calculated using a combination of two existing models (Sezen 2008 and Setzler and Sezen 2008). The shear displacement at peak strength, $\Delta_{v,n}$, is calculated as:

$$
\Delta_{v,n} = \left(\frac{f_y \cdot \rho_l}{5000 \cdot \dfrac{a}{d}\cdot\sqrt{\dfrac{P}{A_g \cdot f'_c}}} - 0.0004\right)\cdot L
\tag{9}
$$

where ρ_l is the longitudinal steel ratio and L is the length of the column.

As described in Sezen (2008), the shear displacement at the onset of shear failure is adopted from Gerin and Adebar (2004). Shear displacement at axial failure is obtained using the procedure given in Setzler and Sezen (2008), which requires the calculation of total lateral displacement. Total lateral drift is calculated using the equation proposed by Elwood and Moehle (2005).

About bar slip model, when a reinforcing bar embedded in concrete is subjected to a tensile force, strain accumulates over the embedded length of the bar. This tensile strain causes the reinforcing bar to slip relative to the concrete in which it is embedded. Slip of column reinforcing bars at column ends (i.e., from the footing or beam-column joint) will cause rigid body rotation of the column. This rotation is not accounted for in flexural analysis, where the column ends are assumed to be fixed. The bar slip model used in this study was originally developed by Sezen and Moehle (2003) and presented in Setzler and Sezen (2008). This model assumes a stepped function for bond stress between the concrete and reinforcing steel over the embedment length of the bar. The bond stress is taken as $1 \cdot \sqrt{f'c}$ MPa for elastic steel strains and as $0.5 \cdot \sqrt{f'c}$ MPa for inelastic steel strains. The rotation due to slip, θ_s, is calculated as $slip/(d-c)$, where slip is the extension of the outermost tension bar from the column end and d and c are the distances from the extreme compression fiber to the centroid of the tension steel and the neutral axis, respectively. The column lateral displacement due to bar slip, Δ_{slip}, is equal to the product of the slip rotation and the column length ($\Delta_{slip} = \theta s \cdot L$).

3.1.2 Total Lateral Response

The total lateral response of a RC column can be modeled using a set of springs in series in OpenSees (where the flexural spring is represented by a fiber section element). The flexure, shear and bar slip deformation models discussed above are each modeled by springs in series. Each spring is subjected to the same lateral force. The total displacement response is the sum of the responses of each spring. The column spring model is shown in Fig. 2. A typical column element includes two zero-length bar slip springs at its ends, one zero-length shear spring and a flexural element with five integration points. The shear behavior is modeled as an uniaxial hysteretic material defined for the spring in the shear direction (i.e., transverse direction of the column or direction 1 in Fig. 2). The longitudinal displacement caused by the bar slip is modeled with two rotational springs at the column ends using an uniaxial hysteretic material (i.e., direction 3 in Fig. 2). Finally, same vertical displacement is maintained between nodes of zero length elements in the vertical direction (i.e., direction 2 in Fig. 2), using the equalDOF option in OpenSees.

The three deformation components are simply added together to predict the total response up to the peak strength of the column (Setzler and Sezen 2008). Rules are established for the post-peak behavior of the springs based on a comparison of the shear strength V_n, the yield strength V_y, and the flexural strength V_p required to reach the plastic moment capacity. By comparing V_n, V_y, and V_p, the columns can be classified into five different categories, as described in Setzler and Sezen (2008): (1) Category I: $V_n < V_y$: the shear strength is less than the lateral load causing yielding in the tension steel. The column fails in shear while the flexural behavior remains elastic; (2) Category II: $V_y < V_n < 0.95 V_p$: the shear strength is greater than the yield strength, but less than the flexural strength of the column. The column fails in shear, but inelastic flexural deformation occurring prior to shear failure affects the post-peak behavior; (3) Category III: $0.95 \cdot V_p < V_n < 1.05 \cdot V_p$: the shear and flexural strengths are very close; (4) Category IV: $1.05 \cdot V_p < V_n < 1.4 \cdot V_p$: the shear strength is greater than the flexural strength of the column. The column experiences large flexural deformations potentially leading to a flexural failure. Inelastic shear deformations affect the post-peak behavior, and shear failure may occur as displacements increase; (5) Category V: $V_n < 1.4 \cdot V_n$: the shear strength is much greater than the flexural strength of the column. The column fails in flexure while the shear behavior remains elastic. Figure 3 shows the three different deformation components and the total lateral displacement for two generic columns of the frame, belonging to two different categories.

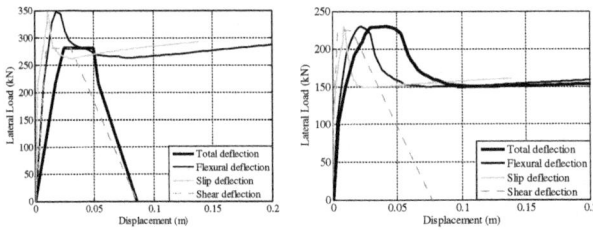

Fig. 3. Three different deformation components and the total lateral displacement for two generic columns of the frame, belonging to Category I (left) and Category III (right).

3.2 Record Selection

A set of 34 strong ground-motion records are selected from the NGA-West2 database (Ancheta et al. 2014). This suite of records covers a wide range of magnitudes between 5.5 and 7.9, and closest distance-to-ruptured area (denoted as RRUP) up to around 40 km. Since the soil shear wave velocity in upper 30 m of soil, Vs30, at the structure's site is around 218 m/sec, all selected records are chosen to be on NEHRP site classes C-D (where C:$360 <$ Vs30 < 760 m/s and D:$180 <$ Vs30 ≤ 360 m/s). The number of records from a single seismic event is limited to one, while only one of the two horizontal components of each recording, with higher spectral acceleration around 1.0 s, is selected. The lowest useable frequency of 0.25 Hz ensures that the low-frequency content is not removed by the ground motion filtering process. The records are selected to be free field or on the ground level without consideration of station housing.

3.3 Cloud Analysis

As explained comprehensively in Sect. 2.2, the Cloud Analysis is a nonlinear dynamic procedure in which the structure is subjected to a set of (un-scaled) ground motion records covering a wide range of IM, herein $Sa(T_1)$, values. Figure 4 shows the Cloud data and the associated Cloud linear regression (fitted to the *NoC* portion of the Cloud data). For each data point (colored squares), the corresponding record ID is shown. It can be see that 7 records out of 34 ground motions cause collapse or global dynamic instability (C data) as shown with red-colored squares. The line $Y = 1$ corresponding to the onset of Near Collapse LS is shown with dashed red line.

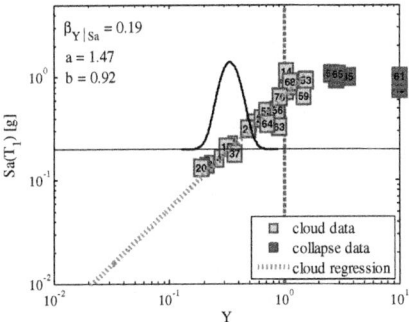

Fig. 4. The cloud regression. (Color figure online)

3.4 IDA Analysis

Each IDA curve herein shows the variation in the performance variable Y for a given ground motion record as a function of $Sa(T_1)$ while the record is scaled-up linearly in amplitude. Each IDA curve has 19 strips. Initially a constant step of 0.1 from 0.1 g to 1.5 g has been adopted. Since the goal is that all IDA curves are able to populate both the $Y < 1$ and $Y > 1$ zones (for the purpose of interpolation of $Sa_{Y=1}$), the records are scaled to four spectral acceleration levels (see the methodology section, in reality one of the levels correspond to the unscaled spectral acceleration). It can be seen that these spectral acceleration levels cover values ranging between 0.4 g and 0.8 g (the zone in which most of the records exceed $Y = 1$). Figure 5 illustrates the complete IDA curves (in thin grey lines) with respect to Y for the suite of 34 ground-motions. The vertical red line plotted at $Y = 1$ demonstrates the dispersion in the spectral acceleration values $Sa_{Y=1}$ plotted as red-star points. The figure also demonstrates the (Log-Normal) probability density function fitted to the $Sa_{Y=1}$ values. The horizontal red-dashed line represents the median of $Sa_{Y=1}$ values (denoted as $\eta_{SaY=1}$) from IDA analysis. In order to facilitate the comparison with Cloud Analysis results, the corresponding Cloud data (the squares) and the regression prediction (blue line) are also plotted. The spectral acceleration value corresponding to $Y = 1$ from the Cloud regression prediction (derived as $\eta_{SaY=1} = (1/a)^{1/b}$, the blue dashed line) represents the median spectral acceleration capacity corresponding to Cloud analysis.

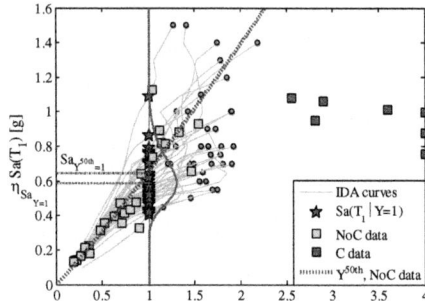

Fig. 5. IDA curves, cloud data, and the regression prediction. (Color figure online)

3.5 Cloud to IDA

As described in Sect. 2.2, *Cloud to IDA* proposes an efficient procedure for performing IDA, which only considers few points (in most cases 4 points) of spectral acceleration for each record, in order to obtain the distribution of $Sa_{Y=1}$. Initially, all the records are scaled to the four points presented in Sect. 2.2. Since, as said, the goal is that all the records are able to populate both the $Y < 1$ and $Y > 1$ zones for producing a good estimation of the distribution of $Sa_{Y=1}$. Just for four records, it was necessary to use a fifth level of scaling equal to $(1/a)^{1/b} \cdot e^{-2\sigma_{Y|Sa,NoC}/b}$. Figure 6 shows the *Cloud to IDA* results.

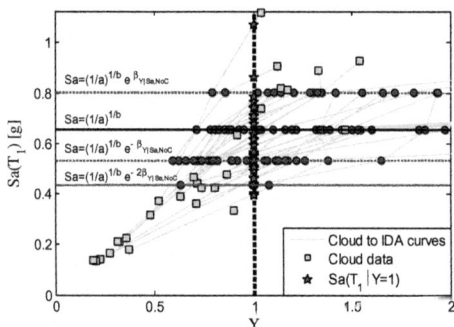

Fig. 6. Cloud to IDA analysis.

3.6 Fragility Curves Results and Comparisons

Figure 7 illustrates a comparison between the three non-linear dynamic procedures (Cloud, *Cloud to IDA* and IDA) discussed herein. Table 1 shows the statistical parameters for the fragility curves, where η is the median value of the fragility curve and β is its standard logarithmic deviation.

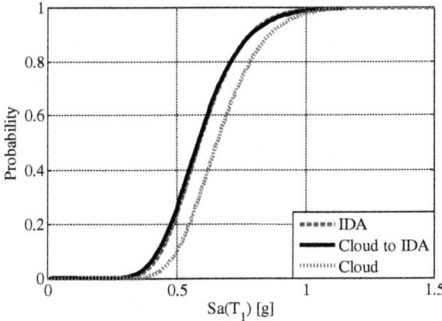

Fig. 7. Fragility curves comparison.

Table 1. Statistical parameters for fragility curves.

Methodology	η(g)	β	Number of analyses
Cloud	0.66	0.22	34
IDA	0.59	0.22	19 • 34
Cloud-IDA	0.59	0.23	4 • 34 + 4

4   Conclusions

Cloud to IDA procedure is proposed herein as an efficient procedure for implementing incremental dynamic analysis (IDA), by exploiting both the data points and the statistic estimates from a simple Cloud Analysis. The transverse frame of a shear-critical seven-storey older RC building in Van Nuys, CA, which is modeled in Opensees with fiber-section considering the flexural-shear-axial interactions and the bar slip, is employed in order to illustrate this procedure. In particular, the procedure of *Cloud to IDA* manages to get the same results (in terms of fragility and for this specific case-study structure) as the complete IDA procedure based on only around 4 data point per ground motion record.

Acknowledgements. This work is supported in part by the executive Project ReLUIS-DPC 2014/2016. This support is gratefully acknowledged.

References

Ancheta TD et al (2014) NGA-West2 Database. Earthq Spectra 30(3):989–1005
Bazzurro P, Cornell CA, Shome N, Carballo JE (1998) Three proposals for characterizing MDOF nonlinear seismic response. J Struct Eng 124(11):1281–1289
Cornell CA, Krawinkler H (2000) Progress and challenges in seismic performance assessment. PEER Cent News 3(2):1–3
Cornell CA, Jalayer F, Hamburger RO, Foutch DA (2002) The probabilistic basis for the 2000 SAC/FEMA steel moment frame guidelines. ASCE J Struct Eng 128:526–533 Special Issue: Steel Moment Resisting Frames after Northridge Part II

Ditlevsen O, Madsen HO (1996) Structural reliability methods. Wiley, New York

Elwood KJ, Moehle JP (2005) Axial capacity model for shear-damaged columns. ACI Struct J 102:578–587

Galanis PH, Moehle JP (2015) Development of collapse indicators for risk assessment of older-type reinforced concrete buildings. Earthq Spectra 31(4):1991–2006

Gerin M, Adebar P (2004) Accounting for shear in seismic analysis of concrete structures. In: 13th world conference on earthquake engineering, Vancouver

Jalayer F (2003) Direct Probabilistic seismic analysis: implementing non-linear dynamic assessments. Ph.D. dissertation, Stanford University, California

Jalayer F, Cornell CA (2003) A technical framework for probability-based demand and capacity factor design (DCFD) seismic formats. Technical report PEER 2003/08, Pacific Earthquake Engineering Research, Berkeley

Jalayer F, Franchin P, Pinto PE (2007) A scalar damage measure for seismic reliability analysis of RC frames. Earthq Eng Struct Dyn 36(13):2059–2079

Jalayer F, Cornell CA (2009) Alternative non-linear demand estimation methods for probability-based seismic assessments. Earthq Eng Struct Dyn 38(8):951–972

Jalayer F, De Risi R, Manfredi G (2015) Bayesian cloud analysis: efficient structural fragility assessment using linear regression. Bull Earthq Eng 13(4):1183–1203

Jalayer F, Ebrahimian H (2017) Seismic risk assessment considering cumulative damage due to aftershocks. Earthq Eng Struct Dyn 46:369–389

Jalayer F, Ebrahimian H, Miano A, Manfredi G, Sezen H (2017) Analytical fragility assessment using unscaled ground motion records. Earthquake Engineering and Structural Dynamics, pp 1–25. https://doi.org/10.1002/eqe.2922

Krawinkler H (2005): Van Nuys hotel building testbed report: exercising seismic performance assessment. Technical report PEER 2005/11, Pacific Earthquake Engineering Research, Berkeley

Islam MS (1996) Analysis of the Northridge earthquake response of a damaged non-ductile concrete frame building. Struct Des Tall Build 5(3):151–182

Miano A, Jalayer F, Ebrahimian H, Prota A (2017) Cloud to IDA: Efficient fragility assessment with limited scaling. Earthquake Engineering and Structural Dynamic, pp 1–24. https://doi.org/10.1002/eqe.3009

Scott MH, Fenves GL (2006) Plastic hinge integration methods for force-based beam–column elements. J Struct Eng 132(2):244–252

Setzler EJ, Sezen H (2008) Model for the lateral behavior of reinforced concrete columns including shear deformations. Earthq Spectra 24(2):493–511

Sezen H, Moehle JP (2003) Bond-slip behavior of reinforced concrete members. In. Proceedings of fib symposium on concrete structures in seismic regions

Sezen H, Moehle JP (2004) Shear strength model for lightly reinforced concrete columns. J Struct Eng 130(11):1692–1703

Sezen H (2008) Shear deformation model for reinforced concrete columns. Struct Eng Mech 28(1): 39–52

Shome N, Cornell CA (1999) Probabilistic seismic demand analysis of nonlinear structures. RMS Program, Stanford University 1999; Report No. RMS35, 320 p

Vamvatsikos D, Cornell CA (2004) Applied incremental dynamic analysis. Earthq Spectra 20(2): 523–553

Lifecycle, Upgrading and Reuse

Lifetime Axial-Bending Capacity
of a R.C. Bridge Pier Cross-Section Subjected
to Corrosion

P. Castaldo[1(✉)], B. Palazzo[2], and A. Mariniello[2]

[1] Department of Structural, Geotechnical and Building Engineering (DISEG),
Politecnico di Torino, Turin, Italy
paolo.castaldo@polito.it
[2] Department of Civil Engineering, University of Salerno, Salerno, Italy

Abstract. Reinforced concrete structures in service may be affected by aging, which may include changes in strength and stiffness assumed in structural design, in particular when the concrete is exposed to an aggressive environment. In this context, this paper provides a computational probabilistic approach to predict the time-evolution of the mechanical and geometrical properties of a statically determinate r.c. structural system (i.e. bridge pier) subjected to corrosion-induced deterioration, due to diffusive attack of chlorides, in order to evaluate its service life. Adopting appropriate degradation models of the material properties, concrete and reinforcing steel, as well as assuming appropriate probability density functions related to mechanical and deterioration parameters, the proposed model is based on Monte Carlo simulations in order to evaluate time variant axial force-bending moment resistance domains, with the aim to estimate the time-variant reliability index. Finally, an application to estimate the expected lifetime of a r.c. bridge pier is described.

Keywords: Reinforced concrete · Time-variant structural reliability
Monte carlo simulations · Corrosion-induced deterioration · Chlorides attack
Lifetime prediction

1 Introduction

Design of reinforced concrete structures is usually aimed at ensuring the performance requirements considering the structural system intact and without taking explicitly into account the progressive deterioration of system properties over time. Reinforced concrete structures in service may be affected by aging, which may include changes over time in strength and stiffness beyond the baseline conditions that are assumed in structural design, in particular when the concrete is exposed to an aggressive environment which may accelerate the risk of structural failure (Verma et al. 2014). Therefore for r.c. structures, the structural performance should be considered time-variant (ISO/DIS 13823 2006, Mori and Ellingwood 1993).

As for a structural system exposed to an aggressive environment, many works (LNEC E-465 2005, Fitch et al. 1995) have been focused on the modelling aspects of diffusion, corrosion and damaging process as well as of the life service prediction. Among the

© Springer International Publishing AG, part of Springer Nature 2018
M. di Prisco and M. Menegotto (Eds.): ICD 2016, LNCE 10, pp. 371–384, 2018.
https://doi.org/10.1007/978-3-319-78936-1_27

several studies and models developed over the years, Life-365 (Thomas and Bentz 2000; Ehlen et al. 2009) is a computer-based model, developed by the American Concrete Institute (ACI), useful to estimate the service life and life-cycle cost analysis of r.c. structures. Biondini et al. (2004, 2006, 2008); Biondini and Vergani (2012) proposed a general approach to the limit analysis of plane framed structures as well as to the probabilistic prediction of the lifetime of r.c. frames with respect to structural collapse. In particular, the structural system has been considered to be exposed to an aggressive environment and the effects of the structural damaging process have been described by the corresponding evolution in time of the axial force-bending moment resistance domains by modelling both mechanical behaviour and aging process of structural elements through a cellular approach and defining appropriate damage indices. Moreover, they also presented a formulation of a three-dimensional r.c. deteriorating beam finite element for nonlinear analysis of concrete structures under corrosion and investigated at a cross-sectional level the time evolution of the uncertainty effects associated with the different parameters defining the probabilistic structural performance of two existing cable-stayed bridges in Italy subjected to deterioration process by means of time-dependent sensitivity factors based on a regression of the simulation results (Biondini et al. 2008). It is noteworthy that, due to the uncertainties in material and geometrical properties, in the magnitude and distribution of the loads, in the physical parameters which define the deterioration process, the time-variant structural safety should realistically be assured only in probabilistic terms.

Within the time-variant reliability assessment of aging r.c. structures exposed to an aggressive environment, this paper provides a reliability-based approach to predict the time-evolution of the mechanical and geometrical properties of a statically determinate r.c. structural system (i.e. bridge pier) subjected to corrosion-induced deterioration in order to evaluate its service life. Following the same approach proposed in some of the abovementioned studies (Biondini et al. 2004, 2006, 2008; Biondini and Vergani 2012), also in this work, through Monte Carlo simulation the behaviour and resistance capacity of the most stressed structural cross sections are analyzed over time considering all the effects due to the chloride-induced deterioration process. Differently, this study provides the reliability assessment by investigating how the material deterioration influences the section mechanical response in probabilistic terms considering different eccentricity values of the axial force. In particular, the structural response of r.c. systems is investigated by modelling the mechanical behaviour through a fiber-approach and considering, within the sectional analysis, some effects of the deterioration process due to the chloride-induced reinforcement corrosion (Biondini and Vergani 2012; DuraCrete 1998; CEB-FIB Task Group 2006; Samson et al. 2005; Marchand 2000), by adopting some of the several models proposed in literature. Note that other degradation effects such as buckling phenomena, shear capacity reduction, loss of steel-concrete bond strength and diffusion-chloride corrosion interaction are herein neglected. The proposed methodology is based on developing Monte Carlo simulations, by assuming appropriate probability density functions related to mechanical and deterioration parameters, in order to define axial force-bending moment resistance domains at different time instants, with the aim to estimate the time-variant reliability of the structural element through the reliability index β (Cornell 1969, Ditlevsen and Madsen 2004, Eurocode 0 2002) both for different axial force eccentricity values and coefficients of variation (COV) values of the probability density functions (PDFs) related to loads and considering different values of the design reliability index β_d. In particular, for each axial force eccentricity value, the probability density function related to axial force/bending moment resistance is considered time-variant whereas the probability density function related to loads (axial force/bending moment actions) is assumed time-invariant. It follows that, assuming a target reliability index β_{target} and COV values of the probability density functions related to actions as well as β_d (design reliability index) values, it is possible to evaluate the service life, or residual service life, of the concrete element subjected to corrosion for each eccentricity value (Palazzo et al. (2015a, b)). Finally, an application of the proposed prediction model to estimate the lifetime of a deteriorating reinforced concrete bridge pier is discussed as example useful to describe the details of the proposed approach.

2 Deterioration Modelling Due to Reinforcement Corrosion

Within reinforced concrete structures, among the most common environmental deterioration factors, reinforcement corrosion, generally associated to carbonation (inducing uniform corrosion) and chloride diffusion (inducing pitting corrosion), is considered the most significant degradation mechanism (Collepardi 2010, Broomfield 1997) and, as described in the model proposed by Tuutti (1982), it is basically a two-phase process consisting of the initiation and propagation phases. Thus, corrosion may affect a r.c. structure inducing many effects such as loss of steel cross-sectional area, loss of steel bars ductility, concrete strength reduction, degradation of concrete cover, cover spalling and loss of steel-concrete bond strength. All the effects of the material deterioration begin only when reinforcement corrosion has begun. In order to model the abovementioned effects, it is necessary to determinate the corrosion initiation time, that is the time required by chloride content at rebar depth to reach the threshold value C_{crit}.

Although in most cases the diffusion process is more properly described by two- or three-dimensional patterns of concentration gradients depending on the geometrical aspect ratio of the cross-section, location of reinforcing steel bars and exposure conditions (Titi and Biondini 2016), the chloride ingress in concrete is herein modeled using the Fick's diffusion equation in a simplified one-dimensional (1D) form through the law proposed by DuraCrete (DuraCrete 1998; CEB-FIB Task Group 5.6 2006), employed also in similar studies (Ghosh and Padgett 2010; Kashani et al. 2012; Pitiliakis et al. 2014) and expressed by Eq. (1):

$$C(x,t) = C_S \left[1 - erf \left(\frac{x}{2\sqrt{k_t \cdot k_e \cdot D_{RCM} \cdot \left(\frac{t_0}{t}\right)^n}} \right) \right] \tag{1}$$

where C(x,t) is the chloride concentration (expressed in kg/m^3) at the cover depth x (expressed in m) and at the time t(s), C_s is the equilibrium chloride concentration at the concrete surface expressed as a percentage by weight of cement (wt% cement), k_t is the transfer variable defined deterministically according to FIB-CEB Task Group 5.6 (2006) equal to 1, k_e is the environmental transfer variable, DRCM is the chloride migration coefficient (expressed in m^2/s), t_0 is the reference point of time and is equal to 0.0767 years (28 days); t is the elapsed time (years), n is the aging exponent and erf is the Gaussian error function. In order to investigate the corrosion effects as functions depending on time, corrosion rate needs to be predicted by adopting the model proposed by Liu e Weyers (Liu and Weyers 1998), that considers several different environmental factors and is described by Eq. (2):

$$i_{corr} = 0.926 \cdot \exp(7.98 + 0.7771 \cdot \ln(1.69 \cdot C_t) + \\ - \tfrac{3006}{T} - 0.000116R_c + 2.24t^{-0.215}) \tag{2}$$

where i_{corr} is the corrosion rate (expressed in $\mu A/cm^2$), C_t is the chloride concentration (wt% cement), given by Eq. (1), T is the temperature (K) at rebar depth; R_c is the electrical resistance of the concrete cover (Ω); t is the time (years). The term R_c is given by the Eq. (3):

$$R_c = \exp[8.03 - 0.549 \cdot \ln(1 + 1.69 \cdot C_t)] \tag{3}$$

3 Lifetime Prediction Methodology

The proposed prediction model is based on modelling both axial-flexural mechanical behaviour and corrosion-induced aging process of a r.c. structural element through a fiber model of the cross section subjected to maximum actions in terms of axial force-bending moment by considering the non-linear behaviour of the materials and taking into account the following degradation-induced effects:

- loss of steel cross-sectional area, evaluated through the model of Val et al. (1998), which takes into account the pitting corrosion;
- loss of steel bars ductility, taken into account through the reduction of the steel ultimate deformation ε_{su} by using the experimental model proposed by Biondini and Vergani (2012);
- concrete strength reduction, degradation of concrete cover and cover spalling, evaluated according to the model proposed by Coronelli and Gambarova (2004). Within the fiber-model, as long as the loss of steel-concrete bond strength cannot be taken into account, a limit strain value of the reinforcing steel bar equal to 1% is assumed as also adopted by Biondini et al. (2008); Biondini and Vergani (2012).

The first step consists of providing the input data, i.e. the geometrical dimensions of the cross-section, the design life, as well as defining the PDFs of each random variable related both to model the corrosion-induced degradation process (i.e., temperature, surface chloride concentration, etc.) and mechanical properties of the materials (i.e., concrete and steel strength).

The second step consists of modelling the linear and non-linear mechanical behaviour of materials by defining the stress-strain relationships both for reinforced concrete and reinforcement steel, also considering effects of lateral confinement for concrete (Saenz 1964; Kent and Park1971; NTC 2008; Okamura et al. 1985).

The third step consists of developing Monte Carlo simulations. In particular, at the initial time instant ($t_0 = 0$ years), the materials mechanical properties and the deterioration modelling parameters are obtained for each sample generated within the Monte Carlo simulations. The number of the generated samples should be high to give validity to the probabilistic model, and, obviously, the accuracy increases with the sample size. It follows that the number of the generated samples should be increased until the statistical values of the results, evaluated at step 5, are not characterized by significant variations anymore.

Subsequently, considering the relations between the different effects due to the degradation, within the fourth step, the time-evolution of the different properties of the material (reinforced concrete) and the variation of the geometry of the resistant section of the structural element are modelled for each time instant along its design life.

The fifth step consists of evaluating the effects of the deterioration process over time by defining the bending moment-curvature diagrams $M–\chi$ for different eccentricity values at each instant time in order to define the resistance domain $N_R(t)–M_R(t)$ (axial force-bending moment) variable over time (time-variant resistance domain). Using appropriate COV values of the PDFs regarding actions, referring to the design reliability requirements β_d provided by the codes and assuming that the structure is designed according to the principle of cost minimization, it is also possible to obtain the actions domain $N_S–M_S$ at time t_0 (initial time), supposed time-invariant.

Finally, in the last step, by comparing actions domains (time-invariant) and resistance domains (time-variant), the reliability index $\beta(t)$ at each time instant within the design life can be estimated for different axial force eccentricity values and considering different COV values of the PDFs related to actions as well as also for different β_d values. It follows that the expected structural lifetime T, associated with a prescribed target reliability level β_{target}, can be evaluated.

4 Case Study

The interlocking cross-section of a bridge pier in Memphis (USA) (Hwang et al. 2001) is only assumed as reference section to apply the proposed approach. The concrete column bent consists of four 4.57 m high, 0.914 m diameter (D) columns. The cross section of the column is shown in Fig. 1.

The vertical reinforcing steel of the column consists of 17, 22 mm diameter and grade 40 vertical bars. The cover depth is 50 mm and the stirrups consist of 10 mm bars with 200 mm spacing. A relatively aggressive atmospheric exposure environment is assumed for an adverse chloride induced deterioration scenario, corresponding to the exposure classes XD3 or XS1 defined by the European Standard (UNI EN 206-1 2006), and a w/c ratio equal to 0.5 is adopted (DuraCrete 1998). It is also assumed that the cross section is characterized by homogeneous properties as well as the parameters related to the environmental exposure are uniform around the circumference of the pier section and that the corrosion uniformly affects each reinforcing bar. From these hypotheses, it follows that the degradation process diffusion and effects such as the cover spalling occur at the same time and way within the concrete cover of the cross section without considering

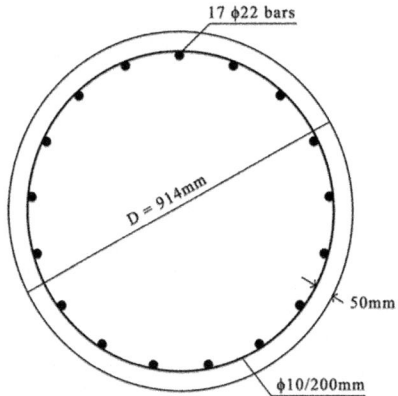

Fig. 1. Cross section of the examined column.

localization effects which can captured by FEM or cellular approaches (Biondini et al. 2004, 2008, Titi and Biondini 2016). The statistical quantification of the model parameters describing the chloride induced corrosion and the mechanical properties of materials adopted in the present study is provided in Table 1. Note that the temperature effects on the concrete behaviour (Ripani et al. 2014, 2016; Vrech et al. 2015; Mroginski et al. 2015; Etse et al. 2013, 2014, 2015, 2016) are not considered. Assuming a design life of 100 years, a usual expected lifetime of bridges (NTC 2008), time intervals of 10 years are analyzed, thus, 11 time instants are examined, from t_0, corresponding to the initial time (end of construction of the structural element), to t_{10}, corresponding to 100 years, in order to develop the time-variant analysis aimed at predicting the lifetime T of the considered bridge pier column.

Table 1. Cumulative distribution functions (CDFs) and corresponding parameters.

Parameter	Distribution	Mean (μ)	COV
Temperature of the structural element or the ambient air T, K	Normal	290.35	0.20
Chloride migration coefficient D_{RCM}, m^2/s	Normal	1.58 E−11	0.20
Critical chloride concentration C_{cr}, %	Beta truncated (limits: 0.2–2.2)	0.4	0.25 (a = 9.2) (b = 13.8)
Aging exponent n	Beta	0.3	0.05 (a = 279.7) (b = 652.6)
Environmental function k_e	Normal	0.67	0.10
Surface chloride concentration C_s, %	Normal	1.2825	0.20
Concrete strength f_c, MPa	Normal truncated (lower limit = 0)	31	0.20
Steel strength f_{sy}, MPa	Lognormal	336	0.11

4.1 Deteriorating Process and Monte Carlo Simulations

For each sample generated through the Monte Carlo simulations, the evolution of the deteriorating process is evaluated during the design life. Therefore, for each time instant, the results are processed in probabilistic terms, by assuming normal or lognormal probabilistic distributions (Fotopolulou et al. 2012; Biondini et al. 2014). The first factor to be determined is the corrosion initiation time, through the Eq. (1), by adopting a lognormal distribution with mean and standard deviation values equal, respectively, 29.8 years and 25.5 years. Once the protective passive film around the reinforcement dissolves due to continued chloride ingress, corrosion and all its induced effects occur. Assuming a normal distribution, within the reinforcing steel of rebars and stirrups, the time evolution of the mean value of the cross-sectional area $A(t)$ is computed. Figure 2 shows the time evolution of the mean value and the PDFs of the cross-sectional area $A(t)$. As expected, the variability in the loss of area $A(t)$ tends to increase with the time due also to the cover spalling which increases the rate of corrosion. At 100 years, the loss of sectional area is about 13%. As shown in Fig. 3, another effect of the pitting corrosion is the loss of steel ductility. It begins at about 40 years and, assuming a normal distribution, there is an average loss of about 22% after 100 years.

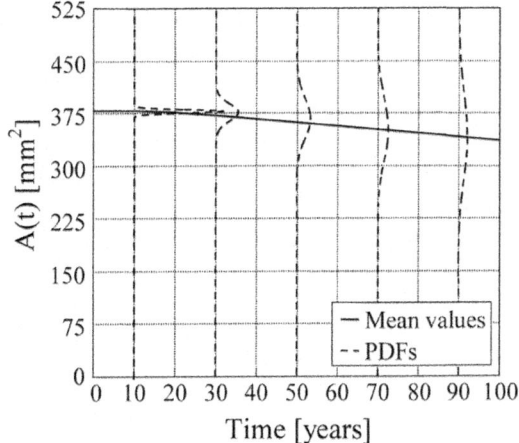

Fig. 2. Time evolution of mean values related to the cross sectional area of the reinforcing steel $A(t)$.

As described in the previous sections, Monte Carlo simulations have been performed in order to obtain the bending moment-curvature diagrams M–χ, at each time instant for different axial force eccentricity values and, then, define the resistance domain $N_R(t)$–$M_R(t)$ (axial force-bending moment resistance) at each time instant. The structural reliability is issued through the reliability index β as explained in the previous section. In order to estimate the lifetime T of the structural element, the reliability index or failure probability, evaluated at each time instant during the aging process, is compared to a

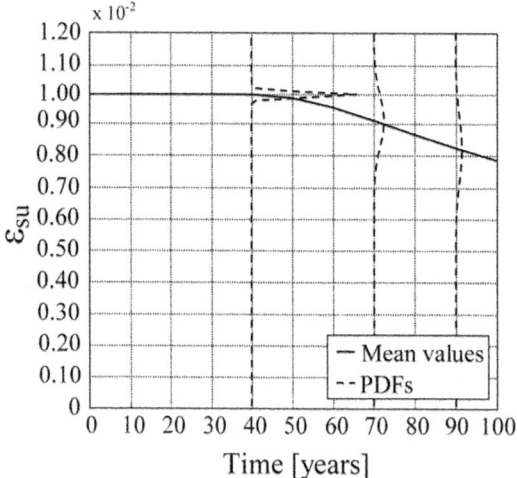

Fig. 3. Time evolution of mean values of the steel ultimate deformation ε_{su}.

prescribed target reliability index or failure probability related to a period of interest equal to 1 year, since the corrosion process is not treated as stochastic process.

Considering the design reliability index β_d equal to 5.2 and 4.2, related to the reliability classes RC3 and RC1 respectively as suggested by Eurocode 0 (2002), two cases of the coefficient of variation COV, equal to 0.2 and 0.1 respectively, according to COV values related to external loads PDFs (Ellingwood et al. 1982), have been considered to define the mean and standard deviation values of the PDFs related to actions for each axial force eccentricity value at the initial instant time. More details may also be found in Palazzo et al. (2015a, b). In Fig. 4, the values of the eccentricity with the corresponding indices are shown.

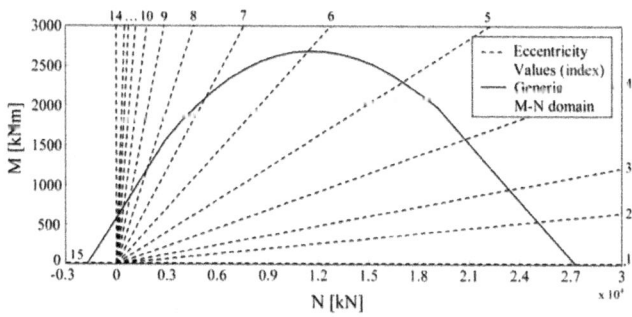

Fig. 4. Straight lines indicating the axial force eccentricity values employed in the analyses.

Assuming the above-mentioned PDFs as time-invariant, it has been possible to evaluate the structural reliability index $\beta(t)$ of the bridge pier over time for the different values of the axial force eccentricity. Figures 5 and 6 show the value of $\beta(t)$ and failure

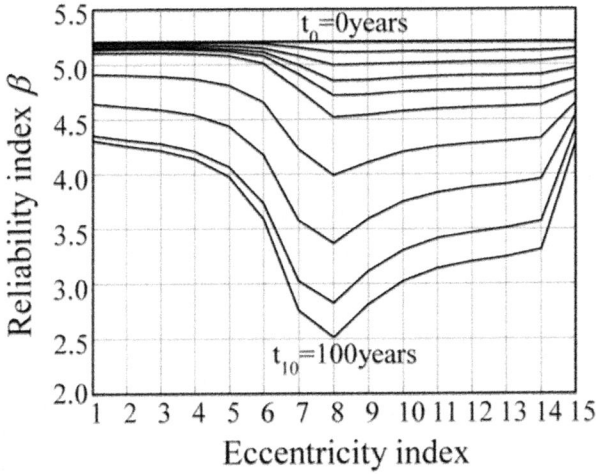

Fig. 5. Time evolution of the reliability index $\beta(t)$ with $\beta_d = 5.2$ and COV = 0.2.

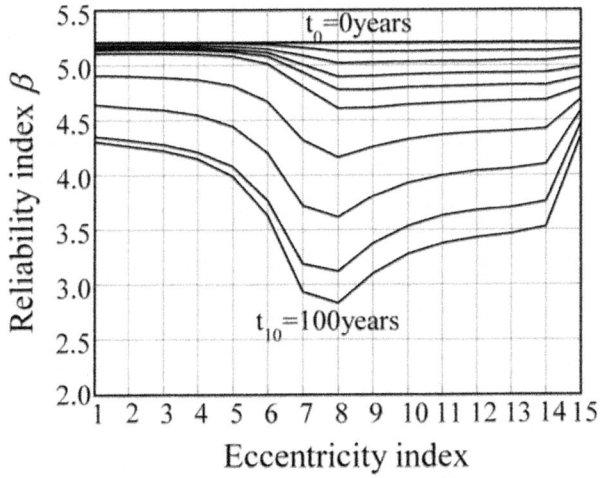

Fig. 6. Time evolution of the reliability index $\beta(t)$ with $\beta_d = 5.2$ and COV = 0.1.

probability for every eccentricity value and time instant, referring to COV value equal to 0.2 and 0.1 respectively. Note that there is a reduction of $\beta(t)$ over time, increase of the failure probability, particularly for average eccentricity values, for which the moments reach higher values. In the worst case, the probability of failure gets down from the initial value of 10^{-7} to about 10^{-3}. By comparing the 2 cases, it is obvious how in the second one there is a more accentuated reduction of the β values. This implies that the reliability decreases with decreasing of the COV characterizing the PDFs of the actions. Finally, the expected structural lifetime T associated to a prescribed reliability level β_{target} is evaluated.

Table 2. Expected lifetime T considering β_d = 5.2 and COV = 0.1.

Expected lifetime T [years] – β_d = 5.2 (P_f = 10^{-7}) – COV = 0.1			
Eccentricity index	β_{target} = 3.09 (P_f = 10^{-3})	β_{target} = 3.72 (P_f = 10^{-4})	β_{target} = 4.27 (P_f = 10^{-5})
1	100	100	100
2	100	100	90
3	100	100	90
4	100	100	80
5	100	100	80
6	100	90	70
7	80	70	60
8	80	70	60
9	90	70	60
10	90	80	60
11	100	80	60
12	100	80	70
13	100	80	70
14	100	80	70
15	100	100	100

Table 3. Expected lifetime T considering β_d = 5.2 and COV = 0.2.

Expected lifetime T [years] – β_d = 5.2 (P_f = 10^{-7}) – COV = 0.2			
Eccentricity index	β_{target} = 3.09 (P_f = 10^{-3})	β_{target} = 3.72 (P_f = 10^{-4})	β_{target} = 4.27 (P_f = 10^{-5})
1	100	100	100
2	100	100	90
3	100	100	90
4	100	100	80
5	100	100	80
6	100	90	70
7	90	70	70
8	90	70	60
9	100	80	60
10	100	80	70
11	100	80	70
12	100	80	70
13	100	80	70
14	100	90	70
15	100	100	100

Tables 2 and 3 show the case related to the value of the design reliability index, $\beta_d = 5.2$ and various thresholds of the target reliability index β_{target}, such as 3.09 ($P_f = 10^{-3}$), 3.72 ($P_f = 10^{-4}$) and 4.27 ($P_f = 10^{-5}$). Analyzing the different results obtained concerning the expected life of the structural element, it can be observed that in the extreme case, with high reliability limit values ($\beta_{target} = 4.27$), the expected life is equal to 60 years. In all other examined cases, however, the prediction of the expected life is equal or exceeds 70 years. The results presented in this study demonstrate that a structural element, such as a bridge pier, subjected to high values of the bending moment actions (average axial force eccentricities) is mainly affected by the chloride-induced deterioration effects which lead to a strong reduction of the safety level. Indeed, structural retrofit aimed at reducing the bending moment actions, such as seismic isolation technique (Castaldo and Tubaldi 2015; Castaldo et al. 2015; 2016a, b, c; Castaldo and Ripani 2016; Palazzo et al. 2014), or protective and maintenance measures are necessary.

5 Conclusions

The structural lifetime of a deteriorating reinforced concrete element (i.e., bridge pier), exposed to an aggressive environment, is investigated by proposing a computational fiber-approach to predict the time-evolution of the mechanical and geometrical properties of the corresponding most stressed cross section considering the damaging process induced by the diffusive attack of chlorides. The prediction model is based on Monte Carlo simulations, by assuming appropriate probability density functions related to mechanical and deterioration parameters, in order to define time-variant axial force-bending moment resistance domains with the aim to estimate the time-variant reliability of the structural element through the reliability index β and investigate how the material deterioration influences the section mechanical response in probabilistic terms considering different axial force eccentricity values. In particular, assuming a target reliability index β_{target} and COV values of the probability density functions related to loads as well as β_d values, it is possible to evaluate the service life, or residual service life, of the concrete element subjected to corrosion for each axial force eccentricity value. On the basis of the hypotheses as well as effectiveness and accuracy of the laws employed to model the deterioration process, the application of the proposed model to estimate the lifetime of a deteriorating reinforced concrete bridge pier, exposed to an aggressive environment corresponding to the exposure classes XD3 or XS1, shows that the proposed approach predicts the time-variant structural resistance and the corresponding remaining lifetime that can be assured under prescribed reliability levels without maintenance. From the obtained results it is possible observe that the structural elements subjected to high values of the bending moment actions (average axial force eccentricities) are mainly affected by the chloride-induced deterioration effects. As for existing r.c. infrastructure in aggressive environments, it follows that the proposed approach can be used to evaluate their service lives and establish the optimal time instant when structural retrofit or protective and maintenance measures are necessary, based on economical analysis, in order to elongate their lifetimes. As for new r.c. infrastructure in aggressive environments, the proposed model can also be used in order to design structural system less vulnerable to corrosion process provided that economic benefits exist.

References

Biondini F, Vergani M (2012) Damage modeling and nonlinear analysis of concrete bridges under corrosion. In: Proceedings of the 6th IABMAS 2012 bridge maintenance, safety, management, resilience and sustainability, Stresa, Italy. CRC Press, Taylor and Francis Group

Biondini F, Bontempi F, Frangopol DM, Malerba PG (2004) Cellular automata approach to durability analysis of concrete structures in aggressive environments. ASCE J Struct Eng 130 (11):1724–1737

Biondini F, Bontempi F, Frangopol DM, Malerba PG (2006) Probabilistic service life assessment and maintenance planning of concrete structures. ASCE J Struct Eng 132(5):810–825

Biondini F, Camnasio E, Palermo A (2014) Lifetime seismic performance of concrete bridges exposed to corrosion. J Struct Eng 10(7):880–900

Biondini F, Frangopol DM, Malerba PG (2008) Uncertainty effects on life time structural performance of cable-stayed bridges. Prob Eng Mech 23(4):509–522

Broomfield JP (1997) Corrosion of steel in concrete: understanding investigat. and repair. E&FN SPON, New York

Castaldo P, Tubaldi E (2015) Influence of FPS bearing properties on the seismic performance of base-isolated structures. Earthq Eng Struct Dyn 44(15):2817–2836

Castaldo P, Ripani M (2016) Optimal design of friction pendulum system properties for isolated structures considering different soil conditions. Soil Dyn Earth Eng 90:74–87

Castaldo P, Amendola G, Palazzo B (2016a) Seismic fragility and reliability of structures isolated by friction pendulum devices: seismic reliability-based design (SRBD). Earthq Eng Struct Dyn. https://doi.org/10.1002/eqe.2798

Castaldo P, Palazzo B, Della Vecchia P (2015) Seismic reliability of base-isolated structures with friction pendulum bearings. Eng Struct 95:80–93

Castaldo P, Palazzo B, Della Vecchia P (2016b) Life-cycle cost and seismic reliability analysis of 3D systems equipped with FPS for different isolation degrees. Eng Struct 125:349–363

Castaldo P, Palazzo B, Ferrentino T, Petrone G (2016c) Influence of the strength reduction factor on the seismic reliability of structures with FPS considering intermediate PGA/PGV ratios. Compos. Part B. https://doi.org/10.1016/j.compositesb.2016.09.072

CEB-FIB Task Group 5.6 (2006) Model Code for service life design. Fédération Internationale du Béton FIB, Bulletin 31

Collepardi M, Collepardi S, Ogoumah Olagot JJ, Simonelli F, Trioli R (2010) Diagnosi del degrado e restauro delle strutture in c.a.. Tintoretto

Cornell CA (1969) A probability-based structural code. ACI-J 66:974–985

Coronelli D, Gambarova P (2004) Structural assessment of corroded reinforced concrete beams: modeling guidelines. J Struct Eng 130(8):1214–1224

Ditlevsen O, Madsen HO (2004) Structural reliability methods. Maritime Engineering, Department of Mechanical Engineering, Technical University of Denmark

DuraCrete (1998) Modelling of Degradation. EU-Project No.BE95-1347, Probabilistic Performance based Durability Design of Concrete Structures, Report 4–5

Ehlen MA, Thomas MDA, Bentz EC (2009) Life-365 service life prediction model version 2.0 – widely used software helps assess uncertainties in concrete service life and life-cycle costs. Concr Int 31(2):41–46

Ellingwood B, MacGregor JG, Galambos TV, Cornell CA (1982) Probability-based load criteria: load factors and load combinations. J Struct Div 108(ST5):978–997

EN 1990 2002: Eurocode 0 - Basis of Structural Design

Etse G, Ripani M, Mroginski JL (2014) Computational failure analysis of concrete under high temperature. Comput Model Concr Struct - Proc EURO-C 2014(2):715–722

Etse G, Vrech SM, Ripani M (2016) Constitutive theory for recycled aggregate concretes subjected to high temperature. Constr Build Mater 111:43–53

Etse G, Ripani M, Caggiano A, Schicchi DS (2015) Strength and durability of concrete subjected to high temperature: continuous and discrete constitutive approaches. American Concrete Institute, ACI Special Publication 2015-January (SP 305), pp 9.1–9.18

Etse GJ, Ripani M, Vrech, SM (2013) Fracture energy-based thermodynamically consistent gradient model for concrete under high temperature. In: Proceedings of the 8th international conference on fracture mechanics of concrete and concrete structures, FraMCoS 2013, pp 1506–1515

Fitch MG, Weyers RE, Johnson SD (1995) Determination of end of functional service life for concrete bridge decks. Transp Res Rec 1490:60–66

Fotopolulou SD, Karapetrou ST, Pitilakis KD (2012) Seismic vulnerability of RC buildings considering SSI and aging effects. In: 15th world conference on earthquake engineering, Lisbon, Portugal

Ghosh J, Padgett JE (2010) Aging considerations in the development of time-dependent seismic fragility curves. J Struct Eng ASCE 136(12):1497–1511 http://www.usclimatedata.com

Hwang H, Liu JB, Chiu YH (2001) Seismic fragility analysis of highway bridges. Mid-America Earthquake Center, Technical report, MAEC RR-4 Project, Center for Earthquake Research and Information, University of Memphis

ISO/DIS 13823 (2006) General principles on the design of structures for durability

Kashani MM, Crewe AJ, Alexander NA (2012) Durability considerations in performance-based seismic assessment of deteriorated RC bridges. In: 15th world conference on earthquake engineering, Lisbon, Portugal

Kent DC, Park R (1971) Flexural members with confined concrete. J Struct Div 97(7):1969–1990

Liu Y, Weyers R (1998) Modeling the time-to-corrosion cracking in chloride contaminated reinforced concrete structures. ACI Mater J 95:675–680

LNEC E-465 (2005) Concrete prescriptive methodology to estimate concrete properties to achieve the design service life under environment conditions XC or XS. National Laboratory of Civil Engineering (LNEC), Lisbon, Portugal

Marchand J (2000) Modeling and behaviour of unsaturated cement systems exposed to aggressive chemical environmental. Mater Struct 34:195–200

Mori Y, Ellingwood BR (1993) Reliability-based service-life assessment of aging concrete structures. J Struct Eng 119:1600–1621

Mroginski JL, Etse G, Ripani M (2015) A non-isothermal consolidation model for gradient-based poroplasticity. In: PANACM 2015-1st pan-American congress on computational mechanics, in conjunction with the 11th Argentine congress on computational mechanics MECOM 2015, pp 75–88

Norme Tecniche per le Costruzioni (2008) D.M. 14 January 2008

Okamura H, Maekawa K, Sivasubramaniyam S (1985) Verification of modeling for reinforced concrete finite element. In: Finite element analysis of reinforced concrete structures, proceedings of the seminar ASCE, pp 528–543

Palazzo B, Castaldo P, Della Vecchia P (2014) Seismic reliability analysis of base-isolated structures with friction pendulum system, In: IEEE workshop on environmental, energy and structural monitoring systems proceedings, Napoli, 17–18 September 2014

Palazzo B, Castaldo P, Mariniello A (2015b) Time-variant reliability of R.C. structures. In: ACE 2015 - the 2nd international symposium on advances in civil and infrastructure engineering, 12–13 June, pp 1–8

Palazzo B, Castaldo P, Mariniello A (2015a) Time-variant structural reliability of R.C. structures affected by chloride-induced deterioration. American Concrete Institute, ACI Special Publication - January (SP 305), pp 19.1–19.10

Pitiliakis KD, Karapetrou ST, Fotopoulou SD (2014) Consideration of aging and SSi effects on seismic vulnerability assessment of RC buildings. Bull Earthq Eng 12(4):1755–1776

Ripani M, Etse G, Vrech S, Mroginski J (2014) Thermodynamic gradient-based poroplastic theory for concrete under high temperatures. Int J Plast 61:157–177

Ripani M, Etse G, Vrech S (2016) Recycled aggregate concrete: localized failure assessment in thermodynamically consistent non-local plasticity framework. Comput Struct 178:47–57. https://doi.org/10.1016/j.compstruc.2016.08.007

Saenz LP (1964) Discussion of "equation for the stress-strain curve of concrete" by Desayi and Krishnan. ACI J 61(9):1229–1235

Samson E, Marchand J, Snyder KA, Beaudoin JJ (2005) Modeling ion and fluid transport in unsaturated cement systems in isothermal conditions. Cem Concr Res 35:141–153

Thomas MDA, Bentz EC (2000) Life-365: computer program for predicting the service life and life-cycle cost of reinforced concrete exposed to chloride. Prod. Manual

Titi A, Biondini F (2016) On the accuracy of diffusion models for life-cycle assessment of concrete structures. Struct Infrastruct Eng 12(9):1202–1215

Tuutti K (1982) Corrosion of steel in concrete. Swedish Cement and Concrete Research Institute, Stockholm Report 4

UNI EN 206-1 (2006) Specificazione, prestazione, produzione e conformità

Val DV, Stewart MG, Melchers RE (1998) Effect of reinforcement corrosion on reliability of highway bridges. Eng Struct 20(11):1010–1019

Verma SK, Bhadauria SS, Akhtar S (2014) Probabilistic evaluation of service life for reinforced concrete structures. Chin J Eng 2014:8 Article ID 648438

Vrech SM, Ripani M, Etse G (2015) Localized versus diffused failure modes in concrete subjected to high temperature. In: PANACM 2015–1st pan-American congress on computational Mechanics, in conjunction with the 11th Argentine congress on computational mechanics, MECOM 2015, pp 225–236

The Effect of Alternative Retrofit Strategies on Reduction of Expected Losses: Evaluation with Detailed and Simplified Approach

M. Gaetani d'Aragona, M. Polese[(✉)], M. Di Ludovico, and A. Prota

Department of Structures for Engineering and Architecture,
University of Naples "Federico II", Naples, Italy
maria.polese@unina.it

Abstract. Limitation of monetary losses due to earthquakes can improve resilience in developed countries. Computation of losses with the PEER performance-based earthquake engineering framework (Porter 2003) normally entails performing a number of Non-linear Response History analyses as a basis for assessment of expected Engineering Demand Parameters and associated losses, that is an elaborate and time-consuming task. In (ATC 2012) an alternative quicker approach relying on simplified modeling and analysis with SPO2IDA is also envisaged. This paper tests the applicability of simplified method using pushover based analysis and CSM method. In particular, it compares the losses obtained for a non-conforming reinforced concrete moment frame building starting from the classical approach, i.e. based on Non-linear Response History analyses, with the one computed with pushover-based assessment. Moreover, it evaluates the reduction of losses after building retrofit with the pushover-based analysis.

Keywords: Economic losses · Pushover · Retrofit

1 Introduction

Seismic risk is a key factor influencing several important decision-making processes from performance-based design of new structures, to investments for rehabilitation of existing buildings and even to seismic mitigation campaigns of large building stocks. Within the PEER performance-based earthquake engineering (PBEE) framework (Porter 2003) the economic losses are adopted as one of the metrics for measuring seismic risk. In the framework, direct economic losses, that include costs for repairing or replacing damaged buildings, are strongly influenced by the probability of replacement of building (i.e. probability of collapse). Further, if a mitigation action is implemented, the earthquake economic losses are generally reduced, while the costs of building retrofit should be included in the evaluation process (Liel and Deierlein 2013; Polese et al. 2017; Gaetani d'Aragona et al. 2017). In FEMA P-58 (ATC 2012), the implementation of the PEER framework is described considering either Non-linear Response History (NLRH) based analyses, or a quicker approach relying on simplified

© Springer International Publishing AG, part of Springer Nature 2018
M. di Prisco and M. Menegotto (Eds.): ICD 2016, LNCE 10, pp. 385–399, 2018.
https://doi.org/10.1007/978-3-319-78936-1_28

modeling and analysis of the building response that involves the adoption of the SPO2IDA (Vamvatsikos and Cornell 2006) for computation of building fragility.

Starting from the idea of simplifying the computational effort normally required for implementation of the PEER framework, the present paper tests the applicability of standard pushover analysis in combination to the capacity spectrum method, CSM, (Fajfar 1999) as a basis for the assessment of engineering demand parameters (EDPs) and related damage and losses. The results of detailed evaluation with standard PEER PBEE approach, entailing the execution of a number of Nonlinear Response History (NLRH) analyses and evaluation of EDPs, are compared with results obtained through pushover based assessment of EDPs for a case study non-ductile reinforced concrete (RC) building. In addition, the effect of alternative retrofit strategies is investigated and the variation of expected losses estimated, showing encouraging results on the possibility to adopt pushover-based analyses for estimation of expected costs.

The following section briefly describe the different methods for computation of economic losses. Next, Sect. 3 presents an application of the detailed and simplified assessment, comparing the results. Finally, Sect. 4 describe the design of possible retrofit schemes and the evaluation of seismic vulnerability reduction as well as of variation of losses with both detailed and simplified approaches.

2   Computation of Losses Within the PEER PBEE Framework

The PEER PBEE framework, that is probably the most refined procedure currently available, allows quantifying different seismic performance measures such as expected number of injuries or Deaths, economic losses (e.g. expressed in terms of Dollars) and time to recovery or Downtime (the 3Ds). The general approach to loss estimation relies on structural analysis to calculate engineering demand parameters (EDPs) such as deformations and accelerations throughout the structure during an earthquake, which are used to predict the damage in structural, non-structural components, and building contents. Damages are the base for computation of expected losses in terms of one or more of the established metrics (e.g., economic losses in dollars or casualties). The framework operates with four analysis stages: hazard analysis, structural analysis, damage analysis and loss analysis, as synthetically shown in Fig. 1.

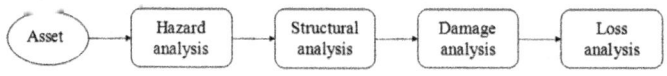

Fig. 1. PEER framework

The outcome of each analysis is then integrated using the Total Probability Theorem, allowing to take into account combined numerical integration of all the conditional probabilities and to propagate the uncertainties from one level of analysis to the next, resulting in probabilistic prediction of performance. As noted in Calvi et al. (2014) the PEER framework is very flexible, since no restrictions are imposed on the approach used to quantify hazard, to perform structural analysis and to relate EDPs to losses and other performance measures. Of course, the results will depend on the assumptions made in applying the procedure and the risk parameters of interest. Different applications of the PEER framework, implementing component-based (e.g. Aslani and Miranda 2009; Ramirez et al. 2012) or storey-based damage assessment (e.g. Ramirez and Miranda 2009), are available. In this study, a component-based approach is adopted.

In Sect. 2.1 the implementation of the PEER framework proposed by Yang et al. (2009) is synthesized, while Sect. 2.2 describes the main differences in the application when pushover based structural analyses for evaluation of EDPs are employed.

2.1 Implementation of the Performance-Assessment Framework for Repair Cost Evaluation

The performance assessment procedure implemented in Yang et al. (2009) starts from the identification within the considered facility of the relevant performance groups (PGs), i.e. groups of structural or nonstructural components whose performance is similarly affected by a particular EDP, e.g. acceleration at a given story or IDR_{max} at another story.

For each PG significant damage states that have a clear correlation with repair actions (and related costs) shall be identified and suitable damage model defined; the latter are expressed in terms of component-level fragility curves, relating EDP (e.g. maximum interstorey drift ratio, IDR_{max}) to the probability of attaining different damage states for the assigned component type (e.g., Fig. 2). In order to allow for quantitative evaluation of the needed repair actions, the number of components or their measurable extension within each PG of the considered facility is estimated.

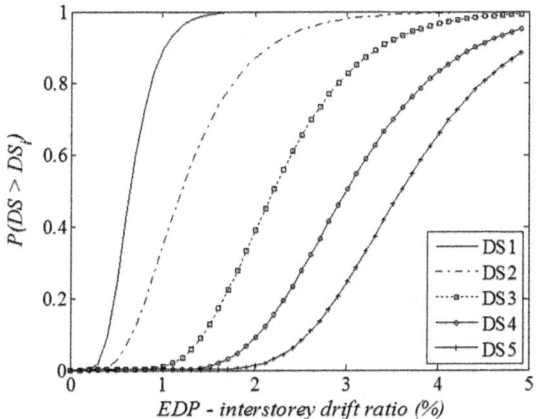

Fig. 2. Example of adopted fragility curve

After selection, e.g. with conventional seismic hazard analysis, of suites of ground motion records, the response of the building is evaluated using NLRH analysis. To allow statistical characterization of EDP at various intensity levels the authors suggest the execution of a number of NLRH analyses with suites of ground motions and for increasing level of hazard. The peak EDP values obtained with NLRH analyses for each record are summarized in an EDP matrix (one matrix for each intensity level); the latter has one column for each considered EDP and one row for each record. Starting from EDP matrices evaluated with structural analysis, additional EDPs are estimated with a simulation-based method that, assuming a joint lognormal distribution among the parameters, enables the generation of large numbers of artificial EDP vectors that have the same statistical distribution as the data that were derived with NRH analyses. This way, large sample of realizations necessary to perform loss evaluations are obtained, and additional sources of uncertainties can be accounted for ATC (2012). The generic EDP realization corresponds to a state of damage. In order to select among possible DSs, given the EDP, a Montecarlo simulation is performed, extracting a random number from the uniform distribution over the interval [0, 1]. For example, if the uniform random number generator produces a number of 0.6 and the EDP value shown in Fig. 2 is 2.5%, the PG is in DS3.

Finally, considering the repair quantities for each item in the PG, multiplying by a unit repair cost (for each DS) and summing over all items, the total repair cost for the building is computed.

2.2   Pushover Based Assessment

A key aspect in the procedure described in Sect. 2.1 is the execution of a number of structural analyses performing NLRH with a suite of ground motion records and for different levels of hazard, i.e. for increasing intensity levels. On the other hand, a simpler approach for practitioners is desirable (Polese et al. 2013).

Static nonlinear pushover PO, under its applicability hypotheses, allows to investigate a building's nonlinear behavior, assessing damage progression and related variation of displacement profile for increasing levels of seismic demand. Moreover, SPO allows investigation of damage distribution within the structural system, hence representing a useful tool in the assessment of post-earthquake condition. SPO may be employed for simplified response calculation; by applying the CSM to the capacity curve resulting from pushover analysis (Fajfar 1999), the building response to earthquakes of assigned spectral shapes may be found. Scaling the demand spectrum, median expected drift profiles for increasing levels of earthquake intensity can be obtained, hence EDPs in terms of IDR_{max} are determined. Median peak floor acceleration (PFA), can be derived using existing proposals in literature (Petrone et al. 2015; Vukobratović and Fajfar 2016; Sullivan et al. 2013; ATC 2012). According to FEMA P-58 (ATC 2012), PFA at floor i can be calculated as a function of peak ground acceleration (PGA) and an acceleration correction factor H_{ai} (see Eq. 1) depending on the strength ratio S, the elastic period of the building T, building height H and the height above the effective base of the building to floor level i, h_i.

$$a_i^* = H_{ai}(S, T, h_i, H) \cdot PGA \quad i = 2 \text{ to } N + 1 \tag{1}$$

Statistical distributions of the relevant EDPs (e.g. IDR_{max} of PFA at the different floors) can be derived hypothesizing they have a lognormal distribution and considering the obtained median values (one for each considered intensity level) and suitable dispersions β. By introducing values of dispersion due to modeling uncertainties β_m and inherent randomness associated to earthquake variability, β_Δ or β_a for drift or acceleration respectively, a global value of β (indicated as β_D) may be calculated and EDP probabilistic distribution obtained for each considered intensity level (see e.g. Eqs. 2 – valid to assess drift dispersion - and 3). In Eq. 3, $\hat{d}_{IM}$ is the median value of drift demand obtained from application of CSM using the demand spectrum scaled at the generic intensity level IM.

$$\beta_D = \sqrt{\beta_\Delta^2 + \beta_m^2} \tag{2}$$

$$P\big[drift > d|IM\big] = \Phi\left[\frac{1}{\beta} \cdot \ln\left(\frac{d}{\hat{d}_{IM}}\right)\right] \tag{3}$$

Suitable values of β_m and of β_Δ or β_a are suggested in FEMA P-58 as a function of available information for modeling (for β_m) or period and strength ratio S (for β_Δ and β_a).

Once the probabilistic distribution of EDP (e.g. IDR_{max}) at each level of considered intensity are obtained, they can be used similarly to the approach described in Sect. 2.1 for computation of expected losses.

3   Application for a Case Study Building

3.1   Building Model

The building selected for this study is a seven-story five-bay frame representative of mid-rise non-ductile reinforced concrete building, see Fig. 3, extracted from the perimeter moment resisting frame of the Van Nuys Holiday Inn located in Los Angeles, California.

The concrete nominal strength is $f_c' = 5$ ksi for the first story columns, 4 ksi for the second story columns and second floor beams, and 3 ksi for columns from third story to the seventh and for beams from the third floor to the roof. Column sections is constant along the height and equal to $14'' \times 20''$. For further details, please refer to Krawinkler (2005).

A two-dimensional finite element multi-degree of freedom model developed using OpenSees (2016) is adopted to simulate the seismic response of the building. Beams and columns are modelled using the nonlinear beam-column element (Spacone et al. 1992). The joints are modeled adopting the "scissor model" by Alath and Kunnath (1995), including a pinching hysteric behavior (Hassan and Mohele 2012) to account for the nonlinear shear deformation of the joint. Shear and axial failures of columns are explicitly modeled using the Limit State material (Elwood 2004). P-Δ effects are not included.

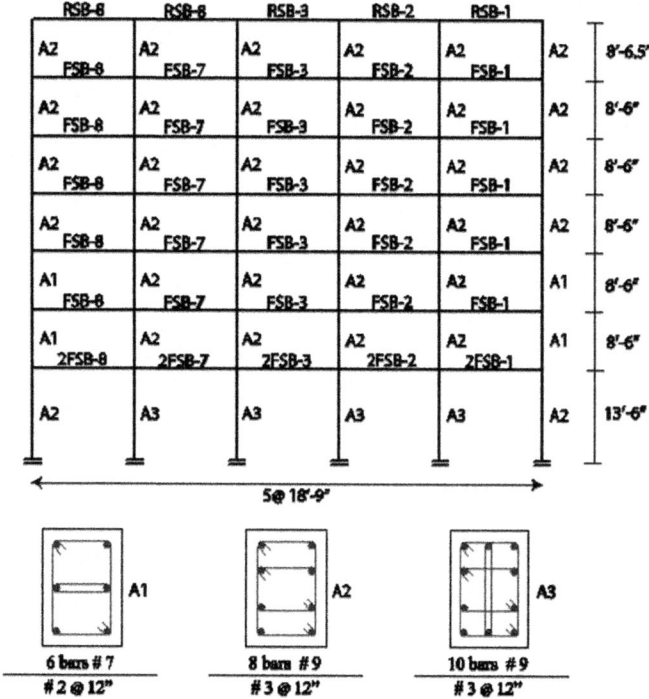

Fig. 3. Schematic view of the frame with element number and column rebar arrangement

The eigenvalue analysis of the original structure provides a fundamental vibration period of 1.05 s. A damping of 2% is assigned to the first and third modes using Rayleigh damping.

To capture the actual capacity for the original non-ductile frame, this study considers two possible system-level collapse mechanisms: Side-sway collapse (SSC) and Gravity load collapse (GLC). The SSC occurs when a single storey has reached its capacity to withstand lateral loads (i.e., when every column in a given floor has reached its residual shear capacity at the same time). GLC occurs when vertical load demand exceeds the total vertical load capacity at a given storey. Collapse is detected based on a comparison of storey-level gravity load demands and capacities (adjusted at each time step to account for member damage and load redistribution). An internal algorithm monitors the dynamically varying capacity of each element and checks the GLC and SSC criteria throughout each nonlinear time history analysis to detect the collapse (with a precision of 0.05 g). The collapse is considered as the first between GLC, SSC.

3.2 NLRH-Based Analysis and Costs Evaluation

In this section, NLRH analyses are initially adopted to estimate building's collapse capacity through successive scaling until collapse (Vamvatsikos and Cornell 2002) and to estimate building response in terms of EDPs. NLRH analyses are performed for increasing intensities of the seismic action appropriately scaling each record to cover the entire range of structural response, from elastic to ultimate response. In order to perform NLRH analyses, the set of 30 ground motions adopted in Vamvatsikos and Cornell (2005) is adopted.

Applying the procedure outlined in Sect. 2.1, and considering a total of 1000 realizations for each intensity level to obtain stable cost estimates, total repair costs are obtained.

Figure 4, gray lines, shows the complementary cumulative distribution function (CCDF) of the total repair cost of the building normalized by building replacement cost (including expected demolition costs), c_r, obtained through the dynamic approach; seven different intensity levels ranging from 0.1 g to 0.7 g are considered.

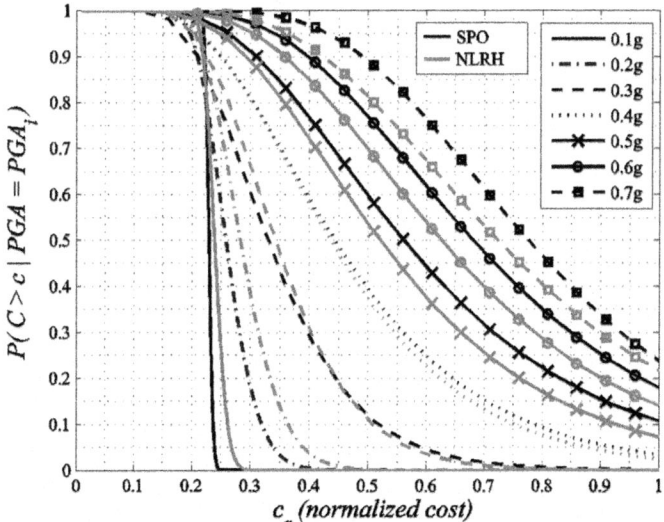

Fig. 4. CCDFs of total normalized repair costs (c_r) for different PGAs obtained through NLRH (gray) and SPO (black) procedure

Figure 4 shows that with the increase of the damaging action the repair cost inflates making the curve translate rightward, while dispersion of results increases. It can be noted that a median c_r of at least 24% is expected even for a PGA = 0.1 g; this can be explained considering the high incidence of nonstructural components and contents that significantly contribute to damage and costs even for lower levels of the seismic action. It is worth to note that the c_r is strongly influenced by the number of collapse cases occurred during the simulation process; consequently, its influence increases as PGA increases (i.e., the probability of collapse increases).

3.3   Pushover Based Analysis and Costs Evaluation

Mass proportional and 1st mode proportional SPO are performed. Both analyses indicate the activation of a 1st story collapse mechanism, with attainment of gravity load collapse at a roof displacement Δ_{Top} of 25.7 and 28.1 cm respectively, corresponding to drifts of 6.24 and 6.84%.

Figure 5 shows the two pushover curves along with the results of the NLRH analysis, expressed in terms of roof displacement vs base shear, for record Imperial Valley scaled at an intensity of PGA = 0.5 g. First story collapse mechanism is attained for 93% of the considered records.

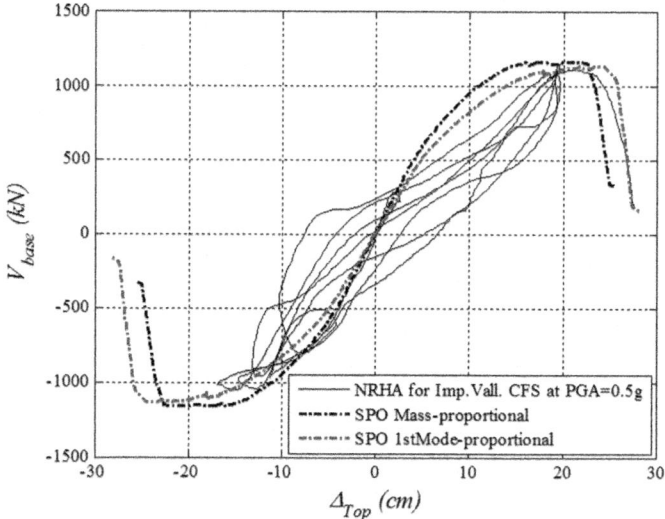

Fig. 5. Comparison between Mass and 1st Mode-proportional SPO and NRHA for Imperial Valley 1979 record scaled at PGA = 0.5 g.

The capacity curves obtained from SPO analyses are transformed in multi-linear curves representing the backbone curve of equivalent SDOF system and IN2 method (Dolšek and Fajfar 2005) is applied to compute building capacity and to obtain seismic demand for earthquakes of varying intensity level. The spectrum utilized to apply IN2 method is the mean spectrum of the 30 records utilized for NLRH analyses. Seismic capacity is expressed in terms of peak ground acceleration and indicated as $PGA_c(g)$.

In particular, the mean result from both SPO analyses is considered, obtaining $PGA_{c,SPO} = 0.61$ g. Median capacity from NLRH analyses (indicated as η in Fig. 5) is $PGA_{c,NLRH} = 0.64$ g.

Figure 6 shows the collapse fragility curves obtained from NLRH analyses and with SPO based approach. The latter curve is built considering $PGA_{c,SPO}$ as median capacity ($=η_{SPO}$) and the logarithmic dispersion $\beta = 0.5$, as indicated in Kosič et al. (2014) for mid-rise old RC frame buildings. It can be noted that the fragility curve representation obtained starting from SPO analyses allows a fair approximation of NLRH based curve.

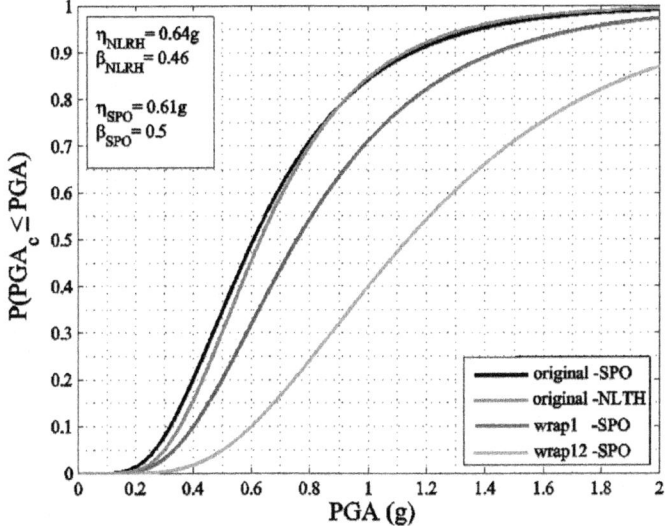

Fig. 6. Collapse fragilities for the original building obtained through NLRH and SPO approach and for the two retrofit configurations (wrap1 and wrap12) through SPO

In Fig. 4, curves obtained with the SPO approach are depicted in black along with c_r curves from NLRH approach. Curves obtained adopting SPO approach show a very similar trend to curves obtained with NLRH. In particular, the SPO-based curves slightly underestimate total repair costs for lower intensities, while lead to an overestimation of costs for higher intensities.

However, the scatter between detailed and simplified approach is very low. In fact, the higher error between two calculations is attained for PGA = 0.7 g, where the ratio between the static and the dynamic c_r is about 9.5%.

4   Building Retrofit and Evaluation of Losses

4.1   Design of Building Retrofit

The retrofit strategy consists in the application of CFRP layers to enhance the structural performance of RC columns. The wrapping is applied to all columns of selected stories and columns are fully wrapped for their entire height to provide an efficient confinement effect. Column wrapping is designed in order to prevent columns shear failure after yielding. In particular, the ACI 369R-11 (2011) is adopted in the design process in order to provide a shear resistance such that flexural failure is ensured (ACI 440 2008). Two different wrapping configurations are adopted: in the first configuration (*wrap1*) columns at 1st story are fully wrapped, while in the second (*wrap12*) columns at 1st and from 2nd story are fully wrapped, allowing a higher increment of displacement capacity.

When CFRP wrapping is applied, an additional collapse criterion is considered: when in that storey a small increase in ground-shaking intensity causes a large increase in lateral drift response, or when IDR_{max} exceeds the 10% threshold.

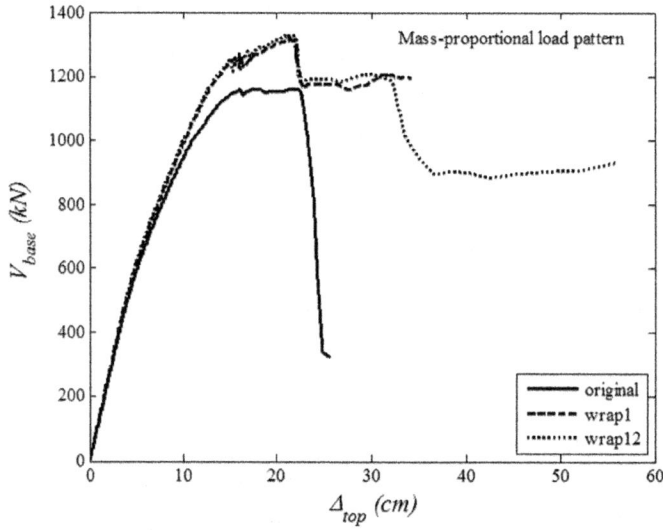

Fig. 7. SPO curves obtained with Mass-proportional load pattern for original building and for the two retrofit configurations (wrap1 and wrap12)

4.2   Reduction of Seismic Vulnerability

Figures 7 and 8 show the SPO curves obtained with Mass-proportional and 1st Mode-proportional load patterns for the three building configurations.

As it can be seen, in both cases the retrofit intervention allows an increase of building capacity, more sensible in the case of wrap12. Coherently, the collapse fragility curves obtained using IN2 (i.e. determining the median capacity with IN2 and considering lognormal standard deviation β), shown in Fig. 6, evidence a higher shift of the curve of wrap12 with respect to original building if compared to the shift of wrap 1.

In fact, the median capacity for wrap1 increases from the original value of $PGA_{c,SPO} = 0.61$ g to $PGA_{c,SPO,wrap1} = 0.76$ g while for the wrap12 configuration it increases up to $PGA_{c,SPO,wrap12} = 1.14$ g.

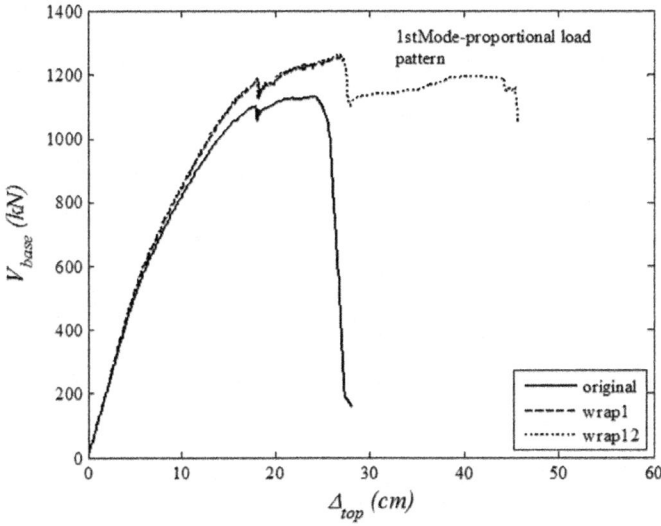

Fig. 8. SPO curves obtained with 1st Mode-proportional load pattern for original building and for the two retrofit configurations (wrap1 and wrap12)

4.3 Expected Losses for Retrofitted Building

Total repair costs are calculated for three building configurations and different earthquake intensities adopting the SPO approach and the procedure outlined in Sect. 2.2. For each simulation, the cost of retrofit is added to total repair costs. Figure 9 shows the CCDFs of the c_r obtained through the SPO approach for three different PGAs and three considered building configurations. Figure 9 shows that the effectiveness of the retrofit strategy varies depending on the earthquake scenario. In particular, wrapping results to be ineffective for lower intensities where repair costs are mainly ascribable to damage in acceleration-sensitive components and the probability of collapse is negligible. Since the applied strategies do not modify the dynamic response of the building very similar results for three building configurations are expected. As the intensity of ground motion increases, the influence of collapse cases on total repair cost becomes more noticeable. For this reason, the effectiveness of retrofit increases since the probability of collapse, conditioned on a given PGA, reduces for wrap1 and wrap12. For instance, for PGA = 0.7 g the median value of normalized repair cost drops from 79% for the original building, to 64% for wrap1 and 57% for wrap12.

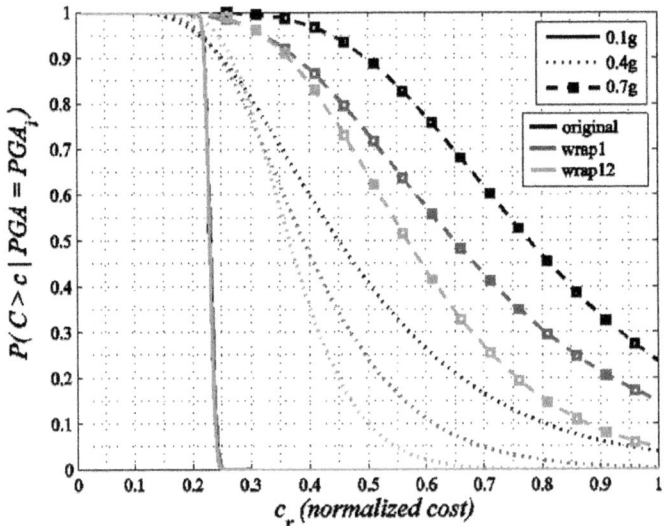

Fig. 9. CCDFs of total normalized repair costs (c_r) for different PGAs obtained through SPO procedure

5 Conclusions

The study presented in this paper aims at contributing in the evaluation of the usability of a simplified SPO-based approach for the computation of total repair costs.

The efficiency of the SPO approach is demonstrated through an application to a case-study building representative of an existing non-ductile reinforced concrete frame. The response of the frame is simulated using a refined 2D finite element model that properly accounts for possible brittle failures of structural members and captures typical non-ductile RC frames failure modes.

The first comparison in terms of collapse fragility shows that the SPO, coupled to the IN2 method (Dolšek and Fajfar 2005), fairly captures the median capacity of the building with an error of just 3%.

Then, the results in terms of total repair costs obtained starting from the classical NLRH approach are compared to those obtained through a SPO-based approach. For both cases, total repair costs are calculated adopting the PEER PBEE framework (Porter 2003) using a Montecarlo simulation. Additional EDPs are generated using a simulation-based method (Yang et al. 2009). In the static approach the IN2 method for increasing intensities is applied and the EDPs in terms of interstorey drift are calculated, peak floor accelerations are obtained adopting simplified formulas (ATC 2012). Suitable dispersions are adopted to account for uncertainties not accounted for in SPO.

The results show that a normalized repair cost (c_r) of at least 24% are expected even for the lower considered intensity, due to the high incidence of nonstructural components and contents even for lower levels of the seismic action. Further, c_r is strongly influenced by the number of collapse cases occurred during the simulation process; consequently, the incidence of collapse increases as the PGA increases.

The SPO approach shows encouraging results leading to very similar losses obtained through the more refined method. In fact, the scatter between two methods is limited, and the higher error is attained for PGA = 0.7 g, where the ratio between the SPO and the NLRH median c_r is of about 9.5%. This suggest that, although the results SPO-based procedure will be inevitably affected by a certain degree of approximation with respect to NLRH analyses, the level of approximation is justified in the view of a significant reduction of computational burden.

Finally, the SPO approach is implemented considering two possible retrofit schemes adopting CFRP wrapping in the first (wrap1) and first two storeys (wrap12). In both cases the retrofit intervention allows an increase of median building capacity, particularly noticeable for wrap12 (about 87%).

The results of cost analysis shows that, the effectiveness of the wrapping scheme varies depending on the earthquake scenario. For lower intensities both retrofit strategies do not lead to a reduction of repair costs since they are mainly ascribable to damage to acceleration-sensitive components, while the building response in the elastic range is not modified by the retrofit (i.e., the peak floor accelerations do not vary). The effect becomes more relevant for higher intensities, since the probability of collapse significantly reduces for these intensities leading to an important reduction of repair costs. For instance, for PGA = 0.7 g the median value of c_r drops from 79% for the original building, to 64% for wrap1 and 57% for wrap12. Consequently, as expected, the adoption of retrofit strategies increasing ductility capacity does not significantly vary c_r, except the portion depending on costs related to collapse.

Additional studies are required to assess the advantages of other alternative strategies in terms of seismic performances and to further simplify the computation of repair costs for an easier application to large building inventories.

Acknowledgements. This study was performed within the framework of the PON METROPO LIS "Metodologie e tecnologie integrate e sostenibili per l'adattamento e la sicurezza di sistemi urbani" grant n. PON03PE_00093_4 and the joint program DPC-Reluis 2014-2016 Task 3.3: Reparability limit state and damage cumulated effects. The S.Co.P.E. computing infrastructure at the University of Naples Federico II was used for the parallel computing.

References

ACI 440.2R-08 (2008) Guide for the design and construction of externally bonded FRP systems for strengthening concrete structures. Report by ACI Committee 440.2R-08. American Concrete Institute, Farmington Hills

ACI 369R-11 (2011) Guide for seismic rehabilitation of existing concrete frame buildings and commentary. Report by ACI committee 369. American Concrete Institute

Alath S, Kunnath SK (1995) Modeling inelastic shear deformation in RC beam-column joints. In: Proceedings of 10th conference on engineering mechanics, Boulder, CO, USA, 21–24 May. ASCE, pp 822–825

Aslani H, Miranda E (2009) Probabilistic earthquake loss estimation and loss disaggregation in buildings, Report No. 157. John A. Blume Earthquake Engineering Center, Stanford University, Stanford

ATC (2012) FEMA P-58-1: Seismic performance assessment of buildings. Volume 1– methodology

Calvi GM, Sullivan TJ, Welch DP (2014) A seismic performance classification framework to provide increased seismic resilience. In: Perspectives on European earthquake engineering and seismology, pp 361–400

Dolšek M, Fajfar P (2005) Simplified non-linear seismic analysis of infilled reinforced concrete frames. Earthquake Eng Struct Dynam 34(1):49–66

Elwood KJ (2004) Modelling failures in existing reinforced concrete columns. Can J Civ Eng 31:846–859

Fajfar P (1999) Capacity spectrum method based on inelastic demand spectra. Earthquake Eng Struct Dynam 28:979–993

Gaetani d'Aragona M, Polese M, Di Ludovico M (2017) Building retrofit prior to damaging earthquakes: reduction of residual capacity and repair costs. In: 16th world conference on earthquake engineering, Santiago, Chile, 9–13 January 2017

Hassan WM, Moehle JP (2012) A cyclic nonlinear macro model for numerical simulation of beam-column joints in existing concrete buildings. In: Proceedings of the 15th world conference of earthquake engineering

Kosič M, Fajfar P, Dolšek M (2014) Approximate seismic risk assessment of building structures with explicit consideration of uncertainties. Earthquake Eng Struct Dynam 43(10):1483–1502

Krawinkler H (2005) Van Nuys hotel building testbed report: exercising seismic performance assessment, PEER 2005/11, University of California, Berkeley, October 2005

Liel AB, Deierlein GG (2013) Cost-benefit evaluation of seismic risk mitigation alternatives for older concrete frame buildings. Earthq Spectra 29(4):1391–1411

OpenSees (2016) Open system for earthquake engineering simulation OpenSees framework-Version 2.5.0. Univ. of California, Berkeley

Petrone C, Magliulo G, Manfredi G (2015) Seismic demand on light acceleration-sensitive nonstructural components in European reinforced concrete buildings. Earth Eng Struct Dyn 44(8):1203–1217

Polese M, Gaetani d'Aragona M, Prota A, Manfredi G (2013) Seismic behavior of damaged buildings: a comparison of static and dynamic nonlinear approach, paper # 1134. In: Proceedings of COMPDYN 2013 4th ECCOMAS thematic conference on computational methods in structural dynamics and earthquake engineering, Kos Island, Greece, 12–14 June 2013

Polese M, Marcolini M, Gaetani d'Aragona M, Cosenza E (2017) Reconstruction policies: explicitating the link of decisions thresholds to safety level and costs for RC buildings. Bull Earthquake Eng 15(2):759–785

Porter KA (2003) An overview of PEER's performance-based earthquake engineering methodology. In: Proceedings of 9th international conference on applications of statistics and probability in civil engineering

Ramirez CM, Miranda E (2009) Building-specific loss estimation methods & tools for simplified performance-based earthquake engineering report no. 171, John A. Blume Earthquake Engineering Research Center, Stanford University, Stanford

Ramirez CM, Liel AB, Mitrani-Reiser J, Haselton CB, Spear AD, Steiner J et al (2012) Expected earthquake damage and repair costs in reinforced concrete frame buildings. Earthquake Eng Struct Dynam 41(11):1455–1475

Spacone E, Ciampi V, Filippou FC (1992) A beam element for seismic damage analysis. Report No. UCB/EERC-92/07. Earthquake Engineering Research Center, College of Engineering, University of California, Berkeley

Sullivan TJ, Calvi PM, Nascimbene R (2013) Towards improved floor spectra estimates for seismic design. Earthq Struct 4(1):109–132

Vamvatsikos D, Cornell A (2002) Incremental dynamic analysis. Earthq Eng Struct Dyn 31(3): 491–514

Vamvatsikos D, Cornell A (2005) Seismic performance, capacity and reliability of structures as seen through incremental dynamic analysis. Report no. 151, John A. Blume Earthquake Engineering Research Center, Stanford University

Vamvatsikos D, Cornell A (2006) Direct estimation of the seismic demand and capacity of oscillators with multi-linear static pushovers through IDA. Earthquake Eng Struct Dynam 35:1097–1117

Vukobratović V, Fajfar P (2016) A method for the direct estimation of floor acceleration spectra for elastic and inelastic MDOF structures. Earthquake Eng Struct Dynam 45(15):2495–2511

Yang TY, Moehle J, Stojadinovic B, Der Kiureghian A (2009) Seismic performance evaluation of facilities: methodology and implementation. J Struct Eng 135(10):1146–1154

R. C. Beams Strengthened on Shear
with FRP and FRCM Composites

J. H. Gonzalez-Libreros[1], C. Pellegrino[1], and G. Giacomin[2(✉)]

[1] Department of Civil, Environmental and Architectural Engineering,
University of Padua, Padua, Italy
[2] G&P Intech S.r.l., Altavilla Vicentina, Italy
amministrazione@gpintech.com

Abstract. This paper presents the results obtained during an experimental campaign carried out in order to investigate the behaviour of reinforced concrete beams strengthened on shear with composite materials. In order to compare their performance, two different strengthening techniques were used: Fibre Reinforced polymers (FRP) and Fibre Reinforced Cementitious Matrix (FRCM) composites. The beams were simple supported and a traditional four point bending scheme was used to execute the tests. Midspan displacements were measured using linear variable transducers (LVDTS) while strain gauges were placed on the stirrups to measure strains.

Keywords: FRP · FRCM · Reinforced · Concrete · Shear · Strengthening

1 Introduction

The need of strengthening existing reinforced concrete (RC) structures arises from their deterioration with age, change in applied loads due to modification of their original use or upgrading to currently available design codes, among other factors. Traditional strengthening techniques, that may include the enlargement of concrete sections and use of steel parts, are still used worldwide but, although they can consider structurally adequate, their execution is usually not time/cost efficient. For this reason, there is a growing interest in researchers to develop strengthening techniques that overcome the drawbacks associated to traditional strengthening systems. Among these techniques, Fibre Reinforced Polymer (FRP) composites have proven to be an adequate solution. FRP composites provide additional strength for flexural, shear, and confinement applications which combined with their low invasiveness, high strength, and ease of application, minimize some disadvantages associated with traditional strengthening techniques such as increase in self-weight of the structure, undesirable change in stiffness, and handling of heavy steel parts.

FRP composites started to be developed during the 80's and applications of this technique can be found worldwide nowadays. The research carried out in the topic during the previous decades has allowed researchers to develop design codes and guidelines that are used today by designers all over the world (ACI440, *fib* bulletin 14, among others). However, an important effort in order to fully understand the behaviour of

© Springer International Publishing AG, part of Springer Nature 2018
M. di Prisco and M. Menegotto (Eds.): ICD 2016, LNCE 10, pp. 400–410, 2018.
https://doi.org/10.1007/978-3-319-78936-1_29

structures strengthened in this fashion is still required. In fact, as pointed out by several researchers, current design equations for predicting the contribution of FRP composites to shear capacity of R.C. elements are not able to replicate the behaviour observed in experimental tests (Sas et al. 2009; Lima and Barros 2011). Most codes assume that the individual contribution of the FRP system should be added to the contribution of concrete and steel reinforcement. However, several researchers (Bousselham and Chaallal 2004; Pellegrino and Modena 2006; Chen et al. 2010) have found that the maximum shear contributions of steel stirrups and FRP are not reached simultaneously implying that their combined contribution is actually less than the simple summation of their respective values and therefore the interaction between internal steel reinforcement and external FRP reinforcement should be taken into account to properly predict the overall shear strength of a strengthened element. Pellegrino and Vasic (2013) also showed that the shear crack angle may have a significant influence in the prediction of the FRP shear strength contribution, which affects the performance of codes such as the ACI440 which specify a constant shear crack angle of 45°.

In addition, some drawbacks of the use of FRP composites, linked mainly to the use of the epoxy resins, such as poor behaviour at temperatures close or above the glass transition temperature, poor compatibility with the substrate, inability to apply the system on wet surfaces or at low temperatures, and inconvenience to carry out post-earthquake assessment of damaged structures have been reported (Al-Salloum et al. 2012). For this reason, advanced composites in which the resin matrix is replaced by a cementitious matrix, known as Fibre Reinforced Cementitious Matrix (FRCM) composites, have recently raised interest in researchers worldwide. Although research performed on the topic is still scarce, the effectiveness of this technique for flexural, shear strengthening and confinement of axially/eccentrically loaded RC elements is confirmed by available experimental evidence (e.g., Triantafillou et al. 2006; Blanksvard et al. 2009; Pellegrino and D'Antino 2013; and Ombres 2015).

Regarding R.C. beams strengthened with shear, Triantafillou and Papanicolaou (2006) carried out test on R.C. beams strengthened in shear with textile reinforced mortar (TRM) and found that the TRM jackets were able to provide additional shear strength comparable to FRP jackets. In fact, depending on the number of layers, the system was able to change the type of failure from shear to a more ductile flexure type.

Blanksvard et al. (2009) observed that the strain in the stirrups in R.C. beams decreased in strengthened elements when compare to unstrengthened beams, for a certain load. This result implies that the interaction between internal and external shear reinforcement is also present for FRCM strengthened elements. A similar behaviour was also witnessed by Ombres (2015).

More recently, Tetta et al. (2015) found that the shear failure in FRCM strengthened beams maybe caused by slippage of the vertical fibre rovings through the mortar, partial rupture of the fibres crossing the shear crack or a combination of both. The slippage phenomenon is more pronounced for side bonded configurations while it tends to be eliminated for fully wrapped elements.

In this paper, preliminary results of an on-going experimental campaign on shear strengthening of R.C. beams with FRP and FRCM composites, carried out at the University of Padua, are presented. Observations regarding the comparison of the performance of the beams depending of the composite used and fibre type are provided as well as the analysis of the cracking pattern and failure mode and the interaction between the strengthening system and the internal shear reinforcement.

2   Experimental Program

Five RC beams with the geometry and configuration of internal longitudinal and transverse reinforcement presented in Fig. 1a were tested at the Construction Materials Testing Laboratory of the University of Padua (Italy).

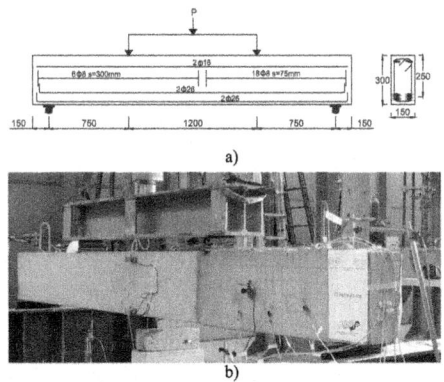

Fig. 1. Beam geometry, reinforcement, and experimental setup.

The beams were designated according to the following convention:

$$S1\text{-}XXXX\text{-}F\#\text{-}UY$$

Where *S1* corresponds to the stirrup spacing in the tested shear span (300 mm, centre to centre), *XXXX* defines the type of strengthening system (control, FRP, or FRCM), # is the type of fibre (1 = carbon mesh, 2 = steel mesh, 3 = carbon net, and 4 = steel net), U corresponds to the strengthening configuration (U wrapped) and *Y* is related to the use of composite anchors (N = no anchors).

The beams were tested using a four-point bending scheme with a total span length of 2.70 m, as shown in Fig. 1. The shear span-to-depth ratio (a/d) was equal to 3.0. The test was controlled by slowly increasing the force. The loading was paused at different increments to mark cracks and take photographs.

The beams were cast in a single batch. The average concrete compressive strength, determined by testing 3 cylinders with dimensions $\phi100 \times 200$ mm in accordance with EN 12390-3 (2009) within +/− 4 days of the day of testing, was 23.3 MPa. The longitudinal reinforcement ratio $\rho_l = A_s/b_w d$ (A_s = longitudinal reinforcement area, b_w = beam width, d = beam effective depth) was 0.057. The $\phi8$ mm, $\phi16$ mm, and $\phi26$ mm deformed reinforcing steel bars had measured yield strength (average of three specimens for each diameter) of 527 MPa, 535 MPa, and 545 MPa, respectively. In Table 1, the main characteristics of the tested beams are shown.

Table 1. Mechanical and geometrical properties of the beams

Beam	f'_c (MPa)	b_w (mm)	d (mm)
S1-CONTROL	23.3	150	250
S1-FRP-F1-UN	23.3	150	250
S1-FRP-F2-UN	23.3	150	250
S1-FRCM-F3-UN	23.3	150	250
S1-FRCM-F4-UN	23.3	150	250

The beams were strengthened with a U-wrapped configuration with one layer of fibres. Before installing the FRP/FRCM system, the beam surface was subjected to mechanical grinding, the corners of the specimens were rounded to a radius of approximately 20 mm, and any loose sand grains were removed. For the FRCM system, the concrete surface was wetted and the first layer of cementitious matrix was immediately applied. The beams were strengthened in shear within just one of the shear spans, with a continuous strip (see Fig. 1b). In order to force the shear failure in the designated span, the opposite shear span of the beam had steel stirrups spaced 75 mm (see Fig. 1a).

The overall area weight (W), elastic modulus (E_f), ultimate tensile strength (f_u), and thickness of the fibres (t) reported by the manufacturer are presented in Table 2.

Table 2. Composite fibre mechanical and geometrical properties

Fibre type	W (g/m²)	E_f (GPa)	f_u (MPa)	t (mm)
F1-Carbon	300	390	3000	0.084
F2-Steel	1910	190	3345	0.24*
F3-Carbon	170	240	4700	0.047*
F4-Steel	2200	190	2400	0.270*

*Equivalent nominal thickness (mm/m).

For specimens strengthened with FRCM composite, the cementitious matrix was applied in layers of approximately 2 mm thickness. Three 40 × 40 × 160 mm samples taken from the cementitious matrix were tested +/− 4 days from the day of testing of the beams. The average flexural strength f_{cf} (EN 12390-5 2009) and average compressive strength f_{cm} (EN 12390-3 2009) were equal to 6.33 MPa (COV = 0.15) and 45.2 MPa (CoV = 0.039), respectively.

3   Results and Discussion

Figure 2 shows the applied load (P) versus midspan displacement (Δ) curves for all specimens.

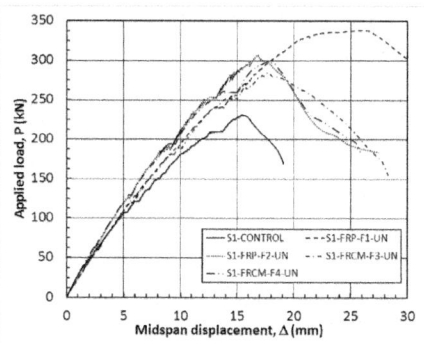

Fig. 2. Applied load (kN) vs. midspan displacement (mm) curves for all specimens.

The initial stiffness of the six beams was similar as shown in Fig. 2. However, for the control beam, there is a reduction of the stiffness after a level of load associated with cracking of the concrete (150 kN) that is not noticeable in the other beams. The curves for the strengthened specimens are similar up to failure, with a significant reduction of the stiffness close to P_{max}. However, beams strengthened with steel fibres, either for FRP and FRCM, showed a slightly higher stiffness than their carbon counterparts.

The maximum load (P_{max}), the nominal shear strength ($V_n = 0.5P_{max}$, neglecting the effects of self-weight), the displacement at midspan corresponding to P_{max} (Δ_{Pmax}), and the failure mode of each beam are summarized in Table 3.

Table 3. Summary of tests results

Beam	P_{max} (kN)	V_n (kN)	Δ_{Pmax} (mm)	Failure mode
S1-CONTROL	230.5	115.2	15.4	Shear
S1-FRP-F1-UN	338.3	169.1	26.0	Flexure
S1-FRP-F2-UN	306.8	153.4	16.8	Shear
S1-FRCM-F3-UN	284.8	142.4	17.8	Shear
S1-FRCM-F4-UN	299.5	149.7	17.7	Shear

The additional shear strength provided by the strengthening system (V_f) and the increase in the nominal shear strength of the control beam ($V_f/V_{n-control}$) are shown in Table 4. V_f is computed as the total shear strength of a given strengthened specimen minus the shear strength of the control specimen (S1-Control). In Table 4, the ratio V_f/t^*, which represents the efficiency of the system in terms of the thickness of the fibres t^*, is also included.

Table 4. Additional strength provided by FRP/FRCM composites

Beam	V_f (kN)	V_f/V_n	V_f/t^* (kN/mm)
S1-FRP-F1-UN	53.9	0.47	641.9
S1-FRP-F2-UN	38.2	0.33	159.1
S1-FRCM-F3-UN	27.2	0.24	577.8
S1-FRCM-F4-UN	34.5	0.30	127.8

*Thickness of the fibres.

3.1   Additional Shear Strength Provided by the Strengthening System

From Tables 3 and 4, it can be seen that using Carbon-FRP composite the failure mode of S1-CONTROL was transformed from shear to flexure, with an increase in strength of $V_f/V_{n\text{-control}} = 0.47$. The value of additional strength provided can be taken as the lower bound of the additional strength provided by the FRP composite with carbon fibres. For Steel-FRP composites, the increase in strength was equal to 0.33, with a shear failure.

The specimens strengthened with FRCM jackets failed also in shear and had an increase in shear capacity ranging from 24% to 30%. Specimen with steel fibres had a slightly larger increase of V_f than specimen with carbon fibres. It is important to highlight that specimens reinforced with FRP composites showed larger values of $V_f/V_{n\text{-control}}$. It is highlighted that specimens S1-FRP-F2-UN and S1-FRCM-F4-UN (strengthened using steel fibres) showed almost the same increase in strength (33% and 30% respectively) independent of the matrix that was used.

In addition to the increase in shear strength, all strengthened beams had an increase of Δ_{Pmax} relative to the control beam. This effect was more important for specimen S1-FRP-F1-UN. For the remaining specimens, the values of Δ_{Pmax} ranged from 16.8 to 17.7 mm, with larger values for specimens with FRCM jackets. In fact, the values for specimens S1-FRCM-F3-UN and S1-FRCM-F4-UN were the same independent of the type of fibre used (17.8 mm and 17.7 respectively).

Regarding the efficiency of the system (V_f/t^*), it is clear from Table 4, that elements strengthened with carbon fibres (S1-FRP-F1-UN and S2-FRCM-F3-UN) have a better behaviour. In fact the value of V_f/t^* for specimen S1-FRP-F1-UN is the higher of the reported values, even if for this specimen V_f corresponds to the lower bound of additional shear strength provided by the strengthening system due to the type of failure mode observed for this specimen, i.e., flexure. For specimens S1-FRP-F2-UN and S1-FRCM-F4-UN, the values of V_f/t^* were approximately a quarter of the value found for specimen S1-FRCM-F3-UN. This implies that having larger values of equivalent thickness, as it is the case of steel fibres, does not imply higher values of additional shear strength.

3.2   Failure Mode and Cracking Pattern

Specimen S1-Control presented a typical beam shear failure characterized by the formation of a main diagonal crack and minor diagonal cracks distributed along the shear span. Specimen S1-FRP-F1-UN failed by flexure caused by concrete crushing. For this specimen, it was not possible to identify cracks in the strengthened shear span, although some shear cracks were recognized in the unstrengthened shear span of the beam. The failure mode of specimen S1-FRP-F2-UN was characterized by debonding of the FRP jacket. Cracks were not visible in the composite before failure. However, a main diagonal crack, similar to the one observed for the control specimen, developed below the FRP jacket and was visible due to the detachment of the FRP composite.

Unlike beams S1-FRP-F1-UN and S1-FRP-F2-UN, cracking was visible on the surface of the FRCM composite for beams strengthened using cementitious matrix. This is an advantage of the FRCM system over the FRP system because it allows for immediate and easy inspection of damaged regions (Triantafillou and Papanicolaou 2006).

Beam S1-FRCM-F3-UN failed by diagonal tension. Local detachment of the entire thickness of the jacket close to the point of application of the load and fibre slippage along the main crack were observed. The inclination of the main diagonal crack was less steep than that of the control beam, and the distribution of the cracking was located in a smaller area.

For beam S1-FRCM-F4-UN, failure was caused by detachment of the composite system. It is important to notice that the cracking pattern reflected in the steel FRCM jacket is different than those observed by the control specimen and the beams strengthened with carbon FRCM composite. Figure 3 shows the cracking pattern observed on the surface of specimens S1-Control and on the composite jackets for specimens S1-FRCM-F3-UN and S1-FRCM-F4-UN.

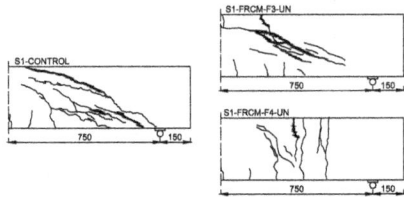

Fig. 3. Cracking patterns for beams S1-Control, S1-FRCM-F3-UN and S1-FRCM-F4-UN

For specimen S1-FRCM-F4-UN, after the FRCM jacket was removed, it was observed that diagonal cracks formed in the beam beneath the composite (see Fig. 4). It is also observed that above the diagonal crack, the fibres and the external matrix layer debonded from the internal matrix layer. However, below the diagonal crack, the entire FRCM jacket detached from the concrete substrate.

Fig. 4. Cracking pattern for beam S1-FRCM-F4-UA.

3.3   Internal and External Shear Reinforcement Interaction

In order to verify the interaction between the internal (stirrups) and external (FRP/FRCM jackets) shear reinforcement, strain gauges were placed on the stirrups as shown in Fig. 5. Considering the spacing of the stirrups (300 mm), two stirrups were located in the shear span (ST1 and St2, see Fig. 5). The strain gauges were located in the middle of the height of the beam.

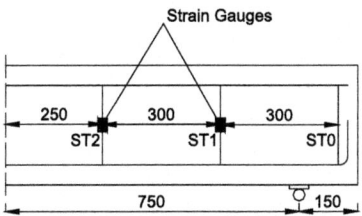

Fig. 5. Location of stirrups and strain gauges

Figures 6 and 7 show the strain distribution for ST1 and ST2 respectively for the 5 beams tested. The graphs also include the yielding strain of the steel links ($\varepsilon_y = 2650 \times 10^6$ mm/mm).

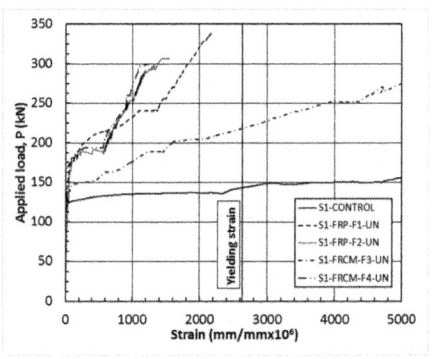

Fig. 6. Strain distribution on stirrup ST1

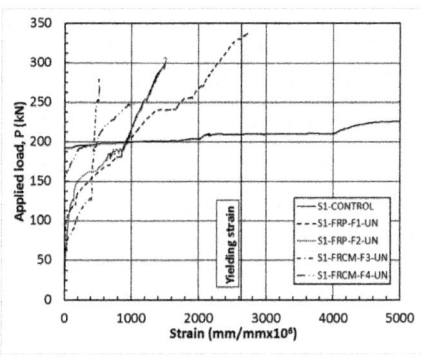

Fig. 7. Strain distribution on stirrup ST2

In Fig. 6, it is possible to see that for the specimen S1-control, ST1 reaches yielding for a value of load of 145 kN approximately. It is important to notice that for this particular specimen, the values of strain reached at peak load were roughly equal to 20000×10^6 mm/mm but they are not shown in the graph for a matter of scale. The remaining beams, with exception of the beam S1-FRCM-F3-UN, did not reach yielding, with lower values of strain found for S1-FRP-F2-UN and S1-FRCM-F4-UN, i.e., beams strengthened using steel fibres. Regarding specimen S1-FRCM-F3-UN, it is important to highlight that even though it reached yielding, the stirrup strain at peak load was approximately 25% of the maximum strain for the control beam.

Figure 7 shows the strain distribution in ST2. For this case, ST2 reaches yielding for a load of 210 kN and the strain at P_{max} is equal to 5436×10^6 mm/mm, which is considerable lower than the strain found for ST1. From the strengthened specimens, only beam S1-FRP-F1-UN showed a strain slightly higher than ε_y. The strain in ST2 for the remaining beams was lower than ε_y. At this point, it is important to notice that the lowest value of strain was witnessed for beam S1-FRCM-F3-UN. It is important to notice that the strain gauge placed on ST2 placed on beam S1-FRCM-F4-UN broke before reaching P_{max}.

In Table 5, the values of strain associated to P_{max} found for ST1 and ST2 for the tested specimens are included. Stirrups that reached yielding are highlighted in bold font.

Table 5. Strains on ST1 and ST2 for P_{max}

Beam	P_{max} (kN)	ε-ST2 (10^6 mm/mm)	ε-ST2 (10^6 mm/mm)
S1-CONTROL	230.5	**20113.9**	**5436.5**
S1-FRP-F1-UN	338.3	2171.8	**2745.4**
S1-FRP-F2-UN	306.8	1549.0	1472.1
S1-FRCM-F3-UN	284.8	**5133.3**	511.8
S1-FRCM-F4-UN	299.5	1241.5	1056.2

*For ST2, the strain gauge broke at a load of 249 kN.

As shown in Figs. 6 and 7 and Table 5, the presence of the strengthening system generates an important reduction of the strain in the stirrups. In fact, for most of the cases, the strengthening avoids the yielding of the stirrups. This implies that a simple summation of the individual contributions of the steel and the FRP/FRCM composites is not adequate. In fact, the contribution of the stirrups is computed based on the yielding stress of the steel and therefore, achieving lower values of strain means that the contribution of the stirrups in a strengthened element might be lower than for the unstrengthened beam. The previous observation is valid for FRP and FRCM composites.

4   Conclusions

In this paper, preliminary results of an on-going experimental campaign carried out in order to study the behaviour of R.C. beams strengthened in shear with FRP and FRCM composites were presented.

The use of carbon-FRP composites was able to change the shear failure of the control beam into a flexural failure for the strengthened beam, providing an increase in the beam capacity of 47%. There was also an increase in the displacement at midspan corresponding to maximum load. For the specimen S1-FRP-F2-UN, strengthened with steel fibres, the increase in strength was equal to 33%.

Beams strengthened with carbon- and steel-FRCM composites showed an increase in shear strength of 24% and 30%, respectively.

Unlike with FRP composites, cracking was visible on the surface of the FRCM jackets. However, for the case of steel-FRCM composites, after the removal of the jacket, it was observed that the cracking pattern on the surface of the jacket was different from that observed in the concrete beneath it.

For FRCM strengthened beams, the type of failure was different depending on the type of fibre used. For carbon fibres, local detachment of the entire thickness of the jacket from the concrete substrate was observed. For steel fibres, detachment of the fibres and external matrix layer or of the entire composite from the substrate was witnessed.

The presence of the strengthening system, either FRP or FRCM composites, implied an important reduction of the strains in the internal stirrups. This finding corroborates the presence of the internal and external shear reinforcement interaction. Based on this observation, the common assumption of adding the individual contributions of the steel and strengthening system in order to find the total shear capacity of the element is not adequate and can generate non-conservative results.

Acknowledgements. The first author would like to acknowledge the technical and economic support from the European Network for Durable Reinforcement and Rehabilitation Solutions (endure), a Marie Skłodowska Curie Initial Training Network. G&P Intech of Altavilla Vicentina (Italy) is gratefully acknowledged for providing the FRP and FRCM composite materials.

References

ACI Committee 440 (2008) Guide for the design and construction of externally bonded FRP systems for strengthening concrete structures, ACI 440.2R-08, Farmington Hills, MI, U.S.A.

Al-Salloum Y, Elsanadedy M, Alsayed S, Iqbal R (2012) Experimental and numerical study for the shear strengthening of reinforced concrete beams using textile-reinforced mortar. J Compos Constr 16:74–90

Blanksvard T, Taljsten B, Carolin A (2009) Shear strengthening of concrete structures with the use of mineral-based composites. J Compos Constr 1(25):25–34

Bousselham A, Chaallal O (2004) Shear strengthening reinforced concrete beams with fibre-reinforced polymer: assessment of influencing parameters and required research. ACI Struct J 101(2):219–227

Chen GM, Teng JG, Chen JF, Rosenboom OA (2010) Interaction between steel stirrups and shear-strengthening FRP strips in RC beams. J Compos Constr 14(5):498–509

EN 12390-3 (2009) Testing hardened concrete – Part 3: Compressive strength of test specimens, European Committee for Standardization, Brussels, Belgium

EN 12390-5 (2009) Testing hardened concrete – Part 5: Flexural strength of test specimens, European Committee for Standardization, Brussels, Belgium

Federation Internationale du Beton, *fib* (2001) Externally Bonded FRP Reinforcement for RC Structures, Task Group 9.3, Bulletin No. 14, Lausanne, Switzerland

Lima J, Barros J (2011) Reliability analysis of shear strengthening externally bonded FRP models. Proc Inst Civ Eng (ICE) – Struct Build 164(1):43–56

Ombres L (2015) Structural performances of reinforced concrete beams strengthened in shear with a cement based fibre composite material. Compos Struct 122:316–329

Pellegrino C, Modena C (2006) FRP shear strengthening of RC beams: experimental study and analytical modeling. ACI Struct J 103(5):720–728

Pellegrino C, D'Antino T (2013) Experimental behaviour of existing precast prestressed reinforced concrete elements strengthened with cementitious composites. Compos Part B Eng 55:31–40

Pellegrino C, Vasic M (2013) Assessment of design procedures for the use of externally bonded FRP composites in shear strengthening of reinforced concrete beams. Compos Part B 45:727–741

Sas G, Täljsten B, Barros J, Lima J, Carolin A (2009) Are available models reliable for predicting the FRP contribution to the shear resistance of RC beams? J Compos Construct 13(6):514–534

Tetta ZC, Koutas LN, Bournas DA (2015) Textile reinforced mortar (TRM) versus fibre-reinforced polymers (FRP) in shear strengthening of concrete beams. Compos Part B 77:338–348

Triantafillou T, Papanicolaou CG, Zissinopoulos P, Laourdekis T (2006) Concrete confinement with textile-reinforced mortar jackets. ACI Struct J 103(1):28–37

Triantafillou TC, Papanicolaou CG (2006) Shear strengthening of RC members with textile reinforced mortar (TRM) jackets. Mater Struct 39(1):85–93

Seismic Retrofitting of RC Frames: A Rational Strategy Based on Genetic Algorithms

R. Falcone, C. Faella, C. Lima, and E. Martinelli[✉]

DiCiv - Department of Civil Engineering, Università degli Studi di Salerno,
Fisciano, SA, Italy
e.martinelli@unisa.it

Abstract. This paper outlines a rational strategy for retrofitting Reinforced Concrete (RC), which is based on combing member- and structure-level techniques in order to achieve optimal design objectives in a Performance-Based approach. Member-level techniques (such as confinement with composite materials, steel or concrete jacketing) are supposed to enhance capacity of single members, whereas structure-level techniques (generally based on introducing steel bracings systems or shear walls) aim to reduce the seismic demand on the existing frame as a whole. A novel procedure, based on a "dedicated" genetic algorithms, is developed by the authors for selecting "optimal" retrofitting solutions, among the technically feasible ones, obtained by combining alternative configurations of steel bracing systems and FRP-confinement of critical members. The main assumptions about the representations of "individuals" and the main information about the genetic operations (i.e. selection, crossover and mutation) are summarised in the paper. Finally, a sample application of the procedure is proposed with the aim to demonstrate its potential in selecting rational retrofitting solutions.

Keywords: RC frames · Seismic retrofitting · Optimal strategy
Genetic algorithm

1 Introduction

Recent earthquake events have induced heavy losses of life and substantial damage to existing reinforced-concrete (RC) buildings (May 2006). Retrofit practices may prove helpful in reducing risks in seismic areas, generally characterised by a significant number of structures often designed for gravitational loads only, without any consideration of Capacity Design rules introduced by modern seismic codes (CEN 2005). Therefore, seismic retrofitting is nowadays as a technical challenge and a societal priority (fib 2003).

Plenty of researches have been conducted in the last two decades on seismic assessment and retrofitting of RC frames: they have been mainly aimed at developing both effective methodologies for quantifying seismic vulnerability and consistent technical solutions for seismic retrofitting (Thermou and Elnashai 2006; Ireland et al. 2007; Kaplan et al. 2011).

© Springer International Publishing AG, part of Springer Nature 2018
M. di Prisco and M. Menegotto (Eds.): ICD 2016, LNCE 10, pp. 411–424, 2018.
https://doi.org/10.1007/978-3-319-78936-1_30

As a matter of principle, feasible retrofitting solutions can be grouped into two broad classes: on the one hand, the so-called *"Member-level"* (or so called *"local"*) techniques and, on the other hand, the *"Structure-level"* (also referred to as *"global"*) techniques (fib 2003).

Member-level techniques are based on strengthening single structural members (Rodriguez and Park 1991): they aim at improving their capacity (i.e. in terms of ductility and/or strength) and, hence, enhancing the capacity of the structure as a whole. Confinement with steel and/or composite (Fiber-Reinforced Polymer, hereinafter denoted as FRP) materials, as well as concrete jacketing, are widely adopted with the aim of realising member-level strengthening of columns (fib 2006).

Conversely, structure-level techniques consist in connecting the existing structure with newly realised sub-systems (such as RC shear walls and steel bracing systems) properly designed for working in parallel with the former, in order to contribute their global capacity to the overall seismic response. In principle, they aim at increasing strength and stiffness in the coupled structural system and, hence, reducing the displacement demand on the existing RC structure and its members (Martinelli et al. 2015).

The aforementioned techniques, considered on their own, represent two somehow "extreme" solutions for retrofitting existing structures: on the one hand, the former may enhance structural capacity in order to meet the requested seismic demand, whereas, on the other hand, the latter aims at reducing displacement demand on the existing members, whose single members' capacity is fairly unchanged.

However, member- and structure-level techniques may be duly combined with the aim to obtain a synergistic action in increasing seismic capacity of under-designed members and reducing demand on the structure as a whole. In fact, each one of the (potentially infinite) combinations of member- and structure-level interventions leads to different direct costs, life-cycle costs, reliability levels and other quantitative/qualitative parameters describing their "fitness" to be considered as a retrofitting solution. In this light, choosing the "fittest" combination of the aforementioned interventions is a task that can be regarded as a constrained optimization problem.

Several researches have been carried out for selecting the most structurally efficient and cost-effective solution for seismic retrofitting, but definition of a rational strategy for obtaining the optimal retrofitting solution is still an open issue. Particularly, no well-established conceptual design procedure is still available and completely accepted by the scientific community for seismic retrofitting; in fact, the possibility of combining member- and structure-level techniques for minimising the initial cost of retrofitting is only conceptually explored in previous papers (Martinelli et al. 2015). Moreover, recent scientific contributions on optimization algorithms for structural design are restricted to new structures (Quan et al. 2012).

In the Authors' best knowledge, no relevant study is currently available for approaching seismic retrofitting of existing RC frames as an optimisation problem. As a matter of fact, in the current practice, as well as in most relevant scientific contributions on this topic (Rodriguez et al. 1991; Kunisue et al. 2000; Thermou and Elnashai 2006; Ireland et al. 2007; Bordea and Dubina 2009; Kaplan et al. 2011), seismic retrofitting of RC frames is addressed as a solely technical problem. Hence, any considerations about optimisation (even under the merely "economical" standpoint) is basically left to engineering judgement.

This is partly due to the complexity of the constrained optimisation problem under consideration, which cannot be duly approached by means of analytical techniques commonly employed in structural engineering. Conversely, it can only be solved by means of meta-heuristic techniques, possibly based on multi-criteria optimisation objectives (Caterino et al. 2009), which are not in the background of common structural engineers.

Therefore, this paper proposes a genetic algorithm capable of selecting the "fittest" solution (in terms of initial costs) obtained by combining structure-level interventions, based on steel bracing systems, and FRP-based member-level techniques. More specifically, the paper outlines a Genetic Algorithm (Gas) inspired to the well-known Darwin's "evolution of species" and the assumption of the "survival of the fittest" rule (Darwin 1859). Although some pioneering applications of these techniques are already available in the field of structural engineering, they are mainly restricted to the design of new structures (Papadrakakis and Lagaros 2002; Fragiadakis et al. 2006; Lagaros et al. 2007; Fragiadakis and Papadrakakis 2008). The following sections summarise the main aspects of the proposed genetic algorithm and its application in the rational design of retrofitting interventions. Finally, as an example, the last section presents an application of the proposed procedure.

2 The Genetic Algorithm

2.1 Problem Statement and Formulation

The conceptual definition of the seismic retrofitting problem under discussion can be generally written as the following Limit State (LS) function g_{LS}:

$$g_{LS,i} = C_{LS,i} - D_{LS,i} \geq 0 \qquad (1)$$

where the capacity $C_{LS,i}$ and demand $D_{LS,i}$ are intended in terms of displacements or forces, respectively for ductile and brittle mechanisms. Consequently, this function is negative in seismically vulnerable structures. Hence, retrofitting aims to enhance their "as built" condition at the i-th relevant LS under consideration and raise the values of $g_{LS,i}$ towards positive values. As a matter of fact, seismic retrofitting of RC structures requires that the structure meet various LSs. In this paper, the LS of Damage Limitation (SLD), related to a seismic event whose Probability of Exceedance (PoE) equal to 63% in 50 years, and the Limit State of Life Safety (SLV), corresponding to an event with 10% PoE in 50 years, are taken into account (M.II.TT., 2008).

The retrofitting objective can be achieved by acting on both capacity and demand and, hence, implementing a combination of member-level and global-level technique (Martinelli et al. 2015). The retrofitting solution can be found by solving the following constrained optimisation problem:

$$\bar{x} = \arg \min_{x}[f(x)]$$
$$g_{LS,i} \geq 0 \quad \forall i = 1. \ldots. n_{LS} \qquad (2)$$

where f(x) is the selected objective function and x is the vector of design variables defining the interventions consisting of both FRP confinement of single RC members and concentric steel bracings installed in parallel with the existing structure.

The total direct cost is assumed herein as f(x): it is defined by adding the costs $C_{loc}(x)$ and $C_{glob}(x)$, respectively referred to local and global interventions:

$$f(x) = \left[C_{loc}(x) + C_{glob}(x)\right] \cdot \Phi\left(\max_i \left[f\left(g_{LS,i}(x)\right)\right]\right) \qquad (3)$$

where the two cost functions take into account both demolition and reconstruction operations needed for realising FRP confinement and installing steel bracings. It is worth highlighting that $C_{glob}(x)$ might also include the costs of foundation strengthening, often requested by the realisation of "collaborative" bracing systems, which result in concentrating vertical reactions at their base. This aspect is not fully developed in the current version of the code.

Finally, $\Phi(\cdot)$ is a generic penalty function, which aims at increasing the nominal cost of intervention for those solutions that do not meet the constraint condition in Eq. (2)

2.2   Representation of "Individuals"

The conceptual flow-chart shown in Fig. 1 depicts the main steps of the procedure proposed for finding the optimal solution of the constrained problem described by Eq. (2).

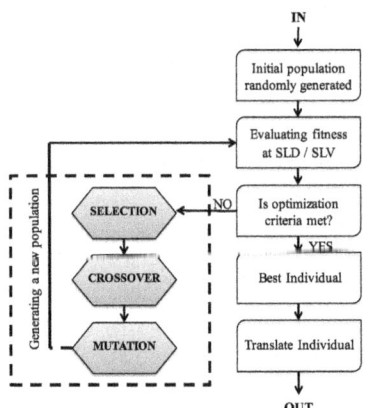

Fig. 1. Flow-chart of the optimization

The procedure starts with the randomly generating an initial population of N_{ind} individuals. They are encoded as strings (chromosomes) composed over some alphabet (s), so that the genotypes (chromosome values) are uniquely mapped onto the decision variable (phenotypic) domain (Biondini 1999).

Several coding methods are currently available, such as binary, gray, non-binary, etc. (Jenkins 1991a, b; Hajela 1992; Reeves 1993). The most commonly used representation in GAs is the binary alphabet {0, 1} although other representations can be used, e.g. ternary, integer, real valued etc. Within the present work, each individual "x" of such a population is represented through a simple "chromosome-like" array of bits. The vector "x" is composed of variables describing both member- and structure-level techniques: the string representing the binary coding (Biondini 1999) of one individual is structured by concatenating the set of variables that represent it.

Figure 2 depicts an example of the binary genotype adopted in the present algorithm for representing a simple three-storey structure with four beams at each floor (one-bay for each direction in plan): the first part of the string describes the member-level techniques considered in the aforementioned individual, whereas the second part is dedicated to describing the structure-level ones.

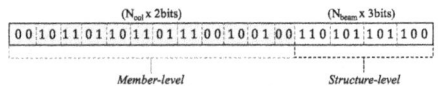

Fig. 2. Example of binary genotype for coding a retrofitting intervention

On the one hand, in the first part of the individual, each couple of bits contains a code related to the number of FRP layers possibly employed for confining the corresponding column of the structure under consideration: hence, a total of $2N_{col}$ bits is contained, N_{col} being the number of column elements in the structural model of the existing frame. Therefore, in the current implementation, the confinement ranges between zero (as-built configuration denoted by the value "00" assumed in the binary code of the corresponding column) and three layers of FRP (denoted by the value "11" of the binary code representing the i-th column).

The presence of a non-zero value for the confinement code of the generic column modifies the original (unconfined) stress-strain relationship describing the material behaviour of concrete. As is well-known, the resulting stress-strain relationship for confined concrete depends on several parameters, such as the number of FRP layers, the shape of transverse section and the type of FRP considered for this application. Specifically, the information collected in this first part of the vector are employed for modifying the mechanical values of the Kent and Park model (1971) available in OpenSEES (Mazzoni et al. 2006) for simulating the behaviour of concrete in columns (Fig. 3).

On the other hand, the following assumption are considered for describing the steel bracings system representing the global intervention codified in the second part of the genotype:

1. only the profiles of the first storey level are reported in the codified "x" vector, whereas those adopted for the upper levels stem out of a consistent design criterion;
2. steel bracings are supposed to be possibly realised between each couple of columns connected by a beam: therefore, the maximum number of bracings is equal to the number of beam N_{beam} at the first floor;

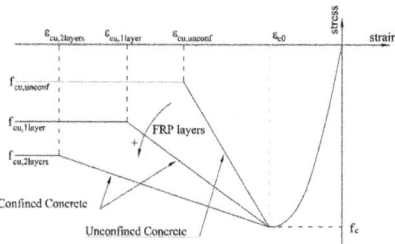

Fig. 3. Model by Kent and Park (1971) for confined concrete.

3. the transverse section of each couple of diagonals is identified by means of a label corresponding to a commercial steel profile.

The aforementioned assumptions are intended at keeping the set of design variables as small as possible. Therefore, the current implementation of the proposed procedure employs three bits for each variable related to steel diagonal bracings and, hence, only $2^3 = 8$ codes are available to describe the section of diagonals possibly installed at the first storey level, in correspondence of a generic beam (see assumption n. 2). They can range from 0 (absence of bracing) to 7: each code included between "001" and "111" points to a position in a commercial steel profile table (assumption n. 3) and, hence, the relevant properties of bracing diagonals are available therein.

The relationship between the section of steel members at the first level and the section of steel bracings at upper floors is defined below:

$$A_{k,des} = \frac{\sum\limits_{j=k}^{n} h_j \cdot W_j}{\sum\limits_{i=1}^{n} h_i \cdot W_i} \cdot A_1 \tag{4}$$

where h_j represents the position in height of the floor with respect to the foundation level, n is the total number of floors, W_j is the seismic mass of the j-th floor. Moreover, A_1 is the area of the cross section of the bracing at the first level and $A_{k,des}$ is the theoretical area of the bracing cross section required at the k-th floor. It is worth highlighting that any other consistent design criteria, defining the upper level sections depending on the first level ones, might be possibly adopted in lieu of Eq. (4). Finally, the knowledge of theoretical areas allows to select the steel commercial section whose area must be greater than $A_{k,des}$.

2.3 Seismic Analysis of "Individuals" and Evolution Criteria

Static Non Linear (Pushover) Analysis and the N2 Method (Fajfar 1999) are considered for determining displacement capacity and demand needed for evaluating the values of function $g_{LS,i}$ according to Eq. (1) for a given "individual" (Fig. 4).

On the one hand, the seismic demand is determined according to the well-known N2-Method (Fajfar 1999): Fig. 5 depicts the main operations requested to calculate the

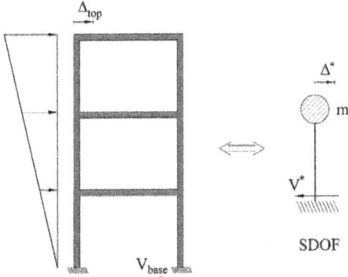

Fig. 4. Existing frame and equivalent SDOF system

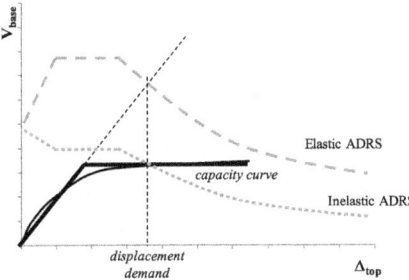

Fig. 5. Evaluating seismic displacement demand via N2 Method

"performance-point" $D_{LS,i}$. On the other hand, the capacity models adopted by current seismic codes (CEN 2005) are considered for determining $C_{LS,i}$.

In addition to the triangular distribution of lateral loads currently considered in the proposed procedure (Fig. 4), further distributions might be easily taken into account for performing the seismic analysis (CEN, 2005a; M.II.TT., 2008).

Finally, for each population, the total cost and the objective function (Limit State function) $g_{LS,i}$ are evaluated for all N_{ind} individuals, hence, the genetic algorithm evolves through three operators until the counter of population reaches a maximum fixed number. Starting from the randomly generated initial population, the *reproduction*, *combination* and *mutation* operators concur in defining the following generation of retrofitting solutions.

2.4 Reproduction Operator

Reproduction, also known as *selection*, selects individuals on the bases of their fitness to the purpose and forms a mating pool, with the aim of realising the "survival of the fittest" principle, which is the basis of the evolution theory (Darwin 1859). The essential idea is that individuals are picked out of the current population and, based on probabilistic assumptions, they are included in the mating pool.

To this end, a fitness function $F(x_k)$ is defined, whose value $F(x_k)$ for the k-th individual of the current generation can be determined as follows:

$$F(x_k) = \frac{\min\limits_{h=1\ldots N_{ind}} f(x_h)}{f(x_k)} \qquad (5)$$

where $f(x_k)$ is the value of the objective function for the of the k-th individual and N_{ind} is the (invariant) number of individual forming each generation (currently assumed equal to 50). The fitness function "measures" the capability of the k-th individual to "compete" with the other ones of the same generation in achieving the objective of the optimisation problem under consideration. As can be easily understood from Eq. (5) it is always lower than the unit, which is determined for the individual characterised by the lower value of the objective function.

The fitness $F(x_k)$ of each individual can be employed in determining the "probability of survival" $p(x_k)$ of the k-th element as part of the current generation; besides other scaling operations (Biondini, 1999), which are not considered in the current implementation, $p(x_k)$ can be determined as follows:

$$p(x_k) = \frac{F(x_k)}{\sum\limits_{h=1}^{N_{ind}} F(x_h)}. \qquad (6)$$

It is apparent that, by definition, the values of $p(x_k)$ range between 0 and 1 and the sum of probabilities for each individual is equal to the unity.

Under the algorithmic standpoint, $p(x_k)$ is the probability that the k-th individual is "selected" within the current generation to be part of the mating pool and, hence, "reproduce" itself into the following generation. In fact, the selection procedure is implemented through the so-called "roulette-wheel" rule (Lipowski and Lipowska 2012), whose circular sections are marked proportionally to the probability of survival of each individual (Fig. 6). If n individuals have to be selected, n random numbers are generated in the [0, 1] range and the individual whose circular segment includes the random number is actually selected for reproduction. It goes without saying that the individuals featuring higher values of $p(x_k)$ have the higher probability to be selected as "parents" for generating "offsprings" in the following generation.

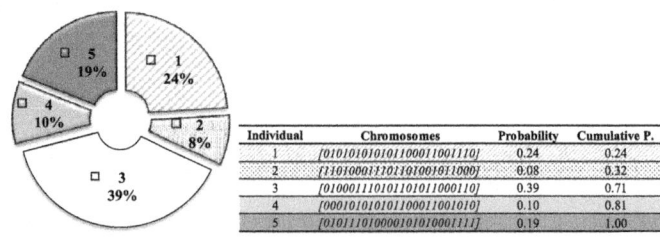

Individual	Chromosomes	Probability	Cumulative P.
1	[0101010101011000110011110]	0.24	0.24
2	[1101000111011010011011000]	0.08	0.32
3	[0100011101011010101000110]	0.39	0.71
4	[0001010101011000110010010]	0.10	0.81
5	[0101110100001010100001111]	0.19	1.00

Fig. 6. Roulette wheel rule in the selection operator

2.5   Crossover Operator

As well as its ideal counterpart in nature, the *crossover* operator produces new individuals that have some parts of both parents' genetic material: the recombination of genotype creates different individuals in the following generations by combining material from two individuals of the previous generation. The two genotype strings of the individual participating in the crossover are "selected" as parents and the resulting strings are defined as "children".

Figure 7 shows the working principle of the crossover operator. Once n individuals are selected according to their $F(x_k)$, crossover "mixes" the segments delimited with dashed lines, into new "offspring" individuals (children): several crossover points can be defined in a process usually referred to as "multi-point" crossover. In the example shown in Fig. 7 crossover operator is applied column-by-column and bay-by-bay.

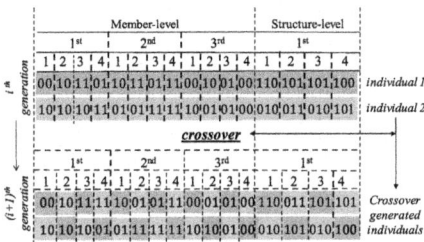

Fig. 7. Crossover operator applied to couples of "parent".

The influence of the number of crossover points on the resulting efficiency is a key aspect, whose discussion is outside the scope of this paper. Readers may refer to De Jong and Spears (1992) for further details.

Finally, it is worth highlighting that, in order to preserve some of the fitter individuals, not all selected strings in the mating pool are used in crossover. Hence, a threshold combination probability p_c is defined: only $N_{ind}*p_c$ individuals of the population are used in the crossover operation, whereas and $N_{ind}*(1-p_c)$ remain unchanged and preserve themselves for following population.

2.6   Mutation Operator

The *mutation* operator introduces diversity in the population whenever the population tends to become homogeneous due to repeated use of reproduction and crossover operators. It operates at the bit level of the genotype code: to this end, the so-called "mutation probability" p_m is defined as the probability of a single bit to be "inverted" in each child individual.

As well as in nature, this probability is usually assumed as a small value: in the current implementation this parameter is set equal to 0.002. Figure 8 shows an example of mutation occurring in a generic individual.

Old chromosomes				Random numbers				New chromosomes			
1	0	1	0	0.702	0.345	0.267	0.433	1	0	1	0
0	0	1	**0**	0.879	0.134	0.029	**0.001**	0	0	1	**1**
Mutation Probaility p_m=0.002											

Fig. 8. Examples of mutation

A higher value might be selected in order to enhance the so-called *ergodicity* of the algorithm, namely its capability to explore the parametric field of the problem under consideration. Under the operational standpoint, mutation has the role of introducing new features in the individuals of the new generation that are not present in the current one. In fact, a coin toss mechanism is employed: if random number between zero and one is less than the mutation probability, then the bit is inverted. If N_{bits} is the number of bits totally allocated in each string, mutation points are randomly selected from the $N_{ind}*N_{bits}$ total number of bits in the population matrix. The importance of this operator is relevant.

3 Sample Application

The working detail and the potential of the presented optimization procedure are shown by a sample application. A simple 3D three-storey RC frame with three bays in x-direction and one bay along y axis is taken as preliminary case for applying the proposed retrofitting strategy: Fig. 9 depicts its in-plane configuration and 3D view. The cross sectional area of beams and columns is 30×50 cm^2. Foundation is not simulated and fixed supports are considered. Rigid joints are used for simulating beam-to-column connections.

The total number of design variables considered in the test example is 34: 3×8 member-level variables (8 columns for each floor), plus 10 structure-level variables (6 bays in x-direction and 4 along the y axes). Each generation includes 50 individuals and the genetic algorithm stops either after 150 generations or if the objective function keeps unchanged for 20 consecutive generations.

Both the LS of Life Safety (SLV) and Damage Limitation (SLD) are considered and, hence, demand and capacity are evaluated in both LSs.

A bilinear stress-strain curve with Young modulus equal to 210 GPa and yield stress $F_y = 350$ MPa is adopted for describing the elasto-plastic behaviour of steel. The Kent-Scott-Park model (Kent and Park 1971) with degraded linear unloading/reloading stiffness and no tensile strength is used in order to describe the constitutive law of existing concrete. The effects of FRP confinement result in increasing the ductility of concrete (Fig. 3), according to the afore mentioned model. The live load is taken as $Q = 2.00$ kN/m^2 and the permanent load is $G = 5.00$ kN/m^2. The gravity loads are contributed from an effective area of 75 m^2.

A Finite Element model is built in OpenSEES (Mazzoni et al. 2006) for simulating the seismic response of the structure under consideration. The well-known fiber approach is used to account for material non-linearity by means of the so-called "nonlinear beam-column elements".

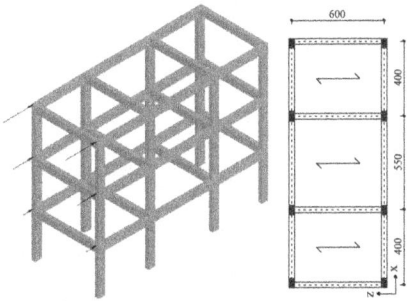

Fig. 9. 3D view and in-plain configuration of the considered structure.

Five integration sections, located at the Gauss-Lobatto quadrature points, are considered for each beam-column element. Each section is divided into a number of fibers. Similar elements are also employed for modelling the concentric steel bracings in structure-level intervention. An accidental eccentricity is assigned in the middle point according to EN 1993-1-1 (2005) in order to obtain the buckling of the bracing in compression.

Figure 10 depicts the outcome of the proposed algorithm throughout the generations: the objective (initial cost) function starts from a cost of 127'866 € and decreases progressively. As expected, the curve shows a very steep slope over the first generations and a slower and slower reduction, often characterised by a staircase shape, towards the final convergence, which is supposed to be achieved after 150 generations, as no further improvement is observed in f(x) over the last 23 iterations.

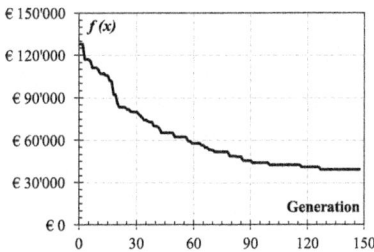

Fig. 10. Convergence history: objective function vs generation

Figure 11 shows the evolution of the maximum demand-to-capacity ratio D/C defined as follows:

$$\frac{D}{C} = \max_i \frac{D_{LS,i}}{C_{LS,i}}. \tag{7}$$

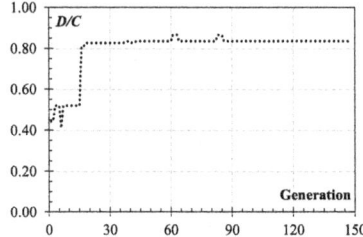

Fig. 11. Evolution of the demand-to-capacity ratio D/C

As can be seen, D/C, strictly related to the LS function defined in Eq. (1) tends to approach the unit, meaning that the solution evolves towards a full utilisation of the available capacity in both existing and newly designed members.

Finally, the optimal phenotype in the 150th population (Fig. 10) has a cost of 39'195 €, the lowest of all previously processed solutions. In the example proposed herein, the optimal solution came up to consist of a concentric steel bracing (realised in the two plain frame along the x-direction and the y-direction) and no local FRP interventions are actually required. The optimal section of steel members for the first bracing level is a HE 120 B, whilst for the two upper floors, according to the relationship (4), the best one is an HE 100 A.

4 Conclusions

This paper outlined the formulation of a rational procedure intended at optimising seismic retrofitting of existing structures by combining member- and structure-level techniques. Although in its current implementation it does not consider some relevant features (such as the cost of foundations) and assumes the actual cost as a simple objective function, the proposed procedure demonstrates its potential in formally approaching seismic retrofitting as a constrained optimisation problem.

The proposed genetic algorithm has the potential to support engineering judgement (being far from the ambition to rule it out) in determining the "fittest" seismic retro-fitting solution for RC frames. In fact, the proposed application demonstrates that the implemented genetic algorithm is capable of finding a solution characterised by a cost significantly lower than the initially assumed trial solution.

Nevertheless, the implementation of this numerical model is still under develop-ment: the work ahead should be primarily intended at including the aspects that are not taken into account yet (i.e. multi-criteria objective function, among the others). Moreover, it should also aim at enhancing the computational efficiency of the computer procedure, whose computational cost is one of the main critical issues to be duly addressed for the proposed method be actually feasible in real applications.

Acknowledgements. This paper is part of the Ph.D. project by Mr. Roberto Falcone, enrolled at the XXX Cycle of the Doctoral Course in "Risk and sustainability in civil, construction and environmental engineering" at the University of Salerno, whose financial support is gratefully acknowledged.

References

Bordea S, Dubina D (2009) Retrofitting/upgrading of reinforced concrete elements with buckling restrained bracing elements. In: Proceedings of the 11th WSEAS international conference on sustainability in science engineering, Timisoara (Romania), 27–29 May 2009, pp 407–402

Biondini F (1999) Optimal limit states design of concrete structures using genetic algorithms. Stud Res 20. http://citeseerx.ist.psu.edu/viewdoc/download?doi=10.1.1.487.3892&rep=rep1&type=pdf. Accessed 15 Apr 2016

Caterino N, Iervolino I, Manfredi G, Cosenza E (2009) Comparative analysis of multi-criteria decision-making methods for seismic structural retrofitting. Comput Aided Civ Infrastruct Eng 24(6):432–445

CEN (2005) EN 1998-1:2005 - Design of structures for earthquake resistance. Part 1: General rules, seismic actions and rules for buildings. European Committee for Standardization (CEN)

CEN (2005) EN 1993-1-1:2005 (Eurocode 3) - Design of steel structures Part 1-1: General rules and rules for buildings. European Committee for Standardization (CEN)

Darwin C (1859) On the origin of species by means of natural selection or the preservation of favoured races in the struggle for life (The origin of species). Murray, London

Fajfar P (1999) Capacity spectrum method based on inelastic demand spectra. Earthq Eng Struct Dynam 28(9):979–993

FEMA356 (2000) Prestandard and commentary for the seismic rehabilitation of buildings. Federal Emergency Management Agency

fib (2003) Seismic assessment and retrofit of reinforced concrete buildings. Bulletin No. 24, 138, ISBN: 978-2-88394-054-3

fib (2006) Retrofitting of concrete structures by externally bonded FRPs, with emphasis on seismic applications, Bulletin No. 35, 220, ISBN: 978-2-88394-075-8

Fragiadakis M, Lagaros ND, Papadrakakis M (2006) Performance-based multiobjective optimum design of steel structures considering life-cycle cost. Struct Multidiscip Optim 32:1–11

Fragiadakis M, Papadrakakis M (2008) Performance-based optimum seismic design of reinforced concrete structures. Earthq Eng Struct Dynam 37(6):825–844

Hajela P (1992) Stochastic search in structural optimization: genetic algorithms and simulated annealing. Chapter 22, pp 611–635

Jenkins WM (1991a) Towards structural optimization via the genetic algorithm. Comput Struct 40(5):1321–1327

Jenkins WM (1991b) Structural optimisation with the genetic algorithm. Struct Eng 69(24):418–422

Kaplan H, Yilmaz S, Cetinkaya N, Atimtay E (2011) Seismic strengthening of RC structures with exterior shear walls. Sadhana Indian Acad Sci 36(1):17–34

Kent DC, Park R (1971) Flexural members with confined concrete. J Struct Div 97(7):1969–1990

Kunisue A, Koshika N, Kurokawa Y, Suzuki N, Agami J, Sakamoto M (2000) Retrofitting method of existing reinforced concrete buildings using elasto-plastic steel dampers. In: Proceedings of the twelfth world conference on earthquake engineering (12WCEE), Paper 0648

Ireland MG, Pampanin S, Bull DK (2007) Experimental investigations of a selective weakening approach for the seismic retrofit of R.C. walls. In: Palmerston north (New Zealand), 30 March–1 April 2007

Lagaros ND, Tsompanakis Y, Fragiadakis M, Papadrakakis M (2007) Soft computing techniques in probabilistic seismic analysis of structures. In: Lagaros N, Tsompanakis Y (eds) Intelligent computational paradigms in earthquake engineering. Idea Group Publishing ltd, Hershey

Lipowski A, Lipowska D (2012) Roulette-wheel selection via stochastic acceptance. Phys A Stat Mech Appl 6:2193–2196

Martinelli E, Lima C, Faella C (2015) Towards a rational strategy for seismic retrofitting of RC frames by combining member- and structure-level techniques. In: SMAR2015 – Third conference on smart monitoring, assessment and rehabilitation of civil structures, Antalya (Turkey), 6–9 September 2015

May PJ (2006) Societal implications of performance-based earthquake engineering. Pacific Earthquake Engineering, PEER Report 2006/12

Mazzoni S, McKenna F, Scott MH, Fenves GL (2006) OpenSEES command language manual. Pacific Earthquake Engineering Research (PEER)

M.II.TT (2008) Norme tecniche per le costruzioni - D.M. 14 gennaio 2008 (in Italian)

Papadrakakis M, Lagaros ND (2002) Soft computing methodologies for structural optimization. Appl Soft Comput 3:283–300

Quan G, Barbato M, Conte JP, Gill PE, McKenna F (2012) OpenSees-SNOP framework for finite-element-based optimization of structural and geotechnical systems. J Struct Eng 138 (6):822–834

Reeves GR (1993) Modern heuristic techniques for combinatorial problems. Wiley, New York

Rodriguez M, Park R (1991) Repair and strengthening of reinforced concrete buildings for seismic resistance. Earthq Spectra 7(3):817–841

Thermou GE, Elnashai AS (2006) Seismic retrofit schemes for RC structures and local–global consequences. Prog Struct Eng Mat 8:1–15

Analytical Prediction of the Flexural Response of External RC Joints with Smooth Rebars

G. Campione[(⊠)], F. Cannella, L. Cavaleri, L. La Mendola,
and A. Monaco

Department of Civil, Environmental, Aerospace and Material Engineering,
University of Palermo, Palermo, Italy
giuseppe.campione@unipa.it

Abstract. An analytical model in a closed form able to reproduce the monotonic flexural response of external R.C. beam-column joints with smooth rebars
is presented. The column is subjected to a constant vertical load and the beam to
a monotonically increasing lateral force applied at the tip. The model is based on
the flexural behavior of the beam and the column determined adopting a concentrated plasticity hinge model including slippage of the main bars of the beam.
A simplified bilinear moment-axial force domain is assumed to derive the
ultimate moment associated with the design axial force. For the joint a simple
continuum model is adopted to predict shear strength and panel distortion.
Experimental data given in the literature are utilized to validate the model.
Finally, the proposed model can be considered a useful instrument for preliminary static verification of existing external R.C. beam-column joints with
smooth rebars.

Keywords: Joint · Beam · Column · Shear · Flexure · Smooth rebars

1 Introduction

Many R.C. buildings currently present in the Mediterranean area (Italy, Greece, Spain,
Turkey, etc.), constructed in the 50's–70's, were designed before the advent of seismic
codes, and show peculiar structural characteristics and material qualities. Among them a
large number of old buildings have been identified as having potentially serious structural deficiencies with respect to earthquake resistance. Typical structural deficiencies of
existing R.C. buildings are the general lack of ductility as well as inadequate lateral
strength: they has been recognized as the fundamental source of deficiency in seismic
performance of gravity- load-designed existing buildings, as a consequence of total
absence of capacity design principles and poor reinforcement detailing (Priestley 1997).

R.C. buildings constructed in the absence of seismic codes have both global
deficiencies, connected to lack of regularity in plan and elevation, as well as to the
possible onset of weak-column mechanism, with a tendency to develop soft-story
mechanisms, and local deficiencies, connected to insufficient transverse reinforcement
of beams, columns, joints, and to insufficient anchorage (Ehsani and Wight 1985;
Hegger et al. 2003). In particular, at the local level, inadequate protection of the panel
zone region within beam-column joint subassemblies is expected as well as brittle

© Springer International Publishing AG, part of Springer Nature 2018
M. di Prisco and M. Menegotto (Eds.): ICD 2016, LNCE 10, pp. 425–439, 2018.
https://doi.org/10.1007/978-3-319-78936-1_31

failure mechanisms of structural elements. In most cases examined experimentally and theoretically failure was due to the shear collapse of external joints producing brittle response; in some other cases failure was due to flexural failure of the columns due to strong beam/weak column design (Attaalla 2004; Hegger et al. 2004).

Generally, it is observed that typical structural deficiencies can be related to: - inadequate confining effects in the potential plastic regions; - insufficient amount, if any, of transverse reinforcement in the joint regions; - insufficient amount of column longitudinal reinforcement, when considering seismic lateral forces; - inadequate anchorage detailing, for both longitudinal and transverse reinforcement; - lapped splices of column reinforcement just above the floor level; - lower quality of materials (concrete and steel) when compared to current practice, in particular use of smooth (plain) bars for both longitudinal and transverse reinforcement and low-strength concrete.

2   Aim of Research and Range of Variation of Case Studies

The main object of the present research is to propose a simple model for hand verification of the flexural behavior of external old type R.C. frames with smooth rebars representative of mid-rise building types designed to support vertical loads, constructed in the pre-70s period. With the aim of understanding the structural behavior of these R. C. frames, detailed experimental investigations on T sub-assemblages were carried out (Calvi et al. 2001; Calvi et al. 2002; Braga et al. 2009).

The range of study cases was that in which buildings were only designed for gravity and so did not respect the criteria of plastic design. The beams were designed with the scheme of continuous beams for gravity loads and columns were designed for gravity loads considered as isolated member. This approach was based on elastic verification of R.C. cross-sections in which the compressive strength of concrete was limited to 6–7 MPa for beams and to 4–5 MPa for columns. The minimum compressive strength assumed, measured on cube specimens at 28 days of curing, was 12 MPa. The smooth steel rebars adopted had 280 MPa yielding stress. The steel rebars were straight at the ends or had hook anchorages. The dimensions of the cross-sections of the columns were at least 300 × 300 mm or 300 × 400 mm (for a number of floors limited to 3 or 4). The span of the floors and beams was at most 5000 mm. The longitudinal reinforcement of columns was constituted by 4 ϕ 12 mm (or 4 ϕ 14 mm) longitudinal bars and stirrups having 6 mm diameter were placed at pitch 250 mm. The beams had recurrent dimensions of 300 × 400 mm and 700 × 200 mm in the floor thickness. At least two bottom and two 12-mm top rebars were adopted. Additional reinforcement in the fixed sections and in the bottom central portion of the beams was adopted to verify the cross-sections (generally 2 or 3 rebars ϕ 14–16 mm diameter). The stirrups in the beams were designed to support 50% of shear, while the other 50% was supported by inclined rebars (inclination 45°). The stirrups were placed at a pitch of 330 mm.

3 Proposed Model

The case examined here is the one shown in Fig. 1. It represents an external beam-joint-column system subjected to a constant vertical load N acting on the column and to a monotonically increasing lateral force F applied at the tip of the beam.

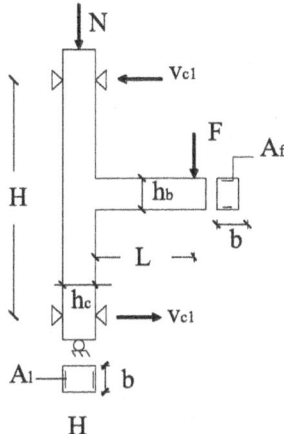

Fig. 1. Beam-column specimen geometry and loading scheme

The analytical model presented below gives the overall beam tip force-displacement (F–δ) curve at the tip of the beam meeting the column at the joint, for any fixed value of the axial load N acting on the column. The force F shown in Fig. 1 is related to the maximum beam moment $M_b = F \cdot L$, while the column is loaded by the abovementioned constant axial load N and by a linear moment whose maximum value is $M_b/2$. Actually, the case in which sections of the column are subjected to a moment equal to $M_b/2$ is a specific case of equal flexural stiffness above and below the joint; otherwise, the moment is distributed according to the relative stiffness. The shear force V_{c1} is related to F by a simple equilibrium equation $(F = (V_{c1} \cdot H)/L)$. The inter-story drift is directly related to the displacements δ which correspond to the applied force F and it is calculated as δ/L.

The overall structure response is given (see Fig. 2), for each force level F, by adding the displacements due to the beam δ_b, to the column δ_c and to the joint δ_j. In particular, the beam displacement is obtained by modeling the beam as a cantilever including slippage of bars as suggested in Fib Bulletin No. 24 (2003), while the column is considered as a beam element simply supported and loaded by a constant axial force and a moment in the nodal region. The joint is modeled as a continuum panel as suggested in Calvi et al. (2002). In the following sections, firstly the beam, column and joint displacements are evaluated and, secondly, they are put together in the F–δ curve of the system. No shear failure is considered in the present model.

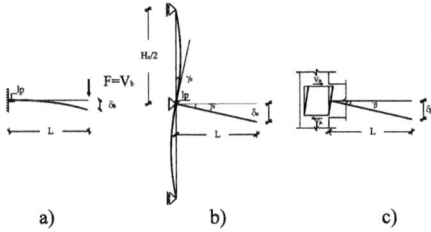

Fig. 2. Contributions of single components in T subassemblages: (a) beam; (b) column; (c) joint.

3.1   Beam Contribution

In flexure, with reference to the rectangular cross-section shown in Fig. 3, representing the section of the beam fixed into the column, using the translational and rotational equilibrium equations of internal forces, the position of the neutral axis c_{cu} and the ultimate flexural strength M_{uy} can be obtained in the following form:

$$c_{cu} = \frac{A_s \cdot f_{yb}}{\alpha \cdot f_c \cdot \beta \cdot b} \tag{1}$$

$$M_{uy} = A_s \cdot f_{yb} \cdot \left(d - \frac{1}{2} \cdot \beta \cdot c_{cu} \right) \tag{2}$$

α and β being the stress block coefficients, assumed equal to 0.85 and 0.80 for normal strength concrete, f_{yb} the working stress of the steel, A_s the area on the main rebars, f_c the compressive strength of concrete and d the effective depth of the beam.

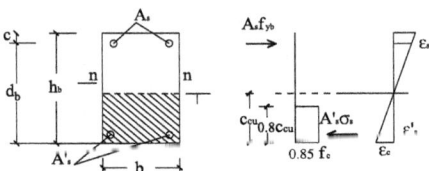

Fig. 3. Design assumptions for analysis of R.C. beams

The working stress f_{yb} of the beam longitudinal rebars present in Eqs. (1 and 2) is related to the bond stress distribution and they are both unknown. Therefore, in order to determine these two terms without relying on iterative procedures, Pauletta et al. (2015), on the basis of experimental results on 61 test specimens, show that it is possible to obtain a single analytical expression giving the tensile stress trend in the beam longitudinal rebars when joint shear failure occurs. Specifically, if smooth rebars with straight anchorages are utilized, the effective stress f_{yb} decreases with an increase in the mechanical percentage of beam tensile reinforcement $\omega_s = (A_s \cdot f_y)/(b \cdot h_b \cdot f_c)$ according to the following analytical law:

$$\chi = \frac{f_{yb}}{f_y} = 0.63 \cdot (\omega_s)^{-0.21} \leq 1 \tag{3}$$

f_y being the yielding stress of the steel, b and h_b the width and depth of the beam and A_s the longitudinal tensile reinforcement area. The value of ω_s that gives $f_{yb}/f_y = 1$ is $\omega_{s\,max} = 0.110$.

By contrast, when smooth rebars had hook anchorage, as shown in Fabbrocino et al. (2004), yield stress is fully attained without loss of bond, but significant slippage occurs. The flexural strength given by Eq. (2) does not exceed the moment capacity M_{uc} for compression failure, which is calculated as:

$$M_{uc} = \alpha \cdot f_c \cdot \beta \cdot c_{cu} \cdot b \cdot \left(d - \frac{1}{2} \cdot \beta \cdot c_{cu} \right) \tag{4}$$

If strain hardening effects in longitudinal rebars are neglected the force F_b corresponding to the ultimate moment is expressed as $F_b = M_u/L$ with M_u the minimum value among those given by Eqs. (3) and (4). In the case of yielding of the main bars if we refer to $c_{cu}/d = 0.2$ Eq. (2) gives $M_{uy}/(b \cdot h_b^2 \cdot f_c) = 0.92 \cdot \omega_s$.

The position of the neutral axis c_{cy} at yielding of the steel rebars can be found with the well-known expression:

$$\frac{c_{cy}}{d} = \sqrt{(\rho \cdot n)^2 + 2 \cdot \rho} - \rho \cdot n \tag{5}$$

with n the ratio between the elastic modulus of steel and concrete E_s and E_c (modulus of elasticity of concrete assumed $E_c = 4200 \cdot \sqrt{f_c}$) and ρ the geometrical ratio of main bars $\rho = A_{sb}/(bd)$.

The yielding and the ultimate curvatures can be expressed as:

$$\varphi_y = \frac{\varepsilon_y}{d - c_{cy}} \tag{6}$$

$$\varphi_{su} = \frac{\varepsilon_{su}}{d - c_{cu}} \quad \text{for } M_u < M_c \tag{7}$$

$$\varphi_{cu} = \frac{\varepsilon_{cu}}{c_{cu}} \quad \text{for } M_u < M_c \tag{8}$$

The deflection δ_c and the force F_c at first cracking neglecting the reinforcement and the slippage of steel bars proves to be:

$$\delta_c = \frac{2 \cdot \sigma_{ct} \cdot L^2}{3 \cdot E_c \cdot h_b}; \quad F_c = \sigma_{ct} \cdot \frac{b \cdot h_b^2}{6 \cdot L} \tag{9}$$

with σ_{ct} the tensile strength of concrete assumed as in ACI 318 (2011) in the absence of experimentation. The deflection of the beam at yielding δ_{by} can be assumed as:

$$\delta_{by} = \frac{M_u \cdot L^2}{3 \cdot E_c \cdot J_n} + \theta_y^{slip} \cdot L \tag{10}$$

the moment of inertia being expressed as:

$$J_n \cong \frac{b \cdot c_{cy}^3}{3} + n \cdot A_f \cdot (d - c_{cy})^2 + n \cdot A_f' \cdot (c_{cy} - c)^2 \tag{11}$$

θ_y^{slip} being the plastic rotation at first yielding due to the slippage of the bar calculated as suggested in Fib Bulletin No. 24. (2003) in the form:

$$\theta_y^{slip} = \frac{\varepsilon_y}{2 \cdot \left(d - \frac{c_{cy}}{3}\right)} \cdot \frac{\varphi}{4} \cdot \frac{f_y}{f_{b,y}} \tag{12}$$

with $f_{b,y}$ the uniform bond stress along the development length L_b assumed $f_{b,y} = 0.2\sqrt{f_c'}$ as in Fib Bulletin No. 24 (2003) and ϕ the diameter of the bar. At rupture we have:

$$\delta_{bu} = \delta_{by} + (\varphi_u - \varphi_y) \cdot l_p \cdot \left(L - \frac{l_p}{2}\right) + \theta_u^{slip} \cdot L \tag{13}$$

l_p being the plastic hinge length assumed as suggested in Thom (1983) in the form $l_p = \alpha \cdot L + c_1$, where c_1 represents a correction for the tension shift effect which occurs due to diagonal cracking and α = the normalized strength increase from yield to ultimate tension steel (e.g. $\alpha = 0.08$ according to Priestley et al. 1997). In Eq. (13) θ_u^{slip} is the plastic rotation at ultimate state due to the slippage of the bar calculated as suggested in Fib Bulletin No. 24 (2003) in the form:

$$\theta_u^{slip} = \theta_y^{slip} + \frac{(\varepsilon_u - \varepsilon_y)}{\left(d - \frac{c_{cu}}{2}\right)} \cdot \frac{\varphi}{4} \cdot \frac{f_u - f_y}{f_{b,u}} \tag{14}$$

with ε_u and f_u the ultimate strain and stress of longitudinal bar and $f_{b,u} = \lambda \cdot f_{b,y}$ and $\lambda = 1.2$.

3.2   Column Contribution

For the cross-section analysis of the column, the scheme adopted is the one shown in Fig. 4. The moment-axial force domain adopted here has been simplified with respect to the effective domain by considering a bilinear domain (this simplification is conservative because this domain is internal to the effective one). The bilinear domain is constituted by the two linear branches, the first one of which connects the case of pure compression with the case of balanced failure and the second one connects the case of balanced failure with the case of pure flexure. Columns have an area of steel in tension equal to that in compression ($A_l = A_l'$). This approach has the advantage that the ultimate moment can be calculated with simple analytical expressions. In the case of pure compression and absence of buckling of longitudinal rebars (a hypothesis justified after a preliminary verification) the ultimate axial force is given by:

$$N_u = \alpha \cdot f_c \cdot h_c \cdot b + 2 \cdot A_1 \cdot f_{yl} \tag{15}$$

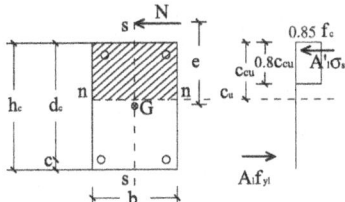

Fig. 4. Design assumptions for analysis of R.C. columns /

Introducing the dimensionless axial force gives:

$$n_u = \frac{N_u}{f_c \cdot h_c \cdot b} = \alpha + 2 \cdot \omega_l \tag{16}$$

$$\omega_s = \frac{A_l \cdot f_{yl}}{b \cdot h_c \cdot f_c} \tag{17}$$

In flexure, neglecting the compressed rebars gives:

$$\frac{M_u}{b \cdot d_c^2 \cdot f_c} \cong 0.9 \cdot \omega_l \tag{18}$$

For balanced failure both the rebars in compression and in tension have to yield and we have:

$$N_b = \beta \cdot c_{cb} \cdot b \cdot \alpha \cdot f_c \tag{19}$$

$$M_b = N_b \cdot e = (\alpha \cdot f_c \cdot \beta \cdot c_{cb} \cdot b) \cdot \left(\frac{h_c}{2} - \frac{\beta \cdot c_{cb}}{2}\right) + 2 \cdot A_l \cdot f_{yl} \cdot \left(\frac{h_c}{2} - c\right) \tag{20}$$

If N_b and M_b are dimensionless we have:

$$n_b = \frac{N_b}{f_c \cdot h_c \cdot b} = \alpha \cdot \beta \cdot \frac{c_{cb}}{h_c} \tag{21}$$

$$m_b = \frac{M_b}{b \cdot d_c^2 \cdot f_c} = \left(0.5 \cdot \alpha \cdot \beta \cdot \frac{c_{cb}}{d_c}\right) \cdot \left(\frac{h_c}{d_c} - \beta \cdot \frac{c_{cb}}{d_c}\right)$$
$$+ 2 \cdot \omega_l \cdot \left(\frac{h_c}{2 \cdot d_c} - \frac{c}{d_c}\right) \cdot \frac{h_c}{d_c} \tag{22}$$

The ultimate moment associated with the design axial force N can be calculated in the following form:

$$M_u(N) = M_u + N \cdot \left(\frac{M_b - M_u}{N_b}\right) \text{ with } N \leq N_b \tag{23}$$

or in dimensionless form

$$m_u(n) = \frac{M_u(N)}{b \cdot d_c^2 \cdot f_c} = m_u + \frac{n}{n_b} \cdot (m_b - m_u) \tag{24}$$

If we assume $c/d_c = 0.05$, $h_c/d_c = 0.95$ and $f_{yl} = 280$ MPa Eq. (24) yields the simplified expressions:

$$m_u(n) = 0.9 \cdot \omega_l - 0.0765 \cdot \omega_l \cdot n + 0.176 \cdot n \tag{25}$$

$$n < n_b = 0.653 \tag{26}$$

For balanced failure we have: $n_b = 0.623$ and $m_b = 0.125 + 0.85 \cdot \omega_l$.

Using Eq. (24) the force in the beam determined by the crisis of the column F_c can be expressed as:

$$F_c = 2 \cdot \left[m_u + \frac{n}{n_b} \cdot (m_b - m_u)\right] \cdot \frac{b \cdot d_c^2 \cdot f_c}{L} \tag{27}$$

To determine the load-deflection contribution due to the column, the rotation of the column in the loaded section γ_e is calculated as:

$$\gamma_{cy} = \frac{\varphi_y \cdot H}{6} \text{ with } \varphi_y = \frac{\varepsilon_y}{h_c - c_{ce}} \tag{28}$$

with the position of the neutral axis given by

$$c_{ce}^3 + 3 \cdot \left(e - \frac{h_c}{2}\right) \cdot c_{ce}^2 + \frac{6 \cdot n}{B_c} \cdot \left[(A_l' + A_l) \cdot \left(e - \frac{h_c}{2} + d\right)\right] \cdot c_{ce} +$$
$$- \frac{6 \cdot n}{B_c} \cdot \left[A_f' \cdot c \cdot \left(e - \frac{h_c}{2} + c\right) + A_l \cdot h_c \cdot \left(e - \frac{h_c}{2} + d\right)\right] = 0 \tag{29}$$

The deflection at the tip of the beam δ_{cy} due to the column at yielding proves to be:

$$\delta_{cy} = \gamma_{cy} \cdot L \tag{30}$$

At rupture the rotation γ_{cu} proves to be:

$$\gamma_{cu} = \gamma_{cy} + (\varphi_u - \varphi_y) \cdot \left(\frac{h_c}{2} - l_p^c\right) \tag{31}$$

with

$$\varphi_u = \frac{\varepsilon_{cu}}{c_{cu}} \tag{32}$$

In Eq. (31) the length l_p^c is assumed as previously done for the beam. Finally, the ultimate displacement δ_{cu} is:

$$\delta_{cu} = \delta_{cy} + \gamma_{cu} \cdot L \tag{33}$$

3.3 Joint Contribution

An important review of existing models for shear strength previsions of external and internal beam-to-column joints can be found in Martinelli et al. (2012). Unfortunately, most of these models refer to R.C. structures with deformed bars, while very few models are available for R.C. members with smooth bars (which mainly refer to old constructions). Among them the model of Calvi et al. (2001) here adopted was. According to this model the shear strength stress can be expressed as:

$$v_{jh} = 0.2\sqrt{f_c'}\sqrt{1 + \frac{f_a}{0.2\sqrt{f_c'}}} \tag{34}$$

with f_a the average compressive stress in the column.

The shear distortion of the panel zone according to the continuum approach proves to be:

$$\gamma_j = \frac{v_{jh}}{G} = \frac{2(1+v)}{E} \cdot v_{jh} \tag{35}$$

The horizontal force V_{jh} represents the shear force that the joint is able to transmit being γ_j the angular distortion of the joint, the V_{jh}-γ_j curve can be calculated on the basis of the deformed configuration. Once γ_j and V_{jh} are known, it is possible to obtain the corresponding force F_j and displacement δ_j at the end of the beam with the following expressions:

$$F_j = V_{jh} \cdot \frac{h_b}{L} \tag{36}$$

$$\delta_j = \gamma_j \cdot L \tag{37}$$

3.4   Load-Deflection Response of the Structure

As done in Campione (2015) after calculating the beam load-deflection curve, the column moment-rotation curve and the joint shear-distortion curve, the total response of the structure, in terms of deflection δ of the beam, can be determined by summing the deflection contribution of each component for a fixed load stage F (in equilibrium with the system) with the following relationship:

$$\delta(F) = \delta_b(F) + \delta_c(F) + \delta_j(F) \tag{38}$$

Figure 5 shows two examples of application of the proposed model in terms of load-deflection curves. The first case examined refers to a beam-to-column sub assemblage with: - beam having cross-section 300 × 400 mm, length L = 2500 mm, bottom reinforced with 2 smooth rebars having 12 mm and top reinforcement 2 having diameter 16 mm; - column having cross-section 300 × 300 mm, length H = 3000 mm, reinforced with 2 smooth rebars having diameter 12 mm, both in tension and compression.

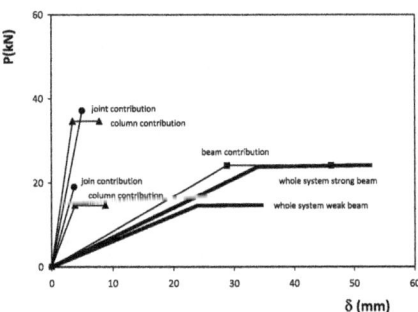

Fig. 5. Theoretical load-deflection response of R.C. external joint

The concrete has compressive strength f_c = 9.95 MPa (0.83 R_{ck} = 12 MPa) and the steel rebars have yielding stress f_y = 280 MPa. The tensile strength of the concrete is neglected. The second example is similar to the first one but the column has the dimensions 300 × 400 mm and is reinforced with 2 smooth rebars having a diameter of 16 mm, both in tension and compression. These examples represent two typical cases of existing R.C. frames designed in the Mediterranean area without seismic detail and referring to a top floor and bottom floor. The comparison shows that in the case of columns and beams both 300 × 400 mm, failure is governed by the beam crisis, the joint and columns having overstrength with respect to the beam. Conversely, for columns with a 300 × 300 mm cross-section, the failure is attained by the column and the lower strength of the joint region. These examples show the effectiveness of the proposed model in predicting the structural response of assemblages in spite of a preliminary verification of the ductility of subassemblages and before more detailed and complex finite element nonlinear analyses.

Figure 6 shows the dimensionless strength domain with variation in the h_c/h_b ratios for two fixed values of $f_c = 15$ and 25 MPa. The strength is the lowest one between F_b, F_c and F_j. The cases examined refer to subassemblies with beams designed for a gravity load with an area of influence of floors of 5 m and a dead load of 60 kN/m^2. The steel grade was 280 MPa and the percentages of steel respected the limit of 0.8% of the area of the concrete transverse cross-section. From the graph it emerges clearly that for h_c/h_b lower than 1 failure is due to the column and the joint. Most existing buildings designed according to these rules do not respect the principles of ductile design and the risk of brittle failure under a seismic attack is very high.

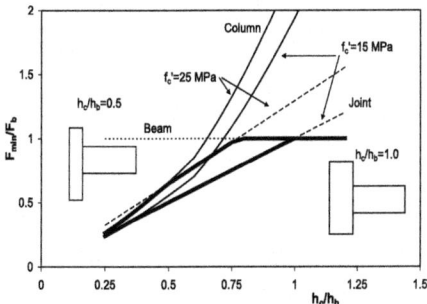

Fig. 6. Dimensionless strength of sub-assemblages with hc/hb variation.

4   Experimental Validation

The cases considered for experimental validation are those shown in Fig. 7, referring to researches by Calvi et al. (2002) and Braga et al. (2009).

The first case examined (Calvi et al. 2002) refers to quasi-static cyclic tests performed on six gravity-load-designed beam-column subassemblies. Six one-way beam-column subassembly specimens, 2/3 scaled, were tested, representing the following typologies: - two exterior knee-joints (specimens L); - two exterior tee-joints (specimens T); - two interior cruciform joints (specimens C). Steel smooth bars, with mechanical properties (allowable stress 160 MPa) similar to those typically used in that period, were adopted for both longitudinal and transverse reinforcements. The main mechanical characteristics of the concrete and reinforcing steel were the following: average cylindrical and cube compression strength equal to 23.9 MPa and 29.1 MPa, respectively; yielding and ultimate stress of the steel reinforcement equal to 385.6 MPa

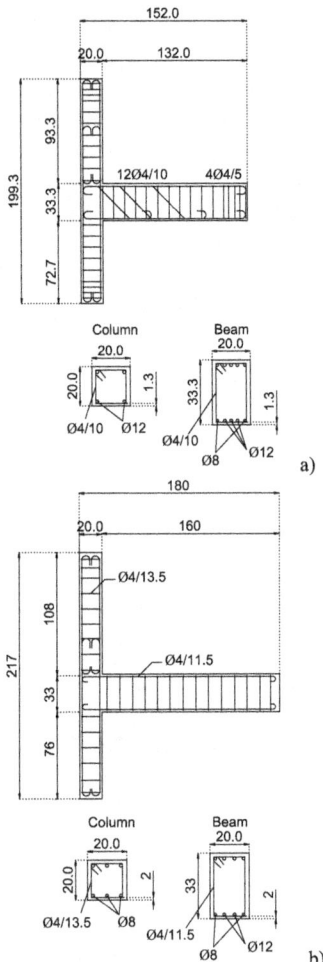

Fig. 7. External RC joint tested in (a) Braga et al. (2009); (b) Calvi et al. (2002).

and 451.2 MPa, respectively, for 8 mm diameter rebars, and 345.9 MPa and 458.6 MPA, respectively, for 12 mm diameter rebar. The test setup for the different specimens was intended to reproduce the configuration of a beam-column subassembly in a frame subjected to reversed cyclic lateral loading. A constant value of axial force due to the gravity loads was applied at 100 kN. The axial load was applied by means of a vertical hydraulic jack, acting on a steel plate connected to the column base plate by vertical external post-tensioned bars. No simulated gravity loads were applied to the beam elements. The second case examined (Braga et al. 2009) refers to an experimental investigation on four internal (C-Joint) and external (T-Joint) R/C beam-column joints built using concrete with low strength and smooth reinforcing bars, without hoops in the panel zone. The tests were performed by increasing cyclic horizontal displacements up to collapse.

The experimental results show that seismic response of this kind of structures is mainly influenced by bond slips of longitudinal bars, and that the shear collapse regards the external joints rather the internal ones. Failure mechanisms observed (column plastic hinging for internal joints, shear failure for external joints) point out the vulnerability of these structures due to the soft story mechanism. All specimens were cast using concrete with average cubic compressive strength equal to 17.5 MPa, while the yield strength of the steel bars was 350 MPa, 325 MPa and 345 MPa, for the 8 mm, 12 mm and 18 mm diameter bars, respectively. The tests were carried out in a quasi-static way applying a time-history of displacements to the upper column. A vertical load was applied at the head of the upper column equal to 120 kN. The proposed model is applied to the reference dataset and the results are shown in Fig. 8. They show good agreement both in terms of mechanisms of failure than of ultimate load and corresponding displacements.

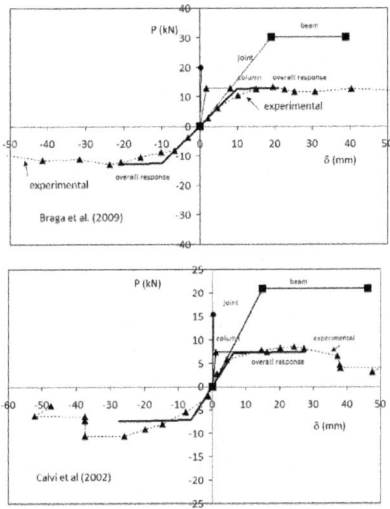

Fig. 8. Theoretical load-deflection response of R.C. external joint.

5   Conclusions

In the present paper an analytical model for predicting the flexural response of external old type reinforced concrete beam-column external joints with smooth bars under monotonic loading is presented.

Yielding of main steel and crushing of concrete in the beam including bars slippage, in the column and in the beam-column joint were identified in order to determine the corresponding loads and displacement and to plot the simplified load-deflection curves of the sub-assemblages subjected at the tip of the beam to monotonically increasing lateral force and in the column to a constant vertical load. The original contribution consists in including in a simple analytical model the main aspects regarding the structural behavior of external beam-joint-column such as slippage on longitudinal bars of beam and brittle failure of joint. The flexural failure of the columns, due to strong beam/weak column design, and the brittle collapse of the external joint in the absence of stirrups due to crushing of the compressed strut are included in the model. The expressions derived, allowing simple hand computation, can be considered a useful instrument for a preliminary verification of the safety state. On the whole, the model gives a physical interpretation of the flexural behaviour of beam-column sub assemblages up to rupture and shows good agreement with the experimental results available in the literature. Finally, prescriptions on mechanical ratios of steel rebars for ductile design of sub-assemblages are derived and discussed.

References

ACI Committee 318 (2011) Building code requirements for structural concrete ACI 318 and commentary. American Concrete Institute, Farmington Hills, Michigan

Attaalla SA (2004) General analytical model for nominal shear stress of type 2 normal-and high-strength concrete beam-column joints. ACI Struct J 101(1):65–75

Bakir PG, Boduroğlu HM (2002) A new design equation for predicting the joint shear strength of monotonically loaded exterior beam–column joints. Eng Struct 24(8):1105–1117

Braga F, Gigliotti R, Laterza M (2009) R/C existing structures with smooth reinforcing bars: experimental behaviour of beam-column joints subject to cyclic lateral loads. Open Constr Build Technol J 3:52–67

Calvi GM, Pampanin S, Moratti M, Ward J, Magened G, Pavese A, Rasulo A (2001) Seismic vulnerability assessment of existing concrete buildings. Experimental tests on a three-story RC frame. Technical report, University of Pavia, Pavia, Italy

Calvi GM, Magenes G, Pampanin S (2002) Relevance of beam-column damage and collapse in RC frame assessment. J Earthq Eng 6(2):75–100

Campione G (2015) Analytical prediction of load deflection curves of external steel fibers R/C beam–column joints under monotonic loading. Eng Struct 83:86–98

Ehsani MR, Wight JK (1985) Exterior reinforced beam-to-column connection subjected to earthquake-type loading. ACI Struct J 82:492–499

Eurocode 8 (2005) Structures in seismic regions, European Committee for Standardization (CEN), ENV 2008

Fabbrocino G, Verderame GM, Manfredi G, Cosenza E (2004) Structural models of critical regions in old type R.C. frames with smooth rebars. Eng Struct 26:2137–2148

Fib Bulletin No. 24 (2003) Seismic assessment and retrofit of reinforced concrete buildings. State-of-art report, 312 pages. ISBN 978-2-88394-064-2

Hegger J, Sherif A, Roeser W (2003) Non seismic design of beam–column joints. ACI Struct J 100(5):654–64

Hegger J, Sherif A, Roeser W (2004) Nonlinear finite element analysis of reinforced concrete beam-column connections. ACI Struct J 101(5):604–614

Hwang SJ, Lee HJ (1999) Analytical model for predicting shear strength of exterior reinforced concrete beam-column joints for seismic resistance. ACI Struct J 96(5):846–858

Lima C, Martinelli E, Faella C (2012a) Capacity models for shear strength of exterior joints in RC frames: state-of-the-art and synoptic examination. Bull Earthq Eng 10:967–983

Lima C, Martinelli E, Faella C (2012b) Capacity models for shear strength of exterior joints in RC frames: state-of-the-art and synoptic examination. Bull Earthq Eng 10:967–983. https://doi.org/10.1007/s10518-012-9340-4

Lima C, Martinelli E, Faella C (2012c) Capacity models for shear strength of exterior joints in RC frames: experimental assessment and recalibration. Bull Earthq Eng 10:985–1007. https://doi.org/10.1007/s10518-012-9342-2

Pampanin S, Calvi GM, Moratti M (2002) Seismic behaviour of R.C. beam-column joints designed for gravity loads. In: 12th European conference on earthquake engineering, Barbican Centre, London, UK, 9–13 September 2002. Paper Reference 726

Park S, Mosalam KM (2012) Parameters for shear strength prediction of exterior beam–column joints without transverse reinforcement. Eng Struct 36:198–209

Park R, Paulay T (1975) Reinforced concrete structures. Wiley, New York, 769 pp

Priestley MJN (1997) Displacement-based seismic assessment of reinforced concrete buildings. J Earthq Eng 1(1):157–192

Pauletta M, Di Luca D, Russo G (2015) Exterior beam column joints – shear strength model and design formula. Eng Struct 94:70–81

Russo G, Pauletta M (2012) Seismic behavior of exterior beam-column connections with smooth rebars and effects of upgrade. ACI Struct J 109(2):225–233

Moment-Axial Domain of Corroded R.C. Columns

G. Campione$^{(\boxtimes)}$, F. Cannella, L. Cavaleri, M. F. Ferrotto,
and M. Papia

Department of Civil, Environmental, Aerospace, Materials Engineering,
Università degli Studi di Palermo, Palermo, Italy
giuseppe.campione@unipa.it

Abstract. In the present paper, a simplified model to determine the moment-axial force domain of the cross-section of reinforced concrete columns subjected to corrosion process is presented. The model considers members with square and rectangular cross-sections and it accounts for cover spalling, buckling of longitudinal reinforcing bars, loss of bond of bar in tension, reduction of confinement pressures (due to the reduction of the area of stirrups and cracking of concrete induced by rust formation). The analytical expressions for prediction of the area reduction of steel, bond strength and critical load of longitudinal bars utilized were verified against experimental data available literature.

Keywords: Moment-axial force domain · Confinement · Buckling
Corrosion · Cover spalling

1 Introduction

It is widely accepted that general and pitting corrosion of reinforcement affects reinforced concrete (R.C.) structures by reducing the cross-sectional area and the mechanical properties of the reinforcement itself, especially when pitting corrosion occurs. Pitting is a localized corrosion type. By contrast, general corrosion is distributed along the bars. Moreover, loss of bond between the steel and concrete and cracking of concrete in the zone of rust formation cause a reduction in the strength and stiffness of the reinforced concrete members.

Based on the literature review it could be seen that individual aspects of reinforced concrete element deterioration are well covered. In addition, literature related to evaluation of the load carrying capacity of deteriorated reinforced concrete columns subjected to axial force and bending moment is available. More work is required to develop a suitable and simple methodology for strength evaluation of deteriorated reinforced concrete columns that incorporates all corrosion-related factors and deterioration scenarios.

For concrete columns cover spalling, buckling of steel bars, reduction of steel area due to rust formation and in some cases opening of the stirrups are the main visible effects due to corrosion.

Experimental researches (Rodriguez et al. 1994, 1996) have been carried out in an attempt to understand the behavior and bearing capacity of corroded R.C. columns. Also, some in situ long term investigations are available (Ismail et al. 2010). At present

© Springer International Publishing AG, part of Springer Nature 2018
M. di Prisco and M. Menegotto (Eds.): ICD 2016, LNCE 10, pp. 440–453, 2018.
https://doi.org/10.1007/978-3-319-78936-1_32

few simple analytical and numerical models have been developed and calibrated with the experimental results to predict the bearing capacity of the moment-axial force domain of corroded R.C. columns (Tapan and Aboutaha 2008, 2011).

In this paper, the effects of reinforcement corrosion, loss of concrete cover, and loss of bond on the structural behavior of R.C. columns with rectangular cross-section are quantified. Moment-axial force domains (M-N) using a simplified model are proposed and verified against experimental data available in the literature.

2 Moment-Axial Forces Domain

The cases examined are those shown in Fig. 1. They refer to short R.C. members having rectangular (or square) cross-sections with side b and height H and reinforced with n longitudinal steel bars with diameter ϕ_l and area A_l, and confined by transverse closed steel stirrups with diameter ϕ_{st} and area A_{st}. Transverse steel is placed in the plane of the cross-section at clear spacing s with a cover δ. In the case of bars distributed up to four layers of reinforcement, an equivalent area of main bars was considered applied to the center of the bar in tension and in compression.

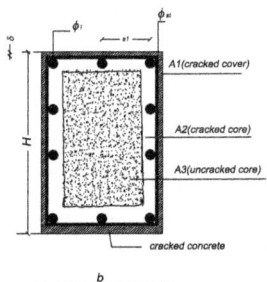

Fig. 1. Cross-sections analyzed

Figure 2 shows the moment-axial force domain of the transverse cross-section obtained numerically using the layer method and indicated with the dashed line. In the same graph the continuous line also shows the simplified model adopted here (Bergmann et al. 1995).

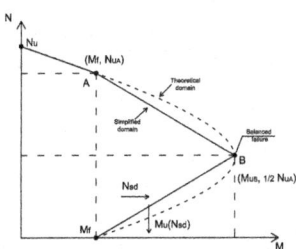

Fig. 2. Moment axial force domain

This model considers three linear branches:- the first one connects the point of pure compression N_c with the point A, the second branch is the segment A-B and the third branch connects B with the point of pure bending M_f. Point A corresponds to a point having the same bending moment of pure flexure M_f and the axial force equal to N_A to be determined. Point B is characterized by an axial force $N_B = 0.5 \cdot N_A$ and a bending moment M_B to be determined. Figure 2 shows the simplified linearized moment-axial force domain fitting the ones obtained numerically. In the next section details for calculation of single points of the interaction diagram are given.

2.1   Case of Pure Compression

As suggested in Campione et al. (2016), the load-carrying capacity of corroded compressed R.C. columns (see Fig. 1) can be determined as the sum of the four different strength contributions due to:- the unconfined concrete cracked cover area (A_{1c}); - the confined cracked concrete of the core area across the bars (A_{2c}); - the internal area of concrete core which is in a triaxial stress state (A_{3c}); - the longitudinal bars $A_{l,red}$ including buckling phenomena.

The expression that gives the load-carrying capacity in compression for the case of a member with a rectangular cross-section is the following:

$$
\begin{aligned}
N_u = &\ \psi \cdot f'_c \cdot \left[(2 \cdot b \cdot \delta + 2 \cdot H \cdot \delta) - 2\delta^2 \right] \\
&+ \psi \cdot f_{cc} \cdot \left\{ b \cdot H - \left[(2 \cdot b \cdot \delta + 2 \cdot H \cdot \delta) - 2\delta^2 \right] - \left[(b - 2 \cdot \delta - 2 \cdot \phi_{st} - 2 \cdot \phi_l) \cdot (H - 2 \cdot \delta - 2 \cdot \phi_{st} - 2 \cdot \phi_l) \right] \right\} \\
&+ f_{cc} \cdot \left[(b - 2 \cdot \delta - 2 \cdot \phi_{st} - 2 \cdot \phi_l) \cdot (H - 2 \cdot \delta - 2 \cdot \phi_{st} - 2 \cdot \phi_l) \right] \\
&+ \beta \cdot \left(A_{l,red} \cdot f_y \right)
\end{aligned}
$$

$$(1)$$

f_{cc} being the compressive strength of the confined concrete, ψ a reduction factor of the compressed concrete in the zone of rust formation and β the reduction factor of the yielding stress of the longitudinal bars due to a buckling effect.

The reduction in ψ, as suggested in Coronelli and Gambarova (2004), can be calculated by considering that the cracking induced by the expansion of the corroded longitudinal and transverse steel bars (rust effect) degrades the strength of the compressed concrete because of the increase in cracking. If the lateral strain, which causes longitudinal micro-cracks, is assumed to be smeared on the cracks, we have:

$$
\psi = \frac{f^*_c}{f_c} = \frac{1}{1 + k \cdot \frac{2 \cdot \pi \cdot n_{bars} \cdot (v_{rs} - 1) \cdot X}{b \cdot \varepsilon_0}}
$$

$$(2)$$

with $k = 0.1$ as suggested in Coronelli and Gambarova (2004), and ε_0 assumed 0.002 for normal strength, normal weight concrete.

X is the thickness of the corrosion attack penetration, which can be measured with the gravimetric method or calculated, as suggested in Val (2007), in a rearranged form of Faraday's law of electrolysis, as:

$$X = 0.0116 \cdot i_{corr} \cdot t \tag{3}$$

where i_{corr} is the corrosion current density in the reinforcing bar expressed in $\mu A/cm^2$ and t the time in years. The compressive strength f_{cc} of the confined concrete was calculated as in Razvi and Saatcioglu (1999) in the form:

$$\frac{f_{cc}}{f_c} = 1 + 6.7 \cdot \left(\frac{f_{le}}{f_c}\right)^{-0.17} \tag{4}$$

with the confinement pressure f_{le} calculated as in Razvi and Saatcioglu (1999), but taking into account, as in Campione et al. (2016), that if corrosion processes are present the confinement pressure has to be reduced. This reduction is due to the reduction of the area of transverse steel bars and to the available stress in the stirrup due to rust formation. It also has to be stressed that although it is true that the expansion of rusted material produces premature loading on the stirrups, we also know that preloading should confine the concrete surrounded by stirrups and longitudinal reinforcement. This effect was not considered in the paper, also because at ultimate stress the concrete cover is spalled off and the longitudinal bars buckle.

The effective confinement pressures prove to be:

$$f_{le} = \left(\frac{2 \cdot A_{st,red} \cdot \sigma_s}{b \cdot s}\right) \cdot \left(0.15 \cdot \sqrt{\frac{b}{s} \cdot \frac{b}{s_1} \cdot \left(\frac{2 \cdot A_{st,red} \cdot \sigma_s}{b \cdot s}\right)}\right) \tag{5}$$

where $A_{st,red}$ is the reduced area of the stirrups, σ_s the available stress in the stirrups, and s_1 the spacing of the laterally supported longitudinal reinforcing.

Considering the free expansion of the side of the cross-section due to rust formation in the external bars and, consequently, the elongation of the stirrup, the stress in the stirrup proves to be:

$$\sigma_s = f_y \cdot \left[1 - \frac{2 \cdot X}{b} \cdot \frac{E_s}{f_y}\right] \tag{6}$$

The reduction of the stirrup area due to rust formation and pitting, if it occurs, can be derived, as in Campione et al. (2016), in the form:

$$A_{st,red}(t) = n_{bar} \cdot \left\{\frac{\pi \cdot [\phi_l - 2 \cdot X]^2}{4} - A_p(t)\right\} \tag{7}$$

A_p is the cross-sectional area of pitting calculated with the expression in Val (2007).

It has to be stressed that, as is well known from the literature, there is no significant reduction of yielding stress due to corrosion processes. Apostolopoulos and Papadakis (2008), Fernandez et al. (2015), Fernandez et al. (2016) and Biondini and Vergani (2015) show that, with an increase in the corrosion level, the yield stress of the bar decreases slowly (less than 15%). In almost all cases, failure of the bar is due to the presence of a brittle region owing to the presence of pitting. Otherwise, the presence of pitting causes a reduction of the ductility of the bars, as can be seen in Stewart (2009).

For longitudinal compressed bars, if the concrete cover is spalled off due to rust formation, the risk of buckling increases. In this case, the maximum allowable stress in the longitudinal bars is the minimum among the yielding stress and the critical stress. The latter can be calculated, as in Campione et al. (2016), in the form:

$$\sigma_{cr} = \frac{3.46 \cdot \sqrt{E_r \cdot I \cdot k_l}}{A_l} \quad (\mathrm{N/mm^2}) \tag{8}$$

$$E_r = \frac{4 \cdot E_s \cdot E_p}{\left(\sqrt{E_s} + \sqrt{E_p}\right)^2} \tag{9}$$

with I the moment of inertia of longitudinal bar, E_p the hardening modulus of steel and k_l a distributed stiffness parameter in the form:

$$k_l = \frac{E_P \cdot A_{st}}{b \cdot s} \cdot \sqrt{2} \tag{10}$$

for a corner bar and

$$k_l = \frac{48 \cdot E_P \cdot I_{st}}{s_1^3 \cdot s} \quad \text{for a mid-face bar} \tag{11}$$

Finally, the reduction factor β of the yield stress in the longitudinal bar due to buckling can be defined as $\beta = \frac{\sigma_{cr}}{f_y}$.

Under corrosion, Eqs. (8, 10, 11) are still utilized adopting the area and the diameter of bars deduced with Eq. (7). If corrosion is due to a pitting effect, the moment of inertia has to be modified. A simplification can be obtained numerically by deriving the diameter of an equivalent bar having the same inertia, giving the following expression:

$$\frac{\phi_{l,red}}{\phi_l} - 1 + 0.0029 \cdot X_p - 0.003 \cdot X_p^2 \tag{12}$$

with X_p the loss of mass in %.

Figure 3 gives experimental results for compressed bars having pitting corrosion. Then a comparison between the analytical results (Eq. 8) and experimental results is given. The scatter between the analytical and experimental results is in the range of 10%. The data utilized of Kashani et al. (2013) refer to compressive tests on single bars with different s/ϕ_l ratios and different levels of loss of mass. The comparison shown in Fig. 3 highlights that in most of the cases examined the model captures the experimental results, emphasizing the importance of including the pitting attack for calculation of the reduced moment of inertia of the cross-section of longitudinal bars.

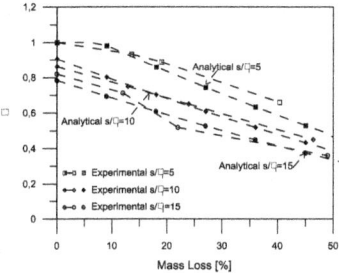

Fig. 3. Normalized reduction of yielding stress in compressed bar with loss of mass in %.

2.2   Case of Pure Flexure

The reference model for prediction of the flexural strength of an uncorroded R.C. section is based on the hypothesis that the plane section theory is applied because of a perfect bond between bars and concrete. In the case of corroded bars, slippage between concrete and steel bars occurs, reducing the strength and the available ductility.

To reproduce the worst condition, it was supposed that pitting and general corrosion occur in all bars and in a section where these effects are very important. This case could be representative of the anchorage zone of longitudinal bars of columns or in zones where high shear forces are present. Moreover, the model is valid only for sections with all reinforcement corroded. Studies in the literature (Tapan and Aboutaha 2008, 2011; Wang and Liang 2008; Guo et al. 2015) analyse cases in which the corrosion level is different along the four sides of the cross-section. The case examined here is the one of greatest interest because it represents the worst condition.

To include the slippage of longitudinal bars due to rust formation a reduction factor γ of the yielding stress of longitudinal bar was introduced. Uniform reduction of the yielding stress of longitudinal bars between two successive flexural cracks was supposed. The expression utilized in Rodriguez et al. (1996) to derive bond strength is:

$$q_{res} = 0.6 \cdot \left(0.5 + \frac{\delta}{\phi_l} \right) \cdot f_{ct} \cdot (1 - \lambda \cdot X^\mu) + \frac{k_q \cdot A_{st} \cdot f_y}{s \cdot \phi_l} \qquad (13)$$

with f_{ct} the tensile strength of the concrete expressed as a function of the characteristic compressive strength f_{ck} in Eurocode 2 (2004) as:

$$f_{ct} = 0.30 \cdot f_{ck}^{2/3} \quad \text{(MPa)} \qquad (14)$$

k_q, μ, λ being empirical constants. According to Rodriguez et al. (1996), k_q and μ are equal to 0.16, 0.1 respectively, while λ is between 0.26 and 0.4 and assumed here to be 0.4.

In the absence of stirrups, assuming $\delta = \phi_l$ Eq. (13) gives a reduction factor of bond strength γ in the form:

$$\gamma = \frac{q_{res}}{q_o} = \frac{1}{1 + \frac{k_q \cdot A_{st} \cdot f_y}{s \cdot \phi_l}} \left[(1 - \lambda \cdot X^\mu) + \frac{\frac{k_q \cdot A_{st} \cdot f_y}{s \cdot \phi_l}}{0.18 \cdot \left(0.5 + \frac{\delta}{\phi_l}\right) \cdot f_c^{\frac{2}{3}}} \right] \quad (15)$$

with X related to X_p through the expression:

$$X_p = 1 - \left(1 - \frac{2 \cdot X}{\phi_l}\right)^2 \quad (16)$$

As can be seen in Fig. 4, the position of the neutral axis c_{cu} and the ultimate flexural strength M_{us} are in the form:

$$c_{cu} = \frac{A_{l,red} \cdot (\gamma - \beta) \cdot f_y}{0.85 \cdot f_{cc} \cdot 0.80 \cdot b} \quad (17)$$

$$M_f = A_{l,red} \cdot (\gamma + \beta) \cdot f_y \cdot (d - \delta) \\ + -\frac{1}{2} \cdot A_{l,red} \cdot \gamma \cdot f_y \cdot (\phi_l + 0.80 \cdot c_{cu}) \quad (18)$$

with $d = H - \phi_{st} - \phi_l/2$ and $A_{l,red}$ the reduced area of a reinforcement layer (tension or compression).

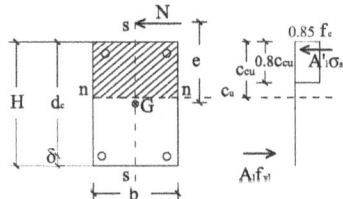

Fig. 4. Design assumptions for analysis of single R.C. column

In Eq. (17, 18) the γ coefficient was introduced to take bond degradation into account. The flexural strength should also not exceed the moment capacity M_{uc} for compression failure of corroded beams (Mac Gregor 1997), calculated as:

$$M_{uc} = 0.125 \cdot \psi \cdot f_c \cdot b \cdot \left(d - \delta - \frac{\phi_l}{2}\right)^2 \quad (19)$$

2.3   Case of Axial Force and Bending Moment

With reference to the simplified moment-axial force domain in Fig. 2 it is possible to obtain the point coordinate of A by solving the following translational and rotational equilibrium equations:

$$N_A = 0.85 \cdot c_{cA} \cdot b \cdot 0.80 \cdot f_{cc} + A_{l,red} \cdot (\beta - \gamma) \cdot f_y \tag{20}$$

$$
\begin{aligned}
M_A = N_A \cdot e &= (0.85 \cdot f_{cc} \cdot 0.80 \cdot c_{cA} \cdot b) \cdot \left(\frac{H}{2} - \delta - 0.4 \cdot c_{cA} \right) \\
&+ A_{l,red} \cdot (\beta + \gamma) \cdot f_{yl} \cdot \left(\frac{H}{2} - \delta - \frac{\phi_l}{2} \right)
\end{aligned}
\tag{21}
$$

By imposing $M_A = M_f$ and solving Eq. (21) with respect to c_{cA}, a second-degree equation is obtained that if solved gives c_{cA}. Substituting the value of c_{cA} in Eq. (20) gives N_A.

Point B in the interaction diagram corresponds to the so-called balanced failure condition. This point is known to be located around $0.4 - 0.6 \cdot A_c \cdot f_c$. However, it depends very much on the reinforcement arrangement. Nevertheless, it is possible to derive rational equations based on plane strain distributions. In this paper the coordinate of point B of Fig. 2 was obtained following the procedure proposed in Bergmann et al. (1995) and Eurocode 4 (2004) in which was set $N_B = 0.5 N_A$. The position of the neutral axis, under these hypotheses, is equal to:

$$c_{cB} = \frac{0.5 \cdot N_A + A_{l,red} \cdot f_y(\gamma - \beta)}{0.8 \cdot f_{cc} \cdot b} \tag{22}$$

and the ultimate moment was obtained from Eq. (21) with substitution of Eq. (22):

$$
\begin{aligned}
M_B &= (0.85 \cdot f_{cc} \cdot 0.80 \cdot c_{cB} \cdot b) \cdot \left(\frac{H}{2} - \delta - 0.4 \cdot c_{cB} \right) \\
&+ A_{l,red} \cdot f_{yl} \cdot \left(\frac{H}{2} - c - \frac{\phi_l}{2} \right) \cdot (\gamma + \beta)
\end{aligned}
\tag{23}
$$

The ultimate moment associated with the design axial force N can be calculated in the following form:

$$M_u(N) = M_u + N \cdot \left(\frac{M_B - M_u}{N_B} \right) \quad \text{with} \quad N \leq N_B \tag{24}$$

or in dimensionless form:

$$m_u(n) = \frac{M_u(N)}{b \cdot d_c^2 \cdot f_c} = m_u + \frac{n}{n_B} \cdot (m_B - m_u) \quad \text{with} \quad n \leq n_B \tag{25}$$

3   Experimental Validation

The data in Tapan and Aboutaha (2008, 2011), Wang and Liang (2008) and Guo et al. (2015) referring to members with square or rectangular sections were utilized in order to validate the proposed model.

Tapan and Aboutaha (2008) analyzed one column with a cross-section of width 1350 mm and height 1830. The longitudinal reinforcement was constituted by nine 36-mm bars for each side with a cover equal to 80 mm. The stirrups, constituted by deformed bars, had a diameter of 12 mm. The authors considered six deterioration cases with two lengths of exposed bars. In this paper, only the results of case with corrosion at all bars are compared with the proposed model.

Figure 5 shows the moment-axial force domain obtained with the proposed model and with the numerical procedure indicated in Tapan and Aboutaha (2008). The comparison is quite satisfactory for the case of flexure and pure compression, while in the case of bending moment and axial force the proposed model slightly overestimated the numerical results with scatter within 5%.

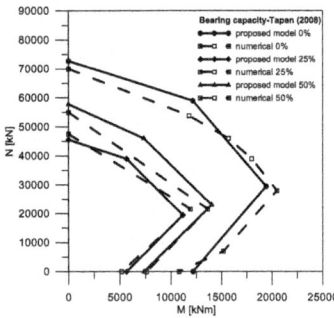

Fig. 5. Comparison between analytical and experimental interaction diagrams

Wang and Liang (2008) analyzed a column with a square cross-section of side 610 mm. The longitudinal reinforcement was constituted by four 28-mm bars for each side, with a cover of 28 mm. The stirrups were constituted by 12-mm bars with a pitch equal to 250 mm. The authors considered six deterioration cases. In this paper, only the results of case with corrosion at all bars are compared with the proposed model.

Figure 6 shows the moment-axial force domain obtained with the proposed model and with the numerical procedure indicated in Wang and Liang (2008). The comparison is also satisfactory in this case. For a loss of mass of 50% the scatter is higher and in the range of 15% for compression.

Figures 5 and 6 clearly also show that for the case in Tapan and Aboutaha (2008) with the increase in the loss of mass the load-carrying capacity decreases significantly. For 10% and 25% the reduction of axial load is almost 35 and 60%. In flexure, similar results are obtained. For the case in Wang and Liang (2008) the analogous decrements of load-carrying capacity for 25 and 50% of loss of mass were 20 and 39%, while in flexure they were 38 and 58%. It is important to stress that if the buckling effect and

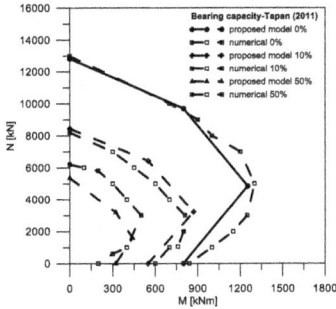

Fig. 6. Comparison between analytical and experimental interaction diagrams.

loss of bond are neglected the reduction of load-carrying capacity and flexural capacity can be addressed only to the area reduction of the steel bars and to the cover spalling, but it gives unsafe results.

For analyses of experimental results, it has to be stressed that failure was controlled by bar buckling, bond deterioration and concrete strength degradation. More specifically, in the case of flexure bond, degradation and buckling of longitudinal bars are the most important phenomena governing failure, while in the case of compression, buckling and reaction of concrete strength were the main phenomena covering the problem. In the case of axial force and bending moment all phenomena occur but for $N < N_B$ buckling and bond failure are predominant, while for $N > N_B$ concrete strength reduction plays an important role too.

Figure 7 shows for the case of pure compression the ratio between the analytical and numerical results given in Tapan and Aboutaha (2008) and Wang and Liang (2008) for different losses of mass, highlighting the good agreement between the numerical and the analytical prediction.

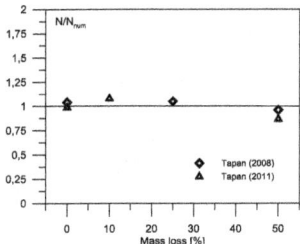

Fig. 7. Comparison between analytical and numerical ultimate axial force

Tapan and Aboutaha (2011) tested twelve columns with a square cross-section of side 200 mm. T longitudinal reinforcement was constituted by four 18-mm deformed bars, one in each corner with a 30-mm cover. The stirrups, constituted by deformed bars, had a diameter of 8 mm and pitch equal to 100 mm. Specimens were corroded by adding 3.5% of calcium chloride ($CaCl_2$) and by connecting the tensile bars with a DC power supply. The columns were tested with two different types of eccentric distances (Type ZX with small eccentricity and Type ZD with large eccentricity), two different positions of the corroded zone (Type L corresponding to a corroded tensile zone and Type Y to a corroded compressed zone) and different types of corrosion levels that were defined in terms of average weight loss of the steel. There were three columns with a high corrosion level (ZDL700-2, ZDL350-3 and ZDY350-3), two with a low level (ZDL700-1 and ZDY700-1) and others with intermediate values.

Guo et al. (2015) tested four columns with cross-section having width equal to 600 mm and height 250 mm. The longitudinal reinforcement was constituted by twelve 16-mm bars, with a cover thickness of 25 mm. The stirrups, constituted by deformed bars, had a diameter of 8 mm and spacing of 150 mm. Four specimens were tested, with a different level of corrosive damage, calculated as mass loss of the longitudinal reinforcement: 0, 5, 10 and 15% for specimen 1, 2, 3 and 4, respectively. The effects of accelerated corrosion were cracks at the corners of the specimen due to corrosion and volume expansion of the corrosion product of the longitudinal rebar and the stirrups. Figure 8a–b shows for the case of flexure under constant axial force the ratio between analytical and experimental results given in Tapan and Aboutaha (2008) and Guo et al. (2015) for different loss of mass, highlighting the good agreement for almost all cases examined between numerical and analytical prediction.

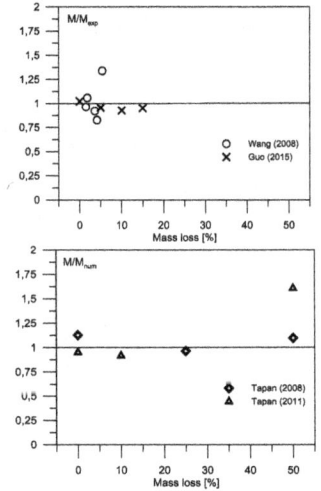

Fig. 8. Comparison between analytical and (a) experimental ultimate bending moment (b) numerical ultimate bending moment

Similarly, Fig. 9 shows for the case of balanced failure the ratio between analytical and experimental results given in Tapan and Aboutaha (2008) and Wang and Liang (2008) for different loss of mass highlighting the good agreement between numerical and analytical prediction.

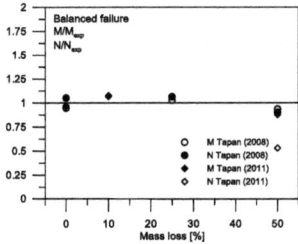

Fig. 9. Comparison between analytical and experimental ultimate axial force bending moment

4   Conclusions

In the present paper, a simplified model to calculate the moment-axial force domain of R.C. corroded columns subjected to corrosion processes was developed and verified against selected experimental data.

The simplified domain is constituted by three branches and four points. To take into account loss of load-carrying capacity due to corrosion processes, the model considers:- cover expulsion; - cracking of the portion of confined core close to the longitudinal bars and stirrups; - reduction of steel area; - buckling of compressed reinforcement; - loss of bond of tensile reinforcement.

The limitations of the proposed model, which covers several cases of practical interest, are mainly that the following: the corrosion process of bars involves them in the same manner; cover spalling occurs along the four sides of the section.

The main results obtained can be summarized as follows:

- The reduction in the load-carrying capacity due to loss of mass of longitudinal bars is not relevant for compressed members designed for static loads;
- The reduction in the load-carrying capacity is between 20% and 30% for severe corrosion conditions because of degradation of the compressive strength of the outer portion of the column.
- In flexure, loss of bond plays a fundamental role with respect to loss of area, and loss of mass of 15–20% produces a loss of strength of 30–40%.
- Under an axial load and bending moment, the reduction of flexural capacity depends on the loss of area of the reinforcement (longitudinal and stirrups) and of the loss of bond and buckling of compressed bars in which pitting plays an important role.
- For low levels of axial force, the main effect is the loss of flexural capacity, which in the case of a loss of mass of 25% can produce a reduction of up to 45%.

Nel seguente lavoro è stato sviluppato un modello semplificato per calcolare i domini in pressoflessione di colonne in cemento armato corrose, attraverso un confronto con i dati sperimentali disponibili in letteratura. Il dominio semplificato è costituito da tre rami e quattro punti. La perdita di capacità portante dovuta ai processi corrosivi è stata calcolata considerando:- espulsione del copriferro; - fessurazione della parte interna del calcestruzzo della colonna vicino le barre longitudinali e le staffe; -

riduzione dell'area dell'armatura; -instabilità delle barre in compressione; -perdita di aderenza acciaio-calcestruzzo.

Le limitazioni del modello proposto, che comunque copre molti casi di interesse pratico sono:- il processo corrosivo riguarda tutte le barre in eguale maniera; - l'espulsione del copriferro si ha in tutti i lati della sezione.

I principali risultati ottenuti con il seguente lavoro sono:

- La riduzione della capacità portante dovuta alla perdita di massa delle barre longitudinali non è importante per gli elementi compressi progettati per carichi statici.
- La riduzione nella capacità portante è tra il 20% e il 30% per condizioni di corrosione elevata, dovuta alla riduzione di resistenza a compressione della parte esterna della colonna.
- In flessione, la riduzione dell'aderenza dovuta alla perdita di massa gioca un ruolo fondamentale. Infatti, con una perdita di masse del 15–20% si ha una riduzione dell'aderenza del 30–40%.
- In presenza di forza assiale e momento flettente, la riduzione della capacità flessionale (legato principalmente al pitting) dipende dalla perdita di area delle barre (longitudinali e staffe), dalla riduzione della tensione di aderenza e dall'instabilità delle barre compresse.
- Per bassi livelli di forza assiale, il principale effetto è la riduzione della resistenza a flessione che, con una perdita di massa del 25% può ridursi fino al 45%.

Acknowledgements. The research was performed within the 2014/18 Research Project "DPC–ReLUIS (Dipartimento Protezione Civile - Rete dei Laboratori Universitari di Ingegneria Sismica)", Linea di Ricerca Cemento Armato. The related financial support was greatly appreciated.

References

Apostolopoulos CA, Papadakis VG (2008) Consequences of steel corrosion on the ductility properties of reinforcement bar. Constr Build Mater 22:2316–2324
Bergmann R, Matsui C, Meinsma C, Dutta D (1995) Design guide for concrete filled hollow section columns under static and seismic loading. Verlag TUV Rheinhald
Biondini F, Vergani M (2015) Deteriorating beam finite element for nonlinear analysis of concrete structures under corrosion. Struct Infrastruct Eng 11(4):519–532
Campione G, Cannella F, Minafo G (2016) A simple model for the calculation of the axial load-carrying capacity of corroded RC columns. Mater Struct 49:1935–1945
Coronelli D, Gambarova P (2004) Structural assessment of corroded reinforced concrete beams: modeling guidelines. ASCE J Struct Eng 130:1214–1224
Eurocode 4 (2004) Design of composite steel and concrete structures. EN 1994-1-1
Fernandez I, Bairàn JM, Marì AR (2015) Corrosion effects on the mechanical properties of reinforcing steel bars. Fatigue and σ-ε behavior. Constr Build Mater 101:772–783
Fernandez I, Bairàn JM, Marì AR (2016) Mechanical model to evaluate steel reinforcement corrosion effects on σ-ε and fatigue curves. Experimental calibration and validation. Constr Build Mater 118:320–333

Guo A, Li H, Ba X, Guan X, Li H (2015) Experimental investigation on the cyclic performance of reinforced concrete piers with chloride-induced corrosion in marine environment. Eng Struct 105:1–11

Ismail M, Muhammad M, Ismail ME (2010) Compressive strength loss and reinforcement degradations of reinforced concrete structure due to long-term exposure. Constr Build Mater 24:898–902

Kashani MW, Crewe AJ, Alexander NA (2013) Nonlinear stress-strain behavior of corrosion-damaged reinforcing bars including inelastic buckling. Eng Struct 48:417–429

MacGregor JG (1997) Reinforced concrete: mechanics and design, 3rd edn. Prentice-Hall, Upper Saddle River

Razvi S, Saatcioglu M (1999) Confinement model for high-strength concrete. J Struct Eng ASCE 125(3):281–288

Rodriguez J, Ortega L, Garcia A (1994) Corrosion of reinforcing bars and service life of R/C structures: Corrosion and bond deterioration. In: Proceedings of international conference on concrete across borders, vol II, pp 315–326

Rodriguez J, Ortega LM, Casal J (1996) Load carrying capacity of concrete structures with corroded reinforcement. Constr Build Mater 11(4):239–248

Stewart MG (2009) Mechanical behavior of pitting corrosion of flexural and shear reinforcement and its effect on structural reliability of corroding RC beams. Struct Saf 31:19–30

Eurocode 2 (2004) Design of concrete structures, EN-1992-1-1

Tapan M, Aboutaha RS (2008) Strength evaluation of deteriorated RC bridge columns. J Bridge Eng 13(3):226–236

Tapan M, Aboutaha RS (2011) Effect of steel corrosion and loss of concrete cover on strength of deteriorated RC columns. Constr Build Mater 25:2596–2603

Val DV (2007) Deterioration of strength of RC beams due to corrosion and its influence on beam reliability. ASCE J Struct Eng 133(9):197–1306

Wang XH, Liang FY (2008) Performance of RC columns with partial length corrosion. Nucl Eng Des 238(12):3194–3202

Correlation of In-Situ Material Characterization Tests and Experimental Performances of RC Members

Ciro Del Vecchio[✉], Marco Di Ludovico, Andrea Prota,
Edoardo Cosenza, and Gaetano Manfredi

Department of Structures for Engineering and Architecture,
University of Napoli Federico II, Naples, Italy
{ciro.delvecchio,diludovi,aprota,ccosenza,
gamanfre}@unina.it

Abstract. Existing RC buildings of the Mediterranean area commonly show high variability in the material mechanical properties because of the construction age and manufacturing processes. Such a variability has been confirmed by number of in-situ tests performed during the reconstruction process in the aftermath of L'Aquila 2009 earthquake. In this context, destructive and non-destructive characterization tests are useful supporting tools to estimate the actual concrete mechanical properties and to improve the accuracy of the seismic capacity assessment. The paper reports of a wide material characterization program consisting of destructive and non-destructive tests carried out on a building severely damaged by the L'Aquila 2009 earthquake. Due to poor concrete mechanical properties and seismic structural weaknesses, the building was demolished. Portions of the structural systems have been extracted before the building demolition, subjected to material characterization tests and then tested in laboratory. The comparison between the two sets of material characterization test programs and the correlation with the experimental performances of an RC column tested under compressive axial load is presented herein.

Keywords: In-situ test · Sonreb · Experimental test · Existing buildings
Reinforced concrete

1 Introduction

Recent devastating earthquakes pointed out the high vulnerability of existing reinforced concrete (RC) buildings. The lack of proper seismic detailing, quality controls and poor mechanical properties of structural materials led to significant damage to structural and non-structural members and high reconstruction costs. In this context, a reliable assessment of the seismic performances of existing RC structures is of paramount importance in order to design proper retrofit solution optimizing the use of economic resources.

Nowadays, the assessment procedures available in current standards or guidelines (CEN 2005; MI 2008; C S LL PP 2009) state to achieve a specific level of knowledge by properly assessing the building geometry, structural details and material properties.

© Springer International Publishing AG, part of Springer Nature 2018
M. di Prisco and M. Menegotto (Eds.): ICD 2016, LNCE 10, pp. 454–466, 2018.
https://doi.org/10.1007/978-3-319-78936-1_33

This is a starting point in the assessment procedure and the adopted level of knowledge will affect the performances of structural members and, in turn, the overall structural capacity.

Different testing methods are available in literature and suggested in available codes to investigate in-situ properties of structural materials. Generally, destructive tests, made on concrete sample properly extracted from structural members, provide reliable estimations of concrete material properties. However, due to the level of disruption and the related costs they can only be performed on a limited number of elements. Indeed, destructive tests can be integrated or replaced by non-destructive measurements, which are assuming a primary role in the assessment procedures.

Nowadays, a reliable quantification of material properties from in-situ characterization tests is very challenging due to sources of uncertainties arising at various levels and caused by: the testing method, the systematic interferences with the environment, random interferences (due to material intrinsic variability), human factor influence and data interpretation, including errors in the model (Breysse 2012). The relevant uncertainties in the testing methods and interpretation of results along with the need for a reliable methodology suitable for application at large a scale promoted the development of a specific guideline for characterization tests on structures and soils (DPC et al. 2009). This guideline provided an useful a supporting tool for technicians involved in the L'Aquila reconstruction process.

The reconstruction process in the aftermath of the L'Aquila earthquake (2009) has been a unique occasion to collect important data on the vulnerability of existing RC buildings, reconstruction costs and details about the repair/retrofit measures (Di Ludovico et al. 2016a; b). It resulted that existing buildings have low quality concrete with average concrete compressive strength ranging between 10 and 15 MPa. Scientific studies (Masi 2005) and guidelines (DPC et al. 2009) outlined that the correlation of results from destructive and non-destructive tests and in-place concrete properties can be very challenging for low quality concrete.

This may significantly affects the seismic performance assessment of existing structures (Del Vecchio et al. 2017). Few studies are available in literature which correlate the in-situ material tests with the performance of existing RC members (Dolce et al. 2007; Vona and Nigro 2015). This is because of the difficulties arising in the member sampling and the specimen conservation during the transportation in a laboratory environment.

The goal of this research work is to quantify the accuracy of destructive and non-destructive test estimating the in-situ mechanical properties for low quality concrete. For this purpose, a RC column, part of the lateral resisting frames, has been extracted from an existing building damaged and then demolished after the L'Aquila 2009 earthquake.

A wide program of destructive and non-destructive tests have been performed in-situ and on the extracted RC members. Core tests and the Sonreb method have bene used to estimate the concrete compressive strength. The estimated strength have been used to predict the axial load capacity of a RC column tested under compressive monotonic load. The experimental results and theoretical predictions have been compared and discussed.

2 Case Study Building

RC buildings damaged by the L'Aquila earthquake were classified according to the AeDES form (Baggio et al. 2007). The buildings with severe damage on structural members were rated E and later subject to a further classification based on a detailed seismic assessments. For 541 out of 2211 buildings classified E, the demolition and reconstruction resulted the most suitable solution. Dangerous structural weaknesses, high residual drift, local or global collapse or economical inconvenience in the structural retrofit were the main reasons for demolition and reconstruction (Di Ludovico et al. 2016a; b). Furthermore, an average concrete compressive cylindrical strength lower than 8 MPa was considered as a criterion for building demolition.

A 5 storey RC building, damaged by the L'Aquila earthquake has been selected for this study. The building reported significant damage to structural and non-structural components (see Fig. 1).

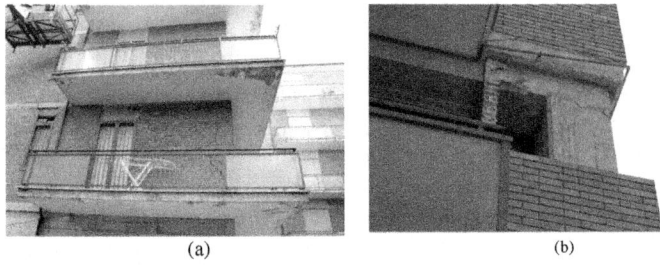

(a)   (b)

Fig. 1. Earthquake damage on the case study building: infill cracking (a) and cracking at the column top end (b).

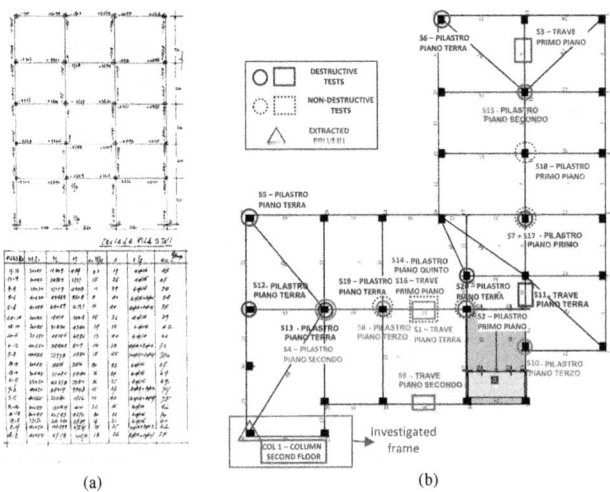

(a)   (b)

Fig. 2. (a) Original design calculations; location of in-situ characterization tests (b)

Table 1. In-situ concrete mechanical characterization tests.

| | | | | FEMA 274, 1997 | | | | | R [-] | V [m/s] | $f_{c,sonreb} = e^a \cdot V^b \cdot R^c$ [Mpa] | | | |
| | | | | | | | | | | | Calibrated S[S1] | Literature | | |
ID	$f_{c,core}$ [MPa]	d [mm]	h [mm]	$F_{l/d}$ [-]	F_{dia} [-]	F_{mc} [-]	F_d [-]	$f_{c,in\text{-}place}^{(S)}$ [MPa]				RILEM[S2]	Gasparik[S3]	Di Leo[S4]
S1	7.31	84	170	1.000	1.019	1.000	1.06	7.90	37.9	3163	7.74	18.94	22.31	20.45
S2	5.81	84	170	1.000	1.019	1.000	1.06	6.28	32.8	2782	6.12	11.08	14.70	12.82
S3	9.22	84	170	1.000	1.019	1.000	1.06	9.96	38.6	3333	9.85	22.27	25.15	23.69
S4	12.21	84	169	1.000	1.019	1.000	1.06	13.19	35.3	3347	13.89	19.86	22.67	21.78
S5	12.01	84	169	1.000	1.019	1.000	1.06	12.97	36.6	3378	12.88	21.40	24.12	23.14
S6	17.82	84	169	1.000	1.019	1.000	1.06	19.25	34.5	3479	18.91	21.27	23.67	23.37
S7	5.10	84	84	0.872	1.019	1.000	1.06	4.81	-	-	-	-	-	-
S8	8.90	84	168	1.000	1.019	1.000	1.06	9.62	-	-	-	-	-	-
S9	10.50	84	168	1.000	1.019	1.000	1.06	11.34	-	-	-	-	-	-
S10	13.50	84	169	1.000	1.019	1.000	1.06	14.58	-	-	-	-	-	-
S11	10.00	84	84	0.874	1.019	1.000	1.06	9.45	-	-	-	-	-	-
S12	9.50	84	167	1.000	1.019	1.000	1.06	10.26	-	-	-	-	-	-
S13	11.20	84	167	1.000	1.019	1.000	1.06	12.10	-	-	-	-	-	-
S14	11.50	84	168	1.000	1.019	1.000	1.06	12.42	-	-	-	-	-	-
S15	10.10	84	168	1.000	1.019	1.000	1.06	10.91	-	-	-	-	-	-
S16	-	-	-	-	-	-	-	-	37.6	3104	7.14	17.84	21.34	19.37
S17	-	-	-	-	-	-	-	-	28.8	2370	3.81	6.09	9.29	7.55
S18	-	-	-	-	-	-	-	-	30.7	2678	6.21	9.15	12.61	10.89
S19	-	-	-	-	-	-	-	-	35.0	3055	8.39	15.48	18.94	17.26
S20	-	-	-	-	-	-	-	-	30.9	3169	16.23	14.30	17.36	16.55
Mean								**11.00**			**10.11**		**17.78**	

where: R is the rebound number; V is the ultrasonic pulse velocity; [S1] is the Sonreb relation calibrated based on in-situ tests (S1–S6) a = –32.1183, b = 5.85229, c = –3.57662; [S2-S3-S4] are the Sonreb relations available in literature in RILEM (1993), Gasparik (1992), Di Leo and Pascale (1994), respectively.

The structural system consists of lateral resisting frames designed with moderate seismic actions. Lack of seismic reinforcement detailing characterizes the structural members. The structural system was designed in the 1963 according to Regio Decreto (1937) and with reference to moderate seismic actions (horizontal force equal to 0.07 the gravity load, according to D.M. 1962). Cement class 730, with concrete compressive strength $R_c = 30$ MPa and smooth bars, AQ50 class, were adopted in the original design. The design scheme of a typical frame and column structural details are reported in Fig. 2a. Geometry, reinforcement details and material mechanical properties were investigated by means of in-situ inspections and characterization tests. A wide program of destructive and non-destructive tests was carried out on the entire structural system (see Fig. 2b). A total of 20 tests (destructive and non-destructive) were carried out to achieve the highest level of knowledge (K.L. 3). Indeed, according to the Italian building code (MI 2008; CS LL PP 2009) 3 destructive tests every 300 m^2 of floor surface are needed for a K.L. 3. The case study building has 5 floors with a floor surface approximately about 400 m^2. Thus, 4 tests per floor (a total of 20 tests) were carried out: 15 destructive tests consisting in specimen sampling and core testing in laboratory; 5 non-destructive tests consisting of rebound hammer and ultrasonic pulse velocity measures. The spatial distribution of the tests is reported in Fig. 2b. The results of the concrete mechanical characterization tests are summarized in Table 1 along with sample properties (diameter, d, and specimen length, h) and strength correction factors ($F_{l/d}$, F_{dia}, F_{mc}, F_d) determined according to FEMA 274 (1997) guidelines to compute the in-place concrete strength, $f_{c,in-place}$, starting from the experimental strength core testing, $f_{c,core}$. The Sonreb method was used to estimate the concrete strength by mean of non-destructive tests. In particular, the concrete mechanical characterization program resulted in three different estimation of the mean concrete compressive strength, f_c, 11.00, 10.11 and 17.78 MPa coming from core tests[S], Sonreb method calibrated on available core tests[S1] and Sonreb method using available literature relationships (Gasparik 1992; Di Leo and Pascale 1994; RILEM 1993) commonly adopted in the practice [S2–S4], respectively. Core tests reported in Table 1 outlined a poor concrete compressive strength (in some portions of the building lower than 8 MPa). This along with economical inconvenience in repair and retrofit (cost 30% higher than demolition and reconstruction cost) resulted in the building demolition.

3   Building Demolition and Specimen Extraction

To investigate on the overall seismic response of the structural system, two beam-column joints and portions of columns and beams were extracted from the structure before the demolition. The specimen were sampled at different floors of a perimetral frame (see Fig. 2b). The absence of proper seismic detailing, commonly, makes these members vulnerable to seismic actions (Calvi et al. 2002; Del Vecchio et al. 2014). Operations for building demolition and specimen sampling are depicted in Fig. 3.

The specimen object of this study is a RC column extracted from the second floor of the case study building, named Col 1. It is 1350 mm tall (about half of the total column height) with a cross-section 350 × 500 mm (see Fig. 2c). Six rebar with

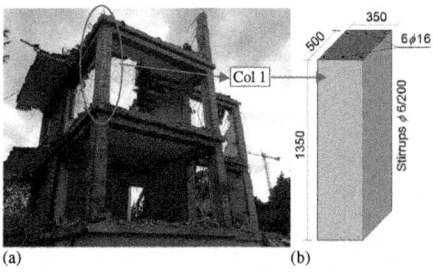

Fig. 3. Building demolition (a) and specimens' sampling (b)

diameter of 16 mm were used as longitudinal reinforcements. Poor transverse reinforcements consists of 6 mm diameter and 200 mm spaced.

4   Material Tests on Extracted Members

A comprehensive material characterization program has been carried out to assess both steel and concrete mechanical properties by means of destructive and non-destructive tests. Number of tests were carried out on the extracted members (two portions of columns and two beam-column subassemblies) in a laboratory environment. Multiple measurements were considered in order to account for their variability and sensitivity of testing instrumentation.

4.1   Destructive Tests

Steel bars employed as internal reinforcement were characterized by means of destructive tests. Number of steel samples representative of longitudinal and transverse reinforcements were taken from the extracted RC members. Test results are summarized in Table 2.

Table 2. Mean mechanical properties of steel internal reinforcement from characterization tests/Proprietà meccaniche medie delle barre d'armatura ottenute da prove a trazione.

	d [mm]	f_y [MPa]	ε_{sy} [-]	E_s [MPa]	ε_{sh} [-]	f_u [MPa]	ε_{su} [-]
Longit. reinf.	16	400	0.0020	203579	0.020	586	0.128
	16	395	0.0022	178424	0.021	587	0.137
	16	374	0.0019	196000	0.024	548	0.135
	Mean	**390**	**0.0020**	**192667**	**0.022**	**574**	**0.133**
Stirrups	6	447	0.0025	181691	0.016	574	0.070
	6	433	0.0027	160459	0.013	548	0.060
	6	392	0.0018	213043	0.011	550	0.150
	6	400	0.0026	153846	0.010	549	0.110
	Mean	**418**	**0.0024**	**177260**	**0.013**	**555**	**0.098**

Table 3. Concrete mechanical characterization tests on extracted RC members.

	ID	$f_{c,core}$ [MPa]	d [mm]	h [mm]	FEMA 274, 1997					$f_{c,sonreb} = e^a \cdot V^b \cdot R^c$					
					$F_{l/d}$ [-]	F_{dia} [-]	F_{mc} [-]	F_d [-]	$f_{c,in\text{-}place}^{(L)}$ [MPa]	$R^{(a)}$ [-]	$V^{(a)}$ [m/s]	Calibrated S + L$^{(a)}$ $f_{c,sonreb}$ [MPa]	$R^{(b)}$ [-]	$V^{(b)}$ [m/s]	Calibrated S + L$^{(b)}$ $f_{c,sonreb}$ [MPa]
Extracted RC members	L1	12.69	143	143	0.875	0.983	1.000	1.06	11.6	25.2	3408	10.36	32.5	3206	10.65
	L2	12.66	143	286	1.000	0.983	1.000	1.06	13.19	27.6	3940	15.49	35.4	3608	15.87
	L3	11.94	143	286	1.000	0.983	1.000	1.06	12.44	-		-	-	-	-
	L4	11.23	143	286	1.000	0.983	1.000	1.06	11.70	-		-	-	-	-
	L5	12.37	143	286	1.000	0.983	1.000	1.06	12.88	-		-	-	-	-
	L6	9.31	143	286	1.000	0.983	1.000	1.06	9.70	-		-	-	-	-
	Col 1	-	-	-	-	-	-	-	-	27.7	3583	**12.33**	31.6	3416	**13.32**
Mean									**11.92**						

where: R is the rebound number; V is the ultrasonic pulse velocity; $^{(a)}$ is the Sonreb relation calibrated on in-situ tests (S1–S6) + laboratory tests (L1a–L3a) a = –25.1915, b = 3.46744, c = –0.124804; $^{(b)}$ is the Sonreb relation calibrated on in-situ tests (S1–S6) + laboratory tests (L1b–L3b) a = –19.1687, b = 2.42054, c = 0.563397.

The yielding strength of ϕ16 bars, f_y, obtained by the experimental tests, satisfactory matches the mean value derived from a large database available in literature for the AQ50 class, $f_{ym} = 371$ MPa (Verderame et al. 2011).

Six concrete cylinders were sampled in laboratory from the extracted RC members. A material characterization program with destructive tests (core testing) was carried out (see Fig. 4a, b). The test results are summarized in Table 3. In particular, the test results on six cores (L1−L6) shows a mean concrete compressive strength of about 11.92 MPa, which do not significantly differ from that obtained from in-situ destructive tests.

4.2   Non-destructive Tests

Non-destructive measurements, rebound number, R, and ultrasonic pulse velocity, V, were measured in correspondence of core drilling (L1−L3) according to practical suggestions reported in available guidelines for in-situ testing (DPC et al. 2009) see Fig. 4. Two sets of measurements (L1a−L3a and L1b−L3b) were collected in laboratory using different instrumentations. Non-destructive measurements were used to predict the concrete compressive strength (i.e. S + L$^{(a)}$, S + L$^{(b)}$). Ordinary Least Squares method was used to estimate the parameters (a, b, c) of the function $f_{c,sonreb} = e^a \cdot V^b \cdot R^c$ [MPa] (DPC et al. 2009).

The concrete characterization program carried out in laboratory on the extracted RC members resulted in three estimations of the concrete compressive strength: 11.92 MPa, 12.33 MPa and 13.32 MPa coming from core tests$^{(L)}$, Sonreb method calibrate on core tests S + L$^{(a)}$ and Sonreb method calibrate on core tests S + L$^{(b)}$, respectively.

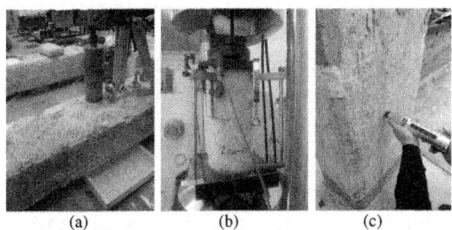

(a)   (b)   (c)

Fig. 4. Laboratory concrete characterization tests: cylinder sampling (a) compression tests (b) and non-destructive tests using rebound hammer (c).

5   Experimental Test on Extracted Column

Column specimen has been prepared for experimental test by levelling the top and bottom face using high strength shrinkage-controlled mortar. In order to avoid local failures, specimen ends have been wrapped with a Carbon Fiber Reinforced Polymers (CFRP) system.

High capacity testing machine (maximum applicable axial load 30000 kN) has been used for the compressive test (see Fig. 5a). A quasi-static monotonic displacement (0.005 mm/s) has been applied on the column until the collapse. Load cell and linear

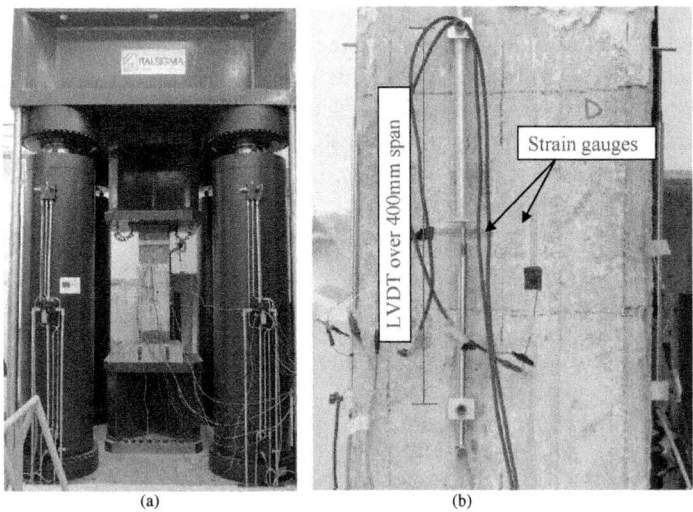

(a)   (b)

Fig. 5. Testing machine (a), test setup and instrumentation (b).

variable displacement transducer (LVDT) have been used to monitor axial load and deformation. LVDTs and strain gauges (s.g.) applied on the concrete surface at the specimen mid-height have been used to monitor specimen elongation and concrete strain at the potential failure surface (see Fig. 5b). Steel strain has been monitored by means of s.g. applied on the longitudinal bars.

Column experimental response in terms of axial load, N, and applied displacement, d, is depicted in Fig. 6a. The first stages of the test have been characterized by an elastic response without significant non-linear phenomena or visible cracks on column surface. At an imposed displacement of about 3 mm (2000 kN) the specimen exhibited minor cracks at the concrete cover. Major cracks on the concrete cover have been detected at the peak strength. (2558 kN at 4.5 mm). At this stage the yielding of longitudinal reinforcement has been achieved as outlined by s.g. measurements. The following stages were characterized by a gradual, almost constant, strength degradation with crack opening, cover spalling and bar buckling (see Fig. 7a).

In the last stages of the test, at 11 mm of imposed displacement, the column exhibited a sudden drop in the axial load capacity due to stirrups failure. The opening of central stirrup, with 90° bent, was observed. This is confirmed by the damage analysis after the removal of damaged concrete (see Fig. 7b). In particular, a cone surface appeared at the mid-height of the column with buckling of longitudinal rein-forcements and stirrups opening.

The comparison of the stress-strain relationship obtained by means of s.g. records at the specimen mid-span and core testing is depicted in Fig. 6c. The stress response of the concrete of the tested column has been derived from the measured axial load reduced by the steel contribution estimated using the stress at the effective strain (monitored by means of s.g. see Fig. 6b). In particular $\sigma_c = N/A - \sigma_s$ where $\sigma_s = \varepsilon_s \cdot E_s$ has been computed, step-by-step, from the effective steel strain depicted in Fig. 6b.

Figure 6c shows a good match between the curves derived from the column test and those obtained from cylindrical samples.

6   Comparison Between Predicted and Experimental Axial Strength

The column peak strength (2558 kN) has been compared with different predictions using concrete compressive strength derived from destructive and non-destructive tests. The comparison is reported in Fig. 6a and Table 4 along with the scatter between experimental peak strength and available predictions.

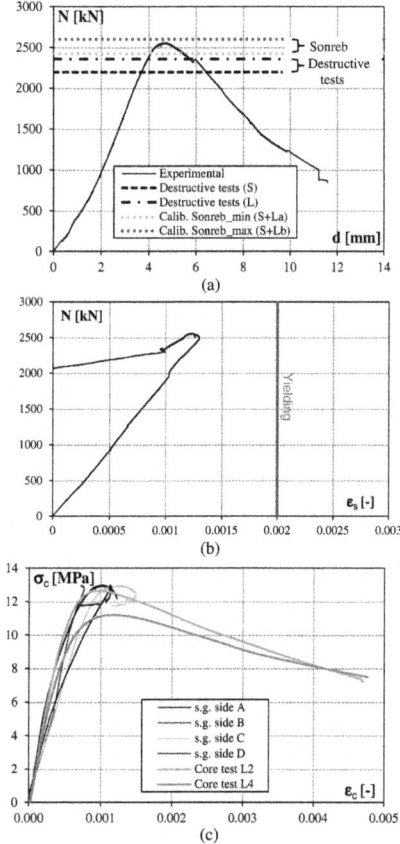

Fig. 6. Column experimental response in terms of: axial load vs. applied displacement (a) strain records on the longitudinal bars (b) concrete stress-strain relationships (c)

In-situ and laboratory tests have been used to estimate the mean concrete compressive strength of the reference specimen (Col 1). In particular, six different estimations have been obtained (see Table 4). Three estimations come from the first set of

(a)   (b)

Fig. 7. Specimen damage at: the end of the test (a), after damaged concrete removal (b).

tests (named in-situ) and three further estimations come from the second set of tests (named Extracted RC members).

The mean concrete compressive strength obtained from destructive and non-destructive tests have been used to predict the axial load capacity of the sample column (Col 1). Basic principles have been adopted for strength predictions. In particular the predicted column axial strength has been computed as $N_{pred} = (A_c \cdot f_c) + (A_s \cdot f_s)$ where: A_c is the concrete area depurated of the area of longitudinal steel bars, A_s; f_s is the effective stress of longitudinal bars at column peak strength ($f_y = 238.6$ MPa at an effective strain of 0.00124, see Fig. 6b) and f_c is the mean concrete compressive strength summarized in Table 4.

Table 4. Comparison of experimental axial strength and theoretical predictions.

			f_{cm} [MPa]	N [kN]	ΔN [%]
Experimental peak strength			-	**2558**	-
In-situ	Destructive tests	$f_{c,in\text{-}place}^{(S)}$	11.00	2200	−14.0
	Calibrated Sonreb	$f_{c,sonreb}^{(S1)}$	10.11	2045	−20.1
	Literature Sonreb	$f_{c,sonreb}^{(S2-S4)}$	17.78	3378	32.1
Extracted RC members	Destructive tests	$f_{c,in\text{-}place}^{(L)}$	11.92	2359	−7.8
	Calibrated Sonreb	$f_{c,sonreb}^{(S+La)}$	12.33	2431	−5.0
		$f_{c,sonreb}^{(S+Lb)}$	13.32	2603	1.8

The comparison outlines that the predicted axial strength using the concrete compressive strength coming from tests on the extracted members ($f_{c,in\text{-}place}^{(L)}$, $f_c^{(S+La)}$, $f_c^{(S+Lb)}$) closely matches the experimental strength with minor differences ranging between the 1.8% and −7.8%. As expected, the best correlation ($\Delta N = 1.8\%$) has been achieved using the concrete compressive strength, $f_c^{(S+Lb)}$, obtained from the Sonreb relationship applied with non-destructive measurements taken on the tested column.

Less accurate resulted the prediction obtained using $f_c^{(S+Lb)}$ derived from destructive tests. This is because the core samples were extracted on the same frame of the sample column, but not exactly on that member.

Significant scatters are attained by using the concrete compressive strength derived from in-situ tests on the entire structural system. In this case, the scatter ΔN ranging between -14.0% and -20.1% for $f_{c,in\text{-}place}^{(S)}$ and $f_{c,sonrev}^{(S1)}$, respectively. Caution should be employed using available literature relationships without any calibration on experimental tests on core samples. In fact, in this case the prediction obtained using the mean concrete compressive strength using available literature relationships, $f_c^{(S2-S4)}$, results not-conservative with a percentage difference of the 32.1%.

7 Conclusions

This study deals with destructive and non-destructive material characterization tests as a supporting tools to predict the strength of existing RC members. A RC column was extracted from an existing building demolished and reconstructed after the L'Aquila 2009 earthquake. The structural system was characterized by a low quality concrete and lack of proper seismic details. The sample specimen was extracted and tested in laboratory to have a reliable benchmark to correlate in-situ material characterization tests and full-scale member experimental performances. A wide material characterization program was carried out by two sets of destructive and non-destructive tests to estimate the mean concrete compressive strength. The study shows that:

- Non-destructive tests combined with destructive tests on core samples represents a reliable tool to perform accurate predictions of the full-scale RC member strength, even in the case of poor quality concrete;
- The best correlation with experimental peak strength has been achieved using the concrete compressive strength, $f_c^{(S+Lb)}$, obtained from the Sonreb relationship applied with non-destructive measurements taken on the tested column.
- Less accurate predictions has been achieved using in-situ characterization tests performed on the entire structural system;
- Caution should be used adopting available literature Sonreb relationships not specifically calibrated for low-strength concrete. In this study, the experimental strength of the tested column has been overestimated of about 32%.

Acknowledgements. The authors would like to express their gratitude to Prof. Francesco Benedettini (University of L'Aquila) for the interest and passion dedicated to these research activities. His premature leaving makes us sad and disappointed. In his memory, we will continue these research activities with the same passion he has conveyed to us.

References

Breysse D (2012) Nondestructive evaluation of concrete strength: an historical review and a new perspective by combining NDT methods. Constr Build Mater 33:139–163

Calvi GM, Magenes G, Pampanin S (2002) Relevance of beam-column joint damage and collapse in RC frame assessment. J Earthq Eng 6(Special Issue 1):75–100

CEN (2005) Eurocode 8: Design of Structures for Earthquake Resistance - Part 3: Assessment and Reofitting of Buildings. European Committee for Standardization, Brussell

C S LL PP (2009) Circolare 617: Istruzioni per l'applicazione delle Norme Tecniche per le Costruzioni (in Italian)

D.M. (1962) Legge n. 1684 (in Italian). G.U. n. 326 del 22/12/1962

D.M. (2008) MI. D.M. 14 Gennaio 2008: Technical codes for structures (in Italian). GU n°29. Rome, Italy

Dolce M, Masi A, Moroni C, Nigro D, Santarsiero G, Ferrini M (2007) Comportamento ciclico-sperimentale di un nodo trave-pilastro estratto da una struttura esistente in c.a. ANIDIS. XII Convegno "L'Ingegneria Sismica in Italia," Pisa, Italy

DPC, Reluis, AGI, ALGI, ALIG (2009) Linee Guida Per Modalita' Di Indagine Sulle Strutture E Sui Terreni Per I Progetti Di Riparazione, Miglioramento E Ricostruzione di edifici inagibili. Doppiavoce

FEMA 274 (1997) NEHRP commentary on the guidelines for the seismic rehabilitation of buildings. Federal Emergency Managament Agency, Washington, D.C., USA, 488

Gasparik J (1992) Prove non distruttive nell'edilizia (in Italian). Quaderno didattico AIPnD, Brescia, Italy

Di Leo A, Pascale G (1994) Prove non distruttive sulle costruzioni in c.a. (in Italian). Il Giornale delle Prove non Distruttive, 4

Di Ludovico M, Prota A, Moroni C, Manfredi G, Dolce M (2017a) Reconstruction process of damaged residential buildings outside the historical centres after L'Aquila earthquake - part I: light reconstruction. Bull Earthq Eng Bull Earthquake Eng 15(2):667–692

Di Ludovico M, Prota A, Moroni C, Manfredi G, Dolce M (2017b) Reconstruction process of damaged residential buildings outside the historical centres after L'Aquila earthquake - part II: heavy reconstruction, Bull Earthquake Eng 15(2):693–729

Masi A (2005) La stima della resistenza del calcestruzzo in situ mediante prove distruttive e non distruttive (in Italian). Il Giornale delle Prove non Distruttive Monitoraggio Diagnostica (1), 1–9

MI (2008) D.M. 14 Gennaio 2008 (D.M. 2008). Technical codes for structures (in Italian). Rome, Italy

R.D. (1937) Regio Decreto Legge n° 2105 (in Italian). G.U. n. 298 del 27/12/1937

RILEM (1993) NDT 4. Recommendations for in situ concrete strength determination by combined non-destructive methods. Compendium of RILEN Technical Recommendations. E&FN Spon, London

Del Vecchio C, Di Ludovico M, Balsamo A, Prota A, Manfredi G, Dolce M (2014) Experimental investigation of exterior RC beam-column joints retrofitted with FRP systems. ASCE J Compos. Constr. 18(4):1–13

Verderame GM, Ricci P, Esposito M, Sansivlero FC (2011) Le Caratteristiche Meccaniche degli Acciai Impiegati nelle Strutture in c.a. realizzate dal 1950 al 1980. XXVI Convegno Nazionale AICAP, Padova

Vona M, Nigro D. (2015) Evaluation of the predictive ability of the in situ concrete strength through core drilling and its effects on the capacity of the RC columns. Materials and Structures, 1043–1059

Del Vecchio C, Del Zoppo M, Di Ludovico M, Verderame GM, Prota A (2017) Comparison of available shear strength models for non-conforming reinforced concrete columns. Eng Struct 148:312–327

ALER Building in Cinisello Balsamo (MI): An Example of Energy Efficient Refurbishment with EASEE Method

R. Brumana[1,2], M. di Prisco[1,2], M. Colombo[1,2], F. Marchi[3],
S. Terletti[4], F. Coeli[5], C. Failla[5], and F. Sonzogni[5(✉)]

[1] Dipartimento ABC, Politecnico di Milano, Milan, Italy
[2] Dipartimento DICA, Politecnico di Milano, Milan, Italy
[3] D'Appolonia S.p.A, Genoa, Italy
[4] Halfen S.r.l, Bergamo, Italy
[5] Magnetti Building S.p.A, Bergamo, Italy
F.Sonzogni@magnetti.it

Abstract. This paper reports the design and the realization of the energetic renovation of a residential building owned by ALER and located in Cinisello Balsamo (MI), by mean of the installation on the existing facades of an innovative insulating prefabricated sandwich panel, for a total of about 520 m^2 covered. The energetic renovation process has been organized in the following steps:

- survey campaign of the existing building's envelope by means of laser scanning technologies and thermographic survey for the identification of non-homogeneous parts of the building structure, to support the technological design as well as the installation of the prefabricated panels;
- use of Building Information Modeling (BIM) aimed at supporting the design of both the prefabricated panels and the anchoring systems, towards the automation in the installation of the elements as well as the energy performance evaluation over time;
- architectural and executive design of the energetic renovation and manufacturing of 186 insulating prefabricated panels, for a total of 28 different typologies, having different sizes, colors and finishing (with and without specific textures);
- panels' installation to the existing building facades by means of steel profiles and realization of the finishing works.

Before and after the installation of the prefabricated insulating panels, a dedicated monitoring campaign was performed for the evaluation of the thermal performance of the building, whose main results are also provided in this paper.

This energetic renovation project has been performed in the framework of the European research project EASEE ("Envelope Approach to improve Sustainability and Energy efficiency in Existing buildings"), funded by the European Union under the 7th Framework Programme for Research and Development. This project, along its four years of duration, was aimed at developing

© Springer International Publishing AG, part of Springer Nature 2018
M. di Prisco and M. Menegotto (Eds.): ICD 2016, LNCE 10, pp. 467–480, 2018.
https://doi.org/10.1007/978-3-319-78936-1_34

innovative solutions for the energy upgrading of multi-storey residential buildings built before 1975, in a historical period in which the focus on energy efficiency was not so pressing, thus being highly energy-consuming buildings.

Keywords: Energy performance · Lightweight modular prefabricated multi-layer insulating panel · BIM approach

1 Introduction

The EASEE (Envelope Approach to improve Sustainability and Energy efficiency in Existing buildings) project - funded under the 7th Framework Programme (FP7), in the framework of the call "Energy saving technologies for buildings envelope retrofitting" - was launched in March 2012 responding to the need to promote buildings' energy efficiency in Europe in order to achieve the EU's 2020 ambitious climate and energy targets. The concept behind the EASEE project was the development of a toolkit for energy efficient envelope retrofitting of existing multi-storey and multi-owner buildings (particularly the project targets residential buildings with cavity walls built before the 70's) allowing a relevant energy demand reduction [1].

Indeed, EASEE proposed both innovations on the technological side, developing different types of advanced insulating components and materials for the three envelope parts (namely for the outer façade, for the cavity wall and for the interior) and innovations on the software side, offering a new consulting service tool for building retrofitting (namely the Retrofitting Planner including the Design Tool). These new technologies, processes and software developed within the project have been integrated towards the development of a new holistic approach to building retrofitting aimed at reducing time and costs associated to this activity while guaranteeing higher energy efficiency, minor burden to building occupants and façade original aesthetic preservation [2].

Concerning the exterior retrofitting, an innovative lightweight modular prefabricated multi-layer insulating panel made of Textile Reinforced Mortar (TRM) with a core made of polystyrene foam obtained without any further interface material has been designed, with a mortar layer thickness below 15 mm and a maximum dimension about 5 m^2 [3]. These panels have been tested first at material level (optimization of mortar mix, adhesion of material, etc.), at prototype level (mechanical, thermal and hygrothermal behavior) and then at real scale level (Ultimate (ULS) and Serviceability (SLS) tests and displacement controlled tests) till the final characterization of the panel's design [4]. Their unique manufacturing process (a dedicated formwork) has been also designed, manufactured and optimized.

During the project have been developed a pre-installed (in the pre-casted panel) and post-installed (drilled in the façade) anchoring systems, able to guarantee a degree of vertical and horizontal adjustability of at least ± 10 mm to compensate on-site tolerances in connections, hold a vertical load between 400 and 500 N for each anchoring, which is compatible with an estimated facade weight of about 600–700 N/m² and made of materials with the lowest possible thermal conductivity and highest strength to minimize thermal bridges.

According to the final design, the delivered panels provided a limited weight (about 75 kg/m²) and were characterized by high durability (expected: 30 years), high insulation properties (reduction of the wall thermal transmittance and minimization of thermal bridges), ensuring an installation without requiring scaffolding with an easy and dry procedure, good finishing and texture in order to reproduce or enhance the original façade appearance and last but not least a quite high customizability in order to easily address every kind of façade. Easiness of replacement was also ensured.

This retrofitting solution has been first installed at test façade level and properly monitored, showing a decrease in heat loss (U-value) of around 65% concerning the exterior envelope.

After testing at small scale, the EASEE retrofitting approach was then implemented at real scale level. Three residential buildings in different countries (Poland, Spain, Italy) have been selected towards the validation of the external retrofitting solutions.

A monitoring campaign showed relevant environmental impacts (in terms of energy savings, CO_2 reduced emissions and increased indoor comfort), societal impacts (in terms of new jobs generation, regeneration of urban areas and safety in installation) as well as economic impacts, namely financial savings and accessible payback periods.

Activities at demo buildings allowed the consortium not only to validate the retrofitting approach, in terms of technologies and software developed and implemented, but also to think about how to propose the EASEE approach to the market, and thus to real clients through dedicated business models and market strategies.

The Project was successfully completed in 2016 (Fig. 2).

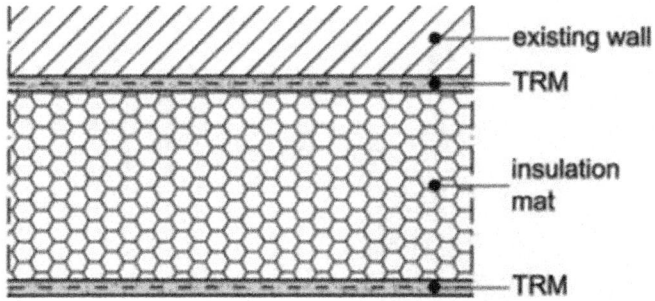

Fig. 1. Sketch of the precast multilayer panel

Fig. 2. Flexural test of the precast multilayer panel

2   Italian Demo Building

The Italian demo building was located in Cinisello Balsamo, in the metropolitan urban area of Milan. The town lies in the high plains of Lombardy, at 154 m above sea level with a temperate climate.

The building (Fig. 1) is a multi-storey and multi-family residential building owned by the Social Housing Agency of Lombardy (Azienda Lombarda Edilizia Residenziale - ALER) Milan division. It was built in 1971 and it is constituted by three floors above ground and a basement where there are garages for cars (Fig. 3).

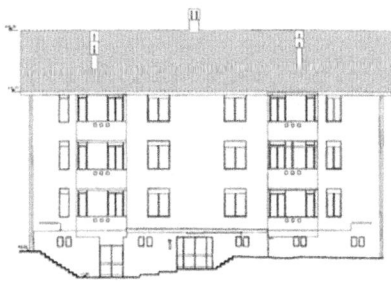

Fig. 3. Italian demo building before the retrofitting

2.1   Goal and Scope

The Italian demo building was targeted to validate the retrofitting approach on a large scale, from assessment of the starting conditions to the manufacturing and installation of the modular elements to the monitoring of performance after retrofitting.

The chosen building was selected among others due to the fact that it perfectly fulfilled the desired specifications towards the demonstration activities (multi-family multi-owner building built in the 1970 s). Moreover the information collected from the survey with a thermal imaging camera revealed and confirmed the needs to apply an energy retrofitting in order to solve the high energy losses from the envelope.

Among the other reasons, the Italian demo building retrofitting, as a social housing, would improve the importance of replicability and impact of the EASEE approach for future improvement and applications. In particular, by retrofitting the overall building (more than 500 m^2) the following aspects have been validated:

- panels' colours and textures;
- methodology for building (geometrical and energetic) assessment;
- BIM approach applied to EASEE retrofitting approach;
- yard preparation and installation process;
- impact on the construction process practice and on the occupants as well as on energy efficiency of the building joints installation and technical details around balconies, corners, doors, windows, etc.;
- finishing activities;
- energy performance of the overall intervention.

Monitoring campaigns have been performed before, during and after the retrofitting of the building in order to know the improvements that the retrofitting solutions provided to the building.

2.2   Building Assessment Before Retrofitting

The building geometrical, structural and energetic assessment has been performed towards a complete evaluation of building boundary conditions and the validation of the building assessment methodology set up by Politecnico of Milan.

Regarding the existing structure, the building was made with pillars and reinforced concrete beams. The floors were made of reinforced cement, while the vertical closures were made of reinforced concrete. The closures had vertical against a wall with a 8 cm air gap of 5.5 cm. Externally there was no plaster but the closures were in exposed concrete.

The survey of the as-built building was devoted to:

- derive plans, elevations and sections (2D);
- derive a BIM model (3D).

Both 2D and 3D design are thought as a support for the design of the retrofitting intervention. In particular, the demo building surveys and data modelling carried out consisted in:

- local datum definition and arrangement of a geodetic network for ground control point measurement;
- laser scanning survey;
- modelling of the building as a support for the design of the external retrofitting and anchoring definition;
- generation of a Building Information Model (BIM) of the building.

2.3 Retrofitting Design

BIM was extensively used at supporting the building design of the Italian building retrofitting, towards the anchoring systems definition, check congruence with irregularities façade, indication for adjustment anchor and related tracking on site.

The envelope was entirely retrofitted: in this way it was possible to identify the building's behavior before and after the application of EASEE solution. Pictures below provide the final design of the facades (final drawing) where the yellow part represents traditional retrofitting, and the gray part the EASEE panels with different finishing surfaces (color and matrix) (Fig. 4).

Fig. 4. Italian demo building: final drawing

During the site work some details have been changed to make better and faster solutions in the connection between EASEE solution and the traditional solution and to finish the covering solution next to balconies and windows (Fig. 5).

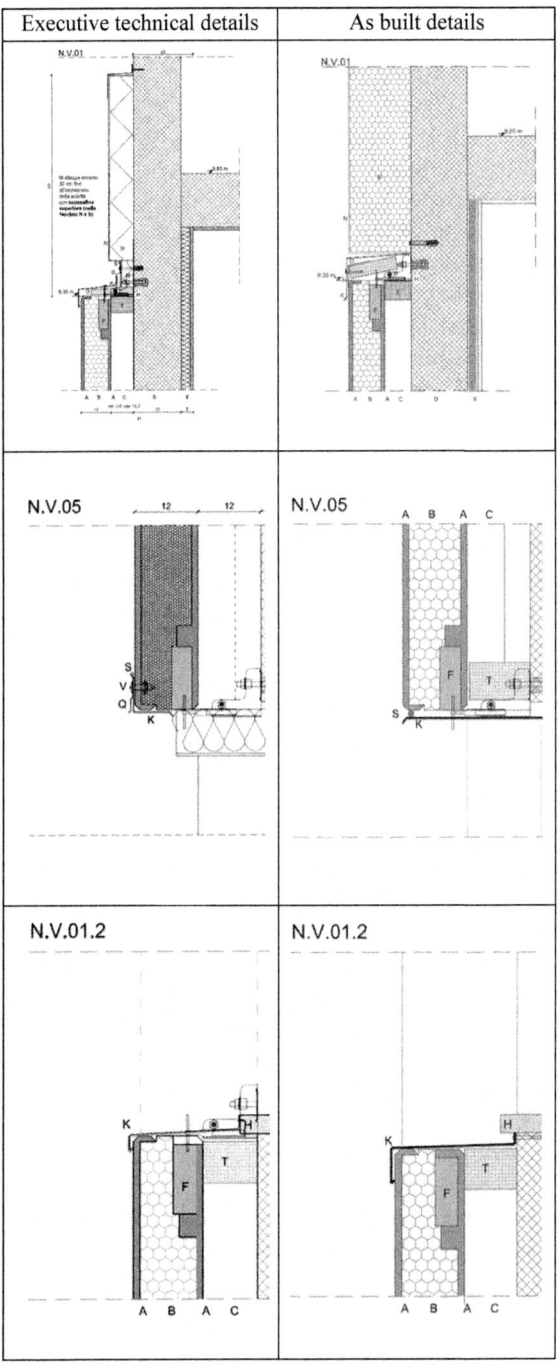

Fig. 5. Technical preliminary details and executive as built details

2.4   External Retrofitting Process

Once the municipality permitting procedure has been accomplished, the site yard has been opened in the second half of October 2015. The various activities were managed by the enterprise by enabling workers to play together to maximize the timing and not to hamper each other. Contingency plan for accelerating the production process for panels manufacturing has been put in place enabling to close in advance the panels production while improving the aesthetic of the panels.

In particular, the contingency plan put in place for respecting the very strict time scheduling foresaw that the production of the panels would have been done both using the vertical and horizontal mould (already in-house at Magnetti). To this aim, the horizontal moulding procedure has been investigated, trial panels produced to evaluate problems and aesthetic of the panels. Tests for understanding the mechanical and bending behavior have been performed.

If in a first phase of material and preparation order of the formwork panel vertically thrown seems winning, the horizontal jet panel is all in all of easier production and with a better surface finish.

The prefabricated panels were thus transported by truck from Carvico (Bg) (where they were produced) to Cinisello Balsamo (Mi) where they were delivered at the test site and unloaded by the crane of the truck. The panel's installation was made without scaffoldings. A small crane vehicle trucks was used for both anchoring and panels' installation.

The joints between panels were made using a low elastic modulus neutral-curing silicone sealant with outstanding ageing resistance. The silicon was placed on polyurethane backfill material in order to reduce the danger of cracking. The joints showed no trace of superficial cracks thanks to the high resistance to UV rays. Around the whole perimeter of the demo building, the cavity between the panels and the existing wall was closed using sealing and metal flashing. Thus, the air permeability was drastically reduced and a close air cavity created.

Below, pictures of the main steps performed during the external retrofitting of the Italian demo building are provided (Fig. 6).

During building retrofitting the occupants have always expressed interest and very positive attitude towards the works. During the construction phase they really appreciated the absence of scaffolding and the possibility to perform daily activities in full freedom. Moreover, they also experience the quick installation of the panels themselves. The widest façade of the building have been indeed retrofitted in approximately 10 days. The total retrofitting lasted less than 3 months.

Finally, building occupants also appreciated the benefits that the EASEE thermal solution was providing, not only in winter but also in summer when they always have bad hot sensation inside their homes.

Fig. 6. Steps of installation processes and technical details

Pictures of the buildings façade are provided below (Fig. 7).

Fig. 7. Pictures before and after retrofitting

3   Monitoring Campaign

The evaluation of the energetic impacts related to the application of the EASEE retrofit solution to the demo building has been done both through simulations and through empirical models based on measured values and real savings.

To evaluate the energy performance, the difference in U-value (W/m^{2}K) before and after the retrofitting has been chosen. A monitoring system, already installed in order to evaluate the building performance before the retrofitting intervention, has been working for months to measure post-installation performance [5].

Figure below provides the stratigraphy of the demo building before retrofitting and the calculation of the thermal transmittance used as benchmarking value, corresponding to 0,80 W/m^{2}K (Fig. 8).

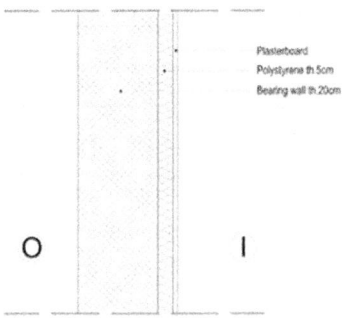

Structure name	Stratigraphy 1
Location	Milan
Structure type	Walls
Number of layers	3
Total thickness	0,2625 m
Total thermal resistance	1,2481 (m^{2}K)/W
Total thermal transmittance	0,8012 W/(m^{2}K)

Fig. 8. Stratigraphic detail before the retrofitting and relative thermal transmittance

After retrofitting through the EASEE panels, the stratigraphy of the wall and results of the transmittance calculated are provided in Fig. 9 and corresponds to 0,24 W/(m^{2}K).

As described, a dedicated monitoring campaign has been performed in situ through the installation of sensors. The data recorded by the sensors have been automatically transmitted by wi-fi connection to a server at Politecnico of Milan and checked and tabulated in order to calculate the thermal transmittance and validate the results previously obtained. The graph below shows the value of the global transmittance step

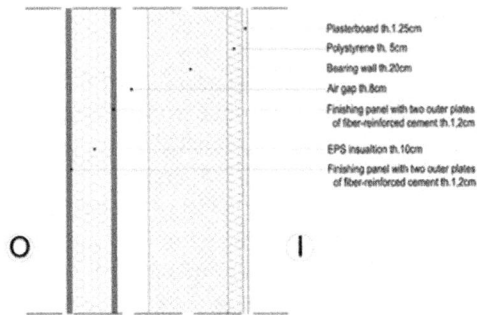

Structure name	Stratigraphy 2
Location	Milan
Structure type	Walls
Number of layers	7
Total thickness	0,4665 m
Total thermal resistance	4,2065 (m^{2}K)/W
Tot thermal transmittance	0,2377 W/(m^{2}K)

Fig. 9. Stratigraphic detail after the retrofitting and relative thermal transmittance

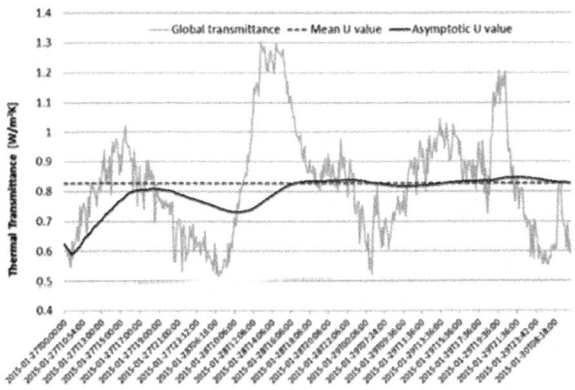

Fig. 10. Thermal transmittance before the retrofitting

by step, the average transmittance and the value of the transmittance obtained by using the progressive average. The transmittance was 0,83 W/(m^{2}K) (Fig. 10).

The transmittance obtained with the new wall configuration was equal to 0,27 W/(m^{2}K). The thermal transmittance reduction between before-after retrofitting was equal to 67%, meaning reduced energy consumption and higher thermal comfort (Fig. 11).

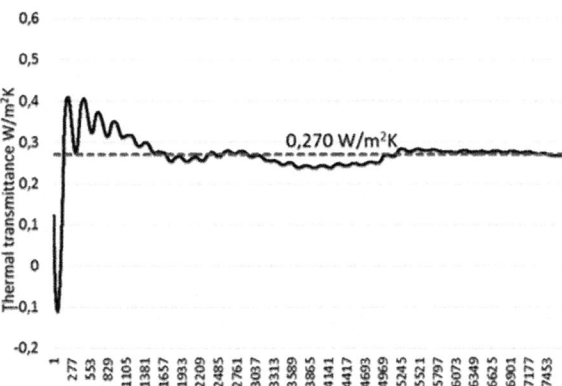

Fig. 11. Thermal transmittance after the retrofitting

4   Conclusions

In this paper an innovative retrofitting process is presented.

More than 500 m^2 (186 panels for a total of 28 different typologies in terms of size, colors and textures) have been retrofitted through the EASEE panels in less than 3 months. The process was entirely carried out without scaffoldings and this was really appreciated by the building occupants.

Another important advantage with the prefabricated EASEE panels was the possibility to cover building imperfection by adjusting the anchoring systems, as for example the non-perfect wall inclination. The prefabricated nature of the EASEE panels and the fact that they do not need any finishing allowed to work also in bad weather conditions due to the dry process used, considerably reducing the dead times especially in cold climates.

The industrial target of EASEE was to achieve the energy performance required by national regulation with an innovative solution whose initial price will be higher than standard ones, but that will reduce or even eliminate the additional costs related to standard retrofitting procedure thanks to the easy installation of modular components on the existing façade without scaffolding. Furthermore, part of the cost could be recovered through the energy savings during the years.

The evaluation of the techno-economic feasibility and success of the EASEE retrofitting approach (including products, processes and services provided to the final user) was performed focusing on the following aspects:

- energetic impacts (energy performances evaluation and related energy savings in terms of building consumptions, thermal comfort, etc.);
- economic impacts (in terms of cost effectiveness during the life cycle of the building);
- indirect industrial impacts (i.e. savings in terms of installation timing and workforce, waste reduction, CO_2 emissions, burden minimization).

Indeed, dedicated monitoring campaign was performed before and after the retro-fitting in order to have data useful for the validation of the panels' performances. Starting from the above numbers and considering as a target market 2% of the 10 million of residential buildings built before 1975 in Europe, a potential annual energy saving of almost 8 million kWh corresponding to 2 billion euros per year can be obtained. Of course these numbers referred to the optimized products, which according to the project partners could be achieved in 2 years from the project end.

Acknowledgment. This research was carried out in the framework of the EASEE "Envelope Approach to improve Sustainability and Energy efficiency in Existing buildings" project, which has received funding from the European Union Seventh Framework Programme (FP7/2007-2013) under Grant Agreement No. 285540 (Call FP7-2011-NMP ENV-ENERGY-ICT-EeB).

References

1. https://www.easee-project.eu/. Accessed 10 July 2016
2. Colombo IG, Colombo M, di Prisco M (2014) Pannello multi-strato prefabbricato di facciata: ottimizzazione strutturale per l'adeguamento energetico e per costruzioni sostenibili. Atti del Convegno CTE, Milano
3. Colombo IG, Colombo M, Magri A, Zani G, di Prisco M (2012) Malte rinforzate con reti in fibra di vetro alcalino-resistenti: un'indagine sperimentale sui parametri di progetto. Atti del Convegno CTE, Bologna
4. Colombo IG, Colombo M, di Prisco M (2014) Comportamento a trazione del calcestruzzo rinforzato con rete in fibra di vetro (TRC) soggetto a cicli di gelo-disgelo in condizione non fessurata e prefessurata. Atti del Convegno CTE, Milano
5. UNI EN 13163 (2009) Thermal insulation products for buildings – Factory made products of expanded polystyrene (EPS) – Specification

Damage Analyses of Concrete Dams Subject to Alkali-Silica Reaction

M. Colombo and C. Comi$^{(\boxtimes)}$

Department of Civil and Environmental Engineering,
Politecnico di Milano, Milan, Italy
claudia.comi@polimi.it

Abstract. The durability of concrete can be reduced by several chemical phenomena, among them the alkali-silica reaction (ASR) plays a fundamental role. Such reaction can have a severe impact on structure safety and functioning. This work deals with the evaluation of the effects of ASR in existing concrete dams. To this purpose, a phenomenological two-phase isotropic damage model, describing the degradation of concrete, is presented and used to simulate the behaviour of concrete dams affected by ASR. This model takes into account the simultaneous influence of both humidity and temperature through two uncoupled diffusion analyses: the heat diffusion analysis and the moisture diffusion analysis. The role of the temperature and humidity fields variations on the development of the deleterious reactions is discussed in the case of an arch dam. The numerical analyses, performed with the proposed thermo-hydro-damage model, allowed to predict the structural behaviour both in terms of reaction extent and increase of crest displacements.

Keywords: Alkali-silica reaction · Damage model · Concrete dams
Finite element analysis

1 Introduction

Concrete is one of the most widely-used materials in the world. The safety, efficiency and durability of concrete structures (many of them with a big social and economic role, e.g. dams) can be endangered by deleterious chemical phenomena like alkali-silica reaction (ASR). Such process occurs between amorphous silica contained in the aggregates and the alkali present in the cement paste, with the formation of a gel. Due to the imbibition of water the gel swells and exerts a pressure that can cause severe expansion and damage. It takes many years, however, for the reaction to occur and the symptoms to become visible.

The material degradation due to ASR was firstly discovered by Stanton (1940). Since then a lot of research effort has been devoted to describe the reactions occurring between aggregates and cement paste, to characterize experimentally the consequent expansion and to develop proper material models to describe the structural effects. These latter should be correctly predicted in order to assess the safety of concrete structures, and in particular dams, affected by ASR. Several examples of concrete dams

© Springer International Publishing AG, part of Springer Nature 2018
M. di Prisco and M. Menegotto (Eds.): ICD 2016, LNCE 10, pp. 481–495, 2018.
https://doi.org/10.1007/978-3-319-78936-1_35

subject to ASR are reported in scientific literature (see e.g. Kladek et al. 1995; Bérubé et al. 2000; Ingraffea 1990).

The influence of temperature and humidity conditions on the ASR was studied through several experimental campaigns (as described in Larive 1998 and Multon and Toutlemonde 2010), but a quantitative assessment of the relative importance of temperature and moisture gradients in concrete dams affected by ASR still lacks.

The present work deals with a phenomenological two-phase isotropic damage model, based on one internal variable of mechanical damage, which depicts the degradation due to the state of stress (tension and compression), caused by the expansion of the reaction products and/or by the external loads (Comi et al. 2012). In the context of the Biot's theory of porous media (Coussy 2004), the concrete subject to alkali-silica reaction is represented as a continuous medium consisting of two phases: the solid skeleton of concrete and the expansive products of the reaction.

The model, accounting for the simultaneous effect of the temperature and humidity, was applied to the analysis of an existing arch concrete dam affected by ASR. A weakly coupled approach was followed: a preliminary heat diffusion analysis and moisture diffusion analysis allowed to compute the varying fields of temperature and humidity, which were the input of the subsequent chemo-damage analysis.

2   Bi-phase Damage Model

At the meso-scale concrete affected by ASR is here described as a two-phase heterogeneous material constituted by the homogenized concrete skeleton (cement paste, aggregates and non-connected porosity) and the homogenized wet gel (combination of the gel produced by the chemical reaction and the adsorbed water).

Let V be the total volume of the representative volume element (RVE), V_s the volume of the solid skeleton, V_g the gel volume and V_w the volume occupied by the water. The degree of saturation of the water S_w is defined by $S_w = V_w/(V - V_s - V_g)$ and used for characterizing the moisture content. The degree of saturation for the gel S_g is defined as the ratio between the volume currently occupied by the gel and the maximum volume that it can occupy. The water and the gel are assumed to fill two different porosities. The water partially fills the initial concrete porosity, while the gel consumes the silica particles necessary for the reaction. Therefore S_g is always set equal to one. Due to the fact that concrete has lower permeability for the gel, it is possible to consider locally drained conditions with respect to the water and locally undrained conditions for the gel.

Assuming small strains and quasi-static conditions, the compatibility and equilibrium equations for the bi-phase solid are

$$\varepsilon = \frac{1}{2}\left(grad\,\boldsymbol{u} + grad^T\,\boldsymbol{u}\right) \tag{1}$$

$$div\,\boldsymbol{\sigma} + \rho\boldsymbol{b} = \boldsymbol{0} \tag{2}$$

where: ε is the small strain tensor of the skeleton; u is the skeleton displacement; σ is the Cauchy stress; ρ is the mass density; ρb is the body force of the solid and the fluid mix.

2.1   Transport Laws

This model takes into account the simultaneous influence of both humidity and temperature through two uncoupled diffusion analyses: the heat diffusion analysis and the moisture diffusion analysis.

As already remarked, the low permeability of concrete with respect to gel allows to neglect the transport of gel, while the migration of the water has to be described by an appropriate transport law for moisture (3). In this paper it is expressed as a function of S_w and it is the combination of the Darcy's law for fluid flow in porous media with the mass conservation law (4).

$$\varphi_w S_w \dot{w}_w = -\frac{k}{\eta_w} \cdot k_{rw} grad p_w \tag{3}$$

$$div(\rho_w \varphi_w S_w \dot{w}_w) + \frac{\partial(\rho_w \varphi_w S_w)}{\partial t} = 0 \tag{4}$$

where: φ_w is the water porosity; $\dot{w}_w$ is the relative velocity of water with respect to the solid; k is the intrinsic permeability of concrete; k_{rw} is the permeability of water; η_w is the dynamic viscosity; p_w is the water pressure.

By substituting (3) in (4) the following non-linear transport law for moisture in its liquid form is obtained

$$div(D_w grad S_w) + \varphi_w \frac{\partial S_w}{\partial t} = 0 \tag{5}$$

where: D_w is the permeability of concrete, dependent on the degree of saturation S_w (for details see Mainguy et al. 2001).

The heat conservation law can be written as

$$C\frac{\partial T}{\partial t} = r - div\,q + \varphi_r \tag{6}$$

where: T is the temperature; C is the volumetric heat capacity of concrete; r is the heat source; q is the heat flux vector in the volume; φ_r is the heat adsorbed or released if the reaction occurs.

The linear and isotropic Fourier conduction law can be adopted for the heat flux

$$q = -D_T grad T \tag{7}$$

where: D_T is the isotropic heat conductivity coefficient.

By substituting (7) in (6) and neglecting the terms φ_r and r, the heat diffusion analysis is governed by the following heat transport law

$$C\frac{\partial T}{\partial t} = div(D_T grad T) \tag{8}$$

2.2   State Equations

In the bi-phase model proposed in Comi et al. (2012) the stress σ is the sum of the effective stress σ' (acting on the skeleton) and of the stress $-bp1$ (acting on the gel), where b is the Biot's coefficient. In order to model the concrete skeleton degradation, the internal isotropic damage variable D is introduced. Finally, the gel formation and expansion are described by the volumetric fraction of the gel ζ_g.

The state equations relating the static variables (stress and gel pressure) to the conjugate kinematic variables (strains and volumetric fraction of gel) are derived from the free energy potential Ψ. They read

$$\sigma = (1-D)2Ge + K(tr\,\varepsilon - \alpha\theta)1 - bp\,1 \tag{9}$$

$$p = -(1-D)M(btr\,\varepsilon - \zeta_g - \alpha_g\theta) \tag{10}$$

where: G and K are the shear and bulk moduli of the homogenized concrete skeleton; e is the deviatoric strain tensor; M is the Biot's modulus; α and α_g are the volumetric coefficients of thermal expansion of the skeleton and of the gel respectively; $\theta = T - T_0$ is the temperature variation with respect to the reference temperature T_0.

The constitutive model is completed by the evolution equations for the variation of the gel volume content, assumed proportional to the rate of the reaction extent ξ, a phenomenological internal variable (varying from 0 to 1)

$$\dot\zeta_g = c\dot\xi \tag{11}$$

$$c = \frac{K + Mb^2}{Mb}\varepsilon_{ASR}^\infty \tag{12}$$

where: ε_{ASR}^∞ is the free asymptotic volumetric expansion due to ASR in the isothermal fully saturated case ($S_w = 1$).

Considering a first order reaction kinetics, the following form for the reaction rate is considered

$$\dot\xi = \frac{\langle f_{Sw} - \xi\rangle^+}{\tilde{t}} \tag{13}$$

$$f_{Sw} = \frac{1 + b_1 exp(-b_2)}{1 + b_1 exp(-b_2 S_w)} \tag{14}$$

where: b_1 and b_2 are material parameters calibrated with experimental data; $\tilde{t}$ is the intrinsic time of the reaction depending on the local histories of temperature $T(t)$ and of degree of saturation $S_w(t)$ and on the reaction extent $\xi(t)$. $\tilde{t}$ is expressed in terms of latency time and characteristic time, both depending on T and S_w

$$\frac{1}{\tilde{t}} = \frac{\xi/f_{Sw} + exp(-\tau_{lat}/\tau_{ch})}{\tau_{ch}(1 + exp(-\tau_{lat}/\tau_{ch}))} \tag{15}$$

where τ_{lat} and τ_{ch} are defined as in Comi et al. (2012).

2.3 Evolution Equations for the Damage

The evolution of the damage variable D, considering possible activation both in tensile (D_t) and in compression case (D_c), is governed by loading-unloading conditions defined in terms of $\boldsymbol{\sigma}$ and p through $\boldsymbol{\sigma}''$

$$\boldsymbol{\sigma}'' = \boldsymbol{\sigma} + \beta p \boldsymbol{1} \tag{16}$$

$$f_i = \frac{1}{2}\boldsymbol{s} : \boldsymbol{s} - a_i(tr\,\boldsymbol{\sigma}'')^2 + b_i tr\,\boldsymbol{\sigma}'' h_i - k_i h_i^2 \leq 0 \tag{17}$$

$$f_i \leq 0 \quad \dot{D}_i \geq 0 \quad f_i \dot{D}_i = 0 \quad i = t, c \tag{18}$$

where: f_t and f_c are the damage activation functions in tension and compression; $\boldsymbol{s}$ is the deviatoric stress; a_i, b_i, k_i (i = t, c) are material parameters governing the shape and dimensions of the elastic domain (see Comi and Perego 2001 for details); h_t and h_c are the hardening-softening functions

$$h_i(D_i) = \begin{cases} 1 - \left[1 - \left(\frac{\sigma_{ei}}{\sigma_{0i}}\right)\right]\left(1 - \frac{D_i}{D_{0i}}\right)^2 \\ \left[1 - \left(\frac{D_i - D_{0i}}{1 - D_{0i}}\right)^{\gamma_i}\right]^{0.75} \end{cases} \tag{19}$$

The model has been validated by simulating the experimental campaign of Gautam and Panesar (2016). Their work presents the development of a new method to apply multiaxial stresses cube specimens of concrete affected by ASR. During these tests the strains were monitored in the three directions. The results demonstrated a strong influence of multiaxial stress states on the ASR expansion of concrete.

Figure 1 shows the comparison between the experimental strains and the model ones in the three directions (X, Y and Z) for specimens subjected to free expansion and to a uniaxial compression load (10 MPa in X direction).

The specimens under free expansion conditions showed similar expansions in the three directions (blue points in Fig. 1(a) and (b)). The small anisotropy is due to the experimental conditions which the model does not take into account (curve in light-blue). When stress is applied in the X-direction, the expansion is suppressed in the same direction, while the remaining two directions have similar expansions, which are

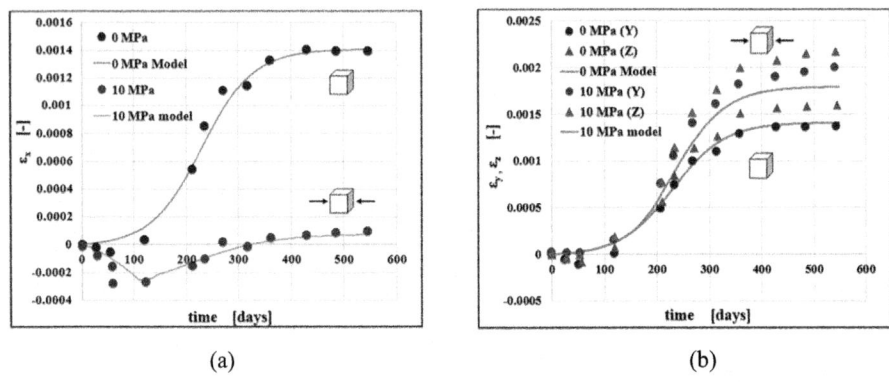

(a)   (b)

Fig. 1. Comparison between experimental data (Gautam and Panesar 2016) and model results in terms of (a) X and (b) Y-Z strains: free expansion and uniaxial loaded specimen (10 MPa) (Color figure online)

greater than the corresponding expansions of the stress-free specimens. The model correctly reproduces the experimental data.

In Fig. 2 the damage evolution corresponding to the two tests is represented.

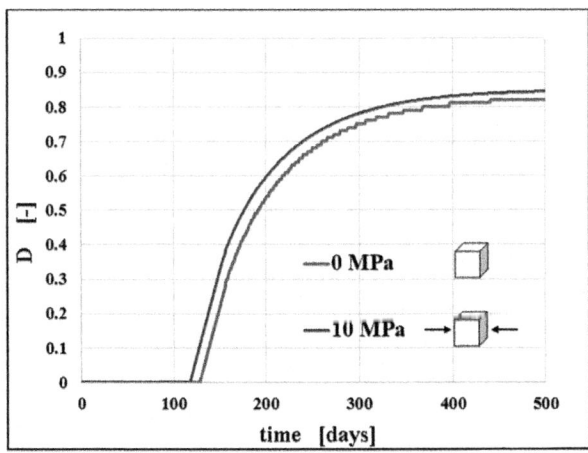

Fig. 2. Damage evolution for the specimens under free expansion condition and compression uniaxial load of 10 MPa.

3   Case Study: An Arch Dam Subject to ASR

In this section a structural analysis of an existing concrete arch dam subject to ASR is presented.

The dam analysed in this work was the object of the XI ICOLD (International Commission on Large Dams) Benchmark Workshop (2011), in which the effects of concrete swelling on the dam equilibrium and displacements were studied. The object of the analysis is a long double curvature arch dam, characterized by a structural height above foundation of 128 m, a developed crest length of 617 m and a crest thickness of 13 m.

3.1   Monitoring System and Data

The water level oscillations from 1963 (year of the completed impoundment of the lake) to 2010 is shown in Fig. 3. During the 1963–1994 operation period, the variation of the water level was limited and 3 periods are identified: 1963–1973 high level (483.8 m), 1974–1981 very high level (486.1 m) and 1982–1994 low level (479.9 m). From 1995, the reservoir level has been more variable.

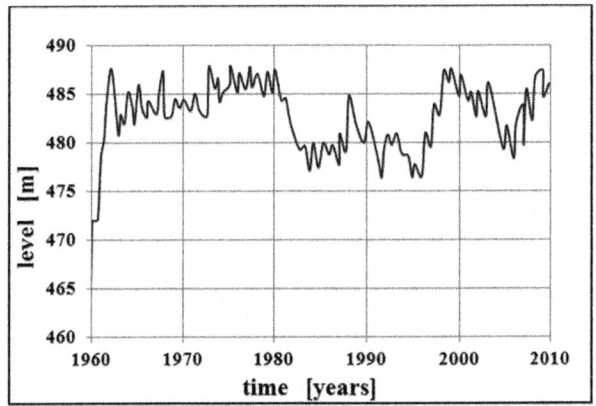

Fig. 3. Reservoir level from 1960 to 2010.

The monitoring system for the dam-foundation system is represented by 6 lines of pendulums and leveling along inspection galleries (Fig. 4). Radial displacements are measured at points marked in blue, while the vertical displacements are measured at points highlighted in red in Fig. 4.

Finally, a special instrumentation consisting of a set of correction vibrating wire sensors is installed which allows to determine correctly the stress inside the hydraulic structure with correction from thermal and creep effects. Indeed, these devices indicate a discrepancy between measured stresses and deformations, which deviate from elasticity. This phenomenon was identified as being due to ASR expansion.

In the seventies long-term laboratory tests have been carried out which demonstrated that a relationship between the local stress tensor and the anisotropic swelling

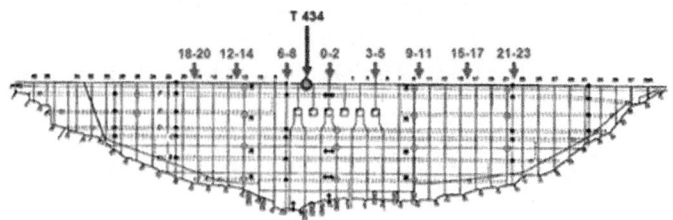

Fig. 4. Location of the monitoring instruments in the arch dam. (Color figure online)

rate tensor exists. The monitoring horizontal and vertical movements confirm that swelling inside the dam body is neither uniform nor isotropic.

3.2 Finite Element Analysis

A tridimensional mesh with brick (20 nodes) and tetrahedron (10 nodes) elements has been employed, as depicted in Fig. 5 (only the hydraulic structure) and Fig. 6 (for the dam-foundation system).

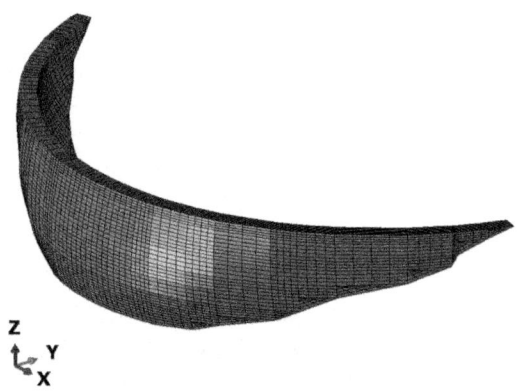

Fig. 5. Finite element model for the arch dam

First the temperature and humidity effects were evaluated, through the diffusion analyses described in Sect. 2.1; then the mechanical analysis of the studied dam was carried out to compare the structural response obtained by the model with the real monitoring data.

3.3 Mechanical Properties of Materials

For the foundation rock and the dam concrete the mechanical properties of Table 1 were assumed: modulus of elasticity E, Poisson's ratio v and unit weight ρ. These properties were considered homogeneously distributed both in the dam body and in the foundation.

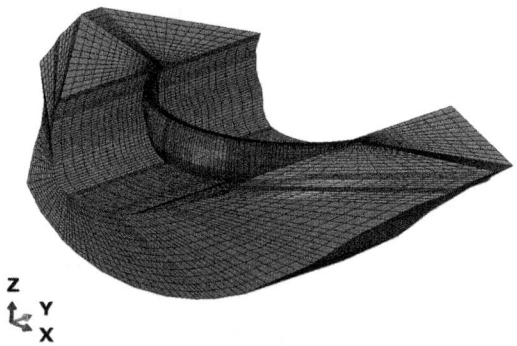

Fig. 6. Finite element model for the dam-foundation system

Table 1. Mechanical properties of the materials

Material	$\frac{E}{GPa}$	$\frac{\nu}{-}$	$\frac{\rho}{kN/m^3}$
Foundation rock	10.0	0.2	-
Dam concrete	22.0	0.2	2350

A compressive strength $f_c = 44$ MPa, a tensile strength $f_t = 2$ MPa and fracture energies (in compression and tension respectively) $G_{fc} = 30$ N·mm^{-1} and $G_{ft} = 0.3$ N·mm^{-1} were assumed for concrete.

The parameter governing the ASR phenomenon were calibrated in order to fit the few available real data.

3.4 Thermal Analysis

To determine the history of the temperature field within the dam, a heat diffusion analysis, solving the Fourier's equation (8) with Dirichlet boundary conditions was performed.

In a first step a steady state analysis allowed to determine the stabilized temperature in the internal nodes of the mesh, starting from the initial uniform field of temperature $T = 288$ K. The following boundary conditions were chosen for this analysis: a constant temperature on the downstream face of the dam and on the dam-foundation contact, equal to the annual average (297 K) and 283 K respectively. Furthermore, a constant water level was considered in this analysis (equal to 482 m, that is the average level from 1970 to 2010) and from this value to 20 m under it (on the upstream face) the temperature was considered linearly varying from the annual air average and 283 K. The boundary conditions just explained are shown in Fig. 7.

Figure 8 highlights some nodes used to present the results in terms of temperature (A = in contact with air, B = influenced by the reservoir level, C = internal node). These nodes will be considered also for the results of the humidity diffusion analyses. In this Figure, also the contour plot of the stabilized temperature is shown.

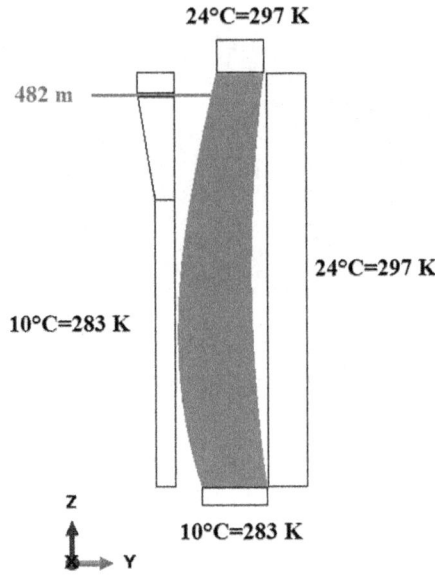

Fig. 7. Initial boundary condition in terms of average temperature field.

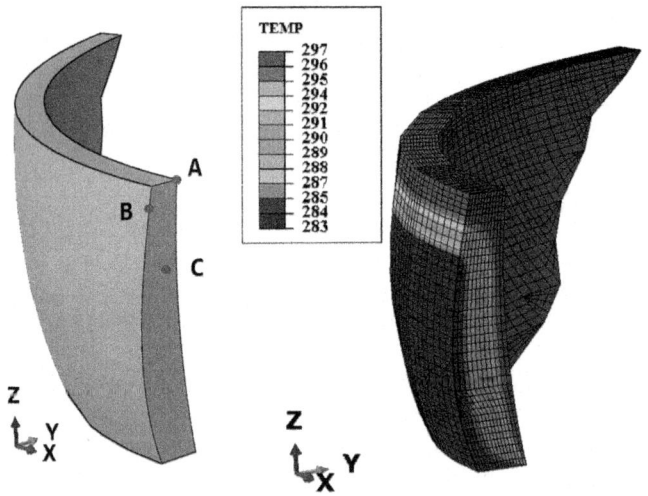

Fig. 8. Location of the nodes considered for the diffusion analyses and contour plot of the stabilized temperature in K.

The second step consisted of a thermal transient analysis, in which the real annual variation of air temperature, shown in Fig. 9, was applied. Furthermore, the real reservoir level variation (Fig. 3) was considered and from the top level to 20 m under it the temperature varied linearly (from air temperature to the constant value of 283 K).

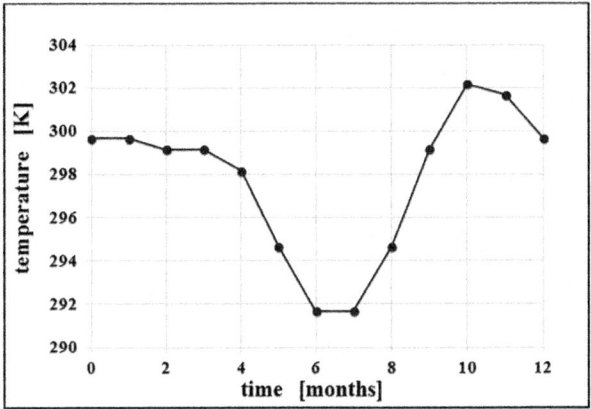

Fig. 9. Annual air temperature in K

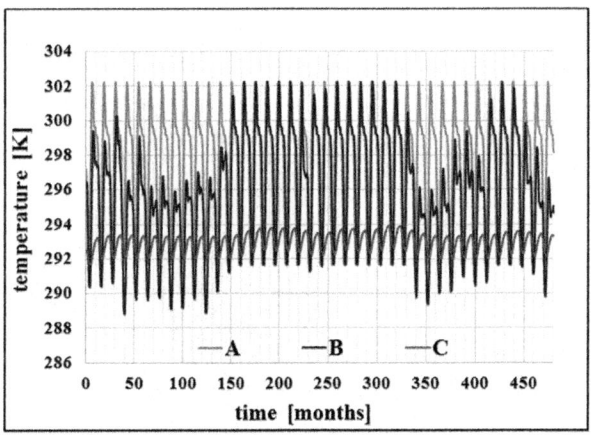

Fig. 10. Histories of temperature of some nodes (A, B, C) of the dam

It is worth noting that the nodes in contact with the rock foundation maintained a temperature of 283 K, as in the first stationary analysis.

In Fig. 10 the histories of temperature at different points of the dam obtained with this second transient analysis are depicted.

3.5   Humidity Analysis

As in the case of thermal analyses, a preliminary moisture diffusion analysis was performed before the transient humidity analysis.

For the first analysis we assumed the initial uniform field of the degree of saturation $S_w = 0.93$ and boundary conditions corresponding to a water level of 482.60 m, so the nodes submerged were characterized by a complete saturation, while at the nodes in

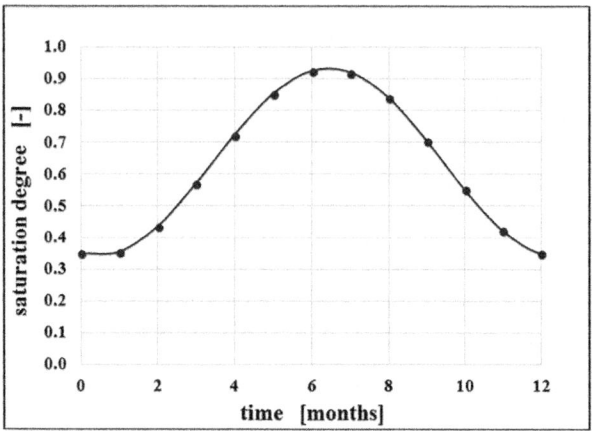

Fig. 11. Annual air humidity in terms of degree of saturation

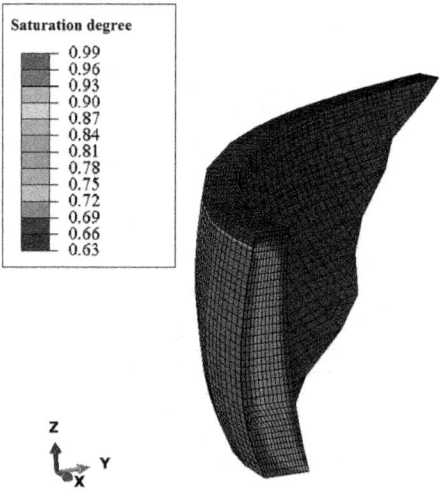

Fig. 12. Stabilized saturation degree pattern

contact with air an average value of 0.64 was applied. This value is the mean of the sinusoidal annual variation of S_w, represented in Fig. 11. The obtained pattern of saturation degree is shown in Fig. 12.

Then, the transient humidity analysis was performed considering the annual oscillation of Fig. 11. Figure 13 reports the obtained histories of humidity of different nodes of the dam mesh. It should be noted that, at difference with what happens for gravity dams, a significant gradient of the saturation degree is present in the whole dam and therefore the humidity effect on the ASR evolution is expected to have a significant role in the structural response.

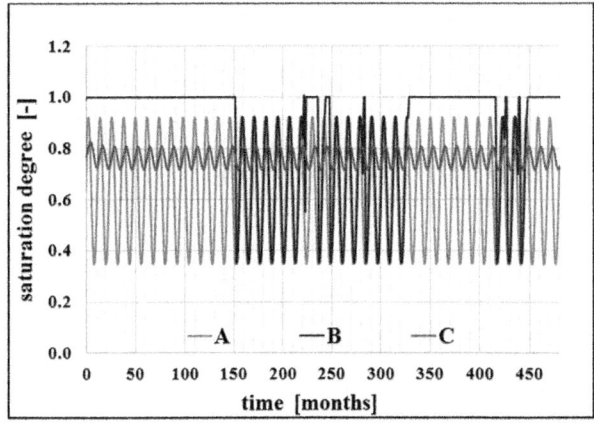

Fig. 13. Histories of saturation degree of some nodes (A, B, C) of the dam

3.6   Mechanical Analysis of the Dam-Foundation System

The results of both the heat and the moisture diffusion problem were used as input for the mechanical analysis of the dam, aimed to evaluate the response due to ASR.

In Fig. 14 the reaction extent evolution in some nodes of the dam is depicted. The reaction extent is different depending on the position of the node in the dam and, differently to the gravity dam cases, it depends mostly to the saturation degree rather than to the temperature.

The development of ASR produced an overall expansion of the dam with a permanent variation of radial and vertical displacements, which was superposed to the normal seasonal oscillation. Figure 15(a) shows the amplified configuration of the dam at t = 4 months, when the reaction has not started yet and the hydrostatic load causes an

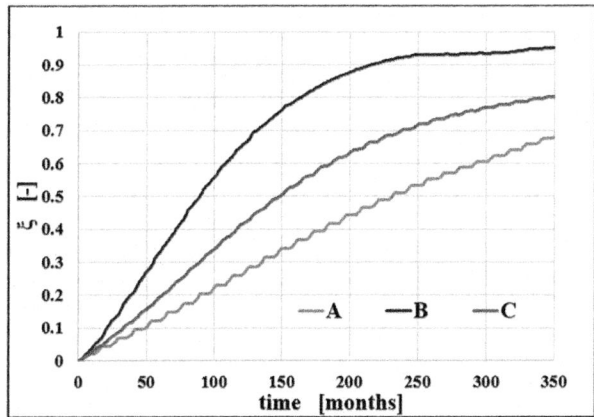

Fig. 14. Reaction extent of some nodes (A, B, C) of the dam.

overall displacement in the upstream-downstream direction. After some years (at t = 120 months) the dam expanded due to the reaction and an overall displacement in the opposite direction is observed, as depicted in Fig. 15(b). These different behaviors, qualitatively in agreement with the available monitored data, are better highlighted in Fig. 15(c), in which the deformed shape is represented with respect to the mid-section of the dam (using an amplification factor = 1000).

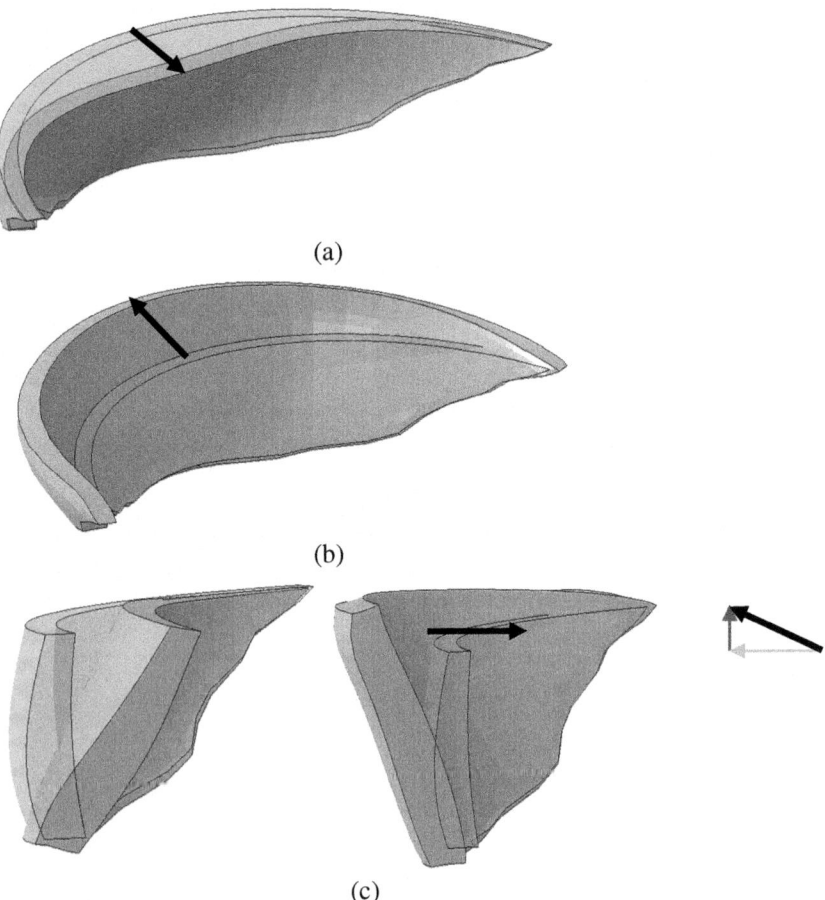

Fig. 15. Deformed configuration of the dam (a) for t = 4 months (due only to hydrostatic pressure), (b) after 10 years (due also to ASR swelling) and (c) corresponding deformed shapes wrt the mid-section

4   Conclusions

In this work a bi-phase damage model for concrete subject to alkali-silica reaction is presented and used to perform a mechanical analysis of an existing concrete arch dam.

Firstly, the effects of temperature and of humidity on the hydraulic structure were evaluated through a heat diffusion analysis and a moisture diffusion analysis respectively. Then, the solution of these two analyses were considered as input for a consequent mechanical analysis, used to define the response due to ASR. The moisture gradient is present in the whole body of the dam and, through the reaction extent, it influenced the structural effect of ASR, contrary to the gravity dams for which the influence is almost absent. The results in terms of displacements are qualitatively in agreement with the available real structural data.

The material parameters used in the model to define the mechanical behaviour and the reaction development have been fixed using the few experimental information available. A further calibration of these parameters, currently in progress, is required to quantitatively reproduce the monitored data.

References

Stanton TE (1940) Expansion of concrete through reaction between cement and aggregate. In: Proceedings of the American society of civil engineers, vol 66, pp 1781–1811

Kladek I, Pietruszczak S, Gocevski V (1995) Modelling of mechanical effects if alkali–silica reaction in Beauharnois powerhouse. In: Proceedings of 5th international symposium on numerical models in geomechanics, Davos, Switzerland, pp 639–645

Bérubé M-A, Durant B, Vézina D, Fournier B (2000) Alkali–aggregate reactivity in Québec (Canada). Can J Civ Eng 27:226–245

Ingraffea AR (1990) Case studies of simulation of fracture in concrete dams. Eng Fract Mech 35:553–564

Comi C, Kirchmayr B, Pignatelli R (2012) Two-phase damage modeling of concrete affected by alkali–silica reaction under variable temperature and humidity conditions. Int J Solids Struct 49:3367–3380

Coussy O (2004) Poromechanics. John Wiley and Sons, New York

Mainguy M, Coussy O, Baroghel-Bouny V (2001) Role of air pressure in drying of the weakly permeable materials. J Eng Mech 127:582–592

Multon S, Toutlemonde F (2010) Effect of moisture conditions and transfer on alkali silica reaction damaged structures. Cem Concr Res 40:924–934

Noret C, Molin X (2011) Effect of concrete swelling on the equilibrium and displacements of an arch dam. In: Proceedings of the 11th ICOLD benchmark workshop on numerical analysis of dams, Valencia, 20–21 October 2011

Gautam BP, Panesar DK (2016) A new method of applying long-term multiaxial stresses in concrete specimens undergoing ASR, and their triaxial expansion. Mater Struct 49:3495–3508

Larive C (1998) Apports combines de l'expérimentation et de la modélisation la comprehension de l'alcali-réaction et de ses effets méchaniques. PhD thesis, LCPC

Comi C, Perego U (2001) Fracture energy based bi-dissipative damage model for concrete. Int J Solids Struct 38:6427–6454

Slot Cutting of the Lago Colombo Dam, Affected by Swelling Deformation, to Bring the Behavior Back from Arch to Gravity

M. Sbarigia[1(✉)], F. Zinetti[2], and M. Hernandez-Bagaglia[2]

[1] Enel – Hydraulic and Civil Engineering, Rome, Italy
matteo.sbarigia@enel.com
[2] Enel – Hydraulic and Civil Engineering, Milan, Italy

Abstract. This paper deals with the rehabilitation and maintenance of the Lago Colombo Dam, owned by Enel S.p.A. since 1962. Comprehensive studies of the dam condition and behaviour pointed out the necessity of important rehabilitation works, to improve the safety conditions and to ensure the efficiency of the dam. Its original gravity behaviour was restored and the arch effect was reduced. The final solution consisted in the crack grouting with special self-compacting mineral mortar and in the carrying out of two new vertical structural joints, by means of slot cutting (2 cm wide) with diamond wire. Others additional improvement works were also carried out. The works started in 2010 and were successfully completed in 2011; in summer 2012 and 2013, the reservoir was re-impounded and the dam behaviour monitored after the rehabilitation. The dam monitoring of the last years has confirmed the change of dam behaviour as required in the rehabilitation project.

Keywords: Rehabilitation · Cutting with diamond wire · Crack grouting
Dam monitoring

1 Introduction

The Lago Colombo Dam is situated in the Orobian Alps (Lombardia - Italy) over 2000 m a.s.l. It was built in 1924–1928 for hydropower generation.

The 31 m high dam was built as a gravity structure, with some particular design features (see Fig. 1):

- no vertical joints;
- a concrete upper part with significant curvature (R_c = 80 m);
- a narrow masonry lower part (about 40 m wide), embedded between rock abutments.

The dam crest is about 140 m long.
The dam foundation consists of a compact sandstone-slate rock.

© Springer International Publishing AG, part of Springer Nature 2018
M. di Prisco and M. Menegotto (Eds.): ICD 2016, LNCE 10, pp. 496–511, 2018.
https://doi.org/10.1007/978-3-319-78936-1_36

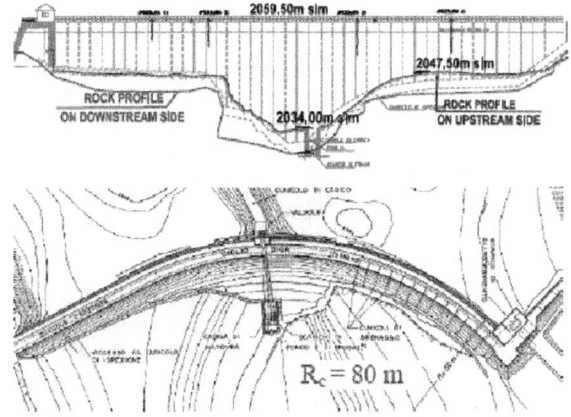

Fig. 1. The dam upstream face and layout

2   Dam Behaviour

2.1   Monitoring

Sub-vertical small fissures arose immediately after the dam construction because of concrete shrinkage. Since early eighties the opening/closing movements of fissures were monitored with manual measures: measured displacements have always been of the order of tenths of a millimeter.

The crest displacements have been monitored by topographic alignment measurements (collimation) since the dam construction. It was observed a regular behavior, with seasonal fluctuations of about 15 mm, until the end of eighties, when the dam crest began to show an irreversible displacement toward upstream. Besides slight cracking started growing on the dam downstream face. A main horizontal crack developed roughly between the concrete upper part and the masonry lower part of the dam. Between 1992–1994 and again between 2003–2004 Enel endowed the monitoring system with implementation, consisting mainly of:

- an inverted pendulum in the maximum cross-section (two measure points, one under the crest and another in foundation);
- 6 extensometers (4 on the main horizontal crack);
- 22 survey points for leveling measures;
- 4 piezometers for uplift control.

All the new measures are both manual and automatic.

The main horizontal crack widened up to 15 mm (as shown by measures, see Fig. 2) in 20 years (1990–2010), extending itself to 100 m onto downstream face (see Fig. 5). The crack was investigated in 2002 and a 4 m depth was estimated, however it never reached the upstream face, even as a fissure.

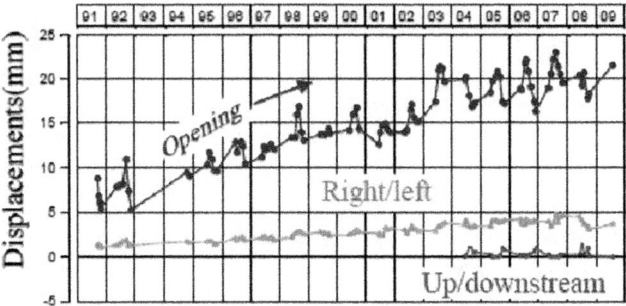

Fig. 2. Crack displacements.

The drift of the crest displacement (as evaluated by means of collimation and inverted pendulum measurements) was about 1.5 mm/year (see Fig. 3). Also the leveling measurements, started in 2003, showed a slight rising (about 0.7 mm/year) of the crest.

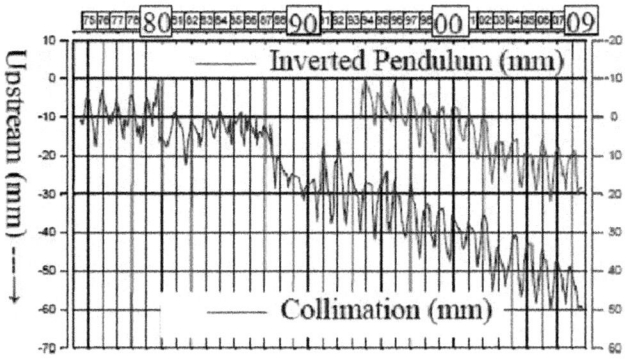

Fig. 3. Crest drift

2.2 Investigations

Extensive investigations were carried out to establish the causes of this behavior. Geological and geomechanical surveys excluded foundation or abutments movements. Diagnostic investigations [1] on concrete, looking for chemical processes causing swelling deformation, found some evidence of slight alkali-silica reactivity and chemical produced gel with high Ca/Si ratio. The laboratory tests evaluated a swelling deformation of 0.05–0.06% in a year.

Besides, the main horizontal crack was investigated [2] by means of 5 corings and video inspections in the central part. Two corings were carried out in the crack plane, other two under this plane and the final one above it. The investigation allowed to discover the crack placement, deepness and size. As shown in Fig. 4 the crack was found sinking inward the dam body for about two meters perpendicularly to the downstream face and for about other two meters horizontally.

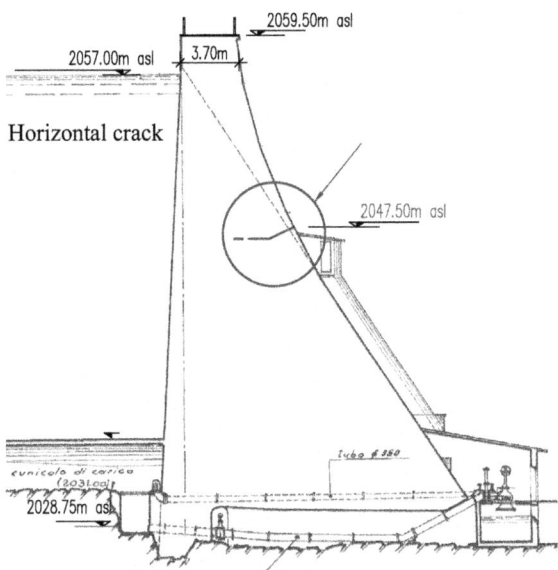

Fig. 4. Extension and direction of the main crack in the highest cross-section.

The investigation showed also the different size of concrete aggregates underlying and overlying the crack. The width of the crack in 2009, before the rehabilitation works, was more than 2 cm.

Statistical and numerical models [3], geological analysis, static and seismic stability calculations [4] were carried out too for the rehabilitation design. The results of the simulation of the dam behavior as monolithic with numerical models showed that in the crack area there were tensile stresses, even if small. Somehow the main crack area ideally divides the wide concrete upper part from the narrow masonry lower one and that area could have been affected by a low tensile strength. The introduction of an artificial load (an increased lapse rate) that simulated the effects of swelling, succeeded in reproducing the measured irreversible displacements, both of the dam crest and of the crack opening.

2.3   Conclusions

In conclusion, particular geometric configuration, different building materials, swelling deformation due to alkali-aggregate reaction, contributed to the crack formation and irreversible opening. The crack and the nearby region became the weakness surface and the plastic hinge for the irreversible rotation of the upper part of the dam. The more the crack opened, the more rotation increased. Finally, winter temperatures and freeze-thaw cycles concurred in undermining the crack region.

3   Rehabilitation Works

The rehabilitation design was carried out by Enel Produzione S.p.A. – Civil and Hydraulic Engineering. It was aimed to restore the structural continuity of the weak and cracked region, to reduce the stress in the dam and to stop the growth of cracking, bringing the dam behavior back from arch to gravity, according to the original structural scheme.

3.1   Design Choices

On the basis of the in-depth analysis and studies described above, two main actions were designed:

- grouting of the crack and nearby region by means of special self-compacting mineral mortar;
- two new vertical structural joints build by means of slot cutting (2 cm wide) with diamond wire.

These works, completed in summer 2010, restored the structural continuity, reinforced that part of the dam body and granted proper strength to the stresses ensuing from the subsequent cutting. Indeed the two cuttings, located at the beginning of the narrow lower part (see Fig. 5) and completed in summer 2011, entailed the splitting of the dam body in three independent blocks, restoring the gravity behavior.

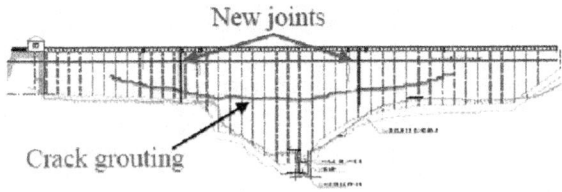

Fig. 5. The rehabilitation design

Two different cement grouts were designed (see Fig. 6):

- a self-compacting, fiber-reinforced, structural mortar, with compensated shrinkage and made of aggregates no larger than 2.5 mm, to fill and plug up the external part of the crack;
- expansive mineral binder for hyperfluid injection slurries, with smaller aggregates (200 μm of maximum size), to clog the deeper and thinner part of the crack.

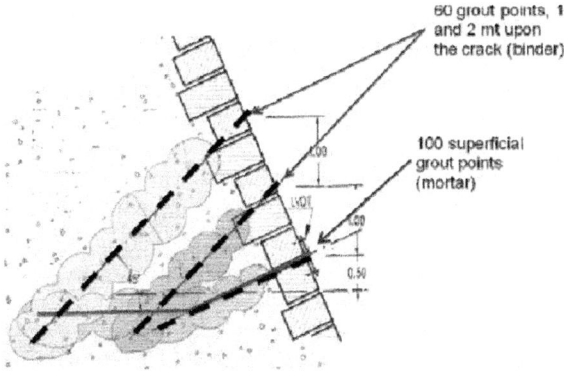

Fig. 6. The grouting scheme of the crack

The width of the slot cuttings was designed 2 cm taking in account the following different factors:

– the residual potential of swelling deformation;
– seasonal opening/closing extent;
– instant elastic release at the cutting.

The 2 cm width should provide the functionality of the new joints for at least 20 years.

The execution of the two cuttings was designed from top to bottom in order to avoid final stress concentration at the narrow crest zone. Therefore an upstream-downstream

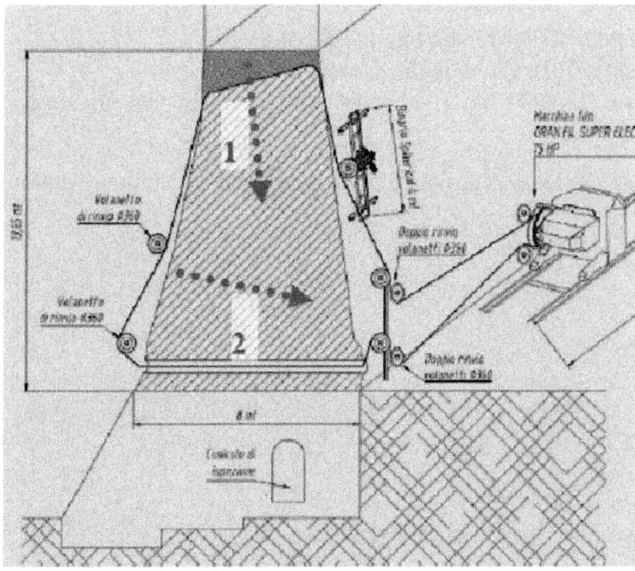

Fig. 7. Cutting scheme.

boring hole was provided to thread the diamond wire. Several gears and a lead mobile track installed upon the upstream face enabled the cutting, first top downwards then downstream-upstream, as long as the wire deformability allowed (see Fig. 7). The machine was located at the dam heel and it moved along a track back and forth, progressively shortening the wire.

The wire, designed expressly for these concrete cutting works, was made of a steel core and a series of steel beads (more than 2 cm diameter), covered of diamond dust (see Fig. 8). The 2 cm width required the carrying out of adequate flywheels, larger than the ones normally used for concrete demolition or quarries activities.

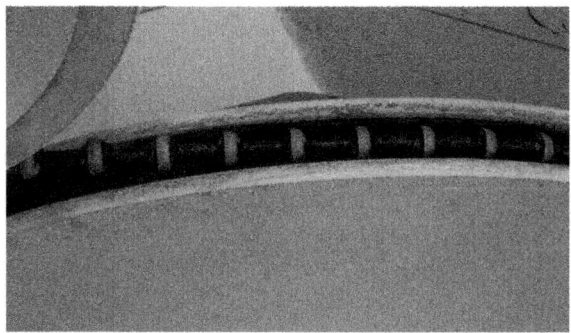

Fig. 8. Diamond wire inside flywheel.

Two reinforced concrete beams were designed to sustain and protect the sealing device of the joint. The sealing of the joints is ensured by the following elements, listed from dam body to upstream face (see Fig. 9):

– Expanded polyethylene foam put into the slot;
– Plastic sheet (TPE), glued on the dam body;
– Plastic sheet (EPDM), blocked between the beams by two structural steel channels, bolted each other.

The sealing device was designed to be removable for future re-cutting.

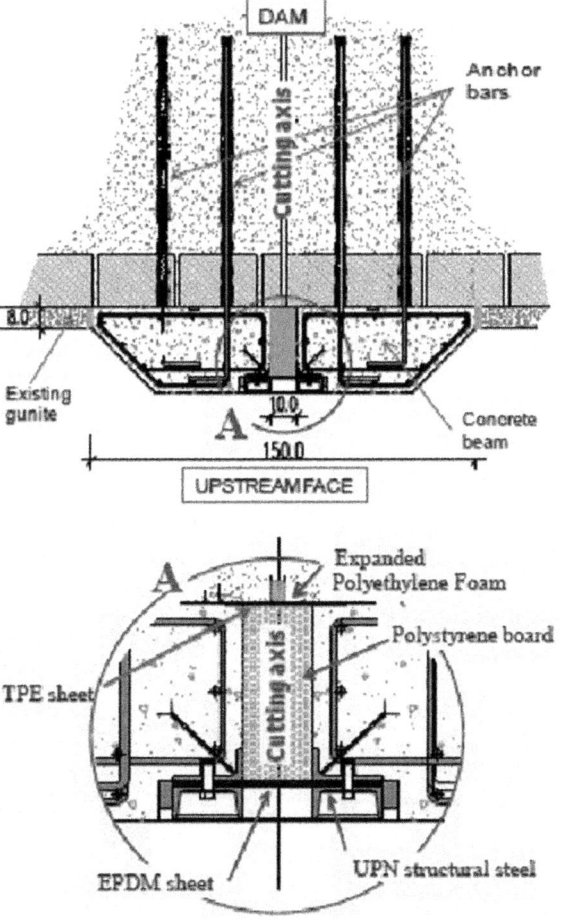

Fig. 9. The designed joint and the detail of the sealing device

3.2 Construction Works

The main works were carried out in summer 2010 and in summer 2011 by Notari S.p.A. It was necessary a constant presence of personnel belonging to the ENEL Construction Supervision Staff, in order to stand over the activities progress and development and to take the most important decision. These building phases were supervised by the Dam Safety Authority (Technical Office of Milan - Ministry of Infrastructure and Transports).

Site logistics
As the dam crest elevation is over 2000 m a.s.l., the working time was concentrated in summer season, from June to October. Furthermore the access to the dam was difficult: building materials had to be carried by means of helicopter, whereas workers reached the site using a cable car and covering a 30 min mountain trail. Fortunately a close

mountain hut could accommodate the workers. Sudden climate changes caused sometimes small floods. Moreover, because of the dam area has great environmental value, a lot of curious tourists interfered with works. All that put a strain on working, but safety has always been ensured.

Crack grouting

In summer 2010 the crack grouting was carried out. A lot of preliminary activities were implemented in order to prepare, clean and seal up the crack surface. The cleaning was made by means of scarification and low-pressure washing. Crack and neighboring region absorbed a mortar quantity three times more than expected. The average absorption of the mortar with greater aggregates was 470 kg/m of crack.

All the injections were carried out by means of a screw feeder, with low pressure (1–2 bar).

After completing the activities, 15 corings (φ80 mm) were carried out in order to check the results and to verify if the crack had been filled. The evidence was positive, as it is shown in Fig. 10:

– Mortar was found near the downstream face, filling the crack (3 cm wide).
– Binder was detected at 2.5 m deep, "connecting" free aggregates.

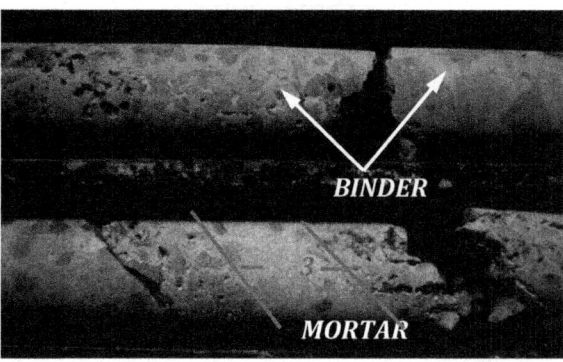

Fig. 10. Evidence of mortar and binder found in one of the corings carried out after crack grouting.

Slot cuttings

The two slot cuttings required preliminary injections of binder in the area of cutting, in order to reinforce the zone and to avoid breakage during cutting.

In summer 2011 each slot was carried out through the following activities:

– horizontal boring hole from upstream to downstream at the foot of the joint;
– installation and setup of the cutting equipments and machines (see Fig. 11);

Fig. 11. The setup of cutting equipments and machines.

- cutting top-downwards for about 2/3 of the height;
- cutting downstream-upstream to completion.

The whole surface of 90 m^2 was cut in about 20–25 h. The cut surface look extraordinarily smooth, as it can be seen in Fig. 12.

Fig. 12. The slot cutting from downstream face. On the left, the kerb of the crest; on the right, the wire in the middle of the cutting.

Several unplanned stops and setbacks happened during cutting: sudden unexpected tears occurred because of a not homogeneous wear of the wire spheres.

Moreover when cutting changed over from top-downward running to down-upstream, friction problems occurred, due to the excessive curvature of the wire. By moving pulleys and flywheels and thanks to the workers' ability, these problems were solved.

Joint sealing device

Referring to the joint sealing device, as shown in Fig. 9, the concrete of the beams, preliminarily tested, was made of the same self-compacting structural mortar used in the crack grouting, adding silica aggregate size 6/10 mm (35%), water (15.5%) and a special anti-shrinkage agent to avoid micro-cracking. During concrete casting, 6 test samples per day were taken. Tests after 28 days (n°3) and 90 days (n°3) were carried out: average cube strength for both beams was over 65 N/mm^2.

One of the completed joints can be seen in Fig. 13.

Fig. 13. The new joints. On the right, the detail of the superficial part of the complex removable sealing device.

3.3 Monitoring During Works

The monitoring system described above was temporarily integrated with:

- 15 digital extensometers LVDT, straddling the horizontal crack before the grouting (measures from July 2010), in order to control its opening/closing movement.
- 16 digital extensometers LVDT, gradually installed during the cuttings, following the passage of the wire and operating during the technical stops (measures from July 2011); 12 of them straddled each slot and the other 4 the existent vertical fissures, to monitor their opening/closing movements.

In summer 2010 the crack extensometers didn't show any significant movements during grouting. Also in summer 2011 nothing special was recorded during cuttings.

The installation of the 12 extensometers following cutting was difficult. Although anchors were prearranged, the instruments should be carefully installed immediately after the passage of the wire, to catch all the following movements. Indeed an elastic instantaneous release was observed and the slot partially closed. The closing deformation decreased from the highest part of the dam (see Fig. 14) to the lowest one.

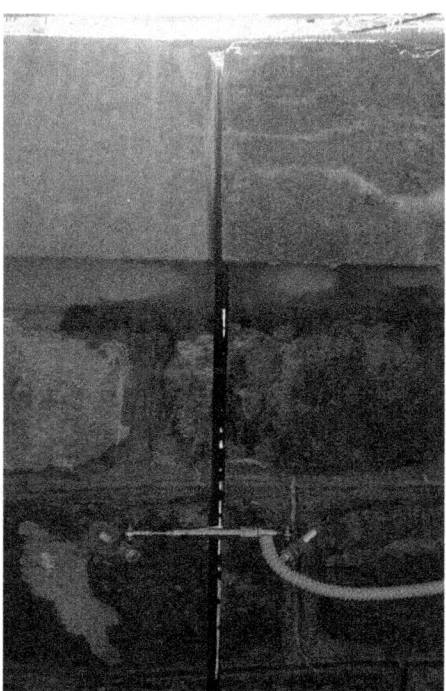

Fig. 14. Provisional extensometer just below the crest on downstream face, installed following the cutting.

As shown in Fig. 15, during the first right slot cutting (completed in about 4 days) the elastic release could be estimated in 5.5 mm at most. Indeed the instantaneous closing deformation of the slot followed exactly the cutting progress (the progress percentages are shown at the left top of the Fig. 15).

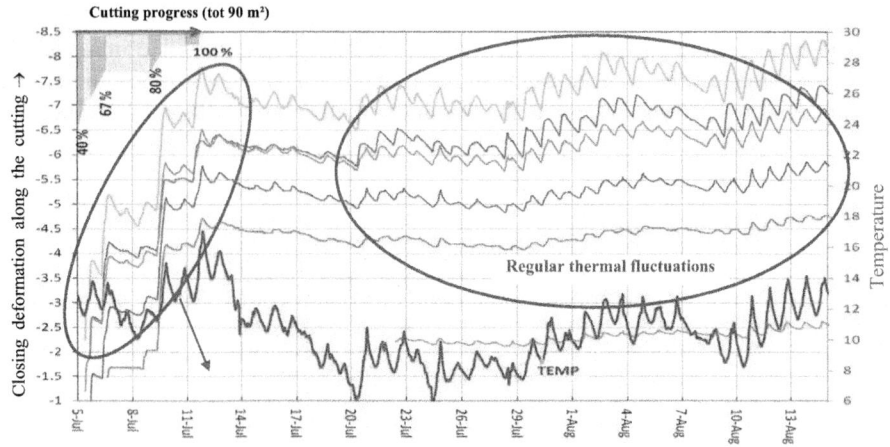

Fig. 15. Measures of the 6 digital extensometers straddling on the right slot.

4   Re-impounding and Dam Behaviour

Since April 2012, in agreement with the Commission for the Final Testing, the reservoir has been re-impounded. In 2012 and 2013 the water level was kept on the normal operating level for long time, in order to test the dam behavior after the rehabilitation with the maximum load. In 2015 the Commission finally certified the goodness of the rehabilitation of the dam.

4.1   Dam Monitoring

At the end of summer 2011, after new joints execution, 6 new definitive extensometers (3 for each slot) were installed on the crest, to monitor the behavior of the joints in the 3 directions of relative movements.

As expected, there have been observed: the opening of joints in the cold season and their closing in the warm season; the relative displacement towards upstream of the central higher block in respect to the lateral ones, a substantial stability in the vertical direction.

Looking at opening/closing displacements (Fig. 16), it is to point out that the seasonal total range of variation is 9–10 mm.

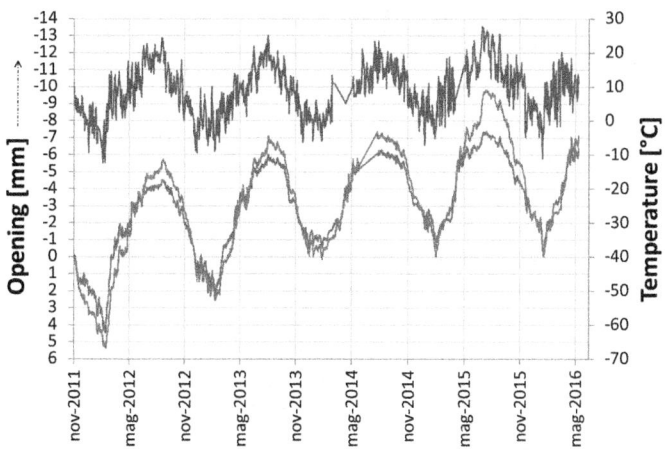

Fig. 16. The opening/closing displacements of the joints.

The summer cutting of the slots allowed, in spite of the instantaneous elastic release, to "adsorb" the thermal movements of the joints, keeping the opening between 13–16 mm in summer and 23–26 in winter.

After 5 years, the expected closing trend is confirmed, about 1–1.5 mm per year. If it was to continue with this rate, it will be necessary a re-cutting in 20 years, as expected by the project as well.

As far as the horizontal crack, the diagram of Fig. 17 shows its absolute stability and closing. At present no trend can be observed.

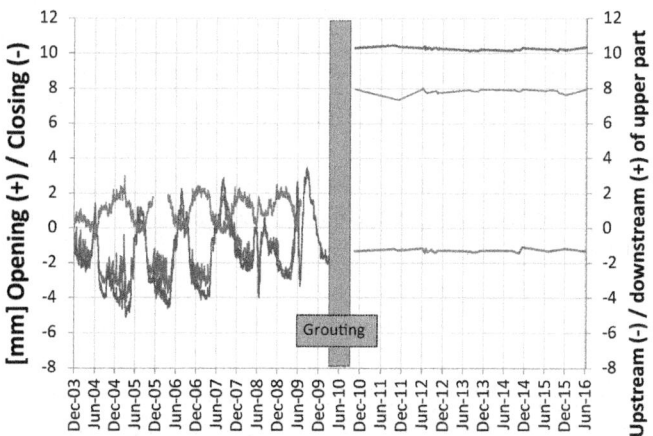

Fig. 17. Extensometers straddling the crack.

Since 2011, the measures of the inverted pendulum (see Fig. 18) have been showing a marked change in the dam behavior, as well as the collimation (see Fig. 19).

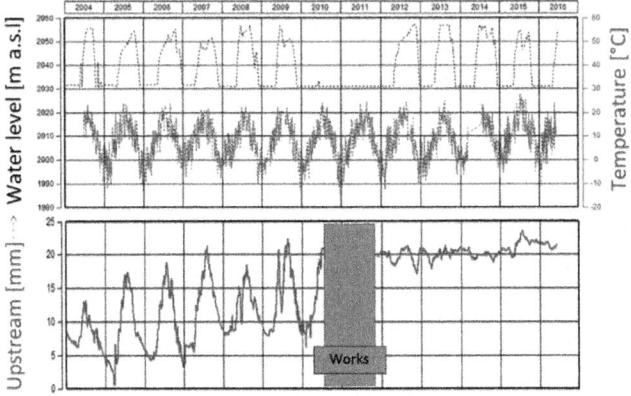

Fig. 18. The crest displacements (inverted pendulum).

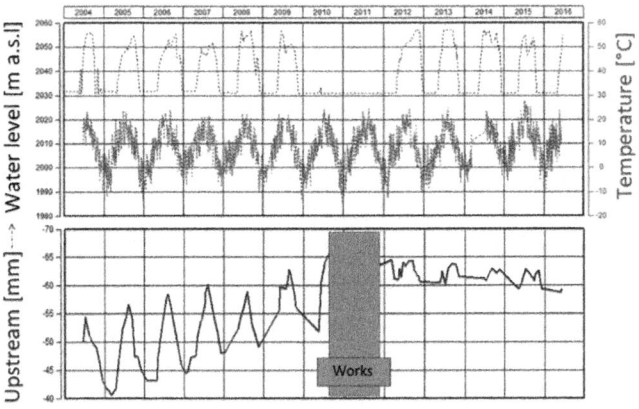

Fig. 19. The crest displacements (collimation)

After the cutting of the dam into three independent blocks, the range of seasonal displacements has been reduced by more than 5 times (from 15–17 mm before the rehabilitation works to 3 mm after).

5   Conclusions

The Lago Colombo gravity dam, built in 1928, was affected by important cracking. The main horizontal crack was related to swelling deformation, due to alkali-silica reaction and tensile stresses not compatible with the building material strength in the involved area. The progressive crack opening, made worse by the freeze-thaw cycles, increased the rotation of the upper part of the dam.

A rehabilitation project was developed to restore the safety conditions and the structural behavior of the dam as a gravity structure, reducing the arch effect. The project consisted in the crack grouting and in the building of two new vertical joints by means of slot cuttings about 2 cm width. This width has shown itself right and proper, considering the residual expansion potential and the elastic release observed. Specifically, after 5 years, the closing rate is in accordance with the project, that supposed a possible re-cutting in 20 years, providing the two joints with a removal sealing device.

The rehabilitation works started in 2010 and completed in 2012. The works were carried out successfully on time and within the design choices.

The dam behavior after rehabilitation confirms the fulfillment of the design targets. The measures of the crest displacements no longer show the historical irreversible drift towards upstream (Fig. 20).

Fig. 20. The Lago Colombo dam at normal operating water level (November 8th 2012). The spillway can be seen on left side

References

1. Enel.Hydro (2002) Diga di Lago Colombo. Indagini diagnostiche sul calcestruzzo
2. Enel.Hydro (2002) Diga del Lago Colombo. Esecuzione di carotaggi e rilievi con sonda televisiva nel corpo murario della diga
3. CESI (2005) Diga del lago Colombo. Analisi numeriche a supporto dell'interpretazione e della valutazione del comportamento strutturale osservato
4. Enel (2007). Diga del lago Colombo. Verifiche di stabilità
5. Marini Quarries Group (2011) Diga del Lago Colombo. Impianto di taglio
6. Enel (2003) Diga del lago Colombo. Risultati delle indagini e prove per la caratterizzazione dello stato fessurativo e della muratura

Spalling Sensitivity Test on Concrete

F. Lo Monte and R. Felicetti[✉]

Department of Civil and Environmental Engineering, Politecnico di Milano, Milan, Italy
roberto.felicetti@polimi.it

Abstract. Concrete sensitivity to spalling in fire is still a critical issue, as no reliable predictive model is currently available. Hence, so far, experimental testing is the most effective means of investigation. This is the reason why an experimental setup has been designed (and discussed in the RILEM TC 256 SPF) by the authors, based on 800×800 mm concrete slabs installed in a steel frame, aimed at applying a biaxial membrane compression. Load and slab thickness can be adjusted in order to simulate the actual service conditions of concrete elements such as tunnel lining segments. The loading system is placed on a horizontal furnace powered by a propane burner fitted with an automatic control system, allowing to follow the prescribed heating curve. This setup allows comparing different concrete mixes as regards their sensitivity to spalling in realistic service conditions and can be of considerable help in initial material testing for strategic infrastructures such as tunnels.

Keywords: Concrete · Fire spalling · Test setup · Tunnels

1 Introduction

Explosive spalling in fire is the more or less violent detachment of concrete pieces from the heat-exposed surface of R/C members. Such phenomenon can be particularly detrimental to the fire resistance of R/C structures both in terms of structural behaviour and repair cost.

As regards the former aspect, spalling progression can lead to a significant reduction of the sectional geometry and to the possible direct exposure of the reinforcing bars to the flames (so dramatically speeding up the mechanical decay of steel). This can be relevant for columns, beams and slabs.

On the other hand, structural performance is seldom a big concern in structures like tunnels, where, however, repair time and cost are critical issues. This is a key point, since the costs connected to traffic disruption must also be considered. Hence, avoiding spalling for this kind of structures is of primary importance. Such result can be achieved with the choice of the right concrete mix in the design phase.

Spalling phenomenon is influenced by both material factors (moisture content, porosity, tensile strength and fibre content) and structural factors (heating rate, external loads and restraints), this making rather difficult predicting the spalling sensitivity of a mix. Even though the exact mechanisms leading to spalling are still not completely clear,

© Springer International Publishing AG, part of Springer Nature 2018
M. di Prisco and M. Menegotto (Eds.): ICD 2016, LNCE 10, pp. 512–523, 2018.
https://doi.org/10.1007/978-3-319-78936-1_37

it is commonly agreed that two are the main actors: (a) stress induced by thermal gradients and external loads and (b) pore pressure rise caused by water vaporization.

Thermal gradients induced by heating introduce compressive stress in the hot layers and tensile stress in the inner core. The increase of compression next to the exposed face favours the formation of cracks parallel to the isothermal surfaces with the consequent decrease of the local mechanical stability. Concrete cracking is also fostered by kinematic incompatibility between aggregate and cement paste, release of absorbed and chemically bound water and cement dehydration (Yufang and Lianchong 2010).

Furthermore, the rise of pressure in concrete pores takes place, mainly due to the vaporization process and to the dilation of liquid water (if saturation occurs). Moisture migration induced by pressure gradients plays an important role, since water is pushed towards both the hot face and the inner core. In the latter case, the build-up of liquid water and the possible condensation of vapour (Khoury 2000) favour the formation of a quasi-saturated layer.

When porosity is low, as the case of High-Performance Concrete (HPC), pores can be completely filled by water (saturation), this leading to the so-called moisture clog. In this region, material permeability is strongly reduced and pressure raise is fostered, leading to values up to 5 MPa (Kalifa et al. 2000). This makes HPC generally more prone to spalling than Normal-Strength Concrete – NSC (Khoury 2000; Kalifa et al. 2000).

The most effective way to reduce concrete spalling sensitivity is to add polypropylene fibre. Limited amounts of polypropylene fibre ($1–2$ kg/m^3) are sufficient to dramatically reduce the probability of spalling, thanks to three main processes: (a) further porosity because of fibre melting at 160–170 °C (Khoury 2008), additional microcracking in the cement paste due to (b) the thermal dilation of melting fibre (Khoury 2008) and (c) to the stress intensification in the cement paste embedding the fibre (Pistol et al. 2012; Tsimbrovska et al. 1997).

The strong interconnection between stress and pore pressure makes necessary to carry out investigation considering both aspects, in conditions close to those of service life. In the literature, both numerical and experimental studies can be found.

In the last decades, numerical models have been developed aimed at simulating the heat and mass transfer taking place in concrete when exposed to high temperature. This involves the solution of a complex set of coupled differential equations. Several approaches based on different simplifying assumptions have been proposed over the past thirty years (Gawin et al. 2011). In these models, however, the mutual interaction between pore pressure and the mechanical response of the material is a critical problem, for which very few experimental evidences are so far available in the literature.

The main point is to define the hydrostatic tensile stress in the solid skeleton which balances the pressure rising in the cement matrix (Lo Monte et al. 2014). Moreover, the characterization of material properties (porosity, permeability) as a function of the temperature is very difficult and not yet completely reliable.

On the other hand, experimental investigation can be performed. For instance, in Mindeguia et al. (2010) pressure was monitored investigating its development in plain and/or fibre concretes, though no external load was considered.

On the other hand, Sjöström et al. (2012) studied concrete sensitivity to spalling in heat-exposed slabs subjected to compressive membrane load applied by means of post-tensioning. This method has the drawback of making difficult the application of a constant load, since the thermal strain affects both the slab and the tendons.

In this paper, the experimental setup developed at the Politecnico di Milano to investigate spalling sensitivity of concrete is discussed. The research project was supported by CTG-Italcementi Group (Bergamo, Italy) and Fondazione Lombardi Ingegneria (Minusio, Switzerland).

Since no standardized test for the assessment of spalling sensitivity has been defined yet, the experimental setup has been designed by the authors (Lo Monte et al. 2015), and is being considered by RILEM TC 256 SPF (Spalling of concrete due to fire: testing and modelling) as a basis for developing a common reference benchmark.

2 Test Setup

The test is based on 800×800 mm concrete slabs installed in a steel frame, aimed at applying a biaxial compressive membrane load up to 1000 kN per axis. Compression is induced by 8 hydraulic jacks (2 per side) and can be kept constant or varied during heating (Fig. 1).

Load and slab thickness (generally between 100 and 200 mm) can be adjusted in order to simulate the actual service conditions of concrete elements such as tunnel lining segments.

The whole loading system is placed on a horizontal furnace powered by a propane gas blow torch fitted with an automatic valve control system, allowing to follow the prescribed heating curve (generally ISO 834).

It is worth noting that the external concrete perimeter in contact with the hydraulic jacks should remain colder to preserve the loading apparatus. Thus, the slab is heated only on its central part of dimensions 600×600 mm. The confining effect provided by the surrounding cold rim is limited by 16 radial cuts whose objective is interrupting its mechanical continuity (Fig. 1a).

In Fig. 2 some details of the load application system are shown. The hydraulic jacks are rigidly mounted to the external restraining frame, made of welded flat steel bars. The thrust is exerted via spherical heads and thick steel plates working as load dividers. The spherical heads also allow to follow the rotation of the slab edges induced by thermal curvature.

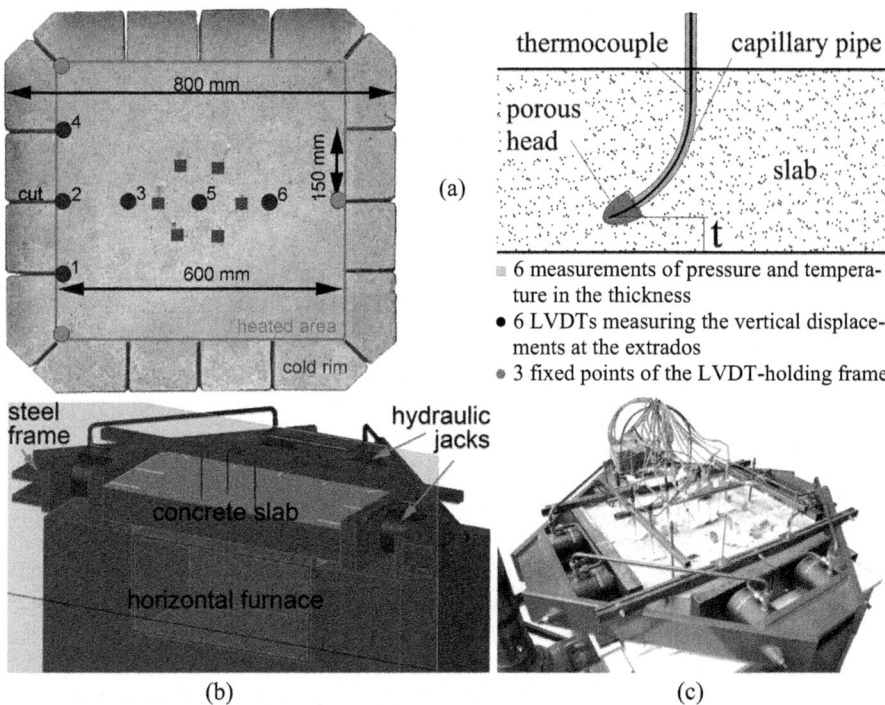

Fig. 1. (a) Concrete slab and measurement points; (b) concrete slab within the loading system (steel frame + hydraulic jacks) positioned on the horizontal furnace; and (c) picture of the instrumented slab, ready for testing.

Fig. 2. Application of the load.

In Fig. 2, the dashed lines represent the centroidal axes of the elements forming the restraining frame. The geometry was studied in order to minimize the eccentricity of the applied loads, this allowing to minimize the sectional size of the members, since they are mainly subjected to tension.

During casting, pressure sensors and thermocouples can be embedded at different depths (Fig. 1a). Thermal and pressure profiles in the slab during heating can give useful information, since the thermal stress is connected to temperature distribution, while pore pressure highlights the ability of polymeric fibre in creating a further connected porosity promoting pressure release. During the test, the flexural behaviour can be monitored by using displacement transducers in order to evaluate the overall effects of thermal curvature and mechanical damage. At the end of the test, laser scanning allows to determine the spalling profile.

The support provided by this setup in comparing different concrete mixes as regards their sensitivity to spalling is shown in the next section. The results reported and the following discussion aim at highlighting the useful information that can be provided in initial material testing for strategic infrastructures like tunnels.

3   Results and Discussion

3.1   General Considerations

So far, more than 30 tests have been carried out on different concretes, with compressive strength ranging from 40 to 125 MPa. Generally speaking, experimental campaigns were planned focusing on two main variables, namely concrete mixes and sustained membrane load. Slab thickness (100 mm) and fire curve (ISO 834) were the same for all tests. In Fig. 3 the typical spalling profile at the heated face of the slab after testing is shown through the laser scanner profilometry.

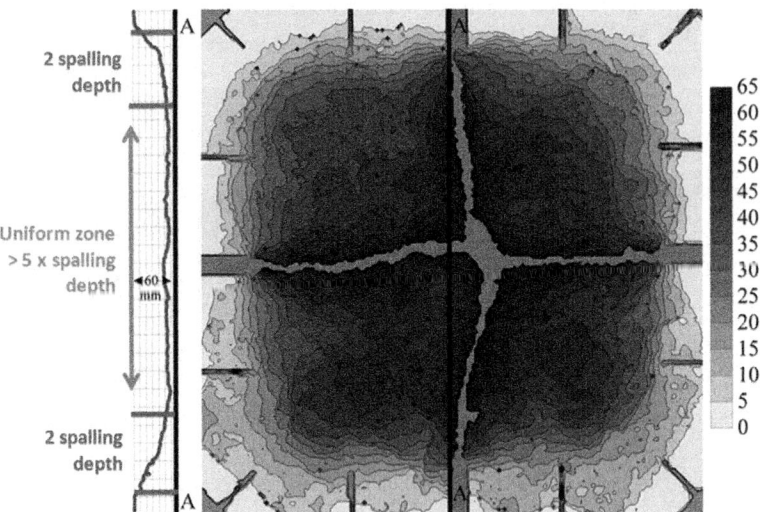

Fig. 3. Laser scanner profilometry of the heated face of the slab after testing.

The two main orthogonal cracks that can be observed in the picture occurred at the end of the test when the bending bearing capacity of the slab was reached. As better described in the following, the rise of the stiffness centroid during heating due to the mechanical damage in the hot layer translates into an eccentricity of external loads and then a hogging bending moment. For high external compression and severe spalling, failure may occur. The slab in Fig. 3 was subjected to 10 MPa and the maximum spalling depth exceeded half of its thickness.

The picture highlights how the presence of the cold rim originates two regions in the spalling profile. There is a transition zone in which the spalling depth goes from zero (in the cold rim) to values close to the maximum (central part of the slab). This region has a width close to two times the spalling depth. Afterwards there is a region in which the spalling depth is rather homogeneous, since the effect of the cold rim becomes negligible. For the deepest spalling observed the uniform region is almost square with a minimum side of about 400 mm.

The regularity of the spalled surface is a very interesting feature, because it can be inferred that, in substantially uniform heating and loading conditions, the size of spalled splinters is mainly governed by the gradients of stress and pore pressure along the depth. Such consideration is of primary importance, since it allows to consider spalling depth as the main parameter in comparing different concrete mixes in terms of sensitivity to the phenomenon.

In the following a few considerations will be made on the role played by load and concrete mix.

3.2   Load Effect

The membrane load applied during fire exposure has a two-fold effect: (1) increase of compression in the hot layer, this reducing its mechanical stability, and (2) decrease of tensile stress (and cracking) in the core, this increasing the likelihood of saturation and fostering higher pore pressure. A simplified scheme of thermal stress and cracking pattern is shown in Fig. 4.

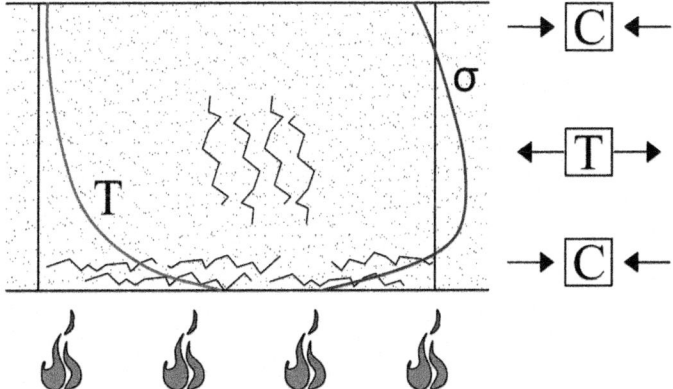

Fig. 4. Thermal stresses in the thickness: compression in the hot layer and tension in the core.

In general, both effects lead to an increase of spalling risk and severity, inducing higher spalled depths and volumes (Figs. 5 and 6). However, to what extent this occurs strongly depends on the variation of the material properties (porosity, permeability, elastic modulus) with temperature. Hence, different concrete mixes can exhibit rather far behaviours.

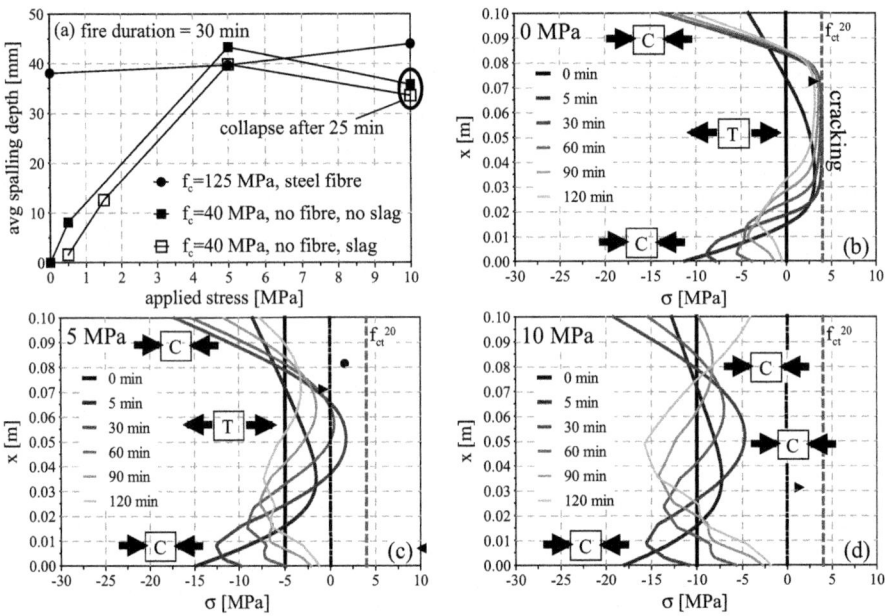

Fig. 5. (a) average spalling depth measured after heating for three different concrete mixes; (b, c, d) stress profiles for different fire durations during heating for three values of external compressive stress, namely 0, 5 and 10 MPa, respectively.

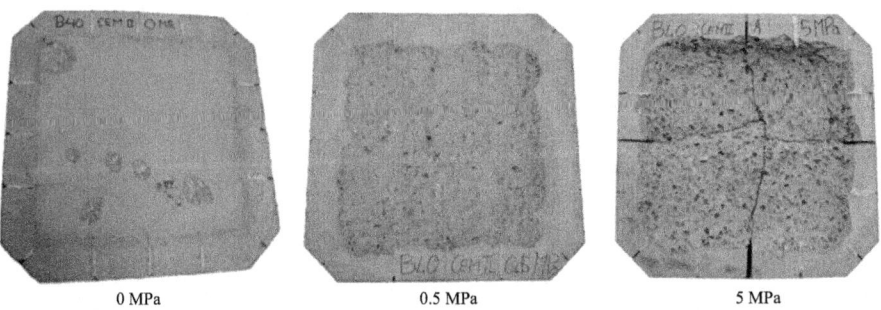

| | 0 MPa | 0.5 MPa | 5 MPa |

Fig. 6. Heated face of NSC slabs after testing for external compressive stress of 0, 0.5 and 5 MPa.

In Fig. 5a the average spalling depth is shown as a function of the applied external compressive stress, for three different concrete mixes. Each symbol in the plot represents an experimental test under constant load during heating.

After the test, the average spalling depth in the uniform square region was measured. It is worth noting that the results can be compared only if they pertain to the same fire duration. In the present examples the fire duration was set as 30 min, but in two cases collapse occurred at a slightly shorter time (about 25 min).

Based on non-linear numerical analyses, the stress profiles along the slab depth are outlined in Fig. 5b–d for different fire duration and applied external compression. As clearly shown, cracking may occur in the core of unloaded slabs due to the tensile stress induced by thermal strain. This leads to a local increase of permeability that favours vapour migration and may prevent moisture clog. For higher values of compression (5 or 10 MPa) tensile strength is never achieved, and cracking is prevented.

Such region is separated from the hot face by a compressed layer, where permeability under transversal load and inherent damage due to strain incompatibility between aggregate and cementitious matrix govern the development of vapour pressure and microcracks. Here the influence of compressive stress is expected to be more pronounced in coarse grained low grade concrete than in inherently homogeneous low-porosity concrete.

Generally speaking, in the plot of average spalling depth versus external compression, two regions can be highlighted: (1) *low compression range*, where increasing external load leads to higher values of compression in the hot layer and to a reduction of cracking in the core, both effects increasing the severity of spalling; (2) *high compression range*, where increasing the external load leads only to higher values of compression in the hot layer, with no cracking in the core.

In region (1) spalling depth rate at increasing load is expected to be higher than in region (2). UHPC (aggregate size = 3 mm, including 20 kg/m^3 of crimped steel fibre 0.30×18 mm) is less liable to cracking both at macro- and meso-scale and the rate difference between regions (1) and (2) is not evident.

Load plays a key role also on the thermal curvature of the slab. In Fig. 7 the vertical displacement at midspan is shown for four nominally identical concrete slabs (NSC, f_c = 40 MPa, no fibre), under different values of biaxial membrane load (0, 0.5, 5 and 10 MPa). When no – or very low – load is applied, the slab sags towards the flames due to the thermal dilation of the hot layers. If a substantial external load is applied, when the decay of stiffness in the hot layers becomes significant, the rise of the stiffness centre produces a negative eccentricity and then a hogging bending moment. Hence, the displacement trend is reversed. Such effect is even more pronounced if spalling occurs.

Since the kinematic behaviour is strictly connected to the bending stiffness, which in turns depends on elastic modulus decay with temperature, monitoring the vertical displacement gives meaningful information about the heat-induced damage.

3.3 Role of Fibre Type

This kind of test proved also to be very effective in highlighting the beneficial effect of polymeric fibre.

In one of the experimental campaigns, different concrete mixes were tested ($f_c \approx$ 60 MPa), sharing the same type of cement, aggregate and very similar water to cement ratios, with the only difference in the presence of fibre.

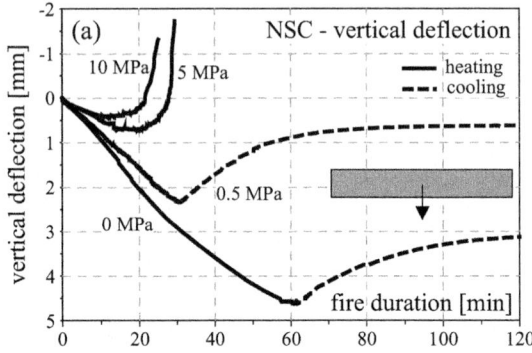

Fig. 7. Centroid vertical displacement for different external loads.

In particular 3 mixes were studied: plain concrete, and concrete with polypropylene or steel fibre.

The results showed that polypropylene fibre is very effective in limiting or avoiding spalling, while the effect of steel fibre is more difficult to predict (Fig. 8). In the experimental tests, plain concrete and steel fibre reinforced concrete (40 kg/m^3 of 0.55 × 35 mm hooked end fibre) underwent very similar amount of spalling, while the addition of 2 kg/m^3 of polypropylene fibre (L = 12 mm; $\varnothing_{eq}$ = 20–48 μm) was sufficient to prevent this phenomenon. As already discussed in the introduction, the reason behind the beneficial effect brought in by the polypropylene fibre is not fully understood, but it is generally accepted that it is mainly due to the additional interconnected porosity.

plain concrete   steel fibre-reinforced concrete   polypropylene fibre concrete

Fig. 8. Heated face of HPC slabs after testing for external compressive stress of 10 MPa.

As regards steel fibre, a possible beneficial effect is the more stable fracture behaviour, though this may foster higher values of pore pressure, due the more limited vapour flow through the cracks.

In these cases, a deeper insight into the possible beneficial effect of any kind of fibre can be gained if pore pressure is monitored. This is the reason why, pressure sensors (fitted with thermocouples, Fig. 1a) are embedded in concrete during casting. Such sensors can be placed at different concrete depths in order to outline the temperature and pressure profiles through the slab thickness. In the test setup at issue, up to six sensors can be installed.

The results in terms of pressure versus time at the six investigated depths and the pressure profiles for different fire durations are shown in Fig. 9 for three different slabs: plain concrete, polypropylene and steel fibre reinforced concretes.

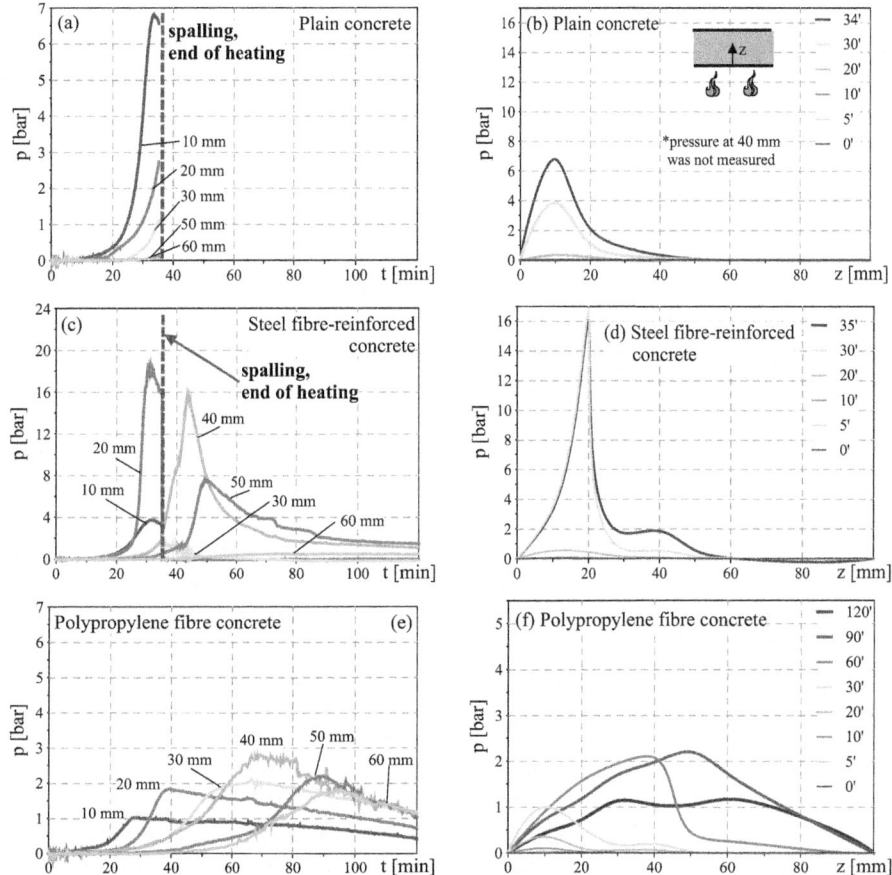

Fig. 9. (a, c, e) pore pressure development at six different depths and (b, d, f) pressure profiles for different fire durations in HPC slabs.

Two plain and two steel fibre reinforced concrete slabs were tested. Spalling was observed in both slabs for plain concrete, while only in one slab for steel fibre reinforced concrete. Spalling occurred after 19 and 34 min for plain concrete and after 35 min in steel fibre reinforced concrete. The final spalling depth was very similar for the three specimens (about 45–50 mm, involving all the heated area).

Polypropylene fibre, on the other hand, proved to be very effective in reducing pore pressure, this being sufficient to avoid spalling in the case at issue.

An explanation can be given by looking at the pressure profiles shown in Fig. 9b, d, f at different time steps for the whole fire duration. In the first 10 min the different slabs showed very similar pressure profiles, since polypropylene fibres were not melted yet. Just before spalling occurred, pressure in plain and steel fibre reinforced concrete slabs was significantly higher with pressure profiles characterized by sharp gradients. In the tests herein presented, after 25 min pore pressure in steel fibre reinforced slab grows faster compared to plain concrete. Though no strong conclusion may be drawn from a single test, this first indication shows that the benefits of steel fibre in terms of enhanced mechanical properties may be offset by the higher pore pressure developed during fire exposure.

4   Conclusions

The present paper introduces the test setup developed by the authors aimed at studying the sensitivity to explosive spalling of different concretes in fire.

So far experimental investigations are recognized as the most reliable way for assessing the spalling sensitivity of concrete, but no standardized test has been defined yet. Within this context, the authors are active in the RILEM 256 SPF Technical Committee (Spalling of concrete due to fire: testing and modelling), with the scope of developing a standardized reference benchmark.

Different experimental campaigns have been carried out, with two main objectives: (a) comparing different concrete mixes as regards their sensitivity to spalling, and (b) assessing spalling severity for a given concrete mix as a function of the external load.

The test proved to be very effective in highlighting differences among concrete mixes, thanks to the possibility of controlling the main parameters such as the fire curve and the external load.

A remarkable sensitivity also to little levels of the membrane compression was observed in Normal-Strength Concrete. The test is also aimed at studying the spalling behaviour of a given concrete mix in service load conditions, this being possible once slab thickness and external load are properly designed.

The most relevant outcome in this experimental procedure is obviously spalling depth. In comparing the beneficial effects of fibres, however, monitoring pore pressure can be important, since it directly shows their effectiveness in reducing vapour pressure in the pores.

Further screening tests on different concrete mixes are in progress in cooperation with concrete manufacturers involved in tunnel lining construction.

Acknowledgements. The Authors are grateful to CTG-Italcementi Group (Bergamo, Italy) and Fondazione Lombardi Ingegneria (Minusio, Switzerland) for the financial support given to this research project.

The tests campaign on NSC under different membrane loads was performed in cooperation with Jihad Md Miah and Pierre Pimienta (CSTB, France).

References

Felicetti R, Lo Monte F, Pimienta P (2012) The influence of pore pressure on the apparent tensile strength of concrete. In: Proceedings of the 7th International Conference on Structures in Fire – SIF 2012, Zurich, Switzerland, 6–8 June, pp 589–598

Gawin D, Pesavento P, Schrefler BA (2011) What physical phenomena can be neglected when modelling concrete at high temperature? A comparative study. Part 2: comparison between models. Int J Solids Struct 48:1945–1961

Kalifa P, Menneteau FD, Quenard D (2000) Spalling and pore pressure in HPC at high temperatures. Cem Concr Res 30:1915–1927

Khoury AG (2000) Effect of fire on concrete and concrete structures. Prog Struct Eng Mater 2:429–447

Khoury GA (2008) Polypropylene fibres in heated concrete. Part 2: pressure relief mechanisms and modelling criteria. Mag Concr Res 60(3):189–204

Lo Monte F, Miah JM, Aktar S, Negri R, Rossino C, Felicetti R (2014) Experimental study on the explosive spalling in high-performance concrete: role of aggregate and fibre types. In: Proceedings of the 8th International Conference on Structures in Fire – SIF 2014, Shanghai, China, 11–13 June, pp 1219–1226

Lo Monte F, Rossino C, Felicetti R (2015) Spalling test on concrete slabs under biaxial membrane loading. In: Proceedings of the 4th International Workshop on "Concrete Spalling due to Fire Exposure", Leipzig, Germany, 8–9 October

Mindeguia JC, Pimienta P, Noumowé A, Kanema M (2010) Temperature, pore pressure and mass variation of concrete subjected to high temperature – experimental and numerical discussion on spalling risk. Cem Concr Res 40:477–487

Pistol K, Weise F, Meng B, Schneider U (2012) The mode of action of polypropylene fibres in high performance concrete at high temperatures. In: Proceedings of the 2nd International Workshop Concrete Spalling due to Fire Exposure, Delft, the Netherlands, 5–7 October, pp 289–296

Sjöström J, Lange D, Jansson R, Boström L (2012) Directional dependence of deflections and damages during fire tests of post-tensioned concrete slabs. In: Proceedings of the 7th International Conference on Structures in Fire – SIF 2012, Zurich, Switzerland, 6–8 June 2012, pp 589–598

Tsimbrovska M, Kalifa P, Quenard D, Daïän, JF (1997) High performance concrete at elevated temperature: permeability and microstructure. In: Transactions of the 14th International Conference on Structural Mechanics in Reactor Technology, Lyon, France, pp 475–482

Yufang F, Lianchong L (2010) Study on mechanism of thermal spalling in concrete exposed to elevated temperatures. Mater Struct 44:361–376

Seismic Loss Analysis of a Non-ductile Infilled RC Building

F. Romano, M. Faggella$^{(\boxtimes)}$, R. Gigliotti, and F. Braga

Department of Structural and Geotechnical Engineering,
Sapienza University of Rome, Rome, Italy
marco.faggella@uniromal.it

Abstract. Mean annual financial losses due to seismic events in Italy are about 2–3 billion euro. For this reason, in recent years increasing attention has been placed on strategies to reduce the seismic risk of the national building stock. In this work, a comparative seismic loss analysis of an infilled R/C building is performed using the FEMA P-58 probabilistic framework and the tool PACT. The objective is to evaluate how the structural modeling and the characterization of structural and nonstructural elements fragility can affect the loss estimation. Fragility and consequence functions for discrete damage states are assumed for structural and non-structural components. A case study prototype typical of Italian pre-1970 R/C infilled buildings is chosen. Nonlinear Incremental Dynamic Analyses (IDA) are performed for three 2D modeling configurations. Financial losses are expressed as median values of repair costs at different hazard levels or in terms of Expected Annual Loss (EAL).

Keywords: Loss analysis · FEMA P-58 · Infilled R/C structures
Shear failure · Fragility functions

1 Introduction

Increasing the seismic safety of the national building stock and managing the risk reducing the monetary impact of earthquakes is an issue more and more ad-dressed in modern building codes, guidelines for mitigation policies and international cooperation (Dolce 2012; Braga et al. 2014; Braga et al. 2015; Faggella et al. 2016; Rossi et al. 2016; Faggella et al. 2012).

In order to design effective intervention measures it is important to accurately identify the probable building collapse mechanisms. Several studies investigated the possible local failure mechanisms in R/C buildings. Among these, Braga et al. (2009) focused on the column flexural failure and shear failure of external joints in older non-ductile structures, designed for gravity loads only. Other research works done by Sezen (2004) and then Elwood and Moehle (2003) focused on shear column failure in bare frames. Another very important issue is frame-infill interaction: observations after seismic events and experimental analyses demonstrated that the frame failure could anticipate adjacent infill failure, usually at the end of columns or beams (Crisafulli 1997).

© Springer International Publishing AG, part of Springer Nature 2018
M. di Prisco and M. Menegotto (Eds.): ICD 2016, LNCE 10, pp. 524–534, 2018.
https://doi.org/10.1007/978-3-319-78936-1_38

2 FEMA P-58 Methodology

The next-generation seismic PBD guidelines employ a probabilistic framework to characterize the building performance. The objective is not only the life safety, but also to evaluate more immediate and useful variables relevant for the decision-making pro-cess. Indeed performance is expressed in terms of median and dispersions values of casualties, repair costs and repair time. Incremental Dynamic Analyses (IDA) are generally used in this approach.

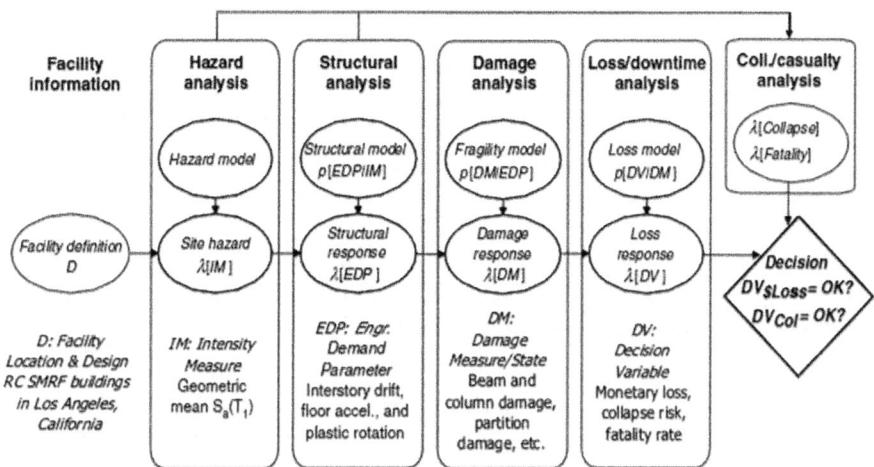

Fig. 1. FEMA P-58 methodology scheme (Goulet et al. 2007)

The FEMA P-58 PBEE process (Moehle and Deierlein 2004, FEMA 2012a, b), illustrated in Fig. 1, can be broken down into logical elements that can be studied and resolved in a rigorous and consistent manner. The performance is estimated as a function of the damage level sustained by the individual building components. Finally the DVs obtained enter into risk management decisions. The en-tire process can be expressed in terms of a triple integral that is an application of the total probability theorem, which is numerically resolved through the PACT tool:

$$vD(V) = \iiint P[DV|DM] dP[DM|EDP] dP[EDP|IM] d\lambda[IM] \qquad (1)$$

3 Case Study

The PBEE framework has been applied to a three-story non-ductile R/C prototype building, typical of the Italian building stock. The basic building dimensions are 10 m (X-direction) by 8 m (Y-direction). Unreinforced masonry infills without openings are present in the X-direction external facades.

This prototype building has been used in the past for several experimental studies (Braga et al. 2009) and numerical studies (Valleriani 2011; Sellitto 2013; Mohammad et al. 2014a, b; Mohammad et al. 2016) to investigate the impact of different models and sources on nonlinearity on the structural response, as well as plan configurations (Faggella et al. 2015a, b) and theoretical regularity assumptions (Faggella 2013; Faggella 2014a, b, c, d). The structure is located in Reggio Calabria, a site represen-tative of the high seismic hazard in Italy (site class A). This structure is representative of structures designed for vertical loads only.

3.1   Numerical Models

Comparative seismic risk analyses are carried out on three 2D structural configurations, developed using OpenSees. All nonlinear beam and column elements in the model are force-based fiber section elements, with different model parameters assigned to unconfined and confined concrete. Infill panels are modelled by strut elements, whose backbone curve is determined according to FEMA 356/ASCE41-06 pro-visions (FEMA 2000).

The 2D computational models are a Bare Frame (BF), a Uniformly Infilled Frame with concentric in-fill strut elements (UIF) and an Uniformly Infilled Frame with shear columns and eccentric infill struts (UIF-s). To capture the local nonlinear shear behaviour of frame

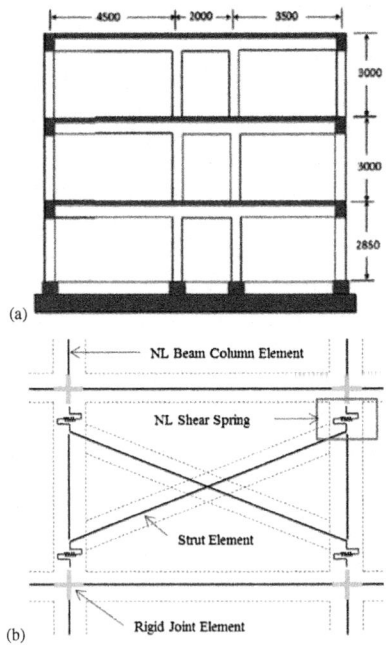

Fig. 2. (a) Case study prototype building; (b) modeling details of infill-frame interaction: nonlinear shear law in nodal short columns (Mohammad et al. 2016)

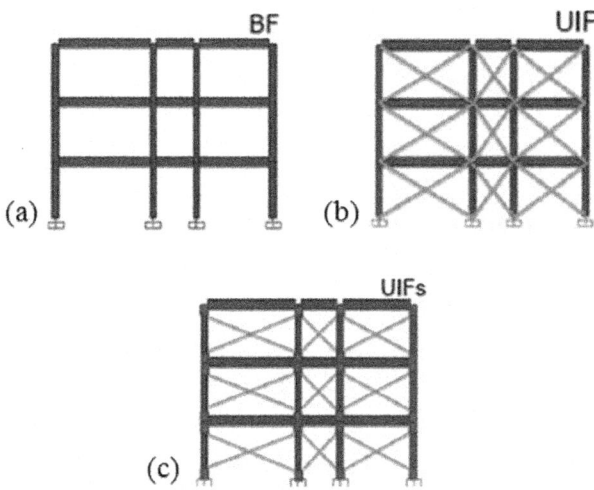

Fig. 3. 2D computational models: (a) Bare Frame, BF; (b) Uniformly Infilled Frame, UIF; (c) Uniformly Infilled Frame with shear columns, UIF-s (Mohammad et al. 2016)

Table 1. Fundamental periods of 2D models, after application of vertical loads.

Model	T1 [s]
Bare Frame (BF)	0.590
Uniformly Infilled Frame (UIF)	0.202
Uniformly Infilled Frame- shear column (UIF-s)	0.253

members, the shear law proposed by Marini and Spacone (2006) is imposed to short column elements next to the nodal regions (Fig. 2b). The three configurations are represented in Fig. 3. The fundamental periods of the three configurations are represented in Table 1.

4 Incremental Dynamic Analyses (IDA)

Incremental Dynamic Analyses (Vamvatsikos and Cornell 2002) are performed to characterize the structural behaviour resulting from structural and nonstructural elements, obtaining EDP distributions at different hazard levels. Eleven ground motions records are extracted from the European strong motion database and scaled to

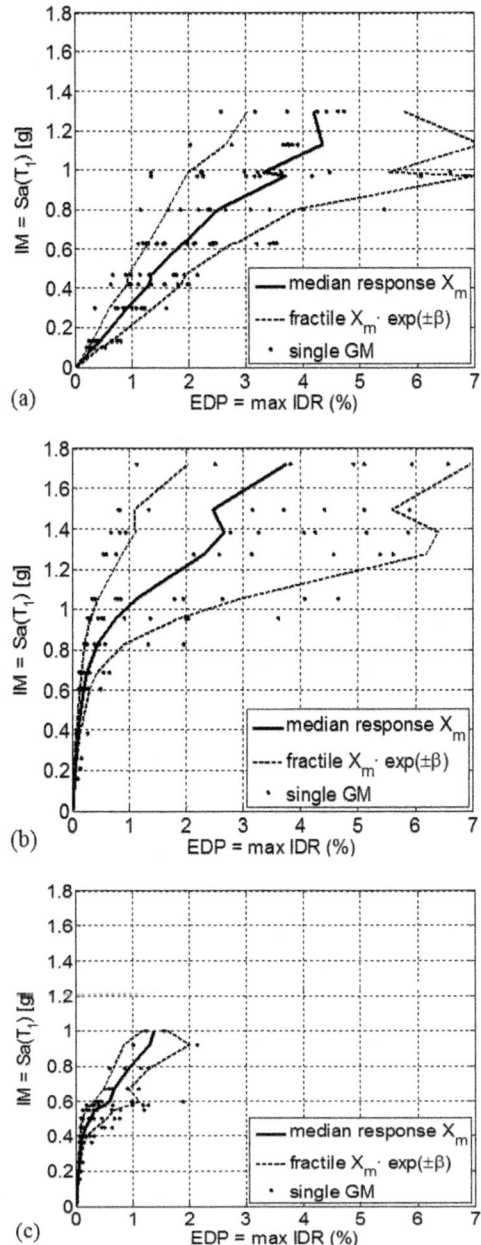

Fig. 4. IDA curves, with 16%, 50% (solid thick line), and 84% fractiles for lognormal drift distribution: (a) Bare Frames; (b) Uniformly Infilled Frames; (c) Uniformly infilled Frames with shear column

$IM = S_a(T_1)$ at 13 hazard levels using different scale factors for each GM and for each hazard level. The analyses were then repeated by scaling the GMs to IM = PGA. The NEEShub cloud-computing platform is used to perform the large number of analyses in parallel (https://nees.org/). The analyses are performed comparing the response obtained in the different modelling configurations. Results highlight the impact of infills contribution and nonlinear shear strength (in the eccentric strut arrangement) on the building behavior, as apparent in Fig. 4.

It is worth noting that analysis results with GMs scaled to IM = PGA are not represented in this paper: max IDR median values and dispersions are essentially lower than in the case of $IM = S_a(T_1)$ (Faggella et al. 2013).

5 Loss Analysis

Structural analysis results are used as input parameters in the damage/loss analysis. The comparative loss analysis is performed to evaluate the importance of the structural modeling and of fragility/consequence functions selection in the building performance estimate. This was done implementing different performance models in the calculation tool PACT, provided by FEMA P-58. It is also necessary to define a Total Replacement Cost (TRC) value in case of building collapse or irreparability. Fragility functions are used in a probabilistic way to correlate EDP to structural and nonstructural element damage; consequence functions are used to associate a probabilistic loss distribution to different element damage states.

5.1 Fragility and Consequence Functions Assembly

In this work, fragility and consequence functions are not selected from the current PACT database. This is due to the difference between Italian and American construction types, especially in frame-infill interaction or the lack of masonry infill characterization in the FEMA P-58 methodology, and differences in the element repair cost libraries. Therefore, median EDP values for the probabilistic curves were derived assembling results from research works in literature and other code references. Nevertheless, the dispersions values used remain those provided by the tool PACT.

The column fragility functions deduced by literature research are represented in Fig. 5: for BF and UIF models, the ductile behavior allows defining three sequential damage states in a wide range of drift; for the UIF-s, only two damage states in a limited range reflect the brittle shear failure due to frame-infill interaction. Consequence functions are then developed considering all the necessary repair activities that would be required to restore each component to its pre-earthquake undamaged condition. A bill of quantities is made using the price list 'Elenco Regionale dei Prezzi delle Opere Pubbliche della (Regione Emilia-Romagna 2012)'.

5.2 Loss Analysis Results

Two types of assessments were performed: an intensity-based and a time-based assessment. Intensity-based assessment provides repair costs for each performance model at five hazard levels (63%, 50%, 10%, 5% and 2% in 50 years). In the

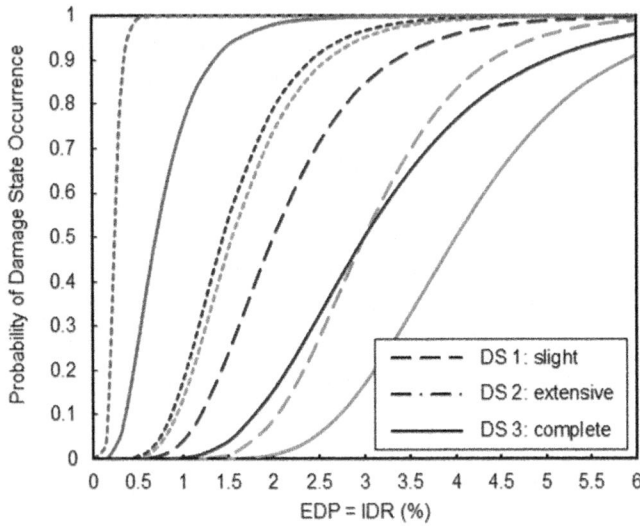

Fig. 5. Fragility functions for the column damage states (DS)

time-based assessment, the DV values are weighted according to the probability of occurrence for each hazard level and then integrated to determine a total life-cycle performance, expressed as Expected Annualized Loss (EAL).

Time-based assessment is performed at 13 intensity levels accounting for the complete site seismic hazard. Figure 6 represents the loss hazard curves of the three models plotting the Repair costs (% of TRC) against the MAFE λ. They are obtained using an approach provided directly by the tool PACT, ac-counting for logarithmic dispersions β. The tool solves numerically the triple integral in Eq. (2). The areas underlying the loss hazard curves represent the EAL (Eq. 3).

$$\nu D(V) = \iiint P[DV|DM]dP[DM|EDP]dP[EDP|IM]d\lambda[IM] \tag{2}$$

$$EAL = \int E[DV|\lambda(DV)]d\lambda(DV) \tag{3}$$

Note that the BF model damage is gradual and hence this is reflected also in the loss progression. In the UIF configuration the presence of concentric in-fill strut underestimate building response in terms of damage and repair costs. For the UIF-s model, the transition from an undamaged to a collapse condition is fast, confirming the occurrence of a brittle collapse mechanism.

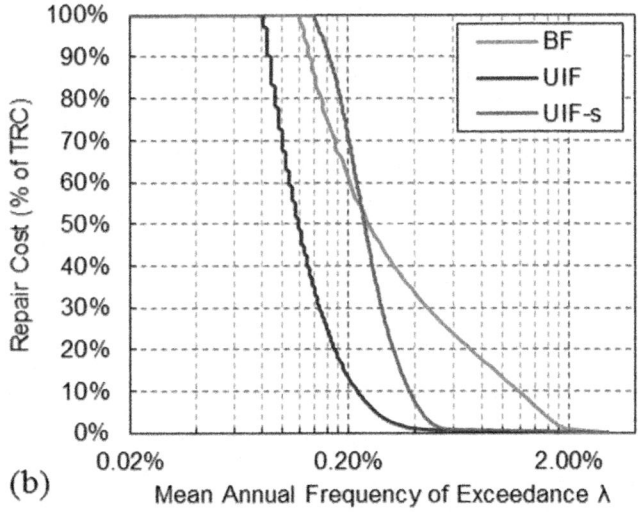

(b)

Fig. 6. Time-based assessment: loss hazard curves comparison between (a) the simplified method and (b) tool PACT

Table 2 shows that the highest EAL value is en-countered for the BF model, despite collapse occurs for very rare seismic events, with TR = 4000 years (λ = 0.02%). This is due to the greater EAL contribution from nonstructural damage in more frequent seismic event (lower seismic intensities). This high nonstructural dam-age in BF performance models is because infill panels are neglected in the computa-tional model, whereas they are incorporated in the BF performance models.

Table 2. Expected Annual Loss (EAL) from tool PACT, accounting for dispersions.

Model	EAL (% of TRC)
[BF]	0.45%
[UIF]	0.14%
[UIF-s]	0.25%

6   Conclusions

In this paper, the probabilistic framework of FEMA P-58 is applied to an older R/C infilled building typical of the Italian stock, to compare the life-cycle seismic performance of different performance models. Structural analysis results are expressed in terms of IDA curves; performance is expressed in terms of EAL (time-based assessment).

Results highlight the importance of using a suitable computational model in predicting the building performance: concentric infill strut modeling can underestimate the overall damage state and then the economic loss value, whereas Bare Frame performance models can overestimate it at lower seismic intensities.

Intensity Measure selection is also important: using the PGA instead of spectral acceleration at the fundamental building period, S_a (T_1), the repair costs are reduced at higher hazard levels for the BF and UIF models, contrary to the UIF-s model for which there is no reduction. The results of time-based assessment show the importance of dispersions in EAL estimate: a probabilistic complete life-cycle analysis should be always preferred.

References

Braga F, Gigliotti R, Laterza M (2009) R/C existing structures with smooth reinforcing bars: experimental behaviour of beam-column joints subject to cyclic lateral loads. Open Constr Build Technol J 3:52–67

Braga F, Gigliotti R, Monti G, Morelli F, Nuti C, Salvatore W, Vanzi I (2014) Speedup of post earthquake community recovery: the case of precast industrial buildings after the Emilia 2012 earthquake. Bull Earthq Eng 12(5):2405–2418

Braga F, Gigliotti R, Monti G, Morelli F, Nuti C, Salvatore W, Vanzi I (2015) Post-seismic assessment of existing constructions: evaluation of the shakemaps for identifying exclusion zones in Emilia. Earthq Struct 8(1):37–56

Crisafulli FJ (1997) Seismic behaviour of reinforced concrete structures with masonry infills. PhD thesis, Department of Civil Engineering, University of Canterbury, New Zealand

Dolce M (2012) The Italian national seismic prevention program. In: Proceedings of 15th World Conference on Earthquake Engineering, 24–28

Elwood KJ, Moehle J (2003) Shake table test and analytical studies on the gravity load collapse of reinforced concrete frames. PEER 2003/01 PEER report 2003/01, Berkeley: Pacific Earthquake Engineering Research Center, University of California, 346p, November 2003

Faggella M, Barbosa AR, Conte JP, Spacone E, Restrepo JI (2013) Probabilistic seismic response analysis of a 3-D reinforced concrete building. Struct Saf 44:11–27

Faggella M, Mezzacapo G, Gigliotti R, Spacone E (2015a) Significance of earthquake incidence on response plan-irregular infilled R/C buildings. In: COMPDYN 2015, 5th ECCOMAS thematic conference on computational methods in structural dynamics and earthquake engineering, Crete Island, Greece, 25–27 May 2015

Faggella M, Gigliotti R, Mezzacapo, G, Spacone E (2015b) Graphical dynamic trends for earthquake incidence response of plan asymmetric systems. In: COMPDYN 2015, 5th ECCOMAS thematic conference on computational methods in structural dynamics and earthquake engineering, Crete Island, Greece, 25–27 May 2015

Faggella M (2014) The ellipse of elasticity and mohr circle-based graphic dynamic modal analysis of torsionally coupled systems. In: Proceedings of the 9th International Conference on Structural Dynamics, EURODYN 2014, Porto, Portugal, 30th June–2nd July 2014

Faggella M (2014) Graphical dynamic earthquake response analysis of one-way asymmetric systems. In: Proceedings of the 10th national conference in earthquake engineering. Earthquake Engineering Research Institute, Anchorage, AK

Faggella M (2014) Graphic dynamic earthquake response analysis of linear torsionally coupled 2DOF systems. In: Proceedings of the 9th international conference on structural dynamics, EURODYN 2014, Porto, Portugal, 30th June–2nd July 2014

Faggella M (2014) Graphical dynamic earthquake response of two-ways asymmetric systems based on directional modal participation radii. In: CST2014, The twelfth international conference on computational structures technology, Naples, Italy, 2–5 September 2014

Faggella, M (2013) Graphical modal analysis and earthquake statics of linear one-way asymmetric single-story structure. In: Papadrakakis M, Papadopoulos V, Plevris V (eds) COMPDYN 2013, 4th ECCOMAS thematic conference on computational methods in structural dynamics and earthquake engineering, Kos Island, Greece, 12–14 June 2013

Faggella M, Laguardia R, Gigliotti R, Morelli F, Braga F, Salvatore W (2016) Performance-based nonlinear response history analysis framework for the PROINDUSTRY project WP2 case studies. In: ECCOMAS congress 2016 VII european congress on computational methods in applied sciences and engineering, Crete Island, Greece, 5–10 June 2016

Faggella M, Monti G, Braga F, Gigliotti R, Capelli M, Spacone E, Laterza M, Triantafillou T, Varum H, Safi MD, Subedi J, Dixit A, Lodi S, Rahman Z, Limkatanyu S, Xiao Y, Yingmin L, Kumar H, Salvatore W, Cecchini A, Lukkunaprasit P (2012) A development cooperation erasmus mundus partnership for capacity building in earthquake mitigation science and higher education. In: 4th international disaster and risk conference IDRC Davos 2012, Davos, Switzerland, 26–30 August 2012

Federal Emergency Management Agency (2000) Prestandard and commentary for the seismic rehabilitation of buildings (FEMA 356 - ASCE/SEI 41–06)

Federal Emergency Management Agency (2012a) Seismic Performance Assessment of Buildings Volume 1 – Methodology (FEMA P-58)

Federal Emergency Management Agency (2012b) Seismic Performance Assessment of Buildings Volume 2 – Implementation Guide (FEMA P-58)

Goulet CA, Haselton CB, Mitrani-Reiser J, Beck JL, Deierlein GG, Porter KA, Stewart JP (2007) Evaluation of the seismic performance of a code-conforming reinforced-concrete frame building—from seismic hazard to collapse safety and economic losses. Earthq Eng Struct Dynam 36(13):1973–1997

Marini A, Spacone E (2006) Analysis of reinforced concrete elements including shear effects. ACI Struct J 103(5):645–655

Moehle J, Deierlein GG (2004) A framework methodology for performance-based earthquake engineering. In: 13th world conference on earthquake engineering, Vancouver, B.C., Canada, Paper No. 679, 1–6 August 2004

Mohammad AF, Faggella M, Gigliotti R, Spacone E (2016) Seismic performance of older R/C frame structures accounting for infills-induced shear failure of columns. Eng Struct 122:1–13

Mohammad AF, Faggella M, Gigliotti R, Spacone E (2014) Probabilistic seismic response sensitivity of nonlinear frame bending-shear and infill model parameters for an existing infilled reinforced concrete structure. In: CST2014, The twelfth international conference on computational structures technology, Naples, Italy, 2–5 September 2014

Mohammad AF, Faggella M, Gigliotti R, Spacone E (2014) Influence of bond-slip effect and shear deficient column in the seismic assessment of older infilled frame R/C structures. In: EURODYN 2014 9th international conference on structural dynamics 30, Porto, Portugal

Regione Emilia-Romagna (2012) Elenco regionale dei prezzi delle opere pubbliche della regione Emilia-Romagna. BUPERT (Official Journal of Regione Emilia-Romagna), 31 July 2012

Rossi E, Ventrella M, Faggella M, Gigliotti R, Braga F (2016) Performance-based earthquake assessment of an industrial silos structure and retrofit with sliding isolators. In: ECCOMAS Congress 2016 VII European Congress on Computational Methods in Applied Sciences and Engineering, Crete Island, Greece, 5–10 June 2016

Sellitto G (2013) Risposta sismica degli edifici esistenti in c.a. Analisi non lineari e modellazione dei fenomeni di aderenza. MS Dissertation, Sapienza Università di Roma. (Advisors: Gigliotti R, Faggella M)

Sezen H, Moehle JP (2004) Shear strength model for lightly reinforced concrete columns. J Struct Eng 130:1692–703. https://doi.org/10.1061/(ASCE)0733-9445(2004)130:11(1692)

Valleriani D (2011) Edifici esistenti in c.a.: confronto tra metodi e modelli per analisi lineari e non lineari. Master Degree Dissertation. Department of Structural and Geotechnical Engineering, Sapienza Università di Roma

Vamvatsikos D, Cornell CA (2002) Incremental dynamic analysis. Earthq Eng Struct Dyn 31 (3):491–514

Influence of Materials Knowledge Level on the Assessment of the Characteristic Value of the Shear Strength of Existing RC Beams

D. Lavorato[1]([⊠]), A. V. Bergami[1], A. Forte[1], G. Quaranta[2], C. Nuti[1], G. Monti[2], and S. Santini[1]

[1] Department of Architecture, University of Roma Tre, Rome, Italy
davide.lavorato@uniroma3.it
[2] Department of Structural and Geotechnical Engineering,
Sapienza University of Rome, Rome, Italy

Abstract. This study deals with the assessment of the shear strength characteristic value (V_{rck}) limited by the concrete strut crushing for existing reinforced concrete (RC) beams on the base of the available material data. A procedure to determine V_{rck} is presented and applied to two existing RC beams extracted from an old structure built in the early 1900s. This procedure requires: (i) an analytical or numerical model for the beam shear strength, (ii) mean and standard deviation values for each uncertain basic variable and (iii) a proper formulation for the tolerance factor depending on amount of data, assumed fractile and given confidence level. The shear strength model adopted in this study is the one proposed by the Italian Code NTC 2008 for beams with transversal steel reinforcement. Several non-destructive tests (namely, rebound and sonic tests) and destructive tests on concrete were performed, and the experimental outcomes showed a great variability of the compressive strength along the same beam, as usual for many old concretes. For that reason, V_{rck} values for each beam were calculated assuming different knowledge levels about the concrete compressive strength. Each level of knowledge is defined taking into account a different combination of available data about the compressive strength carried out from destructive tests. A comparison among the V_{rck} values obtained for each knowledge level is shown to draw useful considerations about the beam shear strength assessment based on materials test data.

Keywords: Existing RC structures · Characteristic value · Capacity assessment

1 Introduction

Capacity assessment of an existing reinforced concrete (RC) structure is a key point for the evaluation of proper repair and/or retrofitting interventions (Albanesi et al. 2008, 2009; Lavorato et al. 2010a, b, 2011, 2015a, b; 2017; Marano et al. 2017) after seismic damage (Fiorentino et al. 2017a). Uncertainty about building materials characteristics has effects on the evaluation of the seismic response of structure (Fiorentino et al. 2017b).

Within this framework, this paper presents a procedure that aims at estimating the capacity of structural elements or systems on statistical basis, providing mean and

© Springer International Publishing AG, part of Springer Nature 2018
M. di Prisco and M. Menegotto (Eds.): ICD 2016, LNCE 10, pp. 535–547, 2018.
https://doi.org/10.1007/978-3-319-78936-1_39

standard deviation values for its probabilistic distribution with minimum computational effort. This procedure starts from a structural capacity model, which is function of some random variables (related to geometry, materials, reinforcing steel configuration, etc.). The capacity model can be obtained by means of analytical approaches (see for instance Trentadue et al. 2014, 2016) or numerical methods (see for instance Scodeggio et al. 2016 or Quaranta et al. 2012). The procedure provides mean and standard deviation of the structural capacity once the first two statistical moments of each uncertain basic random variable are estimated from experimental tests. The characteristic value of the structural capacity is then obtained using the estimated mean and standard deviation, and assuming a suitable probability distribution (e.g., the Lognormal distribution). This task also involves a proper formulation for the tolerance factor, which depends on the amount of data as well as on the assumed fractile and confidence level. The proposed procedure is applied to a real case–study.

(Imperatore et al. 2012a, b, 2013) in order to compute the shear strength limited by the concrete strut crushing of two RC beams (T1 and T2) extracted from an old structure built in the early 1900s. The shear strength model selected in this study is the one proposed in NTC 2008 for beams with transversal steel reinforcement. In doing so, it is assumed that the concrete compressive strength is the only random variable. Beam geometry and steel reinforcement configurations are known with sufficient accuracy, and thus they are assumed as deterministic data. A large number of non-destructive tests (namely, rebound and sonic tests) and destructive tests was performed to determine the concrete compressive strength. The experimental results showed a great variability of the compressive strength along the same beam, as it is the case for many old concretes. The variability of the concrete strength is different for the two tested structural members, even if they were near each other in the structure at the same floor level. Mean ($V_{rc,mean}$) and characteristic (V_{rck}) value of the shear strength limited by the concrete strut crushing are calculated assuming different knowledge levels about the concrete compressive strength. Each level of knowledge is defined by the available experimental test results. Different working hypotheses (WHPs) are formulated, i.e. WHPs that consider a different number of available concrete specimens (2, 3, 5, 8) and WHPs that consider the same number of samples with concrete specimens extracted from different points along the beams. Preliminary considerations about the variation of $V_{rc,mean}$ and V_{rck} calculated for the two beams and assuming different levels of knowledge (WHP) are given at the end of the paper.

2   Procedure to Evaluate the Characteristic Value of a Capacity Indicator

Let C be a performance indicator through which an existing structure is assessed:

$$C = G(X) \tag{1}$$

where $G(\mathbf{X})$ is an analytical or numerical capacity model whereas $\mathbf{X} = [X_1, ..X_i, X_n]$ is the vector of the random basic variables (i.e., related to geometry, materials, etc.). The proposed methodology requires that $s_i = \partial G(X_i)/\partial X_i$ exists and is continuous for each basic variable. Without loss of generality, it is assumed that the basic variables X_i for $i = 1, \ldots, n$ are independent each other. Each basic variable X_i is described by the sample mean μ_{X_i} and the sample variance $\sigma_{X_i}^2$. The mean value of the capacity indicator C is given by Eq. (2):

$$\mu_C = G(\mu_X) \tag{2}$$

whereas the approximated variance of C is obtained from Eq. (3), in which the well-known Taylor expansion-based estimate for the variance is used.

$$\sigma_C^2 = \sum_{i=1}^{n} \left(\frac{\partial G(X)}{\partial X_i} \Big|_{\mu_X} \right)^2 \sigma_{X_i}^2 \tag{3}$$

The characteristic value of the capacity indicator C_k is given by Eq. (4) assuming a Lognormal distribution (which is a reliable assumption for a strength variable of a RC structure):

$$C_k = \frac{\mu_C e^{-k\sqrt{\ln\left(1 + \left(\frac{\sigma_C}{\mu_C}\right)^2\right)}}}{\sqrt{\left(1 + \left(\frac{\sigma_C}{\mu_C}\right)^2\right)}} \tag{4}$$

where μ_C and σ_C are given from Eqs. (2) and (3), respectively. The tolerance factor $k_{\alpha,p,\nu}$ related to a given fractile p and an assigned confidence level $100(1-\alpha)\%$ with ν degrees of freedom is (Monti and Petrone 2016):

$$k_{\alpha,p,\nu} = \frac{z_p}{A} \frac{4\nu - 2}{4\nu - 1} + z_\alpha \sqrt{\frac{1}{\nu} + \frac{1}{2\nu}\left(\frac{z_p}{A}\right)^2} \tag{5}$$

$$A = 1 - \frac{1}{2\nu}(1 + z_\alpha^2) \tag{6}$$

where $z_p = \Phi^{-1}(p)$, $z_\alpha = -\Phi^{-1}(\alpha)$, and $\Phi^{-1}()$ is the inverse cumulative standardized Normal function. Moreover, $\nu = N - 1$ (where N is the number of samples).

3   Beams Geometries and Materials

Two RC beams (Fig. 1) were extracted from an old structure in Rome dating back to 1900s. These beams had very similar geometries and steel reinforcements (see Fig. 2 and Table 1). They were probably built using the same materials (casted concrete and steel rebars) since they were placed near each other at the same floor level. However, concrete compressive strength and steel yielding stress obtained from destructive tests on concrete specimens (cores) and steel specimens (rebar samples) exhibited a large variability, as shown in Tables 2 and 3.

Fig. 1. Extracted beams T1 and T2 from existing RC buildings (1900s)

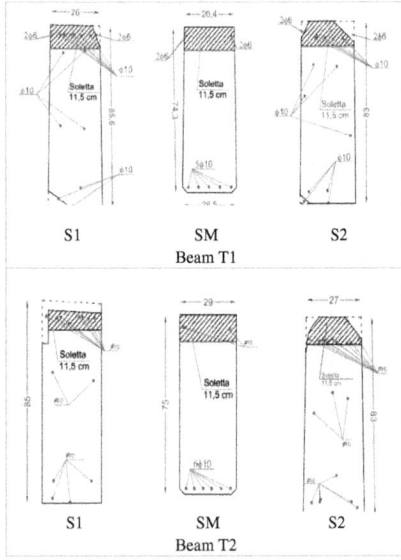

Fig. 2. Extracted beams T1 e T2: concrete geometries and steel reinforcement configurations for sections S1, SM and S2 (Fig. 3)

Table 1. RC beams T1 e T2: concrete geometries and steel reinforcement configurations

	Beam T1	Beam T2
Length	343 cm	389 cm
Midspan height	74.3 cm	75.0 cm
Terminal height	83/85.6 cm	83/85 cm
Upper reinforcement	4Φ6	2Φ10
Bottom reinforcement	2Φ10	2Φ10
Midspan reinforcement	5Φ10	6Φ10
Traversal reinforcement	16Φ6	15Φ6
Spacing of stirrup	15/30 cm	15/32 cm

Table 2. RC beam T1: concrete compressive strength (f_c) and steel stirrups yielding stress (f_y) by destructive tests

	Beam T1	
	f_c [MPa]	f_y [MPa]
Point ST7	16.25	RT1-1[*] 384.55
Point ST8	9.77	RT1-2[*] 393.54
Point ST10	16.51	RT1-3[*] 439.33
Point ST11	20.53	
Point ST11-inf-1	13.94	
Point ST11-inf-2	7.68	
Point ST14	16.02	
Point ST16	16.93	

* Note: positions of steel rebars are not known

Table 3. RC beam T2: concrete compressive strength (f_c) and steel stirrups yielding stress (f_y) by destructive tests

	Beam T2	
	f_c [MPa]	f_y [MPa]
Point ST2	10.72	RT2-1[*] 344.36
Point ST3	10.73	RT2-2[*] 636.51
Point ST8	10.08	RT2-3[*] 322.77
Point ST10	12.27	RT2-4[*] 304.93
Point ST11	10.65	RT2-5[*] 313.44
Point ST13	8.86	
Point ST22	12.91	

* Note: positions of steel rebars are not known

4   Capacity Model

The selected capacity model for the shear strength is the one proposed within NTC 2008 (Italian Construction Technical Code) for beams with transversal steel reinforcement (see Eq. 7a, 7b, 7c):

$$V_{Rd} = \min(V_{Rsd},\ V_{Rcd}) \tag{7a}$$

$$V_{Rsd} = 0,9 \cdot d \cdot \frac{A_{sw}}{s} \cdot f_{yd} \cdot (ctg\alpha + ctg\theta) \cdot \sin\alpha \tag{7b}$$

$$V_{Rsd} = 0,9 \cdot d \cdot b_w \cdot \alpha_c \cdot f'_{cd} \cdot (ctg\alpha + ctg\theta)/(1 + ctg^2\theta) \tag{7c}$$

where V_{Rsd} is the shear strength limited by the stirrups contribution and V_{Rcd} is the shear strength limited by the crushing of the concrete strut while f_{yd}, A_{sw} and s are steel yielding stress, cross section area of shear reinforcement and spacing of the stirrups, respectively. Moreover, d and b_w are effective depth and width of the beam transversal section, $\alpha_c f_{cd}$ is the reduced concrete compressive strength, α is the angle between the shear steel reinforcement and the beam axis whereas θ is the angle of the concrete strut. In this paper, the shear strength limited by the concrete strut crushing is calculated assuming that the concrete compressive strength is the only random variable. This assumption is basically motivated by the fact that the uncertainty affecting the compressive strength of the concrete is much larger than the randomness of the other basic variables of the capacity models.

5   Knowledge Levels About the Beam Materials

The strengths of the materials were obtained by destructive tests on concrete and steel specimens. The concrete specimens are cores extracted from the beams. It is important to underline that the extraction of concrete cores produces effects on the beam that have to be considered carefully. Furthermore, it is usually difficult to extract a concrete core from a loaded beam of an existing structure. In fact, the extraction points should be placed in concrete tension zones where the concrete contribution is not relevant, and this operation has not to damage the steel reinforcement during the core extraction. For that reason, the number of the available concrete cores from the same beam is usually rather limited (one or two concrete cores are typically extracted).

The beams T1 and T2 considered in this study were already tested to evaluate their shear strength during a previous research campaign (Imperatore et al. 2012a, b, 2013). Therefore, it was possible to extract eight and seven concrete cores from T1 and T2 beams, respectively. The core extraction was performed in that zones where the concrete resulted undamaged after the tests and far from the reinforcing bars (extraction positions among the stirrups and the longitudinal rebars). It is important to highlight that the extraction of this large number of cores from the same loaded beam in existing structures is unrealistic. However, using a large number of concrete specimens is interesting to evaluate how quality and amount of experimental data – together with their spatial variability – influence the estimation of the structural capacity.

Table 4. Work hypotheses (WHP) for beam T1: number of concrete cores for each WHP and core label for obtaining the core position along the beam (Fig. 3)

Beam T1	
WHP1	All the available concrete core (8 cores)
WPH2	[ST7] – [ST11inf-1] – [ST11inf-2] – [ST14] – [ST16]
WPH3a	[ST7] – [ST14] – [ST16]
WPH3b	[ST7] – [ST11inf-1] – [ST16]
WPH3c	[ST7] – [ST11inf-1] – [ST11inf-2]
WPH4a	[ST7] – [ST16]
WPH4b	[ST7] – [ST11inf-1]

Table 5. Work hypotheses (WHP) for beam T2: number of concrete cores for each WHP and core label for obtaining the core position along the beam (Fig. 3)

Beam T2	
WHP1	All the available concrete core (7 cores)
WPH2	[ST2] – [ST3] – [ST11] – [ST13] – [ST22]
WPH3a	[ST2] – [ST13] – [ST22]
WPH3b	[ST2] – [ST11] – [ST22]
WPH3c	[ST2] – [ST11] – [ST13]
WPH4a	[ST13] – [ST22]
WPH4b	[ST11] – [ST22]

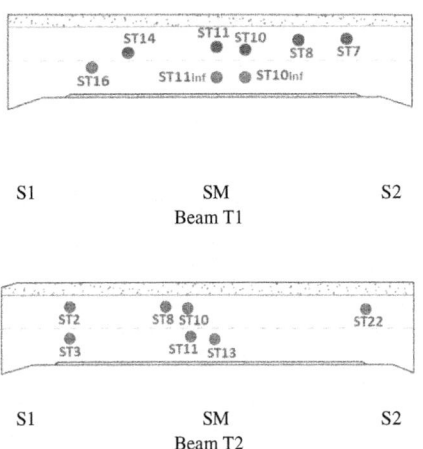

Fig. 3. Extracted beams T1 e T2: positions of the concrete core extraction points along the beams

The WHPs for the beam T1 and T2 are presented in Tables 4 and 5, respectively. In these tables, the number of concrete cores for each WHP and their labels are listed. The concrete core label permits the identification of the concrete core extraction point position in Fig. 3.

The WHP1 considers eight cores for beam T1 and seven cores for beam T2; i.e., all the available cores for each beam. The WHP2 considers the compression strength of five concrete cores for each beam. The WHPs labeled as WHP3a, WHP3b and WHP3c consider different combination of the compression test results carried out from three concrete cores. Similarly, WHP4a and WHP4b consider different combinations of the compression test results from two concrete cores. The main motivation is to evaluate the variation of the selected capacity indicator as function of the inherent uncertainties into the experimental data.

The statistical evaluation of mean and standard deviation values of the concrete compressive strength (f_c) is a key point of the proposed procedure to estimate $V_{rc,mean}$ and V_{rck}. The reliable assessment of the standard deviation values for f_c usually requires at least five compression strength measures. For that reason, WHP2 is a limit case study from a statistical point of view. All the WHPs that consider three concrete cores (WHP3a, WHP3b, WHP3c) as well as those that take into account two concrete cores (WHP4a, WHP4b) have an insufficient number of compression strength measures from a statistical point of view. However, it is important to point out that extracting more than two samples from the same beam is unfeasible for many cases. In fact, a few cores can be usually extracted from a modest number of structural elements of the structure because of technical issues and in order to preserve the structural integrity after the extraction operation. The use of non-destructive tests (i.e., rebound and sound tests) can be a viable strategy to increase the number of estimates about the concrete compressive strength. A great number of nondestructive tests can be performed at low cost on the same element, without jeopardizing its integrity. The results of non-destructive tests are usually calibrated by means of a few outcomes available from destructive tests. As a consequence, there is a correlation among these different types of data, which has to be considered through proper statistical procedures (Giannini et al. 2014). This problem is still topic of ongoing studies and is not addressed herein.

6 Shear Strength of the Beams T1 and T2

The beam shear strength limited by the crushing of the concrete compressed strut (V_{rc}) is evaluated using the model described by Eq. 7b. This model is function of compressive strength of concrete (f_c) and beam geometry (b, d). In this study, the geometrical data of the two beams (T1 and T2) are measured with high accuracy (Fig. 2), and thus they are assumed as deterministic parameters in Table 1. Therefore, the V_{rc} capacity model is function of only one random variable, that is the compressive strength of the concrete ($X = f_c$). The Lognormal distribution is assumed for f_c. The values of mean (μ_x) and standard deviation (σ_x) of f_c for each WHP are given in Tables 6 and 7 for beam T1 and beam T2, respectively.

Table 6. RC beam T1: mean (μ_x) and standard deviation (σ_x) of the Lognormal distribution of the concrete compressive strength $f_c = X$ for each WHP

	Beam T1	
	μ_x [MPa]	σ_x
WHP1	14.87	4.96
WHP2	14.40	4.87
WHP3a	16.40	0.22
WHP3b	15.74	2.61
WHP3c	13.01	5.36
WHP4a	16.59	0.23
WHP4b	15.14	2.71

Table 7. RC beam T2: mean (μ_x) and standard deviation (σ_x) of the Lognormal distribution of the concrete compressive strength $f_c = X$ for each WHP

	Beam T2	
	μ_x [MPa]	σ_x
WHP1	10.90	1.84
WHP2	10.79	2.08
WHP3a	10.89	4.29
WHP3b	11.45	1.57
WHP3c	10.10	1.20
WHP4a	11.08	9.02
WHP4b	11.83	2.62

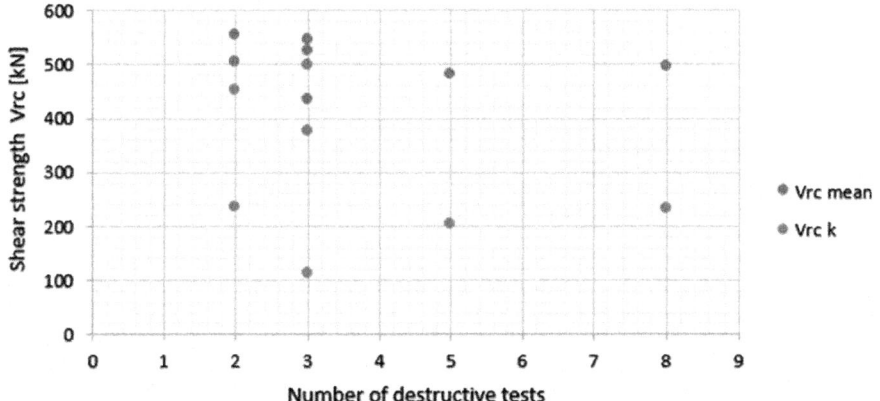

Fig. 4. Extracted beam T1: characteristic values (V_{rck}) and mean values ($V_{rc,mean}$) of the shear strength limited by the crushing of the concrete compressed strut for each WHP

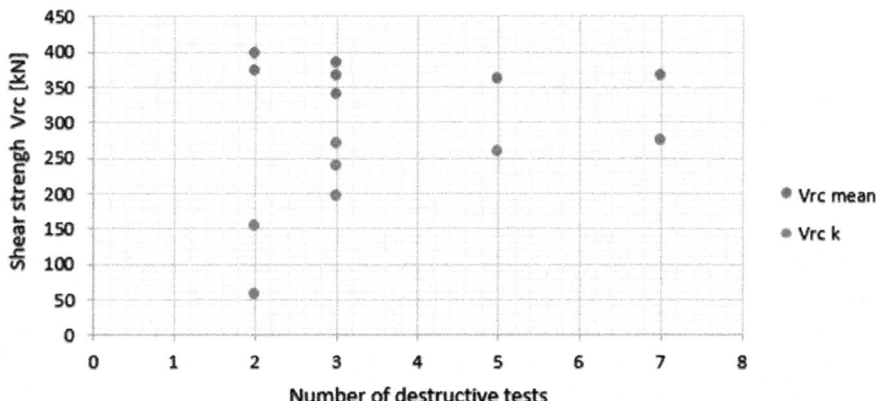

Fig. 5. Extracted beam T2: characteristic values (V_{rck}) and mean values ($V_{rc,mean}$) of the shear strength limited by the crushing of the concrete compressed strut for each WHP

Figures 4 and 5 show mean ($V_{rc,mean}$) and characteristic (V_{rck}) value of the shear strength limited by the concrete strut crushing calculated for each WHP as function of the number of samples for beam T1 and beam T2, respectively.

Tables 8 and 9 show, for each WHP, the number of the concrete cores considered to evaluate f_c and the corresponding values of $V_{rc,mean}$ and V_{rck} for beam T1 and beam T2, respectively.

Table 8. RC Extracted beam T1: Characteristic values (V_{rck}) and mean values ($V_{rc,mean}$) of the shear strength limited by the crushing of the concrete compressed strut for each WHP and number of available concrete cores (n tests)

	Beam T1		
	n tests	$V_{rc\ mean}$ [kN]	$V_{rc\ k}$ [MPa]
WHP1	8	495	231
WHP2	5	480	202
WHP3a	3	547	498
WHP3b	3	524	375
WHP3c	3	433	112
WHP4a	2	553	451
WHP4b	2	505	235

Table 9. RC Extracted beam T2: Characteristic values (V_{rck}) and mean values ($V_{rc,mean}$) of the shear strength limited by the crushing of the concrete compressed strut for each WHP and number of available concrete cores (n tests)

	Beam T2		
	n tests	$V_{rc\ mean}$ [kN]	$V_{rc\ k}$ [MPa]
WHP1	7	365	274
WHP2	5	362	258
WHP3a	3	365	196
WHP3b	3	384	269
WHP3c	3	338	238
WHP4a	2	371	56
WHP4b	2	397	152

The following conclusions can be drawn:

- the value of $V_{rc,mean}$ is almost constant. Indeed, it slightly depends on the size of the considered subset of concrete cores, anywhere the samples are extracted.
- the difference among the $V_{rc,mean}$ values of the two beams is about 20%, even if the two beams are casted at the same time using the same materials.
- V_{rck} values calculated using two or three concrete cores for each beam are very different with respect to the ones obtained considering five or eight cores for T1 and seven or five cores for T2.
- V_{rck} values calculated using five or eight cores for T1 and five or seven cores for T2 appears much similar.

7 Conclusions

A simplified statistical procedure for the structural capacity assessment is presented in this paper. The proposed methodology relies on the calculation of mean and standard deviation of the considered capacity indicator, from which the characteristic value is finally obtained as function of available number of samples, assigned fractile and selected confidence level. This procedure is applied to a real case of study in order to determine the mean ($V_{rc,mean}$) and the characteristic (V_{rck}) value of the shear strength limited by the concrete strut crushing for two existing beams. The obtained results show that:

- concrete strength of old RC structures is highly uncertain. Final experimental results can show significant spatial variability, either when samples are extracted from close beams, or from the same beam;
- as far as the calculation of $V_{rc,mean}$, two samples can be sufficient. As regards the reliable statistical estimation of V_{rck}, no less than five samples are recommended. Since extracting more than two samples from the same beam is unfeasible in many cases, attention must be paid in the identification of statistically homogenous zones.

The proposed procedure will be applied on existing RC columns (Albanesi et al. 2008, 2009; Lavorato and Nuti 2010a, b, 2011, 2015a, b, Lavorato et al. 2017; Marano et al. 2017) considering different concrete and steel materials including stainless-steel, little studied in literature (Zhou et al. 2014, 2015).

Acknowledgements. This work was partially supported by the Italian Consortium of Laboratories ReLUIS, funded by the Italian Federal Emergency Agency, with partial funding from PE 2015–2018, joint program DPC-ReLUIS. This research is also supported by the Structural Laboratory College of Civil Engineering (Fuzhou University), the Sustainable and Innovative Bridge Engineering Research Center of Fuzhou University (SIBERC) and the International Joint-Lab (Universities of Roma Tre and Fuzhou).

References

NTC (2008) Nuove norme Tecniche per le costruzioni, Gazzetta Ufficiale della Repubblica Italiana, n. 29 del 4 febbraio 2008 – Suppl. Ordinario n. 30

Ang AHS, Tang WH (1975) Probability concepts in engineering planning and decision, basic principles. Wiley, New York

Albanesi T, Lavorato D, Nuti C, Santini S (2009) Experimental program for pseudodynamic tests on repaired and retrofitted bridge piers. Eur J Environ Civ Eng 13(6):671–683. https://doi.org/10.1080/19648189.2009.9693145

Albanesi T, Lavorato D, Nuti C, Santini S (2008) Experimental tests on repaired and retrofitted bridge piers. In: Proceedings of the International FIB Symposium 2008 - Tailor Made Concrete Structures: New Solutions for our Society, p 151

Fiorentino G, Forte A, Pagano E, Sabetta F, Baggio C, Lavorato D, Nuti C, Santini S (2017a) Damage patterns in the town of Amatrice after August 24th Central Italy earthquakes. Bull Earthq Eng 16:1399–1423. https://doi.org/10.1007/s10518-017-0254-z

Fiorentino G, Lavorato D, Quaranta G, Pagliaroli A, Carlucci G, Nuti C, Sabetta F, Monica GD, Piersanti M, Lanzo G, Marano GC, Monti G, Squeglia N, Bartelletti R (2017b) Numerical and experimental analysis of the leaning tower of Pisa under earthquake. Procedia Eng 199:3350–3355. https://doi.org/10.1016/j.proeng.2017.09.559

Giannini R, Sguerri L, Paolacci F, Alessandri S (2014) Assessment of concrete strength combining direct and NDT measures via Bayesian inference. Eng Struct 64:68–77

Imperatore S, Lavorato D, Nuti C, Santini S, Sguerri L (2012a) Shear performance of existing reinforced concrete T-beams strengthened with FRP. In: Proceedings of CICE 2012 6th International Conference on FRP Composites in Civil Engineering, Rome, Italy, 13–15 June 2012

Imperatore S, Lavorato D, Nuti C, Santini S, Sguerri L (2012b) Numerical modeling of existing RC beams strengthened in shear with FRP U-sheets. In: Proceedings of CICE 2012 6th International Conference on FRP Composites in Civil Engineering, Rome, Italy, 13–15 June 2012

Imperatore S, Lavorato D, Nuti C, Santini S, Sguerri L (2013) Shear behavior of existing RC T-beams strengthened with CFRP. In: International Association for Bridge and Structural Engineering, IABSE Symposium Report 2013, Assessment, Upgrading and Refurbishment of Infrastructures, vol 99(18), pp 958–965

Lavorato D, Bergami AV, Nuti C, Briseghella B, Xue J, Tarantino A, Marano G, Santini S (2017) Ultra-high-performance fibre-reinforced concrete jacket for the repair and the seismic retrofitting of Italian and Chinese RC bridges. In: COMPDYN 2017 6th ECCOMAS Thematic Conference on Computational Methods in Structural Dynamics and Earthquake Engineering, Rhodes Island, Greece, 15–17 June 2017

Lavorato D, Nuti C (2015) Pseudo-dynamic tests on reinforced concrete bridges repaired and retrofitted after seismic damage. Eng Struct 94:96–112. https://doi.org/10.1016/j.engstruct.2015.01.012

Lavorato D, Nuti C, Santini S, Briseghella B, Xue J (2015b) A repair and retrofitting intervention to improve plastic dissipation and shear strength of Chinese RC bridges. In: IABSE Conference, Geneva. Structural Engineering on Providing Solutions to Global Challenges - Report, pp 1762–1767

Lavorato D, Nuti C (2011) Pseudo-dynamic testing of repaired and retrofitted RC bridges. In: Proceedings of Fib Symposium PRAGUE 2011 on Concrete Engineering for Excellence and Efficiency, vol 1, pp 451–454

Lavorato D, Nuti C (2010a) Seismic response of repaired bridges by pseudodynamic tests. In: Bridge Maintenance, Safety, Management and Life-Cycle Optimization - Proceedings of the 5th International Conference on Bridge Maintenance, Safety and Management, pp 2368–2375

Lavorato D, Nuti C, Santini S (2010b) Experimental investigation of the seismic response of repaired r.c. bridges by means of pseudodynamic tests. In: Large Structures and Infrastructures for Environmentally Constrained and Urbanised Areas, pp 448–449

Marano GC, Pelliciari M, Cuoghi T, Briseghella B, Lavorato D, Tarantino AM (2017) Degrading bouc-wen model parameters identification under cyclic load. Int J Geotechn Earthq Eng 8 (2):60–81. https://doi.org/10.4018/IJGEE.2017070104

Monti G, Petrone F (2016) Test-based calibration of safety factor for capacity model. J Struct Eng 142:04016104

Quaranta G, Kunnath SK, Sukumar N (2012) Maximum-entropy meshfree method for nonlinear static analysis of planar reinforced concrete structures. Eng Struct 42:179–189

Scodeggio A, Quaranta G, Marano GC, Monti G, Fleischman RB (2016) Optimization of force-limiting seismic devices connecting structural subsystems. Comput Struct 162:16–27

Trentadue F, Quaranta G, Marano GC (2016) Closed-form approximations of the interaction diagrams for assessment and design of reinforced concrete columns and concrete-filled steel tubes with circular cross-section. Eng Struct 127:594–601

Trentadue F, Quaranta G, Greco R, Marano GC (2014) New analytical model for the hoop contribution to the shear capacity of circular reinforced concrete columns. Comput Concr 14 (1):59–71

Zhou Z, Lavorato D, Nuti C, Marano GC (2015) A model for carbon and stainless-steel reinforcing bars including inelastic buckling for evaluation of capacity of existing structures. In: 5th ECCOMAS Thematic Conference on Computational Methods in Structural Dynamics and Earthquake Engineering, COMPDYN 2015 pp. 876–886

Zhou Z, Nuti C., Lavorato D (2014) Modeling of the mechanical behavior of stainless reinforcing steel. In: Proceedings of the 10th fib International Ph.D. Symposium in Civil Engineering, pp. 515–520

Author Index

© Springer International Publishing AG, part of Springer Nature 2018
M. di Prisco and M. Menegotto (Eds.): ICD 2016, LNCE 10, pp. 549–550, 2018.
https://doi.org/10.1007/978-3-319-78936-1